Plant Genetic Engineering

Plant Genetic Engineering

Editor: Herbert McCoy

R CALLISTO REFERENCE

www.callistoreference.com

Callisto Reference,
118-35 Queens Blvd., Suite 400,
Forest Hills, NY 11375, USA

Visit us on the World Wide Web at:
www.callistoreference.com

ISBN: 978-1-64116-197-8 (Hardback)

Cataloging-in-Publication Data

Plant genetic engineering / edited by Herbert McCoy.
    p. cm.
Includes bibliographical references and index.
ISBN 978-1-64116-197-8
1. Plant genetic engineering. 2. Genetic engineering. 3. Plant biotechnology.
4. Transgenic plants. I. McCoy, Herbert.
QK981.5 .P53 2019
581.35--dc23

Table of Contents

Preface

Genetic engineering is concerned with manipulating the genetic makeup of cells, by transferring genes either within or across species, to produce improved organisms. Recombinant DNA methods or artificially synthesizing DNA are some methods to obtain new DNA. It is then inserted into the host organism. Crops are genetically engineered to exhibit characteristics of increased tolerance to abiotic stress factors, higher yields and better quality of crops. Research in modern plant genetics has led to the sequencing of plant genomes. The "Gene gun" method, Agrobacterium method, electroporation and microinjection are common techniques of modifying genes in organisms. This book explores all the important aspects of plant genetic engineering in the present day scenario. It strives to provide a fair idea about this discipline and to help develop a better understanding of the latest advances within this field. For all those who are interested in genetic engineering of plants, this book can prove to be an essential guide.

The researches compiled throughout the book are authentic and of high quality, combining several disciplines and from very diverse regions from around the world. Drawing on the contributions of many researchers from diverse countries, the book's objective is to provide the readers with the latest achievements in the area of research. This book will surely be a source of knowledge to all interested and researching the field.

In the end, I would like to express my deep sense of gratitude to all the authors for meeting the set deadlines in completing and submitting their research chapters. I would also like to thank the publisher for the support offered to us throughout the course of the book. Finally, I extend my sincere thanks to my family for being a constant source of inspiration and encouragement.

Editor

Ice-binding proteins confer freezing tolerance in transgenic *Arabidopsis thaliana*

Melissa Bredow[1,*], Barbara Vanderbeld[1] and Virginia K. Walker[1,2]

[1]*Department of Biology, Queen's University, Kingston, ON, Canada*
[2]*Department of Biomedical and Molecular Sciences and School of Environmental Studies, Queen's University, Kingston, ON, Canada*

*Correspondence

email 11mb95@queensu.ca

Seed stock: *Lolium perenne* (Pacific Seed diploid variety; Premier Specific Seeds, Surrey, BC, CA); *Arabidopsis thaliana* (Col-0).
Accession Numbers: EU680848; EU680849; EU680850; EU680851; AJ277399.

Keywords: *Lolium perenne*, antifreeze, ice-binding proteins, ion leakage, freezing survival, *Arabidopsis thaliana*.

Summary

Lolium perenne is a freeze-tolerant perennial ryegrass capable of withstanding temperatures below −13 °C. Ice-binding proteins (IBPs) presumably help prevent damage associated with freezing by restricting the growth of ice crystals in the apoplast. We have investigated the expression, localization and *in planta* freezing protection capabilities of two *L. perenne* IBP isoforms, *Lp*IRI2 and *Lp*IRI3, as well as a processed IBP (*Lp*AFP). One of these isoforms, *Lp*IRI2, lacks a conventional signal peptide and was assumed to be a pseudogene. Nevertheless, both *Lp*IRI2 and *Lp*IRI3 transcripts were up-regulated following cold acclimation. *Lp*IRI2 also demonstrated ice-binding activity when produced recombinantly in *Escherichia coli*. Both the *Lp*IRI3 and *Lp*IRI2 isoforms appeared to accumulate in the apoplast of transgenic *Arabidopsis thaliana* plants. In contrast, the fully processed isoform, *Lp*AFP, remained intracellular. Transgenic plants expressing either *Lp*IRI2 or *Lp*IRI3 showed reduced ion leakage (12%–39%) after low-temperature treatments, and significantly improved freezing survival, while transgenic *Lp*AFP-expressing lines did not confer substantial subzero protection. Freeze protection was further enhanced by with the introduction of more than one IBP isoform; ion leakage was reduced 26%–35% and 10% of plants survived temperatures as low as −8 °C. Our results demonstrate that apoplastic expression of multiple *L. perenne* IBP isoforms shows promise for providing protection to crops susceptible to freeze-induced damage.

Introduction

Many overwintering temperate plants are susceptible to freeze injury during the coldest months. At subzero temperatures, ice crystals form in intracellular spaces, or the apoplast, creating an osmotic gradient that can result in cellular dehydration, expansion-mediated lysis of plasma membranes and even death of the plant (Thomashow, 1998). In order to better adapt to freezing temperatures, some overwintering plants induce the expression of a family of protective proteins, designated ice-binding proteins (IBPs). IBPs are members of a highly diverse family of proteins that have been identified in certain organisms including fish (Davies and Hew, 2013), insects (Duman, 2001), bacteria (Gilbert *et al.*, 2004) and plants (Sidebottom *et al.*, 2000). The activity of IBPs stems from their ability to irreversibly adsorb to ice crystals, resulting in the 'shaping' of ice as they become incorporated into the ice crystal lattice (Bar-Dolev *et al.*, 2012). IBPs are known to enhance freezing tolerance through two distinct properties: ice recrystallization inhibition (IRI), which prevents the growth of ice crystals at high subzero temperatures (Sandve *et al.*, 2008), and thermal hysteresis (TH), or the depression of the freezing point in relation to the equilibrium melting point (Raymond and DeVries, 1977). While freeze-avoidant organisms produce IBPs that depress the freezing point by several degrees, often referred to as antifreeze proteins (AFPs) (Davies and Hew, 2013; Duman, 2001), plants encode IBPs with low TH activity and rely on restricting ice crystal growth as a primary survival strategy (Sandve *et al.*, 2011).

Although IBPs are not commonly found in plants and are absent in *Arabidopsis thaliana* for example, they have been identified and purified from more than a dozen plants including bittersweet nightshade (*Solanum dulcamara*) (Huang and Duman, 2002), carrot (*Daucus carota*) (Smallwood *et al.*, 1999), winter rye (*Secale cereale*) (Hon *et al.*, 1995) and perennial ryegrass (*Lolium perenne*) (Pudney *et al.*, 2003). As ice crystal growth is commonly propagated in the apoplast, secretion of IBPs from the cytoplasm would prevent the recrystallization of extracellular ice, protecting cells from the effects of freeze-induced cellular dehydration. Thus, it is not surprising that most of the IBPs that have been studied have been recovered from the apoplastic extracts of cold-acclimated leaf tissue (Antikainen and Griffith, 1997; Griffith *et al.*, 1992; Hon *et al.*, 1994; Marentez *et al.*, 1993). The presence of an N-terminal signal peptide in most IBPs suggests secretion through the endoplasmic reticulum (ER) secretory pathway. An IBP from *S. dulcamara* has been reported to lack a signal peptide and remains intracellular (Huang and Duman, 2002), suggesting that IBPs might also function to prevent damage associated with intracellular ice nucleation.

The freeze-tolerant perennial grass, *L. perenne* (*Lp*), is native throughout Europe and Eastern Asia where it can survive at temperatures as low as −13 °C (Thomas and James, 1993). Their IBPs have been termed ice recrystallization inhibition proteins (*Lp*IRIPs) due to their low TH activity (~0.3 °C at 1 mg/mL) (Lauersen *et al.*, 2011), but relatively high IRI activity (Sidebottom *et al.*, 2000). Four *Lp*IRIPs have been identified in *L. perenne*: *Lp*IRI1, *Lp*IRI2, *Lp*IRI3 and *Lp*IRI4 (Sandve *et al.*, 2008). A partial protein product, named *Lp*AFP, was identified in another *L. perenne* cultivar (Sidebottom *et al.*, 2000) and has been the subject of much *in vitro* characterization (Lauersen *et al.*, 2011; Middleton *et al.*, 2009).

LpIRIPs have two distinct domains: a leucine-rich repeat (LRR) domain and a carboxyl (C)-terminal IRI domain, which consists of a series of repeated 'ice-binding' motifs (NXVXG/NXVXXG, where X represents an outward-facing residue). The IRI domain of these proteins has been predicted to fold into a β-helix and the crystal structure of LpAFP has verified this fold (Middleton et al., 2009). Specific residues on the ice-binding face (IBF) allow LpIRIPs to adsorb to ice crystals on the basal and primary prism planes resulting in hexagonal bipyramidal crystals (Kumble et al., 2008). N-terminal to the IRI domain is a varying number of LRR motifs, likely derived from phytosulfokine LRR receptor tyrosine kinase sequences (Sandve et al., 2008). The retention of the LRR domain in three of the IBP isoforms may simply have allowed for the presence of the N-terminal signal peptide for secretion to the apoplast.

Notably, one isoform, LpIRI2, lacks most of the region upstream of the IRI domain, including an identifiable signal sequence, and likely evolved through duplication of LpIRI4 and the subsequent deletion of the N-terminal domain (Sandve et al., 2008). The absence of a signal peptide has led to the hypothesis that LpIRI2 is a pseudogene. However, the modest sequence divergence following duplication of LpIRI4 suggests that the gene may still be under selective pressure and thus retain some function (Sandve et al., 2008). Therefore, whether LpIRI2 is a pseudogene, acts as an intracellular IBP or is secreted via a nonclassical pathway is not known.

Biotechnological applications of IBPs for the enhancement freeze tolerance have been tested in tobacco (Nicotiana tabacum) (Holmberg et al., 2001), potato (Solanum tuberosum) (Wallis et al., 1997) and tomato (Solanum lycopersicum) (Hightower et al., 1991). Generally, such transgenic plants have shown IBP accumulation in the apoplast, but few efforts have reported significant differences in freezing tolerance as shown by 50% lethality (LT_{50}) assays. An exception is the enhanced freeze survival following the transfer of IBP sequences to Arabidopsis thaliana (Zhang et al., 2010). However, these studies have focused on the expression of a single IBP from a protein family, and we considered that the isoforms could work synergistically or cumulatively to restrict ice crystal growth. As well, we were particularly interested in determining whether LpIRI2 has retained ice-binding activity and thus function in planta, despite the loss of the N-terminal signal peptide. We have now addressed these outstanding concerns through the expression of various LpIRIP isoforms in A. thaliana, alone and in combination, in order to provide insight into the mechanisms underlying IBP-mediated freezing tolerance.

Results

Bioinformatics analysis of LpIRIP isoforms

Amino acid alignment of LpIRIP isoforms shows high conservation of the C-terminal residues, while the N-terminal domains are more divergent (Figure 1). The N-terminal domains of LpIRIP isoforms have retained few LRR motifs (0–5 motifs across isoforms), with some isoforms having large deletions, or in the case of LpIRI2, having not retained this domain all together. An N-terminal signal peptide was identified for LpIRI1, LpIRI3 and LpIRI4; however, as previously reported (Sandve et al., 2008), there is no apparent secretion signal within LpIRI2 or LpAFP amino acid sequences. Putative ice-binding amino acids appear to be conserved across four sequences with the unprocessed proteins ranging in size from 151 to 285 residues. The Phyre2 algorithm

predicted that all proteins would fold into a right-handed β-helix. While the LpIRI3, LpAFP and LpIRI1 isoforms were predicted to fold into secondary structures with eight β-helical loops, LpIRI2 and LpIRI4 were predicted to have ten loops, as the result of three additional ice-binding motifs (Figure 1).

Functional activity and transcript analysis of LpIRIP sequences

When constructs encoding LpIRI2 and LpIRI3 isoforms were expressed in E. coli, all purified, recombinantly produced proteins restricted ice crystal growth in a splat assay (Figure 2). However, not all recombinant isoforms were equally effective at IRI. High levels of activity were seen with LpAFP and LpIRI3 at 0.01 mg/mL; however, LpIRI2 only demonstrated mild IRI activity, with some ice crystal growth at the same concentration.

Endogenous L. perenne transcript analysis generated ~500-bp and ~850-bp amplification products for LpIRI2 and LpIRI3, respectively. Low levels of LpIRI3 transcript were produced following incubation at 21 °C, but there was no evidence of the LpIRI2 transcript (Figure 3). However, following a 6-d cold acclimation (CA) period at 4 °C both LpIRI3 and LpIRI2 transcripts were abundant.

Ice-binding activity and localization of LpIRIPs in A. thaliana

Following CA, crude cell extracts taken from all four independently generated A. thaliana lines transgenic for each of LpAFP, LpIRI2, LpIRI3, LpAFP and LpIRI2 (designated A3), and all three LpIRIP sequences (designated 2A3) showed functional ice-binding activity as demonstrated by IRI analysis (Figure 4a). There was no IRI activity in control CA plants, and very low activity in those transgenic plants kept at room temperature prior to assay (not shown). This was expected as, at least for LpAFP, circular dichroism-monitored conformational changes to the β-helical structure at temperatures >16 °C were coincident with a loss of activity (Lauersen et al., 2011). All CA LpIRI2-transformed lines consistently showed IRI, but they had lower activity than LpAFP- and LpIRI3-expressing lines (Figure 4a). Extracts from the transgenic plants expressing LpIRI2 and LpAFP demonstrated hexagonal ice shaping indicative of adsorption to the primary prism plane (Figure 4b). Notably, LpIRI3, A3 and 2A3 transgenic lines showed hexagonal bipyramidal ice shaping, which is seen with more active AFPs that bind both the primary prism and basal planes. TH activity depends on both ice-binding properties and protein concentration. Indeed, TH values obtained from cell lysates of our transgenic plants were similar for all the single expression lines with the multiple A3 and 2A3 lines showing greater activity levels (Table 1a).

In order to localize IBP activity, the ORFs of LpIRIPs fused to fluorescent markers were expressed in A. thaliana. Confocal microscopy confirmed protein expression in all LpAFP, LpIRI2 and LpIRI3 transgenic lines. It was difficult to resolve cytoplasmic and apoplastic locations however, due to the large vacuoles in leaf and root tissues (Figure S1). Nevertheless, apoplastic extracts of transgenic A. thaliana lines expressing LpIRI2-mOrange and LpIRI3-mOrange showed fluorescence and had high levels of IRI activity (Table 2 and Figure 5a). Additionally, LpIRI2-mOrange lines consistently showed hexagonal, primary prism plane ice shaping and LpIRI3-mOrange lines exhibited hexagonal bipyramidal crystals, consistent with basal and primary prism plane adsorption (Figure 5b). Both LpIRI2-mOrange and LpIRI3-mOrange lines also demonstrated TH activity (Table 1b). In contrast,

```
LpIRI3   MAKCLMLLLSFAFLLSAAGTATATPCHRDDLRALRGFAENLGGGGALSLRAAWSGASCCD   60
LpAFP    ------------------------------------------------------------   -
LpIRI1   ---MGLLLLFLGFLLPAA-CAATSSCHPDDLRALRGFAKNVGGGGVL-LRTAWSGTSCCV   55
LpIRI2   ------------------------------------------------------------   -
LpIRI4   MAKCWQLLLLFLALLLPAA---SAASCHPDDLYALRDFAGNLRGGGVL-LRAALPGASCCG   56

LpIRI3   WEGVGCDGASGRVTALWLPRSGLT------------------------GPIPSWICQLH   95
LpAFP    ---------------------------------------------------------   -
LpIRI1   WEGVGCNGASGRITSLWLPRRGLAGTITGASLAGLAGLESLNLANNRLVGTIPSWIGELD   115
LpIRI2   ---------------------------------------------------------   -
LpIRI4   WEGVGCDGASGCVKS------------------------------------------   67

LpIRI3   HLRYLDLSGNALVGEVPKNLQVQLKGIT----------NMPLHVMRNRRSLDEQPNTIS   144
LpAFP    -----------------------------------------------DEQPNTIS   8
LpIRI1   HLLYLDLSHNSLVGELPNRLRIRLKGLTTTGHLLGMTFTNMPLDVKHNRRTLAIQPNTIS   175
LpIRI2   -----------------------------------MPLHVKRSQGTLDEDHNTIT   20
LpIRI4   ----------------FQILLKGLTAAGRSLGKAFTHMPLHVKPSQGTLDEDHNTIT   112
                                                             : ***:

LpIRI3   GSNNTVRSGSKNVLAGNDNTVISGDNNSVSGSNNTVV--SGNDNTVTGSNHVVSGTNHIV   202
LpAFP    GSNNTVRSGSKNVLAGNDNTVISGDNNSVSGSNNTVVFTSGNDNTVTGSNHVVSGTNHIV   68
LpIRI1   GTNNLVLSGRNNVVSGNDNTVISGNNNTVSGSFNTVV--TGSDNILTGSNHVVSGRSHIV   233
LpIRI2   GSHNTVRSGSNNVVSGNDNTVISGNNNVMSGSHNTVIF--GGDNFVSGSYHVVSGNHHVV   78
LpIRI4   GINNTVRSGSNNVVSGNDNTVISGNNNVVSGSHNTVV--FGGDNFISGSYHVVSGNHHVV   170
         *  . *  * **  :**::***********:** :*** ***:   *  ** ::** ***** *:*

LpIRI3   TDNNNNVSGNDNNVSGSFHTVSGGHNTVSGSNNTVSG-----------------N   241
LpAFP    TDNNNNVSGNDNNVSGSFHTVSGGHNTVSGSNNTVSGF-----------------TN   109
LpIRI1   TDNNNSVSGDDNNVSGSFHKVSGSHNTVSGSNNTVSG-----------------N   272
LpIRI2   TDNKNAVSGDHNTVSGTQNTVSGNHQIVSGSHGTVSGNHNTVSGRNNSVYGNKNIVSGSN   138
LpIRI4   TDNKNAVSGDHNTVSGSQNTVSGNHQIVSGSHSTVSGNHNTVSGRNNSVYGNNNIVSGSN   230
         ***:* ***: *.***:  ..*** *:****  ****                    *

LpIRI3   HVVSGSNKVVTDA   254
LpAFP    HTVSGSNKVVTDA   122
LpIRI1   HVVSGSNKVVTGG   285
LpIRI2   HVVYGNNKVVTGG   151
LpIRI4   HVVYGNNKVVTGG   243
         *** *.***** .
```

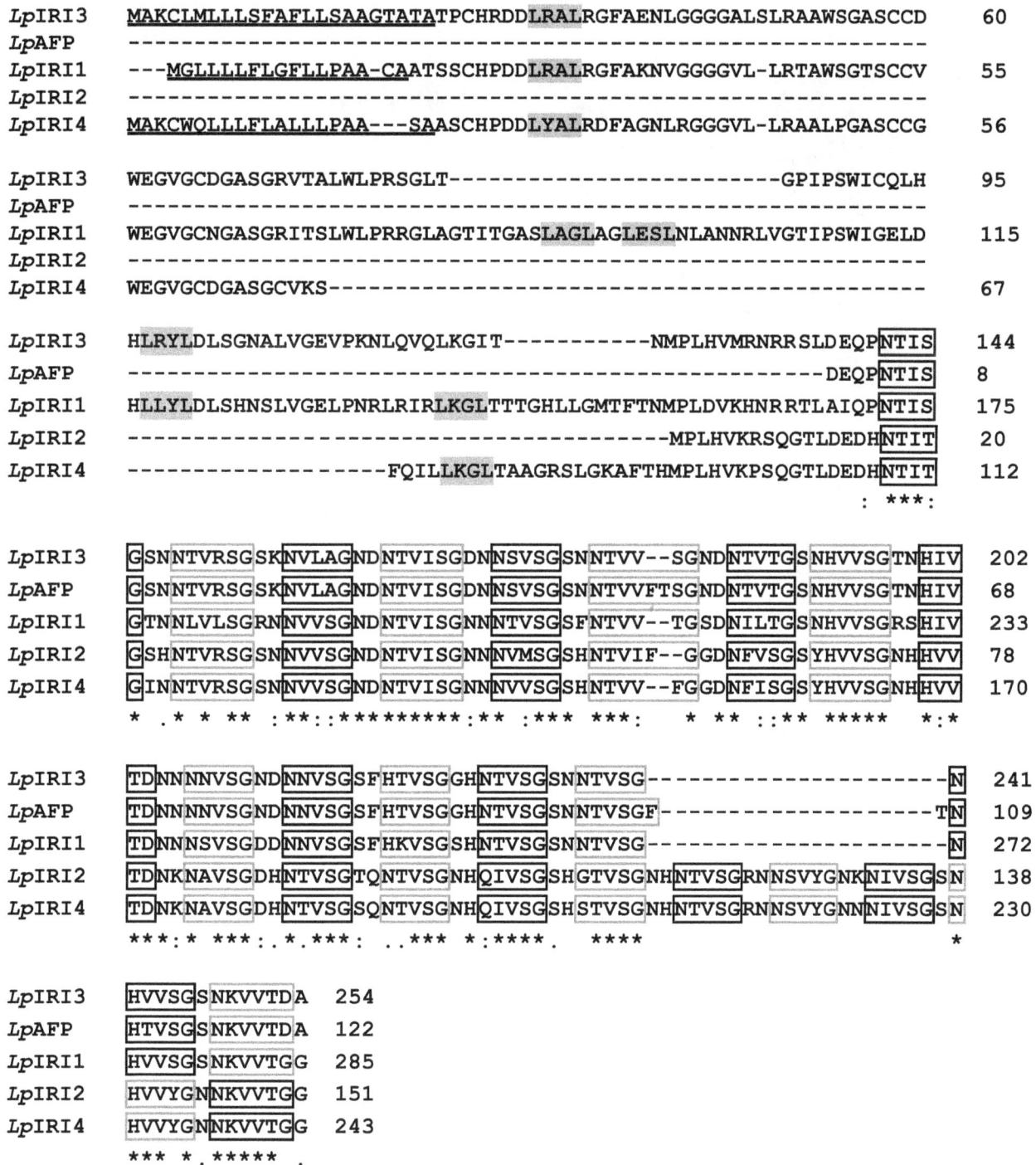

Figure 1 Amino acid sequence alignment of *Lp*IRIP sequences. *Lp*IRI3 (EU680850), *Lp*AFP (AJ277399), *Lp*IRI1 (EU680848), *Lp*IRI2 (EU680849) and *Lp*IRI4 (EU680851) were aligned using the ClustalW2 multiple sequence alignment tool. Putative signal peptides, predicted using the SignalP 4.1 server, are underlined. Leucine-rich repeat motifs (LXXL) are in grey-filled boxes and the amino acids corresponding to the predicted ice-binding and non-ice-binding faces are in black and grey boxes, respectively. Each β-helical turn, predicted by the Phyre2 algorithm, is identified by a distinct colour. (*) denotes a single, fully conserved reside; (:) denotes conservation between groups of highly similar properties (scoring > 0.5 in the Gonnet PAM 250 score), and (.) denotes conservation between groups of weakly similar properties (scoring =<0.5 in the Gonnet PAM 250 score).

apoplastic extracts from *mOrange-LpAFP* lines had similar levels of ice-binding activity to that of *mOrange* transgenic controls, with low levels of fluorescence (Table 2), no observable IRI activity, ice-shaping activity or TH activity (Figure 5 and Table 1b). To determine whether the *mOrange* tag had disrupted the ice-binding activity of *Lp*IRIPs, the crude cell extracts of transgenic plants were assayed for IRI activity (Figure 6). While *mOrange*-expressing lines had no ice crystal inhibition, all *Lp*IRIP-expressing lines retained high levels of IRI activity. Investigations with guttation fluid of transgenic *Lp*IRIP-expressing lines revealed

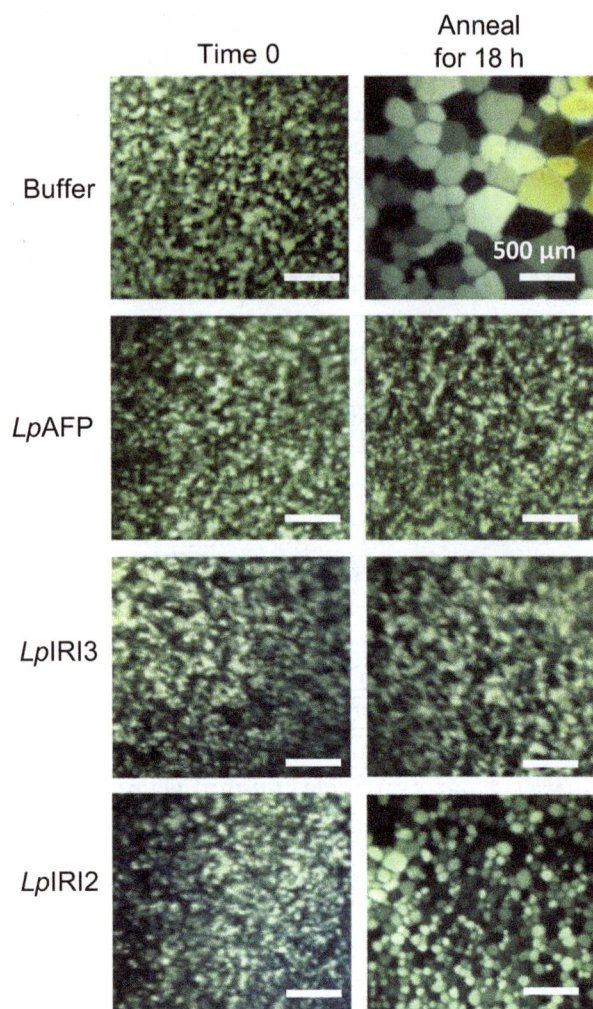

	Time 0	Anneal for 18 h

Buffer

500 μm

LpAFP

LpIRI3

LpIRI2

Figure 2 Ice recrystallization inhibition assessment of recombinantly produced LpIRIPs. Purified LpAFP, LpIRI3 and LpIRI2 proteins were used for a splat assay by diluting samples to a concentration of 0.01 mg/mL and holding ice crystals at −4 °C for 20 h. Shown here are representative images from triplicate assays.

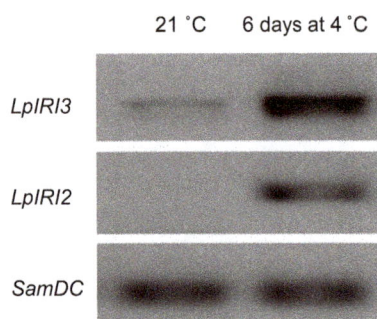

	21 ˚C	6 days at 4 ˚C

LpIRI3

LpIRI2

SamDC

Figure 3 Reverse transcription PCR analysis of endogenous LpIRI3 and LpIRI2 transcript levels. RNA was collected from the leaves of L. perenne grown at 21 °C or cold-acclimated for 6 days at 4 °C. The SamDC transcript was used as a reference loading control. Experiments were conducted in triplicate.

similar results to the apoplastic extracts: transgenic LpAFP-expressing A. thaliana and control plants had no detectable IRI activity with LpIRI3 and LpIRI2 lines with high levels of IRI activity

(Figure 7). It should be noted, however, that the activity observed in LpIRI2-expressing lines was less than in guttation fluid obtained from LpIRI3 lines (Figure 7).

Electrolyte leakage in transgenic leaves

When leaves were collected from CA transgenic A. thaliana plants and incubated at temperatures slowly ramped down to −6 °C, freezing resulted in the leakage of ions, as assessed by conductivity. The expression of LpAFP in transgenic plants appeared to reduce leaf ion leakage by 6%−10% in three of the four independent lines, but these values were statistically insignificant when compared to control plants (Figure 8a). In contrast, expression of LpIRI3 and LpIRI2 decreased ion leakage by 30%−39% and 12%−22%, respectively (Figure 8b and c).

Transgenic plants expressing multiple LpIRIPs also showed reduced electrolyte leakage compared with controls. A. thaliana leaves from plants expressing A3 constructs showed a 28%−35% decrease in electrolyte leakage following a −6 °C treatment (Figure 9a). Transgenic 2A3 lines similarly showed a 26%−35% reduction in electrolyte leakage (Figure 9b).

LpIRIPs and A. thaliana freeze protection

The addition of the various LpIRIP-bearing sequences dramatically enhanced the freeze survival of whole transgenic A. thaliana plants. Freeze survival was significantly increased compared with controls in two of the four LpAFP-expressing lines following freezing at −6 °C (Figure 10a). In one of the LpAFP lines, there was also a significant increase in survival at −7 °C; however, the overall LT_{50} was not changed, remaining at −5.6 °C, not significantly different from the LT_{50} of −5.2 °C seen in control, nontransgenic plants. Significant increases in survival were seen in all four LpIRI3-expressing lines at all tested temperatures between −5 °C and −7 °C (Figure 10b), and this was reflected in a mean LT_{50} of −6.1 °C. Similarly, LpIRI2-transgenic A. thaliana showed enhanced freeze survival at all temperatures between −5 °C and −7 °C (Figure 10c). In these lines, the mean LT_{50} was −6.0 °C compared with −5.4 °C in control plants assayed at the same time.

Transgenic plants expressing multiple LpIRIPs also showed enhanced freezing tolerance. Survival was significantly increased in both the A3 and 2A3 lines at temperatures between −5 °C and −8 °C (Figure 11a and b) with a concomitant significant decrease in LT_{50} to −6.0 °C and −6.4 °C, respectively, compared with the LT_{50} of −5.2 °C and −5.4 °C for the corresponding nontransgenic controls. Notably, none of the lines bearing a single LpIRI sequence or even the A3 lines showed any survival at −8 °C (Figures 10 and 11a), but each of the four 2A3 lines, with all three sequences, showed some survival (10%−17%) at this low temperature (Figure 11b).

Discussion

Freezing tolerance is a complex trait involving biochemical, metabolic and physiological changes. In certain plants, IBPs almost certainly serve as part of a freeze survival strategy to regulate ice crystal growth and to lower the probability of plasma membrane rupture. These proteins have also been shown to lower the activity of bacterial ice nucleation, aiding in freeze survival (Tomalty and Walker, 2014). The L. perenne family of IBPs includes the 'processed' protein sequence, LpAFP, which has been extensively characterized in vitro (Lauersen et al., 2011; Middleton et al., 2009); however, the in planta function and activity of the proteins transcribed and subsequently translated

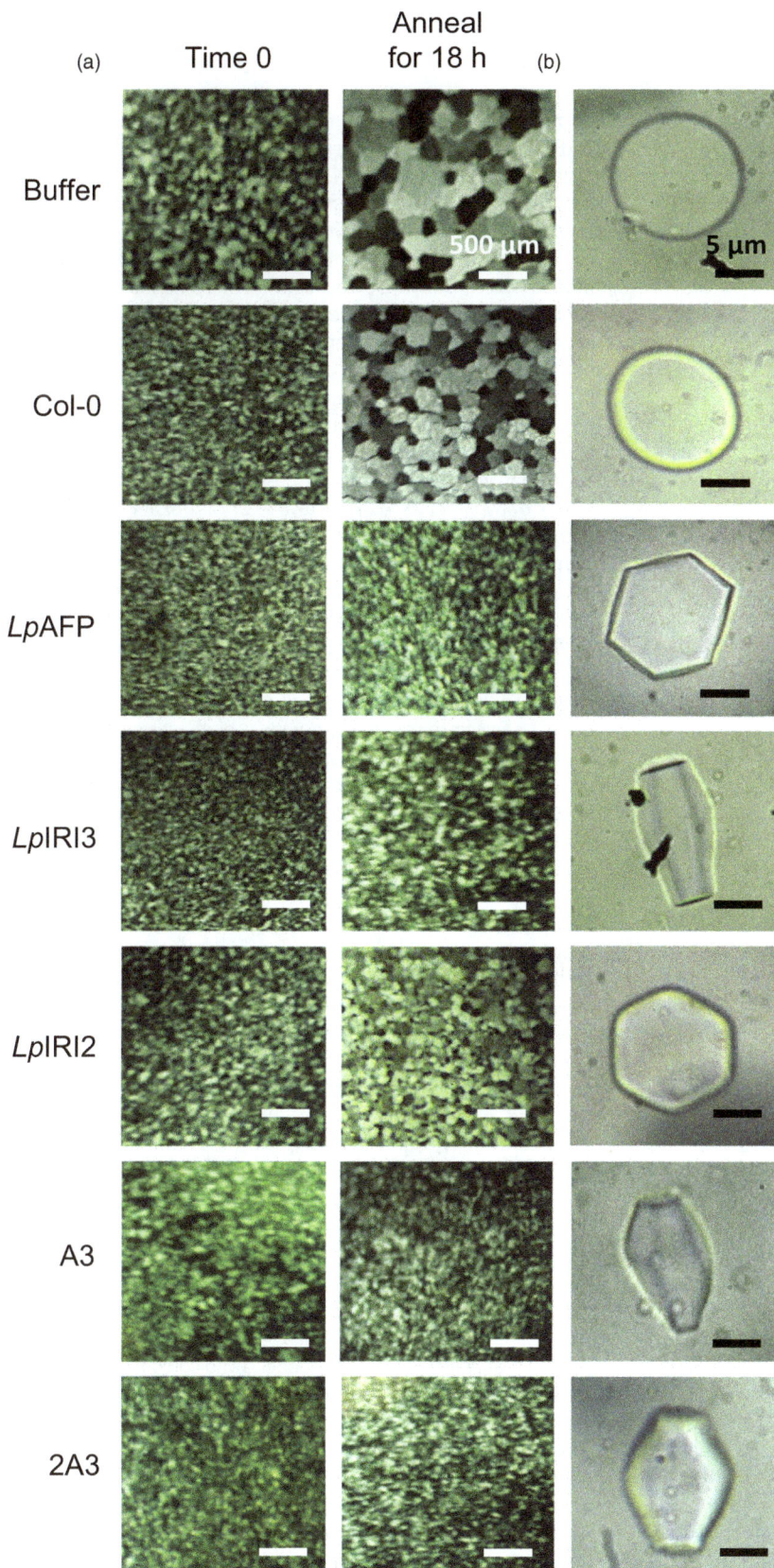

Figure 4 Ice-binding phenotypes in transgenic *A. thaliana*. Ice recrystallization inhibition analysis of crystals annealed at −4 °C for 18 h (a) and ice crystal morphologies (b) using crude cell lysates collected from control (Col-0) and transgenic plants, including lines expressing *LpAFP* and *LpIRI3* (A3) and all three sequences, *LpIRI2*, *LpAFP* and *LpIRI3* (2A3). A total protein concentration of 0.1 mg/mL was used for all assays. Only one representative sample is shown for each of the four *Lp*IRIP-expressing lines. All experiments were performed in triplicate.

from the full-length *Lp*IRIPs are less known. Here, our experiments demonstrate that in transgenic *A. thaliana,* the presence of *LpIRI2* and *LpIRI3* not only reduced electrolyte leakage but also significantly enhanced freeze survival (Figures 8-11).

Importantly, the degree of freeze protection afforded by *Lp*IRIPs in *A. thaliana* was correlated with the ice-binding activity localized to the apoplast (Figure 5). Most ice nucleation occurs outside of the cell (Kajava and Lindow, 1993; Xu *et al.*, 1998).

Table 1 Thermal hysteresis values obtained from crude cell extracts (a) of transgenic lines expressing individual *Lp*IRIP sequences, *Lp*AFP and *Lp*IRI3 (A3), and all three sequences (2A3), compared with wild-type *A. thaliana* (Col-0), as well as apoplast extracts (b) of *mOrange*-tagged *Lp*IRIP lines compared with wild-type *A. thaliana* plants and plants expressing *mOrange* alone

Transgenic line	Thermal hysteresis activity (°C)
(a)	
Col-0	0
*Lp*AFP	0.11
*Lp*IRI2	0.08
*Lp*IRI3	0.13
A3	0.15
2A3	0.18
(b)	
Col-0	0
mOrange	0
mOrange-LpAFP	0
*Lp*IRI2-*mOrange*	0.035
*Lp*IRI3-*mOrange*	0.07

Table 2 Mean fluorescence readings (wavelength emission and excitation of 562 nm and 548 nm, respectively) obtained using apoplast extracts (0.1 mg/mL total protein; triplicate samples) of *mOrange*-tagged *Lp*IRIP-expressing *A. thaliana* lines (as described in Table 1)

Transgenic line	Relative fluorescence units (RFUs)
mOrange	1037
mOrange-LpAFP	1100
*Lp*IRI2-*mOrange*	12 834
*Lp*IRI3-*mOrange*	13 474

Values were normalized based on the level of fluorescence emitted in control, nontransgenic lines.

Intracellular ice nucleation typically only occurs as a result of rapid temperature drops (Siminovitch *et al.*, 1978) and as temperatures were lowered slowly (0.5 °C/h), intracellular ice crystal growth would not be expected in these experiments. In this regard, it may be curious that in two *Lp*AFP-expressing lines, in which the ice activity appeared to have remained intracellular, there was a modest increase in freeze survival, suggesting that intracellular IBPs might additionally contribute to a freeze-tolerant phenotype. Indeed, it is possible that due to the absence of extracellular IBPs and thus the increased probability of plasma membrane rupture, release of intracellular *Lp*AFP facilitated the protection of neighbouring cells in whole plants, explaining the increased survival of one of the *Lp*AFP lines at −7 °C (Figure 10a). Nevertheless, the large increases in freezing stress survival rates observed in *Lp*IRI3-expressing *A. thaliana* lines underscore the dramatic freeze protection conferred by extracellular IBPs. In addition to restricting ice crystal growth, it has also been suggested that IBPs are capable of preventing cell lysis through physical association with the plasma membranes (Beirão *et al.*, 2011; Hays *et al.*, 1996; Rubinsky *et al.*, 1991; Tomczak *et al.*, 2001). Our results showing reduced electrolyte leakage in transgenic *Lp*IRI2 and *Lp*IRI3

A. thaliana lines are consistent with the interpretation that membrane protection is occurring; however, there is no *in planta* evidence of a direct association between IBPs and plasma membranes.

Transgenic expression of multiple *Lp*IRIPs reduced ion leakage and increased freeze survival at lower temperatures than in many of the lines bearing a single *Lp*IRIP sequence (Figures 10 and 11). It is thus likely that multiple transgenes result in a greater accumulation of *Lp*IRIPs, even if this was not reflected in TH values given the hyperbolic relationship between TH activity and protein levels. However, the activity of *Lp*IRIPs can be distinguished on the basis of their adsorption to distinct ice crystal planes. Amino acid alignment suggests that the *Lp*IRI3 isoform has the most regular fold, and highest conservation amongst putative ice-binding residues (Figure 1). In accordance with this observation, transgenic plants with the *Lp*IRI3 sequence either as a single copy, with *Lp*AFP, or with *Lp*AFP and *Lp*IRI2, showed adsorption to both the primary prism and basal planes, characteristic of hyperactive AFPs (Pertaya *et al.*, 2008). It is possible that there could be some synergistic activity with the different transgenes, resulting in reduced ice crystal growth when IBPs with slightly different ice plane affinities are combined. As well, IBPs could have differing affinities for any putative membrane-binding sites, which in turn could enhance freeze survival.

These results are reminiscent of the enhancement of TH activity shown when a low-activity type III AFP isoform was expressed in the Notched-fin eelpout along with a high-activity AFP isoform (Nishimiya *et al.*, 2005). It was suggested that AFP isoforms might act cumulatively to enhance activity levels by high-activity AFPs slowing the growth of ice crystals sufficiently so that less active AFPs would have time to adsorb, or alternatively, that less active AFPs could adsorb at ice crystal sites located between the binding sites of more active AFP isoforms (Nishimiya *et al.*, 2005). Similarly, Burcham *et al.* (1984) hypothesized that antifreeze activity was enhanced in the presence of high- and low-activity antifreeze glycoprotein (AFGP) isoforms, as a result of a cooperative coverage of the initial ice crystal. It is possible that *Lp*IRIPs may function similarly, restricting growth of the initial seed crystal more effectively when more than one isoform is present, providing optimal freeze protection to plants.

Putting aside possible isoform differences in ice or membrane substrate affinity, our results clearly demonstrate that the *Lp*IRI2 isoform with no identifiable signal sequence shows ice-binding activity (TH, ice-shaping and IRI) and CA accumulation in the apoplast, and can confer freeze protection to a host plant. Thus, the evolutionary loss of the N-terminal domain does not render this protein nonfunctional. Similar transcript-level estimates of *Lp*IRI2 and *Lp*IRI3 in CA *L. perenne* leaves, and the reduction in *A. thaliana* leaf electrolyte leakage, further support our contention that *Lp*IRI2 is not a pseudogene and could therefore play a role in *L. perenne* overwintering. As intracellularly localized IBPs are produced in certain plants including the desert evergreen shrub, *Ammopiptanthis mongolicus* (Fei *et al.*, 1994), the flowering shrub, *Forsythia suspense* (Simpson *et al.*, 2005), and *S. dulcamara* (Duman, 1994) as well as the Antarctic microalga, *Chaetoceros neogracile* (Gwak *et al.*, 2014), we initially suspected that *Lp*IRI2 functioned similarly. Of note, intracellularly localized IBPs in addition to restricting ice crystal growth have demonstrated comparably lower ice-binding activity (Duman, 1994). Nonetheless, our experiments indicate that the *Lp*IRI2 isoform is secreted (Table 2 and Figures 5 and 7). Although levels of IRI and TH activities in the apoplast were admittedly lower than

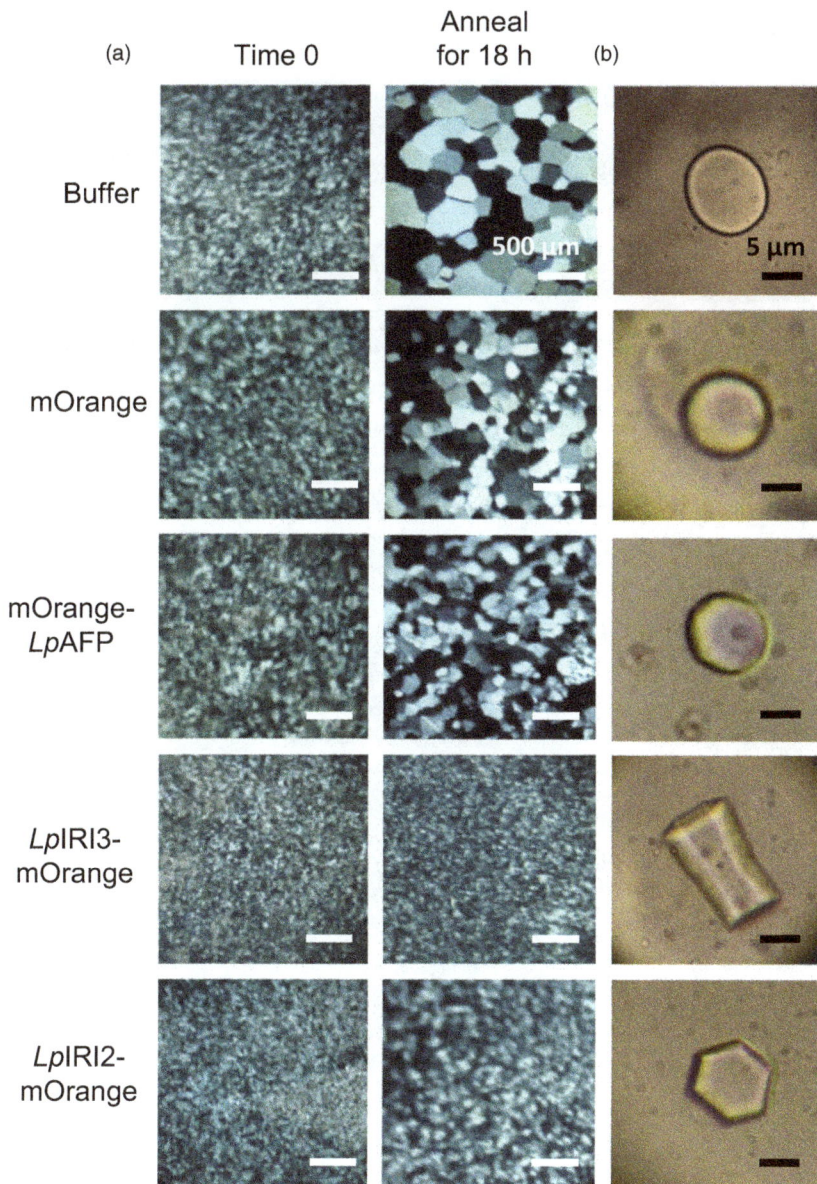

Figure 5 Ice-binding phenotypes in apoplast extracts collected from transgenic *A. thaliana* plants bearing fluorescently tagged *Lp*IRIP sequences. Shown are the ice crystals seen during ice recrystallization inhibition analysis following an 18-h annealing period at −4 °C (a) and the corresponding ice crystal morphologies (b). The apoplast extracts of transgenic lines expressing fluorescently tagged *Lp*IRIP constructs were compared to control (Col-0) nontransgenic plants, at a total protein concentration of 0.1 mg/mL. Only one representative sample is shown for each *Lp*IRIP-expressing line. Both assays were conducted in duplicate.

that observed with *Lp*IRI3-expressing lines, the fact that recombinant *Lp*IRI2 showed a reduced ability to restrict ice crystal growth compared with *Lp*IRI3 (Figure 2) suggests that such differences may be due to varying affinities for ice crystal planes or less efficient ice crystal adsorption. Attempts to localize this isoform, as well as other IBPs, using fluorescently tagged proteins *in planta* did not allow us to unambiguously distinguish between the intracellular plasma membrane and the apoplast (Supplementary 1). Nevertheless, IRI and ice-shaping activity, as well as fluorescence, were present in apoplast extracts and guttation fluid of *Lp*IRI2-expressing *A. thaliana* lines (Figures 5 and 7, and Table 2). Additionally, the level of freeze tolerance observed in *Lp*IRI2 lines was superior to that seen for the intracellular *Lp*AFP lines, in accordance with extracellular localization. Therefore, taken together, these observations support the hypothesis that *Lp*IRI2 is a secreted protein.

Protein secretion using nonclassical pathways is a relatively unexplored area of research. Despite this, it has been estimated that 60% of proteins identified in the secretome of *A. thaliana* are leaderless secreted proteins (LSPs) (Regente *et al.*, 2012). The mechanisms involved in the recognition of LSPs are not well known, and may not be conserved amongst or within secretion systems, suggesting that such mechanisms have evolved independently. Therefore, reliable prediction is not possible. Intriguingly, a nonclassical secretion system appears more common amongst protein families involved in environmental stress responses (Cheng *et al.*, 2009; Gupta and Deswal, 2012; Kim *et al.*, 2008). Thus, it is possible that non-Golgi secretion could provide flexible spatial localization, allowing proteins to take on dual function roles inside and outside of the cell. Perhaps as important, given the large number of proteins up-regulated during the freezing stress response, nonclassical secretion could provide an alternative, and potentially more efficient secretion for critical proteins under these circumstances, allowing other necessarily ER-linked proteins to monopolize the 'traditional' pathway.

Secretion through the ER-Golgi pathway is required for the post-translational modification of proteins. In this regard, a

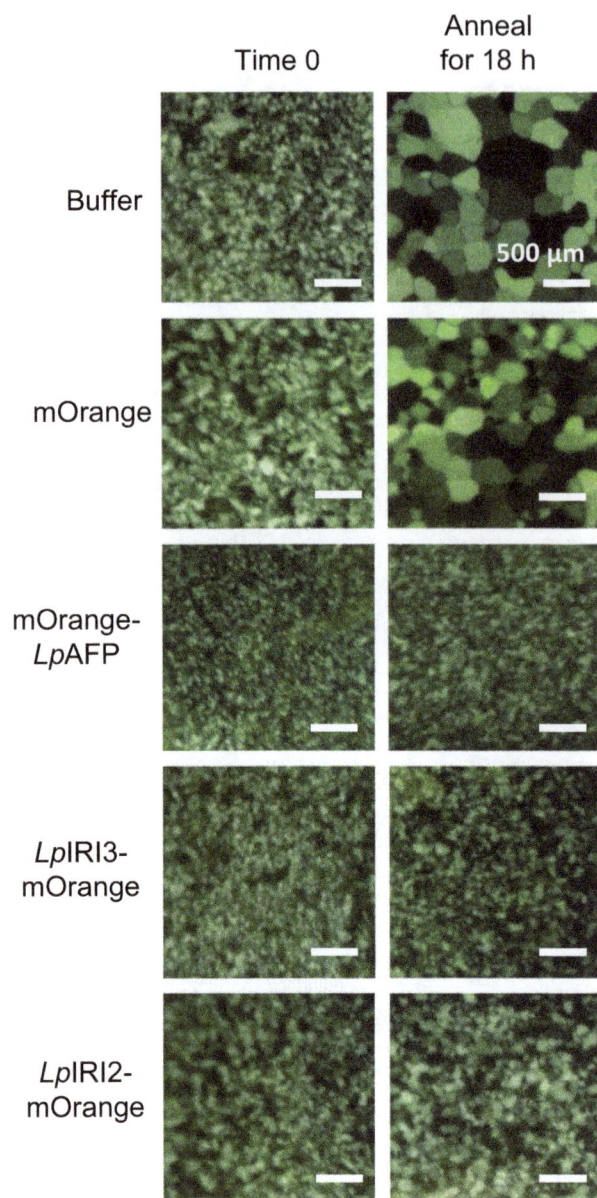

Figure 6 Ice recrystallization inhibition (IRI) analysis of plants expressing fluorescently tagged *Lp*IRIPs. Crude cell extracts were annealed at −4 °C for 18 h. Samples were assayed at a total protein concentration of 0.1 mg/mL. Only one representative sample is shown for each transgenic line. Assays were performed in triplicate.

Figure 7 Ice recrystallization inhibition analysis of guttation fluid collected from transgenic *Lp*IRIP-expressing *A. thaliana* plants. Guttation fluid collected from *Lp*AFP-, *Lp*IRI3- and *Lp*IRI2-expressing lines was used for a splat assay. Samples were held at −4 °C for 24 h and compared to nontransgenic control plants (Col-0). A total protein concentration of 0.1 mg/mL was used for all experiments. Only one representative sample is shown for each *Lp*IRIP-expressing line. Assays were conducted in triplicate.

number of putative glycosylation sites have been identified in *Lp*IRIP family members (Kuiper *et al.*, 2001). Nevertheless, recombinantly produced *Lp*AFP retains IRI and TH activities, indicating that such modifications are not necessary for proper ice-binding activity (Lauersen *et al.*, 2011) and there has been no demonstration that *Lp*IRIPs are post-translationally modified *in planta*. Further experiments regarding the localization of IBPs lacking classical secretion signals, as well as the possible nonclassical mechanisms of protein secretion, could prove useful in understanding the roles that IBPs play in plant freeze tolerance.

We believe that a lack of knowledge regarding the mechanisms underlying IBP-induced freezing tolerance has hindered the

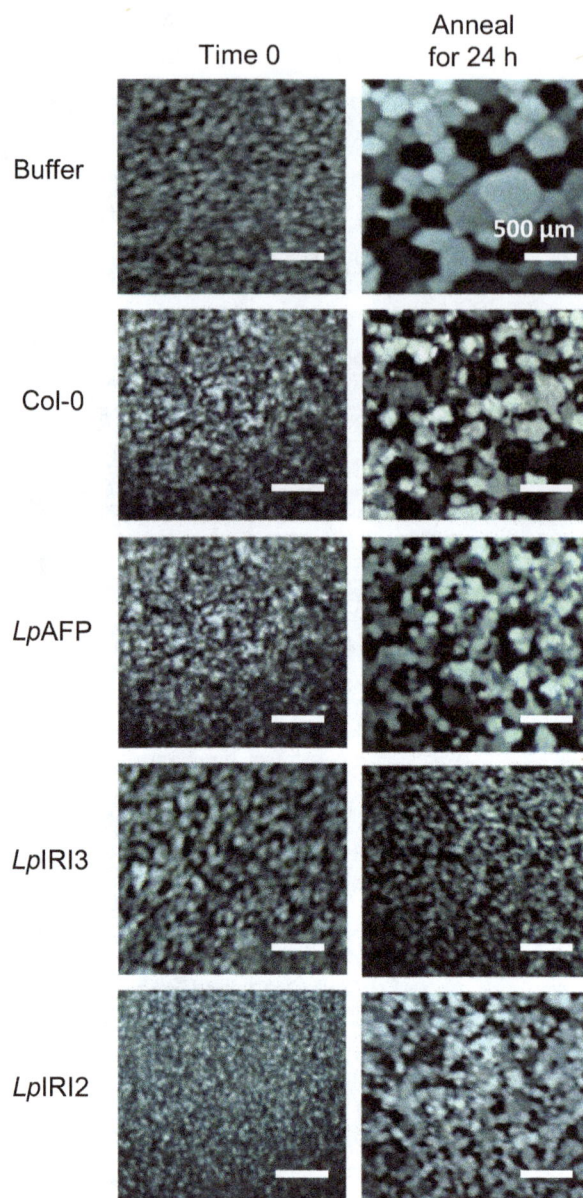

development of freeze-tolerant crops. In the past, a great deal of effort has been invested into the transfer of fish and insect AFPs to plants (e.g. Hightower *et al.*, 1991; Holmberg *et al.*, 2001; Huang *et al.*, 2002; Kenward *et al.*, 1999). These transgenes were considered attractive candidates given the ability of these proteins to depress the freezing point by several degrees. However, the expression of moderately TH-active fish AFPs in plants has not yielded favourable results, likely due to the catastrophic needle-like crystals that are formed once the freezing point is reached. These ice crystal burst patterns are due to

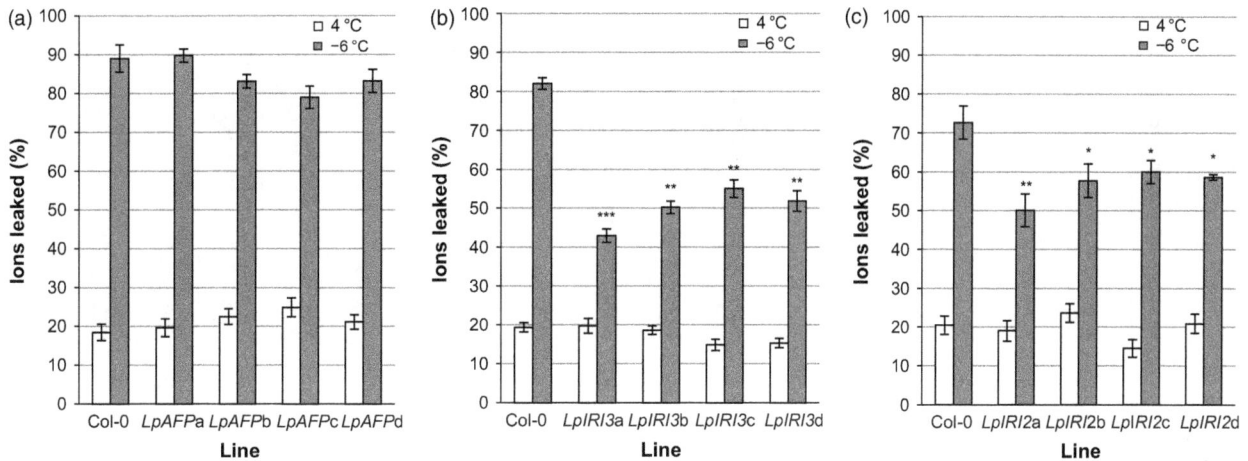

Figure 8 Ion leakage produced by the leaves of transgenic LpIRIP-expressing A. thaliana lines. Shown are the results of A. thaliana plants expressing LpAFP (a), LpIRI3 (b) and LpIRI2 (c) constructs following incubation at 4 °C or freezing to −6 °C. Transgenic plants were compared to nontransgenic A. thaliana plants (Col-0). Ion leakage is represented as the proportion of ions leaked following treatment in relation to the total number of ions in the leaf sample. All experiments were performed in triplicate (n = 12). Error bars represent standard errors of the mean and asterisks denote a significant reduction in ion leakage compared with controls (*P < 0.05, **P < 0.005, ***P < 0.0005; two-tailed t-tests).

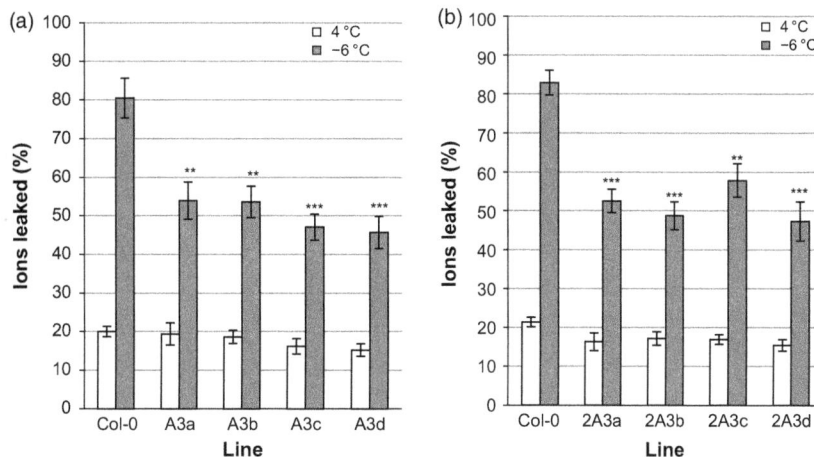

Figure 9 Ion leakage produced by the leaves of transgenic A. thaliana lines expressing multiple LpIRIPs. Plants expressing LpAFP and LpIRI3 (A3) (a) and lines transformed with LpIRI2, LpAFP and LpIRI3 (2A3) (b) were compared to nontransgenic A. thaliana (Col-0) plants. Leaves of transgenic and control A. thaliana plants were incubated at 4 °C or frozen to −6 °C prior to measuring ion leakage, which is represented as the proportion of ions leaked following treatment in relation to the total number of ions in the leaf sample. All experiments were performed in triplicate (n = 12 for each experiment). Error bars represent standard errors of the mean and asterisks denote a significant reduction in ion leakage compared with control plants **P < 0.005, ***P < 0.0005; two-tailed t-tests).

adsorption of AFPs to the primary prism plane exclusively, resulting in growth from the c-axis (Fletcher et al., 2001). Although insect AFPs direct the formation of hexagonal bipyramidal crystals, similar to those produced in the presence of plant IBPs, they have high TH activity. This could allow intracellular freezing at the same time or earlier than extracellular freezing, which would not be compatible with crop survival. We suggest that plant IBPs with their low TH activity, but relatively high IRI activity and 'gentle' burst morphologies (Middleton et al., 2012), should prove more efficacious for such applications. Certainly, given the inevitability that most overwintering plants will freeze, the use of IBPs that have evolved from a freezing tolerant survival strategy would logically appear to be more promising than AFPs that have evolved in species where freezing of the interstitial fluid

is lethal. Here, we have shown that the expression of LpIRIPs in A. thaliana produced a freeze-tolerant phenotype that was enhanced in the presence of more than one isoform. These results strongly suggest that the expression of multiple LpIRIP isoforms in a cold-adapted but freeze-susceptible crop may allow for even more substantial freezing tolerance capabilities than the striking ~2 °C seen here.

Experimental procedures

Bioinformatics analysis

The translated sequences corresponding to the open reading frames (ORFs) of LpIRI1 (GenBank accession no. EU680848), LpIRI2 (EU680849), LpIRI3 (EU680850), LpIRI4 (EU680851) and

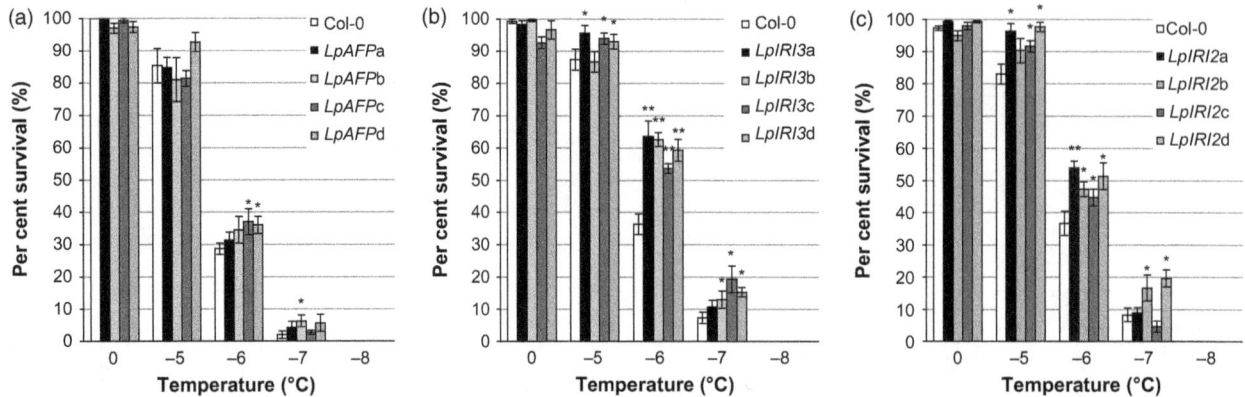

Figure 10 Freezing survival of transgenic *A. thaliana* expressing individual *Lp*IRIP isoforms. Shown are the survival rates of *LpAFP*- (a) *LpIRI3*- (b) and *LpIRI2*- (c) expressing *A. thaliana* following recovery from temperatures between 0 and −8 °C. Transgenic plants were compared to nontransgenic *A. thaliana* plants (Col-0). Error bars represent the standard errors of the mean and asterisks denote lines with significantly enhanced freezing survival compared with Col-0 plants (*P < 0.05, **P < 0.005; two-tailed *t*-tests). Experiments were conducted using 100 seedlings for each independently generated transgenic line, in triplicate.

Figure 11 Survival rates of transgenic *A. thaliana* plants expressing multiple *Lp*IRIPs. Survival was calculated for *A. thaliana* plants expressing *LpAFP* and *LpIRI3* (A3) (a) and *LpIRI2*, *LpAFP* and *LpIRI3* (2A3) (b) following recovery of plants from temperatures between 0 and −8 °C. Error bars represent standard errors of the mean. Asterisks denote a significant increase in freezing survival (*P < 0.05, **P < 0.005, ***P < 0.0005; two-tailed *t*-tests) compared with Col-0 lines. All experiments were performed using 100 seedlings per transgenic line, in triplicate.

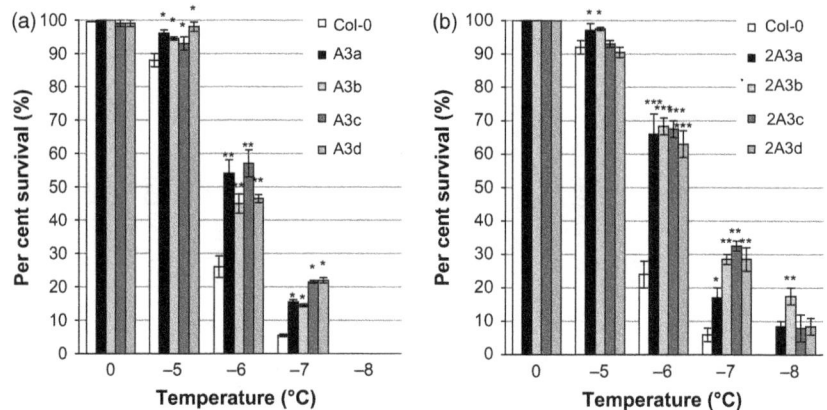

LpAFP (AJ277399) were aligned using ClustalW2 multiple sequence alignment tool (http://www.ebi.ac.uk/Tools/msa/clustalw2/). Predictions regarding the IRI-domain and putative ice-binding residues for *Lp*IRIP sequences were made based on alignment with *Lp*AFP, for which the IBF has been well characterized (Middleton *et al.*, 2009). SignalP 4.1 (http://www.cbs.dtu.dk/services/SignalP/) was used to predict sequences encoding putative signal peptides. The Phyre 2.0 algorithm (http://www.sbg.bio.ic.ac.uk/phyre2/html/page.cgi?id=index) was used to predict secondary protein structure.

Cloning and protein purification of recombinant LpIRIPs

Sequences corresponding to the ORFs of *LpIRI2* and *LpIRI3* were synthesized by GeneArt™ (Invitrogen, Carlsbad, CA), and the stop codons were removed by PCR with the primers LpIRI2NdeIFW/LpIRI2nostopXhoIRV (Table 3a) for *LpIRI2* and LpIRI3NdeIFW/LpIRI3nostopXhoIRV (Table 3a) for *LpIRI3*, in order to incorporate a 6-residue histidine tag to facilitate protein purification. Amplification was performed using Platinum Pfu Taq DNA polymerase (Invitrogen, Carlsbad, CA) using the following program: 2 min at 94 °C followed by 35 cycles of 30 sec at 94 °C, 30 s at 54 °C and 1 min 50 s at 72 °C, with a final extension at 72 °C for 7 min. The amplified

products were then ligated into pET24a(+) vectors (Novagen, Etobicoke, ON, CA) and transformed into ArcticExpress™ *Escherichia coli* cells (New England Biolabs Inc., Whitby, ON, CA) using chemical transformation, with each construct subsequently confirmed by sequencing (Plateforme de séquençage et de génotypage des genomes; Québec City, QC, CA) after each cloning step.

Bacterial cultures were grown to an optical density (OD) of 0.8 (λ = 595) and induced for 48 h at 16 °C, using 0.5 mM of isopropyl β-D-1-thiogalactopyranoside (IPTG). Cells were lysed using a French Press (ThermoFisher Scientific, Nepean, ON, CA) and recombinant proteins purified from soluble lysates using a nickel-NTA agarose column (Qiagen, Toronto, ON, CA) as previously described (Lauersen *et al.*, 2011). Purified proteins were dialysed against 50 mM Tris–HCl and 100 mM NaCl, pH 8.0, for 24 h and used immediately for ice-binding and protein assays or frozen at −20 °C until analysed.

Cloning of LpIRIP constructs for expression in A. thaliana

For expression in *A. thaliana*, the ORFs of *LpAFP*, *LpIRI2* and *LpIRI3* were PCR-amplified using LpAFPBglIIFW/LpAFPPmlIRV, LpIRI2BglIIFW/LpIRI2PmlIRV and LpIRI3BglIIFW/LpIRI3PmlIRV primers, respectively (Table 3a) with the following protocol: 95 °C

Table 3 Primer sequences used for the cloning of *Lp*IRIP constructs prior to transformation of *A. thaliana* plants (a) and reverse transcription PCR analysis of *LpIRI2* and *LpIRI3* transcripts (b), with underlined sequences correspond to internal restriction sites used for cloning purposes

	Primer	Primer sequence	Tm (°C)
(a) Construct			
pET24a(+):*LpIRI2*:His	LpIRI2NdeIFW	5′-TTAACATATGCCATTACATGTGAAGCG-3′	58
	LpIRI2nostopXhoIRV	5′-TTAACTCGAGACCTCCTGTCACGACTTTG-3′	58
pET24a(+):*LpIRI3*:His	LpIRI3NdeIFW	5′-TTAACATATGGCGAAATGCTTGATGCT-3′	58
	LpIRI3nostopXhoIRV	5′-TTAACTCGAGAGCGTCTGTCACGACTTTG-3′	58
pCAMBIA1305.1:*LpAFP*	LpAFPBglIIFW	5′-TTAAAGATCTTATGGATGAACAGCCGAAA-3′	64
	LpAFPPmlIRV	5′-TTAACACGTGTTAAGCGTCTGTCACGACT-3′	76
pCAMBIA1305.1:*LpIRI2*	LpIRI2BglIIFW	5′-TTAAAGATCTTATGCCATTACATGTGAAGCG-3′	58
	LpIRI2PmlIRV	5′-TTAACACGTGTTAACCTCCTGTCACGACTT-3′	58
pCAMBIA1305.1:*LpIRI3*	LpIRI3BglIIFW	5′-TTAAAGATCTTATGGCGAAATGCTTGATGCT-3′	58
	LpIRI3PmlIRV	5′-TTAACACGTGTTAAGCGTCTGTCACGACTT-3′	58
pCAMBIA1305.1:*LpAFP-LpIRI3*	35SBamHIFW	5′-TTTAAGGATCCCATGGAGTCAAAGATTCAAATAG-3′	62
	NOSterHindIIIRV	5′-TTAAAAGCTTGTTTACCCGCCAATATATCCT-3′	60
pCAMBIA1305.1:*LpIRI2-LpAFP-LpIRI3*	35SEcoRIFW	5′-TTTAAGAATTCCATGGAGTCAAAGATTCAAATAG-3′	62
	NOSterSacIRV	5′-TTAAGAGCTCGTTTACCCGCCAATATATCCT-3′	60
pCAMBIA1305.1:*mORANGE-LpAFP*	OFPBglIIFW	5′-AATTAGATCTTATGGTGAGCAAGGGCGAGG-3′	65
	LpAFPPmlIRV	5′-AATTCACGTGTTAAAGCTTTGCAGCGTCTGTCACG-3′	65
pCAMBIA1305.1:*LpIRI2-mORANGE*	LpIRI2NcoIFW	5′-TTAACCATGGTTATGCCATTACATGTGAAGCG-3′	58
	LpIRI2nostopBglIRV	5′-TTAAAGATCTGTACCTCCTGTCACGACTTTG-3′	58
pCAMBIA1305.1:*LpIRI3-mORANGE*	LpIRI3NcoIFW	5′-TTAACCATGGTTATGGCGAAATGCTTGATGCT-3′	58
	LpIRI3nostopBglIRV	5′-TTAAAGATCTGTAGCGTCTGTCACGACTTTG-3′	58
(b) Transcript			
LpIRI2	LpIRI2BglIIFW	5′-TTAAAGATCTTATGCCATTACATGTGAAGCG-3′	58
	LpIRI2PmlIRV	5′-TTAACACGTGTTAACCTCCTGTCACGACTT-3′	58
LpIRI3	LpIRI3BglIIFW	5′-TTAAAGATCTTATGGCGAAATGCTTGATGCT-3′	58
	LpIRI3PmlIRV	5′-TTAACACGTGTTAAGCGTCTGTCACGACTT-3′	58

for 2 min followed by 35 cycles of 95 °C for 45 s, 50 °C for 45 s and 72 °C for 1 min, with a final extension of 7 min at 72 °C. All fragments were ligated into pCAMBIA1305.1 vectors (Cambia, Canberra, ACT, AU) under the control of a cauliflower mosaic virus (CaMV) 35S promoter and a nopaline synthase (NOS) terminator. The construct for the simultaneous expression of *LpIRI3* and *LpAFP* (A3) was generated by PCR amplification of 35S:*LpAFP*:NOS with 35SBamHIFW/NOSterHindIIIRV primers (Table 3a), using the protocol described above, and inserting the amplified fragment into the pCAMBIA1305.1:*LpIRI3* construct. Subsequently, 35S:*LpIRI2*:NOS was amplified using 35SEcoRIFW/NOSterSacIRV (Table 3a) and inserted into the A3 construct for expression of all three isoforms (2A3), again using the same program.

The gene sequence corresponding to *mOrange-LpAFP* was ligated into pCAMBIA1305.1 following PCR amplification using OFPBglIIFW/LpAFPPmlIRV primers (Table 3a) under the following thermocycler program: 94 °C for 2 min, followed by 35 cycles of 94 °C for 30 s, 57 °C for 30 s and 72 °C for 1 min 30 s with a final extension of 72 °C for 10 min. *LpIRI2* and *LpIRI3* were amplified using LpIRI2NcoIFW/LpIRI2nostopBglIRV and LpIRI3NcoIFW/LpIRI3nostopBglIRV primers (Table 3a), respectively, using the PCR conditions described for recombinant *Lp*IRIP constructs, and inserted upstream of the mOrange tag to avoid disruption of the N-terminal signal sequence. In contrast, by placing the mOrange tag upstream of *LpAFP*, we could ensure that this protein remained intracellular, serving as control for cytoplasmic localization. Gene sequences were again confirmed by sequencing.

Plant materials and growth conditions

Lolium perenne seeds (Pacific Seed diploid variety; Premier Specific Seeds, Surrey, BC, CA) used for transcript analysis were grown in potting soil and maintained in a growth chamber (Queen's University, Kingston, ON, CA) on a 20-h/4-h light/dark cycle at 22 °C/18 °C with humidity and light regulated at 70% and 175 μmol/m²/s, respectively. Prior to reverse transcription (RT) PCR analysis, plants were grown for 3 weeks prior to a 6-day acclimation period at 21 °C (no light; control) or a cold acclimation period at 4 °C (no light).

All transgenic expression experiments were conducted using wild-type *A. thaliana* (ecotype: Col-0). For crude cell extractions, apoplast extractions, collection of guttation fluid and electrolyte leakage assays, *A. thaliana* were sown in potting soil and grown in a growth chamber for 3 weeks under standard growth conditions with a 16-h/8-h light/dark cycle at 22 °C/20 °C, 70% relative humidity, and light ~150 μmol/m²/s. Seeds used for survival assays were surface-sterilized using 70% (v/v) ethanol with 0.05% (v/v) Triton X-100 for 5 min, followed by a 95% (v/v) ethanol wash for 5 min, and plated on 0.5× Murashige and Skoog (MS) agar prior to transfer to standard growth conditions.

Prior to experimentation, plants were cold-acclimated (CA) at 4 °C, on a short-day cycle (6-h/18-h light/dark), with ~175 μmol/m²/s light for 48 h. These conditions were imposed for two reasons: first, we sought to prevent *Lp*IRIP misfolding, which occurs above 16 °C (Lauersen *et al.*, 2011), and secondly, we wished to limit the time at 4 °C, because *A. thaliana* reaches optimal freezing tolerance after 1 week of low-temperature exposure (Uemura et al., 1995). Thus, we attempted to balance the accumulation period of well-folded LpIRIPs and at the same time reduce the possible confounding effect of up-regulating other cold-induced freezing tolerance host mechanisms.

Generation of A. thaliana transgenic lines

Tagged and untagged LpIRIP constructs designed for expression in A. thaliana were transformed into GV3101 Agrobacterium tumefaciens cells by electroporation. Agrobacterium-mediated transformation of A. thaliana was carried out using the floral dip method (as described in Middleton et al., 2014). Successfully transformed plants were selected on 0.5× MS medium plates containing hygromycin (50 μg/mL). Four independent, homozygous lines were generated for each construct.

Endogenous LpIRIP transcript analysis

RNA was collected from the leaf tissue of CA L. perenne grown under the conditions described above. Extractions were performed using the RNeasy Plant Mini Kit (Qiagen, Toronto, ON, CA) followed by cDNA synthesis using Superscript® III First-Strand Synthesis System (Invitrogen, Carlsbad, CA) according to the manufacturers' specifications. RT-PCR was carried out using the following cycle conditions: 95 °C for 5 min, 94 °C for 30 s, 53 °C for 30 s and 72 °C for 1 min, for 45 cycles. LpIRI2 was amplified using LpIRI2BgIIIFW/LpIRI2PmlIRV primers (Table 3b) and LpIRI3 was amplified using LpIRI3BgIIIFW/LpIRI3PmlIRV primers (Table 3b). A 'housekeeping gene', s-adenosylmethionine decarboxylase transcript (SamDC), was amplified with primers as previously described (Hong et al., 2008) using the following cycle conditions: 95 °C for 2 min, followed by 35 cycles of 94 °C for 30 s, 55 °C for 30 s and 72 °C for 30 s.

Ice-binding and protein assays

Ice-binding assays were performed using protocols that have been optimized for plant IBPs as described previously. IRI activity was assessed using splat assays and ice-shaping activity and TH activity assayed using a nanolitre osmometer (Middleton et al., 2014). Prior to analysis, crude cell lysates (Lauersen et al., 2011), apoplast extracts (Villers and Kwak, 2013) and guttation fluid (Madsen et al., 2016) were prepared from 4-week-old CA A. thaliana leaf tissue as described previously. Plant extracts were suspended in a native protein extraction buffer (10 mM Tris–HCl, 25 mM NaCl, pH 7.5). Recombinant proteins used for ice-binding assays were prepared as described above. Protein concentration was determined using the Pierce™ BCA Protein Assay Kit (ThermoFisher Scientific, Nepean, ON, CA) following the manufacturer's instructions, with all experiments carried out in triplicate.

Electrolyte leakage assay

Electrolyte leakage assays using 4-week-old CA A. thaliana plants were carried out using a modified protocol from Thalhammer et al. (2014). Two mature rosette leaves were cut and placed in glass tubes containing deionized water (100 μL). Treatments were conducted by placing tubes in a circulating water bath at 0 °C and lowering the temperature to −1 °C, over 30 min. Samples were then nucleated using a single ice chip, and the temperature was decreased (1 °C every 15 min) to −6 °C. Following treatment, experimental samples were removed from the circulating water bath and allowed to recover overnight at 4 °C in the dark. Control samples were left covered in the dark at 4 °C, while treated experimental samples were prepared and

allowed to recover at 4 °C overnight. All cut leaf samples were then transferred to conical tubes (50 mL containing 20 mL of deionized water) and shaken for 18 h at 24 °C, 150 rpm. Initial (C_i) and final conductivity (C_f) measurements were taken before and after autoclaving samples using a direct reading conductivity meter (Bach-Simpson Ltd., London, ON, CA) and presented as a percentage (100 $C_i C_f^{-1}$) with 12 individual plants for each independent line, in triplicate. Significance was evaluated using two-tailed t-tests ($P < 0.05$).

A. thaliana freeze survival assay

Whole plant freeze survival assays were modified from Xin and Browse (1998). A. thaliana seedlings (~2000 per experiment) were grown for 4 weeks on MS agar plates rather than soil to reduce the presence of any ice-nucleating bacteria. Plates were then transferred to −1 °C (no light) for 2 h prior to nucleation using an ice chip in the centre of the plate and kept at this temperature for 12 h to ensure that the agar was frozen. The temperature was then lowered by 1 °C every 2 h, until the temperature reached −8 °C. One plate for each line was removed at temperatures indicated in the Results (between −5 and −8 °C) and allowed to recover at 4 °C (no light) for 24 h, prior to transfer to standard growth conditions for 14 days, when per cent survival and LT$_{50}$ were calculated. Assays were performed using 100 seedlings per independent line, in triplicate, with significant differences evaluated using a two-tailed t-test ($P < 0.05$).

Fluorescence readings

The level of mORANGE-tagged protein present in the apoplast extracts of transgenic A. thaliana plants was determined using the SpectraMax Gemini XS microplate reader (Molecular Devices, Sunnyvale, CA). Emission and excitation wavelengths of 562 nm and 548 nm, respectively, were used based on the specifications of the mORANGE protein product. Readings were taken at a total protein concentration of 0.1 mg/mL.

Confocal microscopy

Two-week-old seedlings grown on 0.5× MS agar plates were cold-acclimated for 48 h at 4 °C (6 h light) prior to visualization. A laser scanning microscope (LSM710; Ziess, Oberkochen, BW, DE) was used to visualize mOrange-tagged proteins (~543 nm). Images were obtained using ZEN 2009 software.

Acknowledgements

We acknowledge Dr. P. Davies for use of his nanolitre osmometer and Dr. Moez Hanin for generously donating the mOrange-LpAFP construct. We also thank Dr. K. Lauersen and Lena Dolman for their efforts in preliminary studies. This work was funded by the NSERC Discovery grant to VKW.

References

Antikainen, M. and Griffith, M. (1997) Antifreeze protein accumulation in freezing-tolerant cereals. Physiol. Plant. **99**, 423–432.

Bar-Dolev, M., Celik, Y., Wettlaufer, J.S., Davies, P.L. and Braslavsky, I. (2012) New insights into ice growth and melting modifications by antifreeze proteins. J. R. Soc. Interface, **9**, 3249–3259.

Beirão, J., Zilli, L., Vilella, S., Cabrita, E., Schiavone, R. and Herráez, M.P. (2011) Improving sperm cryopreservation with antifreeze proteins: effect on gilthead seabream (Sparus aurata) plasma membrane lipids. Biol. Reprod. **86**, 1–9.

Burcham, T.S., Knauf, M.J., Osuga, D.T., Feeney, R.E. and Yeh, Y. (1984) Antifreeze glycoproteins influence of polymer length and ice crystal habit on activity. *Biopolymers*, **23**, 1379–1395.

Cheng, F., Zamski, E., Guo, W.W., Pharr, D.M. and Williamson, J.D. (2009) Salicylic acid stimulates secretion of the normally symplastic enzyme mannitol dehydrogenase (MTD): a possible defense against mannitol-secreting fungal pathogens. *Planta*, **230**, 1093–1103.

Davies, P.L. and Hew, C.L. (2013) Biochemistry of fish antifreeze proteins. *FASEB J.* **4**, 2460–2468.

Duman, J.G. (1994) Purification and characterization of a thermal hysteresis protein from a plant, the bittersweet nightshade Solanum dulcamara. *Biochim. Biophy. Acta*, **1206**, 129–135.

Duman, J.G. (2001) Antifreeze and ice nucleator proteins in terrestrial arthropods. *Annu. Rev. Physiol.* **63**, 327–357.

Fei, Y.B., Sun, L.H., Huang, T., Shu, N.H., Gao, S.Q. and Jian, L.C. (1994) Isolation and identification of antifreeze protein with high activity in Ammopiptanthus mongolicus (in Chinese with English Abstract). *Acta Bot. Sin.* **36**, 649–650.

Fletcher, G.L., Hew, C.L. and Davies, P.L. (2001) Antifreeze proteins of teleost fishes. *Annu. Rev. Physiol.* **63**, 359–390.

Gilbert, J.A., Hill, P.J., Dodd, C.E. and Laybourn-Parry, J. (2004) Demonstration of antifreeze protein activity in Antarctic lake bacteria. *Microbiology*, **150**, 171–180.

Griffith, M., Ala, P., Yang, D.S., Hon, W.C. and Moffatt, B.A. (1992) Antifreeze protein produced endogenously in winter rye leaves. *Plant Physiol.* **100**, 593–596.

Gupta, R. and Deswal, R. (2012) Low temperature stress modulated secretome analysis and purification of antifreeze protein from Hippophae rhamnoides, a Himalayan wonder plant. *J. Proteome Res.* **11**, 2684–2696.

Gwak, Y., Jung, W., Lee, Y., Kim, J.S., Kim, C.G., Ju, J.H., Song, C. et al. (2014) An intracellular antifreeze protein from an Antarctic microalga that responds to various environmental stresses. *FASEB J.* **28**, 4924–4935.

Hays, L.M., Feeney, R.E., Crowe, L.M., Crowe, J.H. and Oliver, A.E. (1996) Antifreeze glycoproteins inhibit leakage from liposomes during thermotropic phase transitions. *Proc. Natl Acad. Sci. USA*, **93**, 6835–6840.

Hightower, R., Baden, C., Penzes, E., Lund, P. and Dunsmuir, P. (1991) Expression of antifreeze proteins in transgenic plants. *Plant Mol. Biol.* **17**, 1013–1021.

Holmberg, N., Farrés, J., Bailey, J.E. and Kallio, P.T. (2001) Targeted expression of a synthetic codon optimized gene, encoding the spruce budworm antifreeze protein, leads to accumulation of antifreeze activity in the apoplasts of transgenic tobacco. *Gene*, **275**, 115–124.

Hon, W.C., Griffith, M., Chong, P. and Yang, D. (1994) Extraction and isolation of antifreeze proteins from winter rye (Secale cereale L.) leaves. *Plant Physiol.* **104**, 971–980.

Hon, W.C., Griffith, M., Mlynarz, A., Kwok, Y.C. and Yang, D.S. (1995) Antifreeze proteins in winter rye are similar to pathogenesis-related proteins. *Plant Physiol.* **109**, 879–889.

Hong, S.Y., Seo, P.J., Yang, M., Xiang, F. and Park, C. (2008) Exploring valid reference genes for gene expression studies in *Brachypodium distachyon* by real-time PCR. *BMC Plant Biol.* **8**, 1–11.

Huang, T. and Duman, J.G. (2002) Cloning and characterization of a thermal hysteresis (antifreeze) protein with DNA-binding activity from winter bittersweet nightshade, Solanum dulcamara. *Plant Mol. Biol.* **48**, 339–350.

Huang, T., Nicodemus, J., Zarka, D.G., Thomashow, M.F., Wisniewski, M. and Duman, J.G. (2002) Expression of an insect (Dendroides canadensis) antifreeze protein in Arabidopsis thaliana results in a decrease in plant freezing temperature. *Plant Mol. Biol.* **50**, 333–344.

Kajava, A. and Lindow, S.E. (1993) A model of the three-dimensional structure of ice nucleation proteins. *J. Mol. Biol.* **232**, 709–717.

Kenward, K.D., Brandle, J., McPherson, J. and Davies, P.L. (1999) Type II fish antifreeze protein accumulation in transgenic tobacco does not confer frost resistance. *Transgenic Res.* **8**, 105–117.

Kim, H.J., Kato, N., Kim, S. and Triplett, B. (2008) Cu/Zn superoxide dismutase in developing cotton fibers: evidence for an extracellular form. *Planta*, **228**, 281–292.

Kuiper, M.J., Davies, P.L. and Walker, V.K. (2001) A theoretical model of a plant antifreeze protein from Lolium perenne. *Biophys. J.* **81**, 3560–3565.

KumbleKumble, K.D., Demmer, J., Fish, S.A., Hall, C., Corrales, S., DeAth, A., Elton, C., Prestidge, R., Luxmanan, S., Marshall, C.J. and Wharton, D.A. (2008) Characterization of a family of ice-active proteins from the Ryegrass, Lolium perenne. *Cryobiology* **57**, 263–268.

Lauersen, K.J., Brown, A., Middleton, A., Davies, P.L. and Walker, V.K. (2011) Expression and characterization of an antifreeze protein from the perennial rye grass, Lolium perenne. *Cryobiology*, **62**, 194–201.

Madsen, S.R., Nour-Eldin, H.H. and Halkier, B.A. (2016) Collection of Apoplastic Fluids from Arabidopsis thaliana leaves. *Methods Mol. Biol.* **1405**, 35–42.

Marentez, E., Griffith, M., Mlynarz, A. and Brush, R.A. (1993) Proteins accumulate in the apoplast of winter rye leaves during cold-acclimation. *Physiol. Plant.* **87**, 499–507.

Middleton, A.J., Brown, A.M., Davies, P.L. and Walker, V.K. (2009) Identification of the ice-binding face of a plant antifreeze protein. *FEBS Lett.* **583**, 815–819.

Middleton, A.J., Marshall, C.B., Faucher, F., Bar-Dolev, M., Braslavsky, I., Campbell, R.L., Walker, V.K. and Davies, P.L. (2012) Antifreeze protein fro freeze-tolerant grass has a beta-roll fold with an irregularly structured ice-binding site. *J. Mol. Biol.* **416**, 713–724.

Middleton, A.J., Vanderbeld, B., Bredow, M., Tomalty, H., Davies, P.L. and Walker, V.K. (2014) Isolation and characterization of ice-binding proteins from higher plants. *Methods Mol. Biol.* **116**, 255–277.

Nishimiya, Y., Sato, R., Takamichi, M., Miura, A. and Tsuda, S. (2005) Co-operative effect of the isoforms of type III antifreeze protein expressed in Notched-fin eelpout, Zoarces elongates. Kner. *FEBS J.* **272**, 482–492.

Pertaya, N., Marshall, C.B., Celik, Y., Davies, P.L. and Braslavasky, I. (2008) Direct visualization of spruce budworm antifreeze protein interacting with ice crystals: basal plane affinity confers hyperactivity. *Biophys. J.* **95**, 333–341.

Pudney, P.D., Buckley, S.L., Sidebottom, C.M., Twigg, S.N., Sevilla, M.P., Holt, C.B., Roper, D. et al. (2003) The physico-chemical characterization of a boiling stable antifreeze protein from a perennial grass (Lolium perenne). *Arch. Biochem. Biophys.* **410**, 238–245.

Raymond, J.A. and DeVries, A.L. (1977) Adsorption inhibition as a mechanism of freezing resistance in polar fishes. *Proc. Natl Acad. Sci. USA*, **74**, 2589–2593.

Regente, M., Pinedo, M., Elizalde, M. and de la Canal, L. (2012) Apoplastic exosome-like vesicles: a new way of protein secretion in plants? *Plant Signal Behav.* **7**, 544–546.

Rubinsky, B., Arav, A. and Fletcher, G.L. (1991) Hypothermic protection: a fundamental property of "antifreeze" proteins. *Biochem. Biophys. Res. Commun.* **180**, 566–571.

Sandve, S., Rudi, H., Asp, T. and Rognli, O. (2008) Tracking the evolution of a cold stress associated gene family in cold tolerant grasses. *BMC Evol. Biol.* **8**, 1–15.

Sandve, S.R., Kosmala, A., Rudi, H., Fjellheim, S., Rapacz, M., Yamada, T. and Rognli, O.A. (2011) Molecular mechanisms underlying frost tolerance in perennial grasses adapted to cold climates. *Plant Sci.* **180**, 69–77.

Sidebottom, C., Buckley, S., Pudney, P., Twigg, S., Jarman, C., Holt, C., Telford, J. et al. (2000) Phytochemistry: heat-stable antifreeze protein from grass. *Nature*, **406**, 256–256.

Siminovitch, D., Singh, J. and De La Rouche, I.A. (1978) Freezing behavior of free protoplasts of winter rye. *Cryobiology*, **15**, 205–213.

Simpson, D., Smallwood, M., Twigg, S., Doucet, C.J., Ross, J. and Bowles, D.J. (2005) Purification and characterization of a novel antifreeze protein from Forsythia suspensa (L.). *Cryobiology*, **51**, 230–234.

Smallwood, M., Worrall, D., Byass, L., Elias, L., Ashford, D., Doucet, C.J., Holt, C. et al. (1999) Isolation and characterization of a novel antifreeze protein from carrot (Daucus carota). *Biochem. J.* **340**, 385–391.

Thalhammer, A., Hincha, D.K. and Zuther, E. (2014) Measuring freezing tolerance: electrolyte leakage and chlorophyll fluorescence assays. *Methods Mol. Biol.* **1166**, 15–24.

Thomas, H. and James, A.R. (1993) Freezing tolerance and solute changes in contrasting genotypes of Lolium perenne L. acclimated to cold and drought. *Ann. Bot.* **72**, 249–254.

Thomashow, M.F. (1998) Role of cold-responsive genes in plant freezing tolerance. *Plant Physiol.* **118**, 1–8.

Tomalty, H. and Walker, V.K. (2014) Perturbation of ice nucleation activity by an antifreeze protein. *Biochem. Biophys. Res. Commun.* **452**, 636–641.

Tomczak, M.M., Hincha, D.K., Estrada, S.D., Feeney, R.E. and Crowe, J.H. (2001) Antifreeze proteins differentially affect model membranes during freezing. *Biochem. Biophs. Acta*, **155**, 255–263.

Uemura, M., Joseph, R.A. and Steponkus, P.L. (1995) Cold acclimation of Arabidopsis thaliana. *Plant Physiol.* **109**, 15–30.

Villers, F. and Kwak, J.M. (2013) Rapid apoplastic pH measurement in Arabidopsis leaves using a fluorescent dye. *Plant Signal. Behav.* **8**, e22587.

Wallis, J.G., Wang, H. and Guerra, D.J. (1997) Expression of a synthetic antifreeze protein in potato reduced electrolyte leakage release at freezing temperatures. *Plant Mol. Biol.* **35**, 323–330.

Xin, Z. and Browse, J. (1998) eskimo1 mutants of Arabidopsis are constitutively freezing-tolerant. *PNAS*, **95**, 7799–7804.

Xu, H., Griffith, M., Patten, C.L. and Glick, B.R. (1998) Isolation and characterization of an antifreeze protein with ice-nucleation activity from the plant growth promoting rhizobacterium Pseudomonas putida GR12-2. *Can. J. Microbiol.* **44**, 64–73.

Zhang, C., Fei, S.Z., Arora, R. and Hannapel, D.J. (2010) Ice recrystallization inhibition proteins of perennial ryegrass enhance freezing tolerance. *Planta*, **232**, 155–164.

2

A comprehensive draft genome sequence for lupin (*Lupinus angustifolius*), an emerging health food: insights into plant–microbe interactions and legume evolution

James K. Hane[1,2,3,#], Yao Ming[4,#], Lars G. Kamphuis[1,5,#], Matthew N. Nelson[5,6,†], Gagan Garg[1], Craig A. Atkins[5,6], Philipp E. Bayer[6], Armando Bravo[7], Scott Bringans[8], Steven Cannon[9,10], David Edwards[6,11], Rhonda Foley[1], Ling-ling Gao[1], Maria J. Harrison[7], Wei Huang[10], Bhavna Hurgobin[6,11], Sean Li[12], Cheng-Wu Liu[13], Annette McGrath[12], Grant Morahan[14], Jeremy Murray[12], James Weller[15], Jianbo Jian[4] and Karam B. Singh[1,5,*]

[1]CSIRO Agriculture, Wembley, WA, Australia

[2]Department of Environment and Agriculture, CCDM Bioinformatics, Centre for Crop and Disease Management, Curtin University, Bentley, WA, Australia

[3]Curtin Institute for Computation, Curtin University, Bentley, WA, Australia

[4]Department of Plant and Animal Genome Research, Beijing Genome Institute, Shenzhen, China

[5]UWA Institute of Agriculture, University of Western Australia, Crawley, WA, Australia

[6]School of Plant Biology, University of Western Australia, Crawley, WA, Australia

[7]Boyce Thompson Institute for Plant Research, Ithaca, NY, USA

[8]Proteomics International, Nedlands, WA, Australia

[9]USDA-ARS Corn Insects and Crop Genetics Research Unit, Crop Genome Informatics Lab, Iowa State University, Ames, IA, USA

[10]Department of Agronomy, Iowa State University, Ames, IA, USA

[11]University of Queensland, Brisbane, Qld, Australia

[12]Data61, CSIRO, Canberra, ACT, Australia

[13]John Innes Centre, Norwich Research Park, Norfolk, UK

[14]Centre for Diabetes Research, University of Western Australia, Crawley, WA, Australia

[15]School of Biological Sciences, University of Tasmania, Hobart, TAS, Australia

---

*Correspondence
e-mail karam.
singh@csiro.au

†Present address: Royal Botanic Gardens Kew, Natural Capital and Plant Health, Ardingly, RH17 6TN, UK.

#These authors contributed equally.

Keywords: Legume comparative genomics, synteny, whole-genome assembly, flowering time genes, polyploidy, Genistoids.

Summary

Lupins are important grain legume crops that form a critical part of sustainable farming systems, reducing fertilizer use and providing disease breaks. It has a basal phylogenetic position relative to other crop and model legumes and a high speciation rate. Narrow-leafed lupin (NLL; *Lupinus angustifolius* L.) is gaining popularity as a health food, which is high in protein and dietary fibre but low in starch and gluten-free. We report the draft genome assembly (609 Mb) of NLL cultivar Tanjil, which has captured >98% of the gene content, sequences of additional lines and a dense genetic map. Lupins are unique among legumes and differ from most other land plants in that they do not form mycorrhizal associations. Remarkably, we find that NLL has lost all mycorrhiza-specific genes, but has retained genes commonly required for mycorrhization and nodulation. In addition, the genome also provided candidate genes for key disease resistance and domestication traits. We also find evidence of a whole-genome triplication at around 25 million years ago in the genistoid lineage leading to *Lupinus*. Our results will support detailed studies of legume evolution and accelerate lupin breeding programmes.

Introduction

Lupins are grain legumes that form an integral part of sustainable farming systems and have been an important part of the human diet for thousands of years (Gladstones, 1970). Planted in rotation with cereal crops, lupins reduce the need for nitrogenous fertilizer, provide valuable disease breaks and boost cereal yields (Gladstones, 1970). Lupins thrive on low-nutrient soils due to their ability to fix atmospheric nitrogen in symbiosis with beneficial bacteria and efficiently take up phosphorus from soils (Gladstones, 1970). Consequently, they are effective ecological pioneers and able to colonize extremely impoverished soils such as coastal sand dunes and new lava soils set down by recently erupted volcanoes (Lambers et al., 2013).

Lupins have emerged as both a human health food and food-additive. Lupin seeds are rich in protein, ranging from 30% to 40% of whole seeds (Williams, 1979), with very little starch compared to other major grain legumes, for example chickpea and soya bean. The narrow-leafed lupin kernel contains 40%–45% protein and 25%–30% dietary fibre, and low fat and carbohydrate content (Lee et al., 2006). An important property of lupin kernel flour is as a food-additive (e.g. in bread); it increases satiety (thus reducing energy intake) and reduces insulin resistance, which are valuable properties in the context of the rising incidence of obesity and diabetes (Lee et al., 2006). Furthermore, lupin flour is increasingly used as a nongenetically modified alternative to soya bean products and is used to produce gluten-free foods such as pasta. Studies on lupin seed proteins have provided valuable information on their number and RNA/protein expression patterns (Foley *et al.*, 2011, 2015) as well as demonstrated that specific members are able to reduce glycaemia to comparable levels as achieved with metformin, a widely used hypoglycaemic drug (Magni *et al.*, 2004).

Lupins belong to a single genus, *Lupinus*, in a legume clade known as the genistoids, which are believed to have diverged early in the evolution of papilionoid legumes (Lavin *et al.*, 2005). There are an estimated 267 species of lupin distributed around the Mediterranean region ('Old World' lupins) and North and South America ('New World' lupins) (Drummond *et al.*, 2012). Andean *Lupinus* species in particular show a rate of speciation unparalleled in the plant kingdom with broad morphological diversity ranging from small prostrate herbs to tall trees (Drummond *et al.*, 2012; Hughes and Eastwood, 2006). Both annual and perennial species have found their niches in a vast array of ecological habitats across 100 degrees of latitude (Drummond *et al.*, 2012). Together, these properties make this genus exceptionally useful for testing hypotheses relating to genome evolution, adaptation and speciation.

While wild lupin species were cultivated as far back as 2000 BC in the Mediterranean and Andean regions, domestication of lupin species was completed only in the 20th century (Gladstones, 1970). The most widely grown domesticated species today is narrow-leafed lupin (*L. angustifolius*; NLL) (Lee *et al.*, 2006). Its domestication was initiated in Germany in the early 20th century and completed in the 1960s in Australia with the development of the first fully domesticated cultivar with low alkaloid content, nonshattering pods, permeable seeds and early flowering. Since then, NLL cultivation has grown to span more than 600 000 hectares in over 20 countries (FAO, 2013).

Over the last decade, various legume genomes that utilized reference genetic maps to order and orient scaffolds for pseudomolecule assembly have been published, and these include those of *Medicago truncatula* (Young *et al.*, 2011), chickpea (*Cicer arietinum*; Varshney *et al.*, 2013), pigeon pea (*Cajanus cajan*; Varshney *et al.*, 2012); common bean (*Phaseolus vulgaris*; Schmutz *et al.*, 2014) and soya bean (*Glycine max*; Schmutz *et al.*, 2010). In the latter two assemblies, synteny-based refinement methods were used in addition to the dense genetic maps to order and orient assembled scaffolds into pseudo-molecules (Schmutz *et al.*, 2010, 2014). A range of genomic resources have also been produced in recent years for the study of lupins, particularly NLL. These include genetic maps for NLL and white lupin (Croxford *et al.*, 2008; Kamphuis *et al.*, 2015; Kroc *et al.*, 2014; Nelson *et al.*, 2010; Yang *et al.*, 2013b) and large genomic insert libraries for NLL (Gao *et al.*, 2011; Kasprzak *et al.*, 2006). Transcriptomic resources have been developed for all four cultivated lupin species (Foley *et al.*, 2015; Kamphuis *et al.*, 2015; O'Rourke *et al.*, 2013; Parra-González *et al.*, 2012; Secco *et al.*, 2014; Wang *et al.*, 2014). Preliminary draft genome data had been generated for NLL and were used to assist molecular marker design (Gao *et al.*, 2011; Kamphuis *et al.*, 2015; Yang *et al.*, 2013a). In this study, we present the first high-quality draft genome for a genistoid legume, narrow-leafed lupin ($2n = 40$), report on a survey of its gene content and provide insights into its genome evolution, symbiotic relationships and host–pathogen interactions. Lupin, as a genus in the early-diverging genistoid lineage in the papilionoid subfamily, serves as an outgroup for the many crop and model species in this subfamily—an outgroup that shares many characteristics of other papilionoid legumes (such as symbiotic nitrogen fixation), but

with sufficient evolutionary distance to make inferences about the timing and histories of important molecular evolutionary events.

Results and discussion

The NLL genome assembly and gene features

The haploid genome size for NLL was previously estimated by flow cytometry to be 924 Mb (Kasprzak *et al.*, 2006; Naga-nowska *et al.*, 2003). K-mer-based estimation of genome size predicted a similar value of 951 Mb (Figure S1). Initial assembly of the Tanjil genome using only paired-end Illumina data produced 191 701 scaffolds in 521 Mb, with an N50 of 10 137 and N50 length of 13.8 kb. The assembly was improved via scaffolding with additional paired-end, mate-pairs and BAC-end data totalling an average coverage of 162.8 X (Table S1). This resulted in a contig assembly with 1 068 669 contigs, totalling 810 Mb or 85% of the K-mer-based estimated genome size. The final scaffold assembly after removing scaffolds less than 200 bp comprised 14 379 scaffolds totalling 609 Mb with a contig N50 length of 45 646 bp and scaffold N50 of 232 and scaffold N50 length of 703 Kb (Table 1).

The NLL genome is highly repetitive (57% of the genome) (Table S2), with over half its repeats (32% of the genome) matching known transposable elements (TEs) (Table 2). Typical of most eukaryotes, TEs were most commonly long terminal repeats (LTRs) retrotransposons (28%), with DNA LTRs, long interspersed nuclear elements (LINE) and short interspersed nuclear elements (SINE) TEs comprising a relatively small proportion (4.8%, 2.7% and 0.1%, respectively). Noncoding RNA was estimated to comprise 0.1% of the genome (Table S3), the majority being ribosomal RNA (0.035%) and transfer RNA (0.012%), with predicted snRNA and miRNA representing 0.009% and 0.006%. Analysis of divergence between known TEs (Jurka *et al.*, 2005) indicated a peak at ~30%, however the same analysis applied to *de novo* repeats produced a bimodal distribution with an additional less diver-gent peak at ~10% (Figure S2).

A total of 33 076 protein-coding genes were annotated (Figure S3) after combining evidence from transcriptome align-ments derived from five different tissue types (leaf, stem, root, flower and seed), protein homology, and *in silico* gene predic-tion (Table 3). Additionally, peptide data from proteomics analysis of leaf, seed, stem and root samples were mapped to both the translated gene annotations and the 6-frame transla-tion of the whole-genome assembly (Bringans *et al.*, 2009) (Table S4). Proteogenomic comparison of peptide-mapping versus gene annotation supported between 94 and 1134 annotations per tissue type (Table S4), and provided valuable information on tissue localization for the products of these genes. InterPro terms were the most informative functional annotation assigned to NLL proteins with 26 580 (80.4%) proteins annotated (Table S5). Comparing gene counts for Interpro terms in NLL to other plant species (source: PLAZA 3.0 (Van Bel *et al.*, 2011)) via Fisher's exact test, numerous Interpro terms were over-represented in NLL and were often significantly higher than most species, excepting *G. max* (Data S1). However, in a few cases, NLL InterPro terms were more abundant versus all species including *G. max*. These included tyrosine protein kinases, photosystem II cytochrome b559, porins and micro-tubule-associated proteins. The NLL assembly was also depleted in genes with InterPro terms corresponding to NBS–LRR proteins,

Table 1 Summary of the narrow-leafed lupin cv. Tanjil genome assembly. The assembly comprises scaffolds, the majority of which have been placed into pseudochromosomes based on genetic map and synteny data

Assembly statistics	Total length (bp)	Average length (bp)	Maximum length (bp)	Minimum length (bp)	N50	N50 length	Total # sequences	Total unknown N bases
Contigs	810 353 784	758	922 429	100	4 246	45 646	1 068 669	0
Scaffolds	609 123 749	42 362	4 089 732	200	232	703 185	14 379	4 078 848
Pseudochromosomes	470 424 067	23 521 203	36 457 581	16 251 777	8	24 697 652	20	3 351 285
Unplaced scaffolds	138 780 182	10 239	1 472 692	200	610	45 366	13 554	808 063

Table 2 Summary of transposon content in the narrow-leafed lupin cv. Tanjil genome assembly

	Repbase TEs		TE Proteins		De novo		Combined TEs	
	Length (bp)	% in Genome	Length (bp)	% in Genome	Length (bp)	% in Genome	Length (bp)	% in Genome
DNA	8 983 926	1.47	7 351 979	1.20	23 429 353	3.83	29 084 889	4.76
LINE	8 299 104	1.35	10 841 081	1.77	13 051 653	2.13	16 438 300	2.69
LTR	79 075 250	12.95	90 533 453	14.83	154 738 027	25.35	172 348 763	28.23
SINE	66 384	0.01	0	0	483 328	0.08	544 025	0.09
Other	3917	0.000642	0	0	0	0	3917	0.000642
Unknown	0	0	2988	0.00049	134 545 040	22.04	134 548 028	22.04
Total	95 943 148	15.71	108 715 411	17.81	319 388 057	52.32	331 905 409	54.37

Table 3 Summary of predicted protein-coding gene annotations of narrow-leafed lupin and their supporting evidence types

Gene set		Number	Average transcript length (bp)	Average CDS length (bp)	Average exon per gene	Average exon length (bp)	Average intron length (bp)
De novo	AUGUSTUS	34 525	2 983.98	1 252.69	5.49	228.06	385.36
	GENSCAN	29 436	10 570.55	1 367.30	6.22	219.52	1 760.19
	A. thaliana	48 717	2 815.40	968.47	3.72	260.12	678.23
Homolog	C. cajan	46 735	2 422.52	929.51	3.90	238.24	514.57
	C. arietinum	42 856	4 349.29	1 125.34	4.04	278.37	1 059.62
	G. max	39 433	3 648.01	1 245.43	4.55	273.19	675.09
	M. truncatula	61 321	2 454.66	843.16	3.10	271.33	764.67
	P. vulgaris	68 168	1 936.06	786.66	3.23	242.93	513.54
EST		1 795	2 134.32	606.22	3.16	191.27	704.40
GLEAN		32 413	3 568.05	1 305.97	5.58	233.78	493.22
RNA-seq		49 946	2 309.00	803.54	4.01	199.94	498.02
Final set		33 076	3 673.44	1 289.14	5.52	233.52	488.41

DNA helicases, peptidase C48, hAT transposases and certain transcription factors.

Chromosome-level analysis of the NLL genome assembly

Enhanced genetic map data were used to place NLL scaffolds in a chromosomal context. Fluidigm assays yielded 469 transcriptome-derived SNPs that were polymorphic in the RIL population ($n = 153$) derived from a domestic (83A:476) by wild (P27255) cross. An additional 8668 DArTSeq molecular markers, including 4767 presence/absence variants (DArT_PAV markers) and 3901 SNPs (DArT_SNP markers) were also applied. When combined with 830 previously reported sequence-associated marker loci and seven trait loci, a total of 9972 loci (Data S2) were used to generate the improved map (Data S3), which comprised 20 linkage groups that correspond to the haploid chromosome complement of NLL (Lesniewska et al., 2011). The genetic map covers 2500.8 cM, with an average interval size of 0.85 cM between 2959 nonredundant framework loci (Table S6). This map incorporated for the first time a small orphan cluster of markers into linkage group 20 (Kamphuis et al., 2015) and has evenly distributed linkage group lengths (cM) (Figure S4A) and average interval sizes (cM) (Figure S4B).

A combination of 7707 markers physically mapped unambiguously to the scaffold assembly, including the following new markers: 3492 DArTSeq markers with presence–absence polymorphism, 2975 DArTSeq markers with SNP polymorphism, 555 Fluidigm markers with SNP polymorphism and 685 other previously reported PCR-based markers (Gao et al., 2011; Kamphuis et al., 2015; Kroc et al., 2014) (Data S2). Twenty pseudomolecule sequences, ranging from 16.2 to 36.5 Mb, were built from 825 scaffolds. The pseudomolecule assemblies total 470 424 067 bp (77.2% of the full assembly length) (Table 1). Of these scaffolds,

820 were anchored to linkage groups and provisionally ordered and oriented using the high-density marker resource, and five were added on the basis of synteny comparisons, using all-by-all dot plot comparisons between the NLL pseudomolecules and remaining unplaced scaffolds, and five other legume genomes (*Glycine max, Lotus japonicas, Medicago truncatula* and *Phaseolus vulgaris*; Data S6). The five added scaffolds comprised 2 004 769 bp or 0.4% of the pseudomolecule length. Additionally, when marker resolution was insufficient to confidently order and orient scaffolds (primarily in pericentromeric regions, where recombination rates are very low), synteny with the species above was considered in the scaffold order and orientation, under the assumption that discontinuities in genomic synteny that occur precisely at NLL scaffold boundaries are likely due to misorientation or local misplacement.

Genome assembly validation and comparison to previous draft assembly

To validate the quality of the genome assembly a CEGMA analysis (Parra *et al.*, 2007) was conducted to identify whether the majority of core eukaryotic genes are present in the assembly. This showed 235 complete and eight partial core eukaryotic genes were present in the assembly which equates to 98.0% or 243 genes of the gene set of 248 genes (for details and the missing protein KOG id's see Data S4). The transcriptome data for five different tissue types was aligned to the NLL assembly and for four of the five datasets 98.5%–99.0% of reads mapped back to the assembly (Table S7), suggesting the majority of the gene-rich space of the NLL genome is captured in the assembly. For the root transcriptome 89.1% aligned back to the assembly, which could be due to contamination from the soil or soil microbes. Furthermore, of the 33 076 genes in the predicted gene set of the current assembly, 1.8% (596 genes) are absent in the previous draft assembly from 2013% and 47.5% (15 703 genes) had partial hits, whereas 50.7% (16 777 genes) had 100% complete alignment in the previous assembly. In conclusion, over 98% of the gene-rich space is captured in our assembly and it is a significant improvement of the fragmented draft assembly from (Yang *et al.*, 2013b) which had a scaffold N50 of 7319 scaffolds compared to 232 scaffolds for this assembly (Table 1).

CoReFinder (Collapse/Repeat Finder) was applied to 20 pseudochromosomes and unplaced scaffolds greater than 10 kb, and a total of 14 923 collapsed regions of 3 462 044 bp (0.58% of the genome) were identified (Data S5). In addition, a total of 66 301 repeated regions of 23 699 757 bp (3.89% of the genome) were identified. A copy number estimate of the repeated regions was also performed and ranged from 1.58 (pseudochromosome NLL-01) to 171.60 (Scaffold_486) (Data S5). In conclusion, the assembly captures the majority of the gene space (~98%) and shows a low level of collapsed genes.

Comparative genomics across legume species

Resequencing of additional NLL lines at 51.5–59.2× coverage (Table S8) allowed comparisons of sequence variation across the NLL lines Unicrop (early domesticated cultivar), 83A:476 and P27255 (wild accession), relative to the pseudochromosomes of the reference cv. Tanjil (Figure 1). This indicated that the wild P27255 was significantly divergent across all regions of the genome with 216 167 indels and 3 053 917 SNPs (Table S8). In contrast, domesticated lines exhibited lower levels of diversity overall with 47 113 indels and 606 035 SNPs for line 83A:476

and 81 375 indels and 1 099 966 SNPs for cultivar Unicrop. Several trait-associated markers (anthracnose and phomopsis resistance, flowering time, bitterness, pod shattering) could be mapped onto pseudochromosomes, facilitating 'reverse-genetic' nomination of candidate genes for disease resistance and domestication traits (Table S9).

Comparison of orthologous gene content across multiple plant species highlighted a significant proportion of proteins that are conserved between NLL and four other legume species (Figure S5, Table S10). Among these species, lupin possesses a relatively high number of expanded paralogous genes (Figure S6), second only to *Glycine max*—likely due in both cases to independent whole-genome duplication (WGD) and whole-genome triplication in the *Glycine* and *Lupinus* lineages, respectively.

We find clear evidence of a whole-genome triplication (WGT) in the genistoid lineage. This is inferred on the basis of synteny comparisons between NLL and itself and between NLL and other sequenced legume genomes (Data S6). Dot plots between NLL and another legume genome frequently show three strong, overlapping synteny blocks when these are viewed with respect to the other legume genome, or two blocks in the NLL self-comparison (with the third copy visible as the NLL self-match on the main diagonal). For the genomes *Lotus japonicus, Medicago truncatula* and *Phaseolus vulgaris*, the proportions of the NLL genome with a 'synteny coverage depth' of three with respect to the other genome are 21.4%, 21.0% and 13.2%, respectively (Table S11), while in comparisons going the 'other way' (with respect to NLL), the proportion of the genome with synteny coverage depth of three is negligible: 0.62%, 1.08%, 1.73%. In contrast, the proportion of those genomes with coverage depth at two (with respect to NLL) is high (14.2%, 27.5% and 28.6%), as expected, due to the papilionoid WGD (Table S11). In comparisons with *Glycine max*, the proportion of the genome with a 'synteny coverage depth' of three is 12.7% with respect to *Glycine*, while going the other way (with respect to NLL), the fourfold synteny coverage depth is greater than the threefold coverage depth (22.9% vs. 12.6%), as expected due to the additional WGD in the *Glycine* lineage.

Divergence times between *Lupinus* and other papilionoid legumes were calculated based on accumulation of synonymous changes between orthologous gene pairs between species (Figure S7), using a known species phylogeny and rooting the tree at the papilionoid WGD. The galegoid clade, containing *Lupinus*, is known to have originated near the base of the papilionoid subfamily (Lavin *et al.*, 2005). If the papilionoid WGD immediately preceded the papilionoid radiation (Cannon *et al.*, 2015), at ~58 Mya (Lavin *et al.*, 2005), then we estimate the genistoid lineage separated from the other papilionoid legumes at ~54.6 Mya, and the whole-genome triplication to have occurred in the genistoid lineage at ~24.6 Mya (Figure S7; Data S7).

These time estimates assume constant rates of synonymous nucleotide changes before and after the WGD. Additional taxon sampling in the Genisteae would be needed to refine the WGT timing; however, it is clear that the genistoid WGT is considerably older than the *Glycine* WGD, as Ks values for the WGT and WGD peaks are more than twofold greater in *Lupinus* than in *Glycine* (0.3 vs. 0.12).

From Ks analyses, we infer that the *Lupinus* lineage has accumulated point mutations at a rate similar to *Lotus* and *Glycine*, but more slowly than for *Phaseolus* or *Medicago*. This is apparent in papilionoid WGD peaks present at ~0.7 to 1.0, in self-comparisons between paralogs (Figure 2). Furthermore, a WGT is

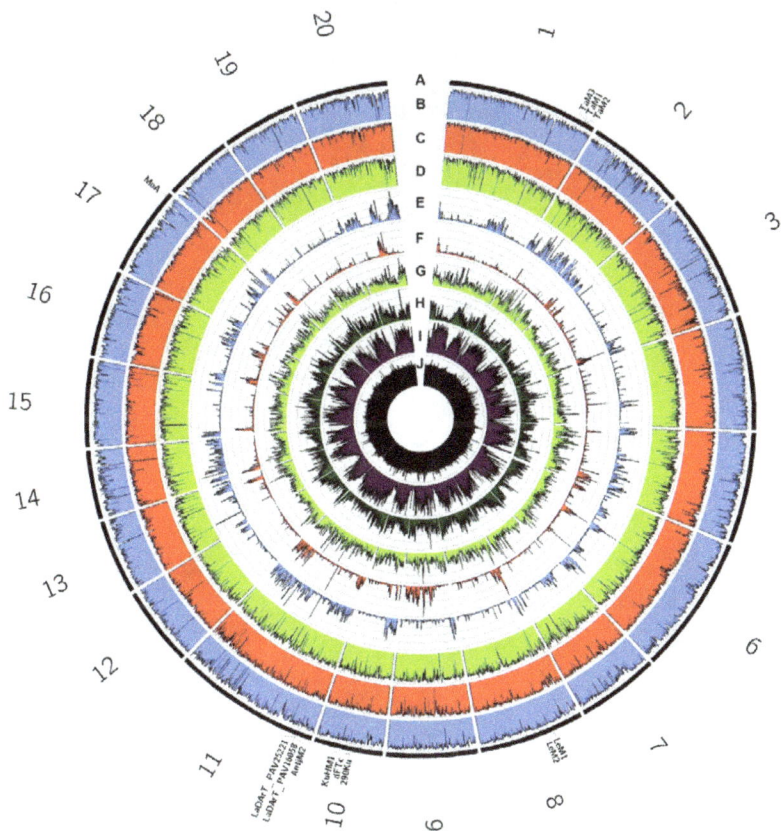

Figure 1 Summary of sequence variability in narrow-leafed lupin lines Unicrop, 83A:476 and P27255, relative to pseudochromosomes (corresponding to linkage groups) of the reference genome of cv. Tanjil. (A) Pseudochromosomes (black), with sequence-based genetic markers relevant to this study highlighted. (B–D) Per cent of 100-Kb windows covered by ≥5× read depth for resequencing data from lines Unicrop (b, blue), 83A:476 (C, red) and P27255 (D, light green). (E–G) Density of polymorphic sequence sites ranging from 0 to 20 000 variants/Mb calculated within 100-Kb windows, for lines Unicrop (E, blue), 83A:476 (F, red) and P27255 (G, light green). (H) Per cent of 100-kb windows representing annotated genes in cv. Tanjil (dark green). (i) Per cent of 100-kb windows represented annotated repetitive DNA in cv. Tanjil (purple). (j) Per cent G:C content ranging from 0% to 50%, calculated in 100-Kb windows, in cv. Tanjil (black).

evident in the genistoid lineage at around Ks ~0.3. This compares with the *Glycine* WGD peak at 0.12 and the papilionoid WGD at ~0.74 in *Lupinus* or ~0.68 in Glycine. If the papilionoid WGD occurred at ~58 Mya (Cannon *et al.*, 2015; Lavin *et al.*, 2005), then, assuming constant rates in this lineage, the genistoid WGT would have occurred at around 24.6 Mya.

Synteny comparisons with other sequenced legume genomes show extended regions of homology on all chromosomes, retained since divergence from the common ancestor of *Lupinus* and the other papilionoid species, which occurred ~55 Mya. For example, blocks spanning more than 6.4 million bases remain between soya bean and NLL (Table S12; Data S6, Data S7). Comparisons between NLL and soya bean generally show at least threefold synteny for NLL synteny viewed on soya bean as the reference, and at least fourfold synteny for soya bean synteny viewed on NLL as the reference, as both soya bean and NLL experienced the papilionoid WGD at ~58 Mya, and independent WGD at ~11 and WGT 24.6 Mya, respectively (see all-by-all chromosome dot plots for the NLL chromosomes compared to other NLL chromosomes and soya bean chromosomes in Data S6, and synteny depth coverage in Table S11). However, blocks are more degraded in NLL than soya bean. From the soya bean self-comparison, in the recent and papilionoid WGDs, the longest remaining blocks are 12.8 million and 3.46 million bases, respectively, while from the NLL self-comparison, in the recent and papilionoid WGDs, the blocks are 5.6 and 1.4 million bases, respectively. Average block lengths follow similar patterns, with the average 'old' (papilionoid) blocks from soya bean being 1.47 times longer than in NLL. The somewhat greater degradation in gene order in NLL is consistent with greater loss of paralogous genes (and decreased total gene count) in NLL than in soya bean.

Relating NLL gene content to industry-relevant phenotypes

Analysis of the annotated gene set using InterPro and Go-terms (Data S8) coupled with the dense reference genetic map (Data S3) allowed the nomination of candidate genes for phenotypes segregating in the recombinant inbred line (RIL) population. A major disease pressure on lupins, including NLL, is anthracnose (caused by *Colletotrichum lupini*). The cultivar Tanjil is resistant to anthracnose, and a single dominant resistance gene (*Lanr1*) maps to linkage group 11 (Kamphuis *et al.*, 2015; Yang *et al.*, 2013b). Using our new genetic map (Data S3), we refined the location of *Lanr1* to a single scaffold (Scaffold_133), between flanking markers LaDArT_PAV20595 and LaDArT_PAV25221 (Table S9). This region spans 388 kb, harbours 5 cosegregating markers and contains 41 predicted genes (*Lup005013.1-Lup005054.1*) including an NLR resistance gene (*Lup005042.1*). Alignment of *Lup005042.1* sequence from the four parents of the two RIL populations used to fine-map the location of *Lanr1* showed complete conservation for resistant lines Tanjil and 83A:476, but considerable divergence to susceptible lines Unicrop and P27255 (Figure S8), thus making *Lup005042.1* a good candidate for *Lanr1*.

Legumes typically undergo important symbiotic relationships with other organisms. This includes associations with beneficial bacteria to form rhizobium–legume symbiosis (RLS) and with beneficial fungi to form arbuscular mycorrhizal symbiosis (AMS). Some of the genes required for a successful association are shared by both types of symbioses, and it is believed that the evolutionary younger RLS recruited part of the genetic programme of the more ancient AMS (Parniske, 2008). Around 80% of land plants can form AMS, but some lineages have lost this ability along with some of the genes required to establish this

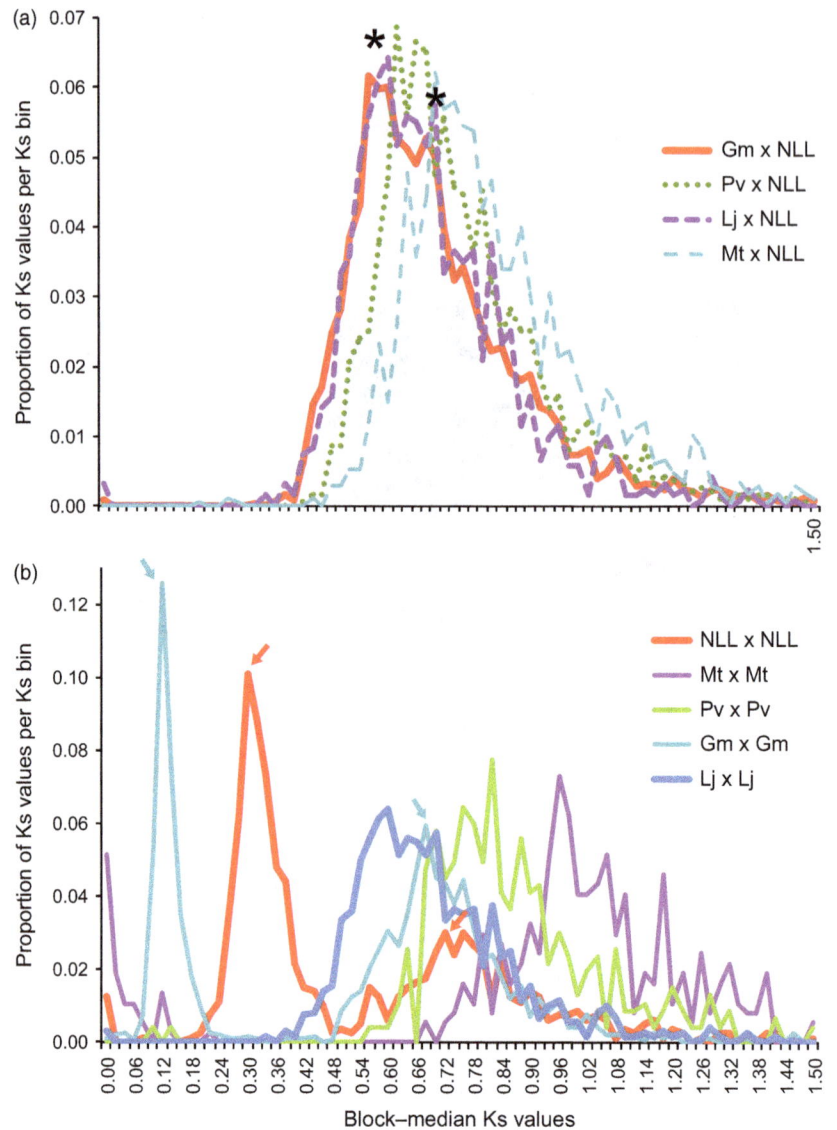

Figure 2 Synonymous substitution (Ks) analysis, showing proportion of values per Ks bin. Ks values are medians from synteny blocks for the indicated comparisons, and values in these plots are scaled to the total number of Ks counts for each comparison. (a) Orthologous comparisons between narrow-leafed lupin (NLL) and *Glycine max* (Gm, red line), *Phaseolus vulgaris* (Pv, green dots), *Lotus japonicus* (Lj, purple dashed line) and *Medicago truncatula* (Mt, blue dashed lines). Asterisks show a primary peak for the speciation-derived orthologs, and a probable smaller secondary peak for the papilionoid whole-genome duplication-derived 'old orthologs'. (b) Paralogous genome self-comparisons for narrow-leafed lupin, *Medicago truncatula*, *Phaseolus vulgaris*, *Glycine max* and *Lotus japonicus*. Coloured arrows show two peaks in both *Glycine max* and narrow-leafed lupin: the first peak in each case represents independent whole-genome duplications in these lineages (*Glycine* at ~11 Mya and *Lupinus* at ~24 Mya), and the second peaks correspond to the shared papilionoid whole-genome duplication. See Figure S7 and Data S7 for additional rate and date estimations.

relationship (Delaux *et al.*, 2014). Among legumes, lupins are unique, because they can form RLS but are unable to form AMS. This has been used to identify genes that are shared between both symbiotic associations (Bravo *et al.*, 2016; Delaux *et al.*, 2014). The NLL genome was screened for the presence of AMS genes and was found to include 20 of 38 characterized mycorrhizal association genes. These included genes involved in rhizobium–legume symbiosis, or the biosynthesis, regulation or transport of plant hormones. However, NLL lacked key genes required specifically for AM symbiosis (in *italics* in Table 4) but not nodulation, including *SbtM1*, *SbtM3*, *HA1*, *EXO70I*, *RAM2*, *PT4*, *STR1*, *STR2*, *RAM1*, *ERF1*, *RAD1*, *DIP1*, *FatM*, *KIN2*, *KIN3*, *KIN5*, *RFCb* and *CYT733A1* (Table 4; Data S9). The only exception was PP2AB'1, which so far is known only to be required for AMS (Charpentier *et al.*, 2014), but may play other, yet to be discovered roles in lupin biology. During nodulation, lupins become infected by rhizobia via intercellular penetration rather than through intracellular infection threads, as do most other legumes (González-Sama *et al.*, 2004). Short-infection-thread-like structures have been observed in cortical cells, but their importance is not clear (González-Sama *et al.*, 2004; James *et al.*, 1997; Tang *et al.*, 1992). Despite this, all genes known to be

required for rhizobial infection were present in NLL (Table 4), suggesting fundamentally conserved mechanisms underlying different infection modes.

We also examined genes involved in flowering time as early flowering is an important trait in NLL (Berger *et al.*, 2013). Most genes and gene families prominent in flowering time control and light signalling in other dicot species were represented in NLL (Table S13), with notable exceptions. These included the FLC clade of vernalization-responsive MADS-domain proteins, which appears to be broadly absent from legume genomes including NLL (Hecht *et al.*, 2005). Other genes appeared absent from NLL despite their presence in other papilionoid legumes (Table S13). These included the red light photoreceptor gene *PHYE*, which is present in Medicago and soya bean but absent in pea (Hecht *et al.*, 2005; Platten *et al.*, 2005; Schmid *et al.*, 2003; Yant *et al.*, 2010), suggesting it may have been lost more than once during legume evolution. A more striking case is the *FT* family which appears to consist of three relatively ancient clades in papilionoid legumes, *FTa*, *FTb* and *FTc*, with the *FTa* clade further divided into *FTa1* and *FTa3* subclades (Hecht *et al.*, 2011). Genes in the *FTa1* and *FTb* clades have significant roles in flowering time control in other legumes (Hecht *et al.*, 2011; Kong *et al.*, 2010; Laurie

Table 4 Overview of genes associated with arbuscular mycorrhizal and rhizobial associations in the genomes of *Medicago truncatula* and narrow-leafed lupin

Symbiotic component	Gene product	Medicago	Lupin	Reference*
NUP85	Nucleoporin	*Medtr1g006690*	*Lup020970.1*	1
NUP133	Nucleoporin	*Medtr5g097260*	*Lup029707.1*	2
NENA	Nucleoporin	*Medtr6g072020*	*Lup022917.1*	3
MCA8	Calcium pump	*Medtr7g100110*	*Lup028615.1 Lup006231.1 Lup028310.1 Lup018698.1*	4
DELLA1	Transcriptional regulators	*Medtr3g065980*	*Lup023873.1 Lup029445.1*	5
DELLA2	Transcriptional regulators	*TC182493*	*Lup007545.1 Lup009138.1*	5
CCD7	Carotenoid cleavage dioxygenase	*Medtr7g045370*	*Lup003751.1*	6
CCD8	Carotenoid cleavage dioxygenase	*Medtr3g109610 Medtr7g063800*	*Lup028507.1*	6
PDR1	ABC transporter	*Medtr3g107870 Medtr1g011640 Medtr1g011650*	*Lup013990.1 Lup001244.1*	7
D27	Carotenoid isomerase	*Medtr1g471050 Medtr7g095920*	*Lup011456.1 Lup018644.1*	8
SUT2	Sucrose transporter	*Medtr8g468330*	*Lup016593.1*	9
DMI1 (Pollux)	Cation channel	*Medtr2g005870*	*Lup014919.1*	10
NSP1	GRAS transcription factor	*Medtr8g020840*	*Lup007304.1*	11
NSP2	GRAS transcription factor	*Medtr3g072710*	*Lup012083.1*	12
DMI3 (CCaMK)	Calcium/calmodulin-dependent protein kinase	*Medtr8g043970*	*Lup001774.1*	13
DMI2 (SYMRK/NORK)	Receptor-like kinase	*Medtr5g030920*	*Lup025527.1*	14
Castor	Cation channel	*Medtr7g117580*	*Lup029273.1*	15
VAPYRIN	MSP and ANK repeat-containing protein	*Medtr6g027840*	*Lup000011.1 Lup001531.1*	16
IPD3 (Cyclops)	Coiled-coil domain containing protein	*Medtr5g026850*	*Lup027672.1*	17
NFP	LysM receptor-like kinase	*Medtr5g019040*	*Lup012981.1*	18
PP2AB'1	Protein phosphatase 2A	*Medtr1g112940*	*Lup024672.1*	19
LYK3	LysM receptor kinase	*Medtr5g086130*	*Lup018960.1*	20
ERN1	Transcription factor	*Medtr7g085810*	*Lup000007.1*	21
ERN2	Transcription factor	*Medtr6g029180*	*Lup009942.1*	22
NIN	Transcription factor	*Medtr5g099060*	*Lup029716.1*	23
NF-YA1	Transcription factor	*Medtr1g056530*	*Lup000323.1*	24
NF-YA2	Transcription factor	*Medtr7g106450*	*Lup019646.1*	25
RPG	Coiled-coil protein	*Medtr1g090807*	*Lup001677.1*	26
LIN	E3 ubiquitin ligase	*Medtr1g090320*	*Lup001700.1*	27
PUB1	E3 ubiquitin ligase	*Medtr5g083030*	*Lup029507.1*	28
SUNN	LRR receptor kinase	*Medtr4g070970*	*Lup003404.1*	29
NPL	Pectate lyase	*Medtr3g086320*	*Lup011017.2*	30
CRE1	Cytokinin receptor	*Medtr8g106150*	*Lup008799.1*	31
FLOT4	Flotillin	*Medtr3g106430*	*Lup030707.1*	32
SYP132A	Syntaxin	*Medtr2g088700*	*Lup029417.1 Lup030298.1*	33
SbtM1	*Subtilisin-like protease*	*Medtr5g011320*		34
SbtM3	*Subtilisin-like protease*	*Medtr5g011340*		34
HA1	*ATPase*	*Medtr8g006790*		35,36
Exo70	*Exocyst complex protein*	*Medtr1g017910*		37
RAM2	*GPAT*	*Medtr1g040500*		38
PT4	*Phosphate transporter*	*Medtr1g028600*		39
STR1	*Half-ABC transporter*	*Medtr8g107450*		40
STR2	*Half-ABC transporter*	*Medtr5g030910*		40
RAM1	*GRAS transcription factor*	*Medtr7g027190*		41
ERF1	*Transcription factor*	*Medtr7g009410*		42
RAD1	*GRAS transcription factor*	*Medtr4g104020*		43
DIP1	*GRAS transcription factor*	*Medtr8g093070*		44
FatM	*Acyl-(ACP) thioesterase*	*Medtr1g109110*		45
KIN2	*Protein kinase*	*Medtr4g129010*		45
KIN3	*Protein kinase*	*Medtr7g116650*		45
KIN5	*Serine–threonine protein kinase*	*Medtr3g104900*		45
RFCb	*Replication factor C*	*Medtr3g118160*		45
CYT733A1	*P450 enzyme*	*Medtr6g034940*		45

*References for each of the functionally characterized genes in relation to either AM symbiosis or Rhizobia symbiosis can be found in Data S10.

et al., 2011; Zhai *et al.*, 2014), but both groups of genes are absent from the NLL draft assembly, which has only duplicated copies of *FTa3* and *FTc* genes. Furthermore, microsynteny analysis between chickpea, common bean, medicago, soya bean and NLL showed conservation of the genes flanking the *Ft* genes in legumes with the *FTa1*, *FTa2* and *FTb* absent in the NLL genome assembly, whereas *FTc1* and *FTa3* are present in the Tanjil assembly (Figure S9). The *FTa1* and *FTb* clade genes are also not found in any of the comprehensive NLL transcriptome datasets (Kamphuis *et al.*, 2015). This implies that the strong vernalization response of NLL (Berger *et al.*, 2012) involves a mechanism distinct from that in *Medicago truncatula* where *FTa1* is the major target (Laurie *et al.*, 2011).

Conclusion

The comprehensive draft assembly of NLL (cultivar Tanjil) is the first representative of the genistoid clade of *Papilionoideae* legumes and will support further whole-genome analysis of other species in this important clade. Resequencing of additional lupin lines and in-depth transcriptome sequencing revealed widespread polymorphisms that were used to generate a dense reference genetic map. These resources are accessible through the Lupin Genome Portal (http://www.lupinexpress.org) which includes interactive BLAST, GBrowse and CMap interfaces (Donlin, 2009; Priyam *et al.*, 2015; Youens-Clark *et al.*, 2009) and provides a platform for genome-wide association studies and genomics-based breeding approaches. The knowledge of germplasm diversity, and capacity for reverse genetics facilitated by the dense genetic map and pseudomolecule assembly can accelerate future breeding of elite cultivars. This will fortify efforts to improve lupins as human health food crops and increase yield stability and productivity of lupins for farmers worldwide.

Experimental procedures

Library preparation and sequencing

Paired-end Illumina gDNA libraries of 100 bp length and 170, 500 and 800 bp insert sizes were generated (82.2× coverage). This was complemented by mate-paired libraries of 50 bp read length and 2, 4, 10, 20 and 40 Kb insert sizes generating a total of 150.41 Gb or 162.8× coverage. Illumina sequence reads were trimmed for adapter and low-quality sequences via CutAdapt v1.1 (min length 25 bp, rounds 3, match length 5 bp) (Martin, 2011). Mate-paired libraries were filtered for contaminating paired-end reads by merging pairs of reads with overlapping 3′ sequences via FLASH v1.2.2 (Magoč and Salzberg, 2011). Additional RNA-seq Illumina data used in this project to complement genome sequence data were described in a previous study (Kamphuis *et al.*, 2015). The total genome size for narrow-leafed lupin was estimated by performing a 17-mer frequency analysis of genomic paired-end libraries via Kmerfreq (Liu *et al.*, 2013), using the following equation: total genome size = (K-mer frequency/ primary peak depth).

Genetic mapping

To assign scaffolds to linkage groups, we developed additional transcriptome-derived SNP markers compatible with the Fluidigm microfluidic array platform as previously described (Kamphuis *et al.*, 2015) to add to the 1475 loci of the previous reference genetic map. These new Fluidigm SNP assays (469) were used to genotype the same 153 recombinant inbred lines (RILs) developed from a cross between 83A:476 (an Australian breeding line) and P27255 (a wild accession from Morocco) used previously to generate a genetic map (Nelson *et al.*, 2006). Additionally, DArTSeq analysis (Diversity Arrays Technology, Canberra) was performed, resulting in a further 3901 SNP-polymorphic markers and 4765 markers polymorphic by allele presence/absence. These new markers, together with the 830 previously reported STS markers, giving a total of 9972 markers and seven phenotypic trait loci (Kamphuis *et al.*, 2015), generated an improved genetic map prepared with the aid of MultiPoint 3.1 (MultiQTL Ltd., Haifa, Israel) using the approach detailed in our previous study (Kamphuis *et al.*, 2015).

De novo genome assembly and validation

Paired-end Illumina data were assembled via SOAPdenovo2 (Luo *et al.*, 2012) producing an initial assembly that was further scaffolded by SSPACE2 v2.0 (Boetzer *et al.*, 2011), progressing iteratively through paired-end (170, 500 and 800 bp) and mate-paired (2, 4, 10, 20 and 40 Kb) sequence libraries in order of increasing insert size. Five rounds of scaffolding were performed for each insert library, followed by five rounds of gap-closing via BGI GapCloser (Luo *et al.*, 2012) using paired-end sequences only. Further scaffolding was performed using BAC-end sequence data (insert size ~100 Kb) (Gao *et al.*, 2011), via Bambus (range 50–400 000 insert) (Pop *et al.*, 2004). The length of assembly 'gaps' (i.e. unknown stretches represented by runs of >10 'N' bases) was corrected to a uniform 100 bp. Scaffolds were screened for simple repeats via RepeatMasker (-no_is -norna - noint) (Smit *et al.*, 1996–2010) and tandem repeats finder (2 7 7 80 10 50 50 -f -d -m -h) (Benson, 1999). Sequences <200 bp length or with >=50% repetitive, simple repeat or unknown N bases were removed from the assembly as per GenBank requirements. The assembly was then validated versus the new genetic map generated in this study. Sequence-based genetic markers were mapped to scaffolds via ePCR (Schuler, 1997) and BLASTN ((Altschul *et al.*, 1990); e ≤ 1-e05). Marker location on scaffolds was determined preferentially by ePCR (max. 2 gaps, 2 mismatches, amplicon range 10–1000 bp for markers designed for the Fluidigm platform, 10–5000 bp for other markers), where *in silico* PCR produced a single amplicon for the match with the minimum possible hamming distance (mismatches+gaps). Where *in silico* PCR could not determine an unambiguous marker location, the locations of unambiguous top BLASTN hits for known marker amplicon sequences to the scaffolds were used instead. Scaffold joins were compared to marker order on the genetic map, and where a conflict was found, preliminary scaffolds were split on all 'gaps' located between conflicted markers. Whole-genome alignments to the *Glycine max* genome assembly using promer and mummerplot (Kurtz *et al.*, 2004), generally filtering at 90% identity and requiring maximum unique matches (–mum), were also used to manually split scaffolds where 'macrosynteny' was observed and this did not conflict with genetic map data. This final filtered and validated set of scaffolds was then assembled into pseudochromosomes based on the 20 linkage groups of the genetic map. Where possible, scaffolds were assigned to linkage groups in the order of their constituent markers on the map, reverse complemented if indicated by two or more markers. Synteny versus *G. max* was also used to manually place scaffolds on the map, particularly where abrupt disruption of synteny corresponded to neighbouring scaffold termini. Scaffolds placed on linkage groups were subsequently joined by uniform unknown gap lengths of 100 bp to form pseudochromosomes.

Annotation of genes and other genome features

Transcriptome sequences for cv Tanjil were previously assembled by Kamphuis et al. (2015), and in this study additional transcriptomes for cv Unicrop and P27255 were generated by the same method. Annotation of gene structure in the cv Tanjil reference genome was predicted de novo using AUGUSTUS (Stanke et al., 2006) and GENSCAN (Burge and Karlin, 1997). Further support for gene annotations was provided through alignment to the genome assembly of EST sequences derived from GenBank EST records listed under the taxon 'Fabaceae' (Benson et al., 2013), and homology to proteins of Arabidopsis thaliana (Initiative, 2000), Cajanus cajan (Varshney et al., 2012), Cicer arietinum (Varshney et al., 2013), Glycine max (Schmutz et al., 2010), Medicago truncatula (Young et al., 2011) and Phaseolus vulgaris (Schmutz et al., 2014). De novo predictions were combined and curated with supporting homology and EST evidence via GLEAN (Elsik et al., 2007). RNA-seq data were aligned to the genome via TopHat (Trapnell et al., 2009), assembled transcripts via Cufflinks (Trapnell et al., 2010) and predicted open reading frames according to transcript alignments. GLEAN results were aggregated with RNA-seq-supported gene models to produce the final gene set. Functional annotations were assigned to genes based on searches against Interpro (Quevillon et al., 2005), KEGG (Kanehisa and Goto, 2000), GO (Ashburner et al., 2000) and UniProt (The UniProt Consortium, 2013).

Repetitive DNA regions were predicted in the genome for both transposable elements (TEs) and tandem repeats. Annotation of TEs was based on homology and de novo methods. The homology approach used RepeatMasker v3.30 (Smit et al., 1996–2010) (with RepeatProteinMasker) to identify repeats matching known repeat sequences in Repbase v16.10 (Jurka et al., 2005). The de novo method predicted repetitive DNA via Repeatmodeler v1.0.5 (Smit and Hubley, 2010). Tandem repeats were predicted using Tandem Repeats Finder v4.04 (Benson, 1999).

Proteogenomics

Samples of NLL were obtained from leaf, seed, stem and root tissues and protein extracted and subjected to iTRAQ by Proteomics International using to the iTRAQ protocol (Sciex, USA).

Spectral data were analysed using ProteinPilot™ 4.0 Software (Sciex) against query and decoy databases generated from both translated gene annotation and six-frame-translated open reading frames. The database of potential open reading frames was generated by obtaining the six-frame translation of scaffolds via EMBOSS getorf (between stop codons, ≥10 aa in length). The spectral data were exported as XML files with proteogenomic mapping of peptides to scaffold and pseudochromosome sequences performed with CDSmapper (http://sourceforge.net/projects/cdsmapper/).

Comparative genomics

Analysis of variation across the cv Tanjil genome

SNP and indel sequence variation was assessed across a panel of cultivars relative to the cv. Tanjil reference genome. NGS reads were aligned to the cv Tanjil reference genome via bowtie v 2.0.5 (–very-sensitive) (Langmead and Salzberg, 2012), and variants were called via the Genome Analysis Tookit 3.4-46 (McKenna et al., 2010). GATK was used to perform read deduplication via Markduplicates,

then variant calling with HaplotypeCaller (–stand_call_conf 20 –stand_emit_conf 20 –min_pruning 5), producing variant data in VCF format (Danecek et al., 2011). Genome comparisons were visualized using Circos v0.67-1 (Krzywinski et al., 2009).

Orthologous gene clusters were predicted via OrthoMCL (Li et al., 2003) comparing translated annotations of NLL to protein datasets from C. cajan (Varshney et al., 2012), C. arietinum (Varshney et al., 2013), G. max (Schmutz et al., 2010), M. truncatula (Young et al., 2011), P. vulgaris (Schmutz et al., 2014) and A. thaliana (Initiative, 2000).

Analysis of rates of silent-site substitutions was carried out by searching all peptides against all others for the species Lupinus angustifolius, Glycine max (v 2.0), Lotus japonicus (v 3.0), Medicago truncatula (v 4.0) and Phaseolus vulgaris (v 1.0). Top respective matches were retained between each species per chromosome pairing (allowing for multiple total hits between two species for a given query gene), and within each species (for analysis of whole-genome duplications). Then in-frame alignments of coding sequences were made for each retained peptide alignment. From alignments of coding sequences, values for K_s, K_a and K_a/K_s were calculated using the 'codeml' method from the PAML package (Yang, 2007). Also from protein alignments, synteny blocks were inferred using DAGchainer (Haas et al., 2004). From the per-gene-pair alignments and the synteny blocks, median K_s values for blocks were calculated and used for K_s histogram peaks (Figure 2).

Ages of species divergences and whole-genome duplications (Figure S7) were calculated from modal K_s peaks (Data S7), by treating initially unknown branch lengths in the known species/duplication tree as variables in a set of equations. The species/duplication tree was rooted at the papilionoid whole-genome duplication, which predated the main papilionoid radiation (Cannon et al., 2015). A time of 58 Mya for the initial papilionoid radiation was assumed (Lavin et al., 2005). There were 11 unknown branch lengths in the tree in Figure S7, and sufficient data from the modal distances between and within species comparisons to solve for these unknowns algebraically.

To evaluate evidence for a whole-genome triplication (WGT), synteny blocks were identified using DAGchainer, and synteny coverage depth was calculated using the BEDTools v2.25.0 (Quinlan and Hall, 2010) 'coverage' function to make comparisons between other genomes and NLL as the reference, or between NLL and each other genome as the reference. Coverage of synteny blocks was calculated at each nucleotide position using the -d option and summarized per coverage depth level.

For visual dot plot assessments of NLL compared with itself and with other legume genomes, we used promer and mummerplot from the MUMmer package (Kurtz et al., 2004), (v3.23) to make comparisons of translated nucleotide sequence, on genomic sequence that was masked for all except exonic sequence. The promer results were filtered to require at least 80% identity.

Genome assembly validation and comparison to previous draft assembly

The quality of the Tanjil draft assembly was evaluated using the default parameters of CEGMA (Core Eukaryotic Genes Mapping Approach) v 2.5 (Parra et al., 2007).

CoReFinder (Collapse/Repeat Finder) is a differential comparative read mapping pipeline, which identifies and discriminates between collapsed and repeated regions in genome assemblies. Paired-end reads with insert sizes of 170 bp, 500 bp and 800 bp totalling a coverage of 138.63× were aligned to the assembly using SOAPaligner v2.21 (Li et al., 2009b), with mapped reads reported in three ways via the '−r' parameter: −r0 (reads that map uniquely), −r1 (reads that map to more than one location, but only one random hit is reported) and −r2 (report all hits) and converted to sorted .bam files using SAMtools v1.2 (Li et al., 2009a). The .bam files were then split into pseudochromosomes/scaffolds using BamTools v2.4.0 (Barnett et al., 2011), such that for each pseudochromosome/scaffold there were three .bam files corresponding to each mapping. The per-base coverage was calculated for each .bam file using BEDTools v2.25.0 (Quinlan and Hall, 2010). The BEDTools output was merged such that each pseudochromosome/scaffold had a single tab-delimited output file consisting of the name of the pseudochromosome/scaffold, the position, the per-base coverage for −r0, the per-base coverage for −r1 and the per-base coverage for −r2.

For each pseudochromosome/scaffold, a custom R script was used on the tab-delimited file to mine for collapsed and repeated regions by iterating through each position in the file. Any region where the median per-base coverage of −r0, −r1 and −r2 was greater than twice the overall median coverage was flagged as 'coll' (collapsed). Any region where the coverage for −r0 was between 0 and 2, the coverage for −r1 was greater or equal to 2, and the coverage for −r2 was 0.5 times the overall median was marked as 'rnc' (repeated, non-collapsed). Regions that were marked as 'coll' or 'rnc' that were within 100 bp of each other were merged using BEDTools.

NGS reads were aligned to the previous draft assembly and the current Tanjil assembly via bowtie v 2.0.5 (−very-sensitive) (Langmead and Salzberg, 2012), and RNASeq datasets for the five tissue types were aligned to the assembly using TopHat v 2.0.9 (−b2-very-fast −r 50 −mate-std-dev 200 -i 20 -I 4000 -g 20 -report-secondary-alignments -m 0 −min-coverage-intron 20 −coverage-search −microexon-search) to determine and compare the coverage of the various paired-end, mate-pair and RNASeq datasets in the two Tanjil draft assemblies.

Comparison of gene function across Lupinus and other plant taxa

To observe general variation in gene function, functional annotations were assigned to the proteins of NLL cv Tanjil via Interproscan (Quevillon et al., 2005) and compared to those assigned in other plant species available from the PLAZA Dicots v3.0 and Monocots v3.0 databases (Van Bel et al., 2011). Fisher's exact test was applied to the number of genes assigned an Interpro term in NLL versus Glycine max, or the average of various groups of species: legumes, dicots, monocots and all available Viridiplantae (Data S1). In the Supplementary Data File provided, further filtering has been applied requiring an expansion in NLL (gene count fold change > 1) and a P-value of ≤ 0.05.

To focus on variation in gene content relevant to arbuscular mycorrhiza and rhizobia association, a protein database was constructed which included the predicted proteins of NLL cv. Tanjil from this study, and 50 other land plant species (Bravo et al., 2016). This database was queried with proteins known to be involved in arbuscular mycorrhizal symbiosis in Medicago truncatula via BLASTP (Altschul et al., 1990), and the top 200 matches were used to create phylogenies. The protein models were aligned using MAFFT v7.205 (Katoh et al., 2002) with default values, and columns of the alignment that contained more than 50% gaps were eliminated. A phylogenetic tree was generated with FastTree v2.1.5 (Price et al., 2010) using the wag

model of amino acid evolution. The presence or absence of NLL true orthologs was assessed through visual analysis of the topology of the phylogenies generated.

Accession code

Genome sequence assembly and annotation data can be found in GenBank under BioProject ID: PRJNA299755 and is also available for download and interrogation via BLAST and GBrowse from the Lupin Genome Portal (http://www.lupinexpress.org).

Acknowledgements

We thank Elaine Smith, Natalie Fletcher and Hayley Casarotto for technical support; James Miller and Paul Lacaze from Millennium Science for support with the development of the Fluidigm SNP markers; and Joel Geoghegan (Centre for Cancer Biology, Adelaide) for performing the Fluidigm genotyping service. We also thank Bevan Buirchell and the Department of Agriculture and Food, Western Australia (DAFWA), for kindly providing the NLL seeds. This research was undertaken with the assistance of resources provided at the Pawsey Supercomputing Centre, and the NCI Specialised Facility in Bioinformatics through the National Computational Merit Allocation Scheme supported by the Australian Government. The mass spectrometry analyses were performed in facilities provided by the Lotterywest State Biomedical Facility-Proteomics node, Harry Perkins Institute for Medical Research. This project was funded by the Grains Research and Development Corporation (GRDC), the Commonwealth Scientific and Industrial Research Organisation (CSIRO) and the University of Western Australia (UWA). The authors declare no conflict of interests.

Author Contributions

JKH, MY, LGK, GG, MNN, SB, RF, L-LG, SL, AM, JJ and KBS contributed to generation of genome sequence, transcriptome sequences, BAC-end sequences, genetic mapping and physical mapping data and development of the genome browser; JKH, MY, LGK, GG, SC and JJ worked on genome assembly; JKH, MY and JJ contributed to the genome annotation; JKH, MY, LGK, MNN, AB, SB, RF, L-LG, MJH, SL, C-WL, AM, JM, JW and JJ contributed to the gene function analysis; JKH, MY, LGK, MNN, SC, GG, WH, PEB, BH, DE and JJ worked on genome analysis and comparative genomics; JKH, LGK and KBS wrote the manuscript with input from MY, MNN, CAA, SB, SC, MJH, AM, GM, JM, JM, DE and JJ; KBS; CAA, GM and KBS conceived and directed the project.

References

Altschul, S.F., Gish, W., Miller, W., Myers, E.W. and Lipman, D.J. (1990) Basic local alignment search tool. J. Mol. Biol. 215, 403–410.

Ashburner, M., Ball, C.A., Blake, J.A., Botstein, D., Butler, H., Cherry, J.M., Davis, A.P. et al. (2000) Gene Ontology: tool for the unification of biology. Nat. Genet. 25, 25–29.

Barnett, D.W., Garrison, E.K., Quinlan, A.R., Stromberg, M.P. and Marth, G.T. (2011) BamTools: a C++ API and toolkit for analyzing and managing BAM files. Bioinformatics, 27, 1691–1692.

Benson, G. (1999) Tandem repeats finder: a program to analyze DNA sequences. Nucleic Acids Res. 27, 573.

Benson, D.A., Cavanaugh, M., Clark, K., Karsch-Mizrachi, I., Lipman, D.J., Ostell, J. and Sayers, E.W. (2013) GenBank. Nucleic Acids Res. 41, D36–D42.

Berger, J.D., Buirchell, B., Luckett, D.J. and Nelson, M.N. (2012) Domestication bottlenecks limit genetic diversity and constrain adaptation in narrowleafed lupin (*Lupinus angustifolius* L.). *Theoret. Appl. Genet.* **124**, 637–652.

Berger, J.D., Clements, J.C., Nelson, M.N., Kamphuis, L.G., Singh, K.B. and Buirchell, B. (2013) The essential role of genetic resources in narrow-leafed lupin improvement. *Crop Pasture Sci.* **64**, 361–373.

Boetzer, M., Henkel, C.V., Jansen, H.J., Butler, D. and Pirovano, W. (2011) Scaffolding pre-assembled contigs using SSPACE. *Bioinformatics* **27**, 578–579.

Bravo, A., York, T., Pumplin, N., Mueller, L.A. and Harrison, M.J. (2016) Genes conserved for arbuscular mycorrhizal symbiosis identified through phylogenomics. *Nat. Plants*, **2**, 15208.

Bringans, S., Hane, J., Casey, T., Tan, K.-C., Lipscombe, R., Solomon, P. and Oliver, R. (2009) Deep proteogenomics; high throughput gene validation by multidimensional liquid chromatography and mass spectrometry of proteins from the fungal wheat pathogen *Stagonospora nodorum*. *BMC Bioinformat.* **10**, 301.

Burge, C. and Karlin, S. (1997) Prediction of complete gene structures in human genomic DNA. *J. Mol. Biol.* **268**, 78–94.

Cannon, S.B., McKain, M.R., Harkess, A., Nelson, M.N., Dash, S., Deyholos, M.K., Peng, Y. *et al.* (2015) Multiple polyploidy events in the early radiation of nodulating and nonnodulating legumes. *Mol. Biol. Evol.* **32**, 193–210.

Charpentier, M., Sun, J., Wen, J., Mysore, K.S. and Oldroyd, G.E. (2014) Abscisic acid promotion of arbuscular mycorrhizal colonization requires a component of the PROTEIN PHOSPHATASE 2A complex. *Plant Physiol.* **166**, 2077–2090.

Croxford, A.E., Rogers, T., Caligari, P.D. and Wilkinson, M.J. (2008) High-resolution melt analysis to identify and map sequence-tagged site anchor points onto linkage maps: a white lupin (Lupinus albus) map as an exemplar. *New Phytol.* **180**, 594–607.

Danecek, P., Auton, A., Abecasis, G., Albers, C.A., Banks, E., DePristo, M.A., Handsaker, R.E. *et al.* (2011) The variant call format and VCFtools. *Bioinformatics*, **27**, 2156–2158.

Delaux, P.-M., Varala, K., Edger, P.P., Coruzzi, G.M., Pires, J.C. and Ané, J.-M. (2014) Comparative phylogenomics uncovers the impact of symbiotic associations on host genome evolution. *PLoS Genet.* **10**, e1004487.

Donlin, M.J. (2009) Using the generic genome browser (GBrowse). *Curr. Protocols Bioinformat.* Chapter 9 Unit 9.9 doi:10.1002/0471250953.bi0909s28.

Drummond, C.S., Eastwood, R.J., Miotto, S.T. and Hughes, C.E. (2012) Multiple continental radiations and correlates of diversification in *Lupinus* (Leguminosae): testing for key innovation with incomplete taxon sampling. *Syst. Biol.* **61**, 443–460.

Elsik, C.G., Mackey, A.J., Reese, J.T., Milshina, N.V., Roos, D.S. and Weinstock, G.M. (2007) Creating a honey bee consensus gene set. *Genome Biol.* **8**, R13.

FAO. (2013) *Statistical Yearbook 2013: World Food and Agriculture*. Rome: Food and Agriculture Organization of the United Nations. 289.

Foley, R.C., Gao, L.-L., Spriggs, A., Soo, L.Y., Goggin, D.E., Smith, P.M., Atkins, C.A. *et al.* (2011) Identification and characterisation of seed storage protein transcripts from *Lupinus angustifolius*. *BMC Plant Biol.* **11**, 59.

Foley, R.C., Jimenez-Lopez, J.C., Kamphuis, L.G., Hane, J.K., Melser, S. and Singh, K.B. (2015) Analysis of conglutin seed storage proteins across lupin species using transcriptomic, protein and comparative genomic approaches. *BMC Plant Biol.* **15**, 106.

Gao, L.-L., Hane, J., Kamphuis, L., Foley, R., Shi, B.-J., Atkins, C. and Singh, K. (2011) Development of genomic resources for the narrow-leafed lupin (*Lupinus angustifolius*): construction of a Bacterial Artificial Chromosome (BAC) library and BAC-end sequencing. *BMC Genom.* **12**, 521.

Gladstones, J. (1970) Lupins as crop plants. *Field Crop Abstracts*, **23**, 123–147.

González-Sama, A., Lucas, M.M., De Felipe, M.R. and Pueyo, J.J. (2004) An unusual infection mechanism and nodule morphogenesis in white lupin (*Lupinus albus*). *New Phytol.* **163**, 371–380.

Haas, B.J., Delcher, A.L., Wortman, J.R. and Salzberg, S.L. (2004) DAGchainer: a tool for mining segmental genome duplications and synteny. *Bioinformatics*, **20**, 3643–3646.

Hecht, V., Foucher, F., Ferrándiz, C., Macknight, R., Navarro, C., Morin, J., Vardy, M.E. *et al.* (2005) Conservation of *Arabidopsis* flowering genes in model legumes. *Plant Physiol.* **137**, 1420–1434.

Hecht, V., Laurie, R.E., Vander Schoor, J.K., Ridge, S., Knowles, C.L., Liew, L.C., Sussmilch, F.C. *et al.* (2011) The pea *GIGAS* gene is a *FLOWERING LOCUS T*

homolog necessary for graft-transmissible specification of flowering but not for responsiveness to photoperiod. *Plant Cell*, **23**, 147–161.

Hughes, C. and Eastwood, R. (2006) Island radiation on a continental scale: exceptional rates of plant diversification after uplift of the Andes. *Proc. Natl Acad. Sci.* **103**, 10334–10339.

Initiative, A.G. (2000) Analysis of the genome sequence of the flowering plant Arabidopsis thaliana. *Nature*, **408**, 796.

James, E., Minchin, F., Iannetta, P. and Sprent, J. (1997) Temporal relationships between nitrogenase and intercellular glycoprotein in developing white lupin nodules. *Ann. Bot.* **79**, 493–503.

Jurka, J., Kapitonov, V.V., Pavlicek, A., Klonowski, P., Kohany, O. and Walichiewicz, J. (2005) Repbase Update, a database of eukaryotic repetitive elements. *Cytogenet. Genome Res.* **110**, 462–467.

Kamphuis, L.G., Hane, J.K., Nelson, M.N., Gao, L., Atkins, C.A. and Singh, K.B. (2015) Transcriptome sequencing of different narrow-leafed lupin tissue types provides a comprehensive uni-gene assembly and extensive gene-based molecular markers. *Plant Biotechnol. J.* **13**, 14–25.

Kanehisa, M. and Goto, S. (2000) KEGG: kyoto encyclopedia of genes and genomes. *Nucleic Acids Res.* **28**, 27–30.

Kasprzak, A., Šafář, J., Janda, J., Doležel, J., Wolko, B. and Naganowska, B. (2006) The bacterial artificial chromosome (BAC) library of the narrow-leafed lupin (*Lupinus angustifolius* L.). *Cell. Mol. Biol. Lett.* **11**, 396–407.

Katoh, K., Misawa, K., Kuma, K.I. and Miyata, T. (2002) MAFFT: a novel method for rapid multiple sequence alignment based on fast Fourier transform. *Nucleic Acids Res.*, **30**, 3059–3066.

Kong, F., Liu, B., Xia, Z., Sato, S., Kim, B.M., Watanabe, S., Yamada, T. *et al.* (2010) Two coordinately regulated homologs of *FLOWERING LOCUS T* are involved in the control of photoperiodic flowering in soybean. *Plant Physiol.* **154**, 1220–1231.

Kroc, M., Koczyk, G., Święcicki, W., Kilian, A. and Nelson, M.N. (2014) New evidence of ancestral polyploidy in the Genistoid legume Lupinus angustifolius L. (narrow-leafed lupin). *Theoret. Appl. Genet.* **127**, 1237–1249.

Krzywinski, M., Schein, J., Birol, I., Connors, J., Gascoyne, R., Horsman, D., Jones, S.J. *et al.* (2009) Circos: an information aesthetic for comparative genomics. *Genome Res.* **19**, 1639–1645.

Kurtz, S., Phillippy, A., Delcher, A.L., Smoot, M., Shumway, M., Antonescu, C. and Salzberg, S.L. (2004) Versatile and open software for comparing large genomes. *Genome Biol.* **5**, R12.

Lambers, H., Clements, J.C. and Nelson, M.N. (2013) How a phosphorus-acquisition strategy based on carboxylate exudation powers the success and agronomic potential of lupines (*Lupinus*, Fabaceae). *Am. J. Bot.* **100**, 263–288.

Langmead, B. and Salzberg, S.L. (2012) Fast gapped-read alignment with Bowtie 2. *Nat. Methods*, **9**, 357–359.

Laurie, R.E., Diwadkar, P., Jaudal, M., Zhang, L., Hecht, V., Wen, J., Tadege, M. *et al.* (2011) The Medicago *FLOWERING LOCUS T* homolog, *MtFTa1*, is a key regulator of flowering time. *Plant Physiol.* **156**, 2207–2224.

Lavin, M., Herendeen, P.S. and Wojciechowski, M.F. (2005) Evolutionary rates analysis of Leguminosae implicates a rapid diversification of lineages during the Tertiary. *Syst. Biol.* **54**, 575–594.

Lee, Y.P., Mori, T.A., Sipsas, S., Barden, A., Puddey, I.B., Burke, V., Hall, R.S. *et al.* (2006) Lupin-enriched bread increases satiety and reduces energy intake acutely. *Am. J. Clin. Nutrit.* **84**, 975–980.

Lesniewska, K., Książkiewicz, M., Nelson, M.N., Mahé, F., Aïnouche, A., Wolko, B. and Naganowska, B. (2011) Assignment of 3 genetic linkage groups to 3 chromosomes of narrow-leafed lupin. *J. Hered.* **102**, 228–236.

Li, L., Stoeckert, C.J. and Roos, D.S. (2003) OrthoMCL: identification of ortholog groups for eukaryotic genomes. *Genome Res.* **13**, 2178–2189.

Li, H., Handsaker, B., Wysoker, A., Fennell, T., Ruan, J., Homer, N., Marth, G. *et al.* (2009a) The sequence alignment/map format and SAMtools. *Bioinformatics*, **25**, 2078–2079.

Li, R., Yu, C., Li, Y., Lam, T.W., Yiu, S.M., Kristiansen, K. and Wang, J. (2009b) SOAP2: an improved ultrafast tool for short read alignment. *Bioinformatics*, **25**, 1966–1967.

Liu, B., Shi, Y., Yuan, J., Hu, X., Zhang, H., Li, N., Li, Z. *et al.* (2013) Estimation of genomic characteristics by analyzing k-mer frequency in *de novo* genome projects. arXiv:1308.2012.

Luo, R., Liu, B., Xie, Y., Li, Z., Huang, W., Yuan, J., He, G. *et al.* (2012) SOAPdenovo2: an empirically improved memory-efficient short-read *de novo* assembler. *GigaScience*, **1**, 18.

Magni, C., Sessa, F., Accardo, E., Vanoni, M., Morazzoni, P., Scarafoni, A. and Duranti, M. (2004) Conglutin γ, a lupin seed protein, binds insulin *in vitro* and reduces plasma glucose levels of hyperglycemic rats. *J. Nutrit. Biochem.* **15**, 646–650.

Magoč, T. and Salzberg, S.L. (2011) FLASH: fast length adjustment of short reads to improve genome assemblies. *Bioinformatics*, **27**, 2957–2963.

Martin, M. (2011) Cutadapt removes adapter sequences from high-throughput sequencing reads. *EMBnet. J.* **17**, 10–12.

McKenna, A., Hanna, M., Banks, E., Sivachenko, A., Cibulskis, K., Kernytsky, A., Garimella, K. *et al.* (2010) The Genome Analysis Toolkit: a MapReduce framework for analyzing next-generation DNA sequencing data. *Genome Res.* **20**, 1297–1303.

Naganowska, B., Wolko, B., Sliwinska, E. and Kaczmarek, Z. (2003) Nuclear DNA content variation and species relationships in the genus *Lupinus* (Fabaceae). *Ann. Bot.* **92**, 349–355.

Nelson, M.N., Phan, H.T., Ellwood, S.R., Moolhuijzen, P.M., Hane, J., Williams, A., Clare, E. *et al.* (2006) The first gene-based map of *Lupinus angustifolius* L.-location of domestication genes and conserved synteny with *Medicago truncatula*. *Theoret. Appl. Genet.* **113**, 225–238.

Nelson, M.N., Moolhuijzen, P.M., Boersma, J.G., Chudy, M., Lesniewska, K., Bellgard, M., Oliver, R.P. *et al.* (2010) Aligning a new reference genetic map of *Lupinus angustifolius* with the genome sequence of the model legume, *Lotus japonicus*. *DNA Res.* **17**, 73–83.

O'Rourke, J.A., Yang, S.S., Miller, S.S., Bucciarelli, B., Liu, J., Rydeen, A., Bozsoki, Z. *et al.* (2013) An RNA-seq transcriptome analysis of orthophosphate-deficient white lupin reveals novel insights into phosphorus acclimation in plants. *Plant Physiol.* **161**, 705–724.

Parniske, M. (2008) Arbuscular mycorrhiza: the mother of plant root endosymbioses. *Nat. Rev. Microbiol.* **6**, 763–775.

Parra, G., Bradnam, K. and Korf, I. (2007) CEGMA: a pipeline to accurately annotate core genes in eukaryotic genomes. *Bioinformatics*, **23**, 1061–1067.

Parra-González, L., Aravena-Abarzua, G., Navarro-Navarro, C., Udall, J., Maughan, J., Peterson, L., Salvo-Garrido, H. *et al.* (2012) Yellow lupin (*Lupinus luteus* L.) transcriptome sequencing: molecular marker development and comparative studies. *BMC Genom.* **13**, 425.

Platten, J.D., Foo, E., Elliott, R.C., Hecht, V., Reid, J.B. and Weller, J.L. (2005) Cryptochrome 1 contributes to blue-light sensing in pea. *Plant Physiol.* **139**, 1472–1482.

Pop, M., Kosack, D.S. and Salzberg, S.L. (2004) Hierarchical scaffolding with Bambus. *Genome Res.* **14**, 149–159.

Price, M.N., Dehal, P.S. and Arkin, A.P. (2010) FastTree 2–approximately maximum-likelihood trees for large alignments. *PLoS ONE*, **5**, e9490.

Priyam, A., Woodcroft, B.J., Rai, V., Munagala, A., Moghul, I., Ter, F., Gibbins, M.A., Moon, H., Leonard, G., Rumpf, W. and Wurm, Y. (2015) Sequenceserver: a modern graphical user interface for custom BLAST databases. *bioRXiv*, doi:10.1101/033142.

Quevillon, E., Silventoinen, V., Pillai, S., Harte, N., Mulder, N., Apweiler, R. and Lopez, R. (2005) InterProScan: protein domains identifier. *Nucleic Acids Res.* **33**, W116–W120.

Quinlan, A.R. and Hall, I.M. (2010) BEDTools: a flexible suite of utilities for comparing genomic features. *Bioinformatics*, **26**, 841–842.

Schmid, M., Uhlenhaut, N.H., Godard, F., Demar, M., Bressan, R., Weigel, D. and Lohmann, J.U. (2003) Dissection of floral induction pathways using global expression analysis. *Development*, **130**, 6001–6012.

Schmutz, J., Cannon, S.B., Schlueter, J., Ma, J., Mitros, T., Nelson, W., Hyten, D.L. *et al.* (2010) Genome sequence of the palaeopolyploid soybean. *Nature*, **463**, 178–183.

Schmutz, J., McClean, P.E., Mamidi, S., Wu, G.A., Cannon, S.B., Grimwood, J., Jenkins, J. *et al.* (2014) A reference genome for common bean and genome-wide analysis of dual domestications. *Nat. Genet.* **46**, 707–713.

Schuler, G.D. (1997) Sequence mapping by electronic PCR. *Genome Res.* **7**, 541–550.

Secco, D., Shou, H., Whelan, J. and Berkowitz, O. (2014) RNA-seq analysis identifies an intricate regulatory network controlling cluster root development in white lupin. *BMC Genom.* **15**, 230.

Smit, A. and Hubley, R. (2010) *RepeatModeler Open-1.0*. http://www.repeatmasker.org.

Smit, A., Hubley, R. and Green, P. 1996–2010. *RepeatMasker Open-3.0*. http://www.repeatmasker.org.

Stanke, M., Keller, O., Gunduz, I., Hayes, A., Waack, S. and Morgenstern, B. (2006) AUGUSTUS: *ab initio* prediction of alternative transcripts. *Nucleic Acids Res.* **34**, W435–W439.

Tang, C., Robson, A., Dilworth, M. and Kuo, J. (1992) Microscopic evidence on how iron deficiency limits nodule initiation in *Lupinus angustifolius* L. *New Phytol.* **121**, 457–467.

The UniProt Consortium. (2013) Update on activities at the Universal Protein Resource (UniProt) in 2013. *Nucl. Acids Res.* **41**, D43–D47.

Trapnell, C., Pachter, L. and Salzberg, S.L. (2009) TopHat: discovering splice junctions with RNA-Seq. *Bioinformatics*, **25**, 1105–1111.

Trapnell, C., Williams, B.A., Pertea, G., Mortazavi, A., Kwan, G., van Baren, M.J., Salzberg, S.L. *et al.* (2010) Transcript assembly and quantification by RNA-Seq reveals unannotated transcripts and isoform switching during cell differentiation. *Nat. Biotechnol.* **28**, 511–515.

Van Bel, M., Proost, S., Wischnitzki, E., Movahedi, S., Scheerlinck, C., Van de Peer, Y. and Vandepoele, K. (2011) Dissecting plant genomes with the PLAZA comparative genomics platform. *Plant Physiol.* **158**, 590–600.

Varshney, R.K., Chen, W., Li, Y., Bharti, A.K., Saxena, R.K., Schlueter, J.A., Donoghue, M.T. *et al.* (2012) Draft genome sequence of pigeonpea (*Cajanus cajan*), an orphan legume crop of resource-poor farmers. *Nat. Biotechnol.* **30**, 83–89.

Varshney, R.K., Song, C., Saxena, R.K., Azam, S., Yu, S., Sharpe, A.G., Cannon, S. *et al.* (2013) Draft genome sequence of chickpea (*Cicer arietinum*) provides a resource for trait improvement. *Nat. Biotechnol.* **31**, 240–246.

Wang, Z., Straub, D., Yang, H., Kania, A., Shen, J., Ludewig, U. and Neumann, G. (2014) The regulatory network of cluster-root function and development in phosphate-deficient white lupin (Lupinus albus) identified by transcriptome sequencing. *Physiol. Plant.* **151**, 323–338.

Williams, W. (1979) Studies on the development of lupins for oil and protein. *Euphytica*, **28**, 481–488.

Yang, Z. (2007) PAML4: Phylogenetic analysis by maximum likelihood. *Mol. Biol. Evol.* **24**, 1586–1591.

Yang, H., Tao, Y., Zheng, Z., Shao, D., Lo, Z., Sweetingham, M.W., Buirchell, B.J. *et al.* (2013a) Rapid development of molecular markers by next-generation sequencing linked to a gene conferring phomopsis stem blight disease resistance for marker-assisted selection in lupin (*Lupinus angustifolius* L.) breeding. *Theoret. Appl. Genet.* **126**, 511–522.

Yang, H., Tao, Y., Zheng, Z., Zhang, Q., Zhou, G., Sweetingham, M.W., Howieson, J.G. *et al.* (2013b) Draft genome sequence, and a sequence-defined genetic linkage map of the legume crop species *Lupinus angustifolius* L. *PLoS ONE*, **8**, e64799.

Yant, L., Mathieu, J., Dinh, T.T., Ott, F., Lanz, C., Wollmann, H., Chen, X. *et al.* (2010) Orchestration of the floral transition and floral development in Arabidopsis by the bifunctional transcription factor *APETALA2*. *Plant Cell*, **22**, 2156–2170.

Youens-Clark, K., Faga, B., Yap, I.V., Stein, L. and Ware, D. (2009) CMap 1.01: a comparative mapping application for the Internet. *Bioinformatics*, **25**, 3040–3042.

Young, N.D., Debellé, F., Oldroyd, G.E., Geurts, R., Cannon, S.B., Udvardi, M.K., Benedito, V.A. *et al.* (2011) The *Medicago* genome provides insight into the evolution of rhizobial symbioses. *Nature*, **480**, 520–524.

Zhai, H., Lu, S., Liang, S., Wu, H., Zhang, X., Liu, B., Kong, F. *et al.* (2014) GmFT4, a homolog of *FLOWERING LOCUS T*, is positively regulated by *E1* and functions as a flowering repressor in soybean. *PLoS ONE*, **9**, e89030.

3

Phosphorus remobilization from rice flag leaves during grain filling: an RNA-seq study

Kwanho Jeong[1,2], Abdul Baten[1], Daniel L. E. Waters[1], Omar Pantoja[1,3], Cecile C. Julia[1,2], Matthias Wissuwa[4], Sigrid Heuer[5], Tobias Kretzschmar[6] and Terry J. Rose[1,2,]*

[1]Southern Cross Plant Science, Southern Cross University, Lismore, NSW, Australia
[2]Southern Cross GeoScience, Southern Cross University, Lismore, NSW, Australia
[3]Instituto de Biotecnología, Universidad Nacional Autónoma de México, Cuernavaca, Morelos, Mexico
[4]Crop, Livestock and Environment Division, Japan International Research Center for Agricultural Sciences, Tsukuba, Ibaraki, Japan
[5]University of Adelaide, School of Agriculture Food and Wine / Australian Centre for Plant Functional Genomics (ACPFG), Adelaide, SA, Australia
[6]Genotyping Services Laboratory, International Rice Research Institute (IRRI), Metro Manila, Philippines

*Correspondence
email terry.rose@scu.edu.au

Keywords: Illumina sequencing, *Oryza sativa*, differential gene expression, phosphorus translocation, senescence.

Summary

The physiology and molecular regulation of phosphorus (P) remobilization from vegetative tissues to grains during grain filling is poorly understood, despite the pivotal role it plays in the global P cycle. To test the hypothesis that a subset of genes involved in the P starvation response are involved in remobilization of P from flag leaves to developing grains, we conducted an RNA-seq analysis of rice flag leaves during the preremobilization phase (6 DAA) and when the leaves were acting as a P source (15 DAA). Several genes that respond to phosphate starvation, including three purple acid phosphatases (*OsPAP3, OsPAP9b* and *OsPAP10a*), were significantly up-regulated at 15 DAA, consistent with a role in remobilization of P from flag leaves during grain filling. A number of genes that have not been implicated in the phosphate starvation response, *OsPAP26, SPX-MFS1* (a putative P transporter) and *SPX-MFS2*, also showed expression profiles consistent with involvement in P remobilization from senescing flag leaves. Metabolic pathway analysis using the KEGG system suggested plastid membrane lipid synthesis is a critical process during the P remobilization phase. In particular, the up-regulation of *OsPLDz2* and *OsSQD2* at 15 DAA suggested phospholipids were being degraded and replaced by other lipids to enable continued cellular function while liberating P for export to developing grains. Three genes associated with RNA degradation that have not previously been implicated in the P starvation response also showed expression profiles consistent with a role in P mobilization from senescing flag leaves.

Introduction

Senescence is the final stage of plant development (Kong *et al.*, 2013). It is typically characterized by visible leaf yellowing and is controlled by both environmental cues and internal elements such as phytohormones and gene regulators (Breeze *et al.*, 2011; Dong *et al.*, 2012). A range of biotic and abiotic environmental cues have been implicated in inducing senescence, including pathogen infection, drought, nutrition limitation and oxidative stress (Buchanan-Wollaston *et al.*, 2003). The yellowing of senescing leaves is related to the degradation of macromolecules like chlorophyll, proteins and RNA, and this degradation ultimately results in a decline in photosynthetic activity (Buchanan-Wollaston, 1997; Guo *et al.*, 2004).

Another key component of senescence in annual crop plants is the remobilization of mineral nutrients from vegetative organs to seeds (Kong *et al.*, 2015). Phosphorus (P) is remobilized to developing seeds with particularly high efficiency, resulting in P harvest indices well above the carbon harvest indices in cultivars of most major crop species (Araújo and Teixeira, 2003; Batten, 1992; Rose *et al.*, 2007, 2010). Although this may be ecologically advantageous because high seed P content improves seedling competitiveness in soils that are naturally low in bioavailable P (White and Veneklaas, 2012), in the context of agriculture it results in the removal of vast amounts of P from fields in harvested products.

The loading of P into grains is of critical importance for the global P cycle because the accumulation of P in the grains of the major cereals, oilseed and pulse crops globally removes the equivalent of over 50% of the P applied as fertilizer each year (Lott *et al.*, 2000). Provided seed germination and seedling vigour can be maintained (see Pariasca-Tanaka *et al.*, 2015 and discussion therein), reducing the amount of P stored in the grains of major crops may be a viable option to reduce the amount of P lost from the P cycle within agricultural production systems (Raboy, 2009; Rose and Wissuwa, 2012; Rose *et al.*, 2013). Although breeding crop cultivars with reduced P levels in grains is an attractive solution, practical advances are hampered by our limited understanding of the physiology, and genetic and molecular regulation of P loading into developing grains, including P remobilization from senescing vegetative tissues (Wang *et al.*, 2016).

The degradation of lipids, nucleic acids and proteins is a key component of the senescence process, and it is possible some of the genes involved in recycling P from phospholipids, RNA and proteins under P starvation in vegetative growth may also play a role in the mobilization of P from senescing tissues, as discussed in a number of recent reviews (Smith *et al.*, 2015; Stigter and Plaxton, 2015). However, to the best of our knowledge, the only genes that have been definitively linked to P loading in developing grains are a putative sulphate transporter which has an endosperm-specific impact on grain P in barley (Raboy et al., 2014; Ye et al., 2011), the ATP-binding cassette (ABC) transporters *ZmMRP4* and *AtMPR5/AtABCC5* which, when inactive, result in low seed P in maize and *Arabidopsis*, respectively (Nagy et al., 2009; Shi et al., 2007), and the purple acid phosphatase (PAP) *AtPAP26* which is involved in remobilization of P from senescing leaves to developing seeds in *Arabidopsis* (Robinson et al., 2012). NAC transcription factors are also involved in nutrient remobilization during senescence (Stigter and Plaxton, 2015); for example, the NAC transcription factor *NAM-B1* plays a role in P loading into wheat grains, but it appears to play a broad role in nutrient remobilization to grains not limited to P (Uauy et al., 2006).

Given the key role of gene families such as P transporters (PTs), PAPs and SPX domain proteins (named after proteins SYG1/PHO81/XPR1) in P recycling under P starvation stress (Wang et al., 2012; Wu et al., 2013), and the role of RNases in remobilization of P from senescing hakea leaves (Shane et al., 2014), we hypothesized some members of these gene families may be involved in the remobilization of P from senescing leaves during grain filling. To test this hypothesis, we first identified a total of 115 P starvation-related (PSR) genes based on recent literature (Lin et al., 2009; Liu et al., 2011; MacIntosh et al., 2010; Secco et al., 2013) (see Table S1). We then investigated the expression of these genes in flag leaves during grain filling in the model cereal rice (*Oryza sativa* L.) using an RNA sequencing (RNA-seq) approach.

Several studies have investigated gene transcript abundance during leaf senescence in crops such as cotton and maize (Lin et al., 2015; Zhang et al., 2014) and in a range tissues during grain filling in rice (e.g. Duan and Sun, 2005; Zhu et al., 2003), including flag leaf tissue, and transcript data are publically available. However, without concurrent data on tissue P dynamics, it is not possible to put these data in context. Here, we quantified P dynamics in flag leaves during grain filling and subsequently examined gene expression at key time points when P concentrations in the flag leaf were stable compared with when the flag leaf was acting as a P source.

Results

Remobilization of phosphorus from flag leaves during grain filling

Flag leaf P concentrations were relatively stable during early grain filling, but declined significantly from 9 days after anthesis (DAA)

(Figure 1). The decline in leaf P concentration from around 3.5 mg/g at anthesis to around 2.5 mg/g at 12 DAA suggests the leaf was acting as a 'P source' by 12 DAA. Two time points were selected for the RNA-seq study: 6 DAA, when the P concentrations were stable (flag leaf was not acting as a P source), and 15 DAA, when the flag leaf was clearly acting as a P source as shown by a 33% reduction in P concentration relative to 0 DAA. The 15 DAA time point was selected because it was the first time point that had a significantly lower P concentration than the 6 DAA time point.

Global gene expression of RNA-seq data

Three biological replicates for each time point (6 DAA and 15 DAA) were analysed using RNA-seq. These six libraries generated 156 million 101-bp paired-end reads after quality control using FASTQC software; 83 million and 73 million from 6 DAA and 15 DAA, respectively (Table 1). Reads were filtered for adapter sequences, poly-N stretches and low-quality reads which resulted in 78 million (94%) and 70 million (95.6%) high-quality reads from 6 DAA and 15 DAA, respectively (Table 1). Of these high-quality reads, around 60 million (77.5%) from 6 DAA and 56 million (79.7%) from 15 DAA were mapped to the reference genome (Table 1). There were 26 788 expressed genes at 6 DAA and 26 747 at 15 DAA (Figure 2a) and 5160 differentially expressed genes (DEGs) between the two time points using a cut-off based on $P < 0.05$ and false discovery rate (FDR) < 0.05 values.

GO enrichment analysis of DEGs

To classify the function of DEGs between two time points, gene ontology (GO) enrichment analysis was performed looking at DEGs at 15 DAA relative to 6 DAA. After filtering for DEGs based on P and FDR values (i.e. <0.05) as well as $\log_2$ fold change (i.e. >1.5), 1643 DEGs were retained. Using parametric analysis of gene set enrichment (PAGE), 1180 DEGs were mapped to the GO term database resulting in a total of 263 enriched GO terms.

Up-regulated DEGs within the term biological process were largely associated with P metabolism, protein modification and transport, which was congruent with up-regulated DEGs within the molecular function term that were found to be associated with phosphotransferase, kinase or transport activity. Interestingly, a high number of up-regulated DEGs could be categorized to nine GO terms related to nucleotide, nucleoside or ribonucleotide binding proteins, although the genes found in each of the respective GO terms are largely identical. No up-regulated

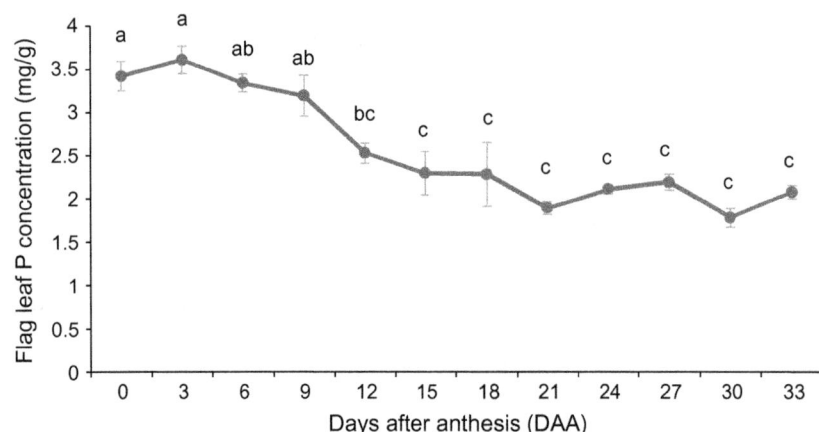

Figure 1 Phosphorus concentration of flag leaves during grain filling. Each value represents the mean ± SE of three biological replicates. Different letters indicate significant difference at $P \leq 0.01$.

Table 1 General information of sequencing reads and mapping statistics

Samples	Replications	Raw reads	High-quality reads		Mapping to genome	
			Number	%	Number	%
6 DAA	1	30 160 726	28 586 476	94.78	22 181 542	77.59
	2	27 979 092	25 551 322	91.32	19 050 896	74.56
	3	25 297 097	24 329 299	96.17	19 612 039	80.61
15 DAA	1	22 787 486	21 747 029	95.43	17 210 696	79.14
	2	25 016 582	23 801 938	95.14	18 800 985	78.99
	3	25 603 135	24 652 300	96.29	19 961 618	80.97

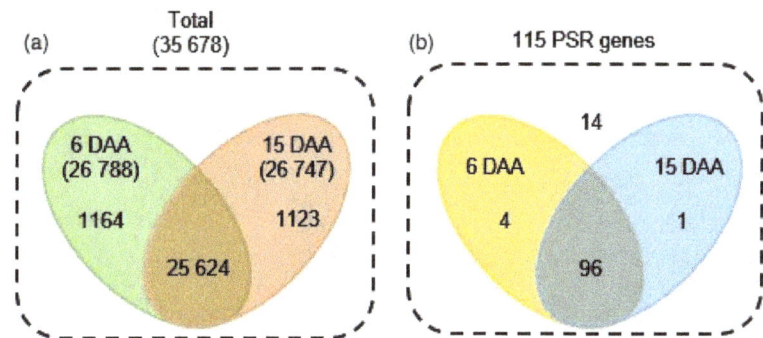

Figure 2 Venn diagram showing (a) the total number of genes expressed between two time points and (b) the expression profile of 115 PSR genes.

DEGs were enriched within the cellular component GO term (Figure 3). Down-regulated DEGS within biological processes were associated with transcription and translation, while, congruently, down-regulated DEGs under molecular functions were associated with RNA binding and ribosome constitution. Several genes related to photosynthesis and photosystems were down-regulated, which was further reflected in a marked down-regulation of cellular component genes associated with plastids and organelles.

Expression of 115 PSR genes in the flag leaf at 6 DAA and 15 DAA

Of the 115 PSR genes, 100 were expressed at 6 DAA and 97 at 15 DAA in flag leaves (Figure 2b). The most highly expressed gene of the 115 PSR genes was the P transporter (*OsPT*) *OsPT21* (Os01g0279700) with expression levels of 1613 FPKM (fragments per kilobase of transcript per million mapped reads) at 6 DAA and 1168 FPKM at 15 DAA. The second most highly expressed of the 115 PSR genes was *OsPT24* (Os09g0570400) with 694 FPKM at 6 DAA and 334 FPKM at 15 DAA (Table S1). Only 14 of the 115 PSR genes were not expressed (FPKM = 0) at either 6 DAA or 15 DAA (Figure 2b).

We identified 38 PSR genes that were differentially expressed between the two time points (Table 2). Twenty-six of the 38 genes were up-regulated at 15 DAA when the flag leaf acts as a source of P for developing grains (Table 2). These 26 genes included *OsPHO1* and *OsPHO2*, which are primary factors in P starvation signalling, as well as *OsSPX1* and *OsSPX2* domain

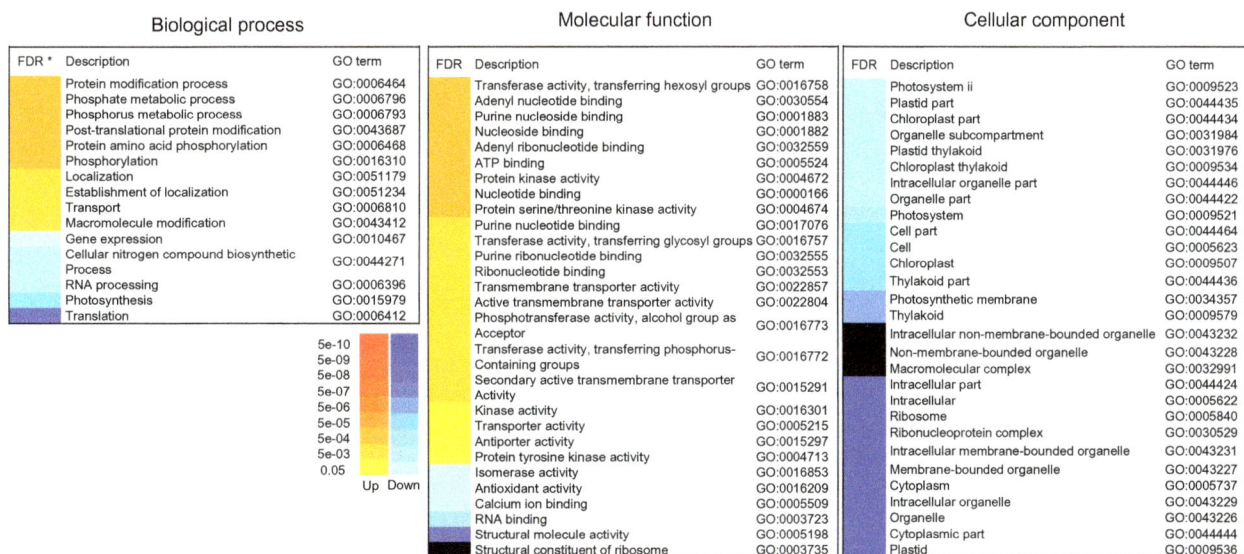

Figure 3 Hierarchical clustering of 72 significant GO terms by PAGE analysis. The adjusted FDR value of the term determines the degree of colour saturation of the corresponding box.

proteins that play a key a role in P homeostasis during vegetative growth (Secco et al., 2012; Wu et al., 2013). Further, three PAPs (*OsPAP1d*, *OsPAP3c* and *OsPAP10a*) which are involved in the release of inorganic P (Pi) from organic P (Wu et al., 2013) and *OsSQD2* which is involved in sugar transport to diacylglycerol (DAG) were among the 26 genes. Three of the 26 known P transporters (Liu et al., 2011), *OsPT5*, *OsPT19* and *OsPT20*, were

also among the 26 PSR genes up-regulated at 15 DAA in the flag leaf (Table 2).

Expression of *OsPAP*, *OsPT* and *OsSPX* domain gene families in the flag leaf at 6 DAA and 15 DAA

Given that several genes in the *OsPT*, *OsPAP* and *OsSPX* gene families were differentially expressed between 6 DAA and 15

Table 2 38 differentially expressed genes among 115 PSR genes

No.	Gene id	MSU id	Name/description	FPKM 6 DAA	15 DAA	Log$_2$
1	OS07G0187400	LOC_Os07g08970	Conserved hypothetical protein	44.01	288.27	2.71
2	OS03G0238600	LOC_Os03g13540	OsPAP3c	10.74	59.36	2.47
3	OS07G0165200	LOC_Os07g07080	Regulator of chromosome condensation/beta-lactamase-inhibitor protein II	8.59	35.30	2.04
4	OS02G0514500	LOC_Os02g31030	Glycerophosphoryl diester phosphodiesterase family protein	6.06	24.77	2.03
5	OS06G0291500	LOC_Os06g18820	Conserved hypothetical protein.	0.34	1.36	1.99
6	OS01G0142300	LOC_Os01g04920	OsSQD2	12.03	37.18	1.63
7	OS04G0185600	LOC_Os04g10690	OsPT5	0.30	0.90	1.60
8	OS01G0110100	LOC_Os01g02000	OsPHO1;1	2.50	7.30	1.55
9	OS02G0802700	LOC_Os02g55910	Similar to MGDG synthase type A	12.18	34.87	1.52
10	OS09G0554000	LOC_Os09g38100	OsPT20	1.80	5.00	1.47
11	OS03G0261800	LOC_Os03g15530	Protein of unknown function DUF3049 domain containing protein	12.71	33.58	1.40
12	OS01G0310100	LOC_Os01g20860	OsPLDzeta2, OsPLDrho2	3.35	8.74	1.38
13	OS01G0776600	LOC_Os01g56880	OsPAP10a	16.63	41.37	1.31
14	OS05G0557700	LOC_Os05g48390	OsPHO2	8.93	18.89	1.08
15	OS07G0134500	LOC_Os07g04210	Similar to hydrolase/protein serine/threonine phosphatase	9.82	20.74	1.08
16	OS08G0156600	LOC_Os08g06010	Major facilitator superfamily protein.	8.95	18.34	1.03
17	OS01G0557500	LOC_Os01g37690	OsCAX1a	19.79	38.98	0.98
18	OS03G0214400	LOC_Os03g11560	Digalactosyldiacylglycerol synthase, chloroplast precursor	5.74	11.04	0.94
19	OS09G0454600	LOC_Os09g28160	OsPT19	3.56	6.68	0.91
20	OS05G0358700	LOC_Os05g29050	OsPLDrho1	9.97	16.72	0.75
21	OS09G0528700	LOC_Os09g35940	Similar to Cytochrome p450 (CYP78A9).	15.78	26.39	0.74
22	OS06G0140800	LOC_Os06g04880	Serine threonine kinase, putative	24.09	39.60	0.72
23	OS06G0603600	LOC_Os06g40120	OsSPX1	31.50	51.15	0.70
24	OS02G0202200	LOC_Os02g10780	OsSPX2	33.67	54.16	0.69
25	OS12G0576600	LOC_Os12g38750	OsPAP1d	12.11	18.76	0.63
26	OS08G0433200	LOC_Os08g33640	Conserved hypothetical protein.	34.26	50.60	0.56
27	OS09G0537700	LOC_Os09g36680	OsRNS4	0.72	0.00	Undefined*
28	OS02G0625300	LOC_Os02g41580	CAMK_CAMK_like.14 – CAMK includes calcium/ calmodulin depedent protein kinases	38.41	25.68	−0.58
29	OS04G0608600	LOC_Os04g51920	Protein disulfide isomerase	14.28	9.11	−0.65
30	OS07G0100300	LOC_Os07g01030	Glycosyl transferase, group 1 domain containing protein	131.17	80.56	−0.70
31	OS04G0598000	LOC_Os04g50970	Seed specific protein Bn15D1B	553.88	297.77	−0.90
32	OS10G0444700	LOC_Os10g30790	OsPT8	13.26	7.11	−0.90
33	OS02G0593500	LOC_Os02g38020	OsPT14	35.75	16.15	−1.15
34	OS07G0187700	LOC_Os07g09000	OsPHF1	28.30	12.13	−1.22
35	OS03G0263400	LOC_Os03g15690	OsPT16	11.83	5.05	−1.23
36	OS01G0852200	LOC_Os01g63290	OsPT22	49.99	18.59	−1.43
37	OS02G0668500	LOC_Os02g44820	Cellular retinaldehyde-binding/triple function, C-terminal domain containing protein	35.42	12.24	−1.53
38	OS05G0387200	LOC_Os05g32140	Similar to UDP-sulfoquinovose synthase, chloroplast precursor (EC 3. 13. 1. 1)	219.30	60.06	−1.87

All the genes were cut off by *P* < 0.05 and FDR < 0.05.

*Undefined = zero expression value for one of the two samples.

DAA, we investigated expression of all known members of these gene families at both time points. Of the 26 *OsPAP* genes in rice (Zhang *et al.*, 2011), in addition to the three *OsPAP* genes included in the 115 PSR genes that were up-regulated at 15 DAA, a further three *OsPAP* genes (*OsPAP9b*, *OsPAP15* and *OsPAP26*) had significantly up-regulated expression (log$_2$ fold change >1.3) at 15 DAA, although the expression of *OsPAP15* was low compared to the other *OsPAP* genes (Figure 4a).

Twenty-one of the 26 P transporters in rice (*OsPT1* to *OsPT26*) were expressed in flag leaves. Three of these—*OsPT5*, *OsPT19* and *OsPT20*—were differentially expressed at 15 DAA (Figure 4b). Expression levels of *OsPT19* and *OsPT20* were relatively low, and the expression level of *OsPT5* was extremely low (Figure 4b). Interestingly, while *OsPT21* and *OsPT24* were not significantly differentially expressed between the time points, their overall expression was the highest (>400 FKPM) at both 6 DAA and 15 DAA (Figure 4b).

Six *SPX* genes (*OsSPX1-OsSPX6*) (Wang *et al.*, 2009) and four *SPX-MFS* (major facility superfamily) genes (*OsSPX-MFS1-4*) are known to be regulated by *OsPHR2* under P deficiency in rice (Lin *et al.*, 2010; Wang *et al.*, 2012; Wu *et al.*, 2013). Examination of all *OsSPX* and *OsSPX-MFS* expression levels and their regulator *OsPHR2* revealed that in addition to *OsSPX1* and *OsSPX2* (see above), *OsSPX-MFS1* and *OsSPX-MFS2* were also significantly up-regulated at 15 DAA with log$_2$ fold changes of 2.55 and 2.80, respectively. The absolute expression level of *OsSPX-MFS2* was greater than 200 FPKM at 15 DAA (Figure 4c). Interestingly, expression of the transcription factor *OsPHR2* was not significantly up-regulated at 15 DAA.

Biological pathways related with P recycling processes during senescence

To investigate the possible involvement of lipid and nucleotide metabolism genes in P remobilization during leaf senescence, we used the KEGG pathway database to retrieve genes associated with these metabolic pathways. We then analysed expression of all DEGs in our RNA-seq data and found that 35 DEGs mapped to lipid metabolism (Table 3). Among these, glycerophosphoryl diester phosphodiesterase (GDPD; PLC-like; Os02g0514500), digalacto-syldiacylglycerol (DGDG) synthase (Os02g0539100), monogalactosyldiacylglycerol (MGDG) synthase (Os02G0802700) and sulphoquinovosyltransferase (SQD2; Os01G0142300), whose products have been proposed to participate in galactolipid synthesis, were up-regulated at 15 DAA (Table 3). Several genes related to lipid degradation were also up-regulated, such as phospholipase D (Os06g0604400) and phospholipase C (Os03g0826600), glycerophosphoryl diester phosphodiesterase (Os02g0514500), aldehyde dehydrogenase (NAD$^+$) (Os12g0178000) and phosphatidate phosphatase (Os05g0462400). In contrast, dihydroxyacetone kinase (Os03g0719300), glycerol-3-phosphate dehydrogenase (NAD$^+$) (Os01g0971600, Os07g0229800), glycerol-3-phosphate O-acyltransferase (Os10g0577900) involved in *de novo* DAG synthesis were down-regulated (Table 3). Also down-regulated were lysophospholipid acyltransferase (Os02g0676000) and 1-acyl-sn-glycerol-3-phosphate acyltransferase (Os10g0497100), gene products involved in phospholipid synthesis.

We also examined the nucleotide metabolism pathway that includes RNA transport and degradation and found 50 DEGs, of which 16 genes were up-regulated and 34 genes were down-regulated at 15 DAA (Table 4). Among the up-regulated genes, six are directly related to RNA degradation (numbers 1, 2, 9, 13, 14 and 15 in Table 4). In general, the down-regulated genes within the nucleotide metabolism pathway are involved in nucleic acid biosynthesis/maturation/processing (19, 26, 33, 36, 38, 39, 41, 44, 47-51; Table 4) or purine biosynthesis (17, 21, 29, 40, 42 and 43; Table 4), which, together with the up-regulated genes, indicates a strong degradation of nucleic acids during leaf senescence.

Discussion

While previous studies have profiled gene expression in a range of tissues during rice grain filling (e.g. Duan and Sun, 2005; Zhu *et al.*, 2003) and leaf senescence in other crops such as cotton

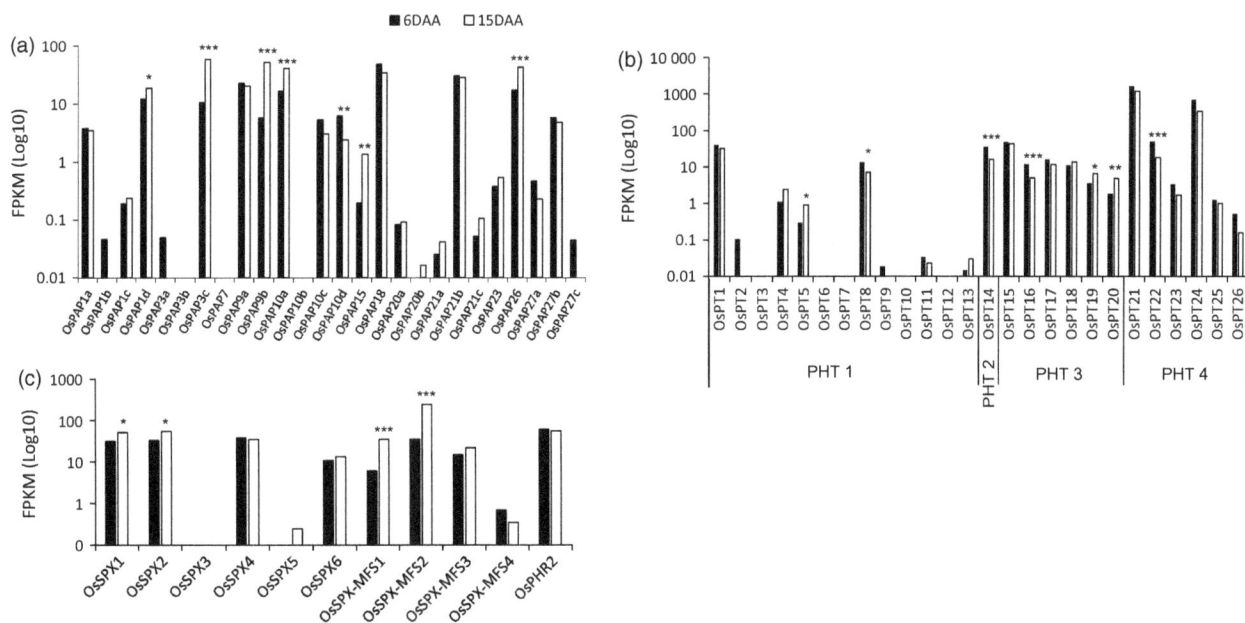

Figure 4 The expression of (a) 26 *OsPAPs*, (b) 26 *OsPTs* and (c) 6 *OsSPXs and 4 OsSPX-MFS* domains and their regulatory transcription factor *OsPHR2* in the flag leaf during grain filling in rice. ***$P < 0.001$; **$P < 0.01$, *$P < 0.05$; no mark = nonsignificant.

Table 3 Differentially expressed genes associated with lipid metabolic pathways

No.	Gene id	KEGG Number and description	FPKM 6 DAA	15 DAA	Log$_2$
1	Os02g0514500	K18696 GDE1; glycerophosphoryl diester phosphodiesterase [EC:3.1.4.46]	6.06	24.77	2.03
2	Os12g0178000	K00128 E1.2.1.3; aldehyde dehydrogenase (NAD$^+$) [EC:1.2.1.3]	1.80	7.29	2.02
3	Os05g0462400	K15728 LPIN; phosphatidate phosphatase LPIN [EC:3.1.3.4]	4.11	15.08	1.88
4	Os10g0473400	K17108 GBA2; nonlysosomal glucosylceramidase [EC:3.2.1.45]	3.36	11.28	1.75
5	Os01g0142300	K06119 SQD2; sulfoquinovosyltransferase [EC:2.4.1.-]	12.03	37.18	1.63
6	Os02g0802700	K03715 E2.4.1.46; 1,2-diacylglycerol 3-beta-galactosyltransferase [EC:2.4.1.46]	12.18	34.87	1.52
7	Os02g0539100	K09480 E2.4.1.241; digalactosyldiacylglycerol synthase [EC:2.4.1.241]	30.23	80.43	1.41
8	Os06g0204400	K00993 EPT1; ethanolaminephosphotransferase [EC:2.7.8.1]	25.77	52.27	1.02
9	Os11g0158400	K09480 E2.4.1.241; digalactosyldiacylglycerol synthase [EC:2.4.1.241]	9.40	18.58	0.98
10	Os03g0214400	K09480 E2.4.1.241; digalactosyldiacylglycerol synthase [EC:2.4.1.241]	5.74	11.04	0.94
11	Os06g0604400	K01115 PLD1_2; phospholipase D1/2 [EC:3.1.4.4]	7.92	15.06	0.93
12	Os11g0242100	K17108 GBA2; nonlysosomal glucosylceramidase [EC:3.2.1.45]	19.35	36.26	0.91
13	Os03g0826600	K01114 plcC; phospholipase C [EC:3.1.4.3]	166.74	303.14	0.86
14	Os12g0121300	K00967 PCYT2; ethanolamine-phosphate cytidylyltransferase [EC:2.7.7.14]	4.83	8.19	0.76
15	Os01g0624000	K12349 ASAH2; neutral ceramidase [EC:3.5.1.23]	58.55	98.81	0.75
16	Os04g0634700	K00901 dgkA; diacylglycerol kinase (ATP) [EC:2.7.1.107]	16.18	26.75	0.73
17	Os11g0186200	K00128 E1.2.1.3; aldehyde dehydrogenase (NAD$^+$) [EC:1.2.1.3]	283.39	464.74	0.71
18	Os11g0516000	K00654 SPT; serine palmitoyltransferase [EC:2.3.1.50]	43.67	69.55	0.67
19	Os01g0931300	K13511 TAZ; monolysocardiolipin acyltransferase [EC:2.3.1.-]	32.54	51.62	0.67
20	Os08g0224800	K00967 PCYT2; ethanolamine-phosphate cytidylyltransferase [EC:2.7.7.14]	22.09	34.10	0.63
21	Os02g0676000	K13519 LPT1; lysophospholipid acyltransferase [EC:2.3.1.51 2.3.1.23 2.3.1.-]	95.28	59.76	−0.67
22	Os07g0100300	K06119 SQD2; sulfoquinovosyltransferase [EC:2.4.1.-]	131.17	80.56	−0.70
23	Os10g0497100	K00655 plsC; 1-acyl-sn-glycerol-3-phosphate acyltransferase [EC:2.3.1.51]	22.15	13.38	−0.73
24	Os01g0175000	K06130 LYPLA2; lysophospholipase II [EC:3.1.1.5]	51.15	30.17	−0.76
25	Os04g0669500	K06130 LYPLA2; lysophospholipase II [EC:3.1.1.5]	21.69	12.41	−0.81
26	Os05g0548900	K05929 E2.1.1.103; phosphoethanolamine N-methyltransferase [EC:2.1.1.103]	18.11	9.72	−0.90
27	Os01g0796500	K01094 GEP4; phosphatidylglycerophosphatase GEP4 [EC:3.1.3.27]	36.87	19.52	−0.92
28	Os06g0649900	K16860 PLD3_4; phospholipase D3/4 [EC:3.1.4.4]	3.63	1.88	−0.95
29	Os03g0719300	K00863 E2.7.1.29; dihydroxyacetone kinase [EC:2.7.1.29]	46.29	21.50	−1.11
30	Os10g0577900	K00630 ATS1; glycerol-3-phosphate O-acyltransferase [EC:2.3.1.15]	56.31	25.82	−1.13
31	Os01g0971600	K00006 GPD1; glycerol-3-phosphate dehydrogenase (NAD$^+$) [EC:1.1.1.8]	4.87	1.75	−1.47
32	Os10g0493600	K07407 E3.2.1.22B; alpha-galactosidase [EC:3.2.1.22]	173.12	61.63	−1.49
33	Os07g0229800	K00006 GPD1; glycerol-3-phosphate dehydrogenase (NAD$^+$) [EC:1.1.1.8]	42.68	14.48	−1.56
34	Os05g0387200	K06118 SQD1; UDP-sulfoquinovose synthase [EC:3.13.1.1]	219.30	60.06	−1.87
35	Os07g0679300	K07407 E3.2.1.22B; alpha-galactosidase [EC:3.2.1.22]	27.76	5.94	−2.23

(Lin *et al.*, 2015) and maize (Zhang *et al.*, 2014), lack of concurrent data on P remobilization precludes conclusions being made about the potential regulation of P remobilization from vegetative tissues during grain filling. In the present study, we identified critical developmental stages during grain filling when flag leaf P concentration was stable (6 DAA) and when the flag leaf was a P source (15 DAA) as the basis for subsequent RNA-seq studies.

Given the dearth of knowledge on specific genes involved in P remobilization from senescing leaves during grain filling (Wang *et al.*, 2016), we investigated whether genes associated with P starvation during vegetative growth were also associated with P remobilization from vegetative tissues during grain filling. Of the 115 PSR genes identified in the published literature, 14 were not expressed in flag leaves at either developmental stage, which is not surprising given that some of these genes have tissue-specific expression. *OsRNS1* and *OsRNS7*, for example, are specifically expressed in root tissue (MacIntosh *et al.*, 2010). As expected, the absolute expression of the 115 PSR genes was low relative to the most highly expressed genes as many of the highly expressed genes (e.g. *OsLIR1*, photosystem subunits I, II and chloroplast precursor) are key genes for photosynthesis and are highly expressed in flag leaves during grain filling (Reimmann and Dudler, 1993; Sugano *et al.*, 2010).

GO enrichment analysis is a common and powerful approach in gene expression analysis (Subramanian *et al.*, 2005). By categorizing treatment or development-dependent global transcript changes into standardized GO terminology, metabolic shifts are revealed, allowing for the interpretation of possible changes that may occur at the molecular and cellular level (Du *et al.*, 2010). Down-regulated DEGs were highly enriched in GO terms related to photosynthesis (GO:0015979) and plastids (45% of the cellular component category), indicating a reduction of photo-assimilatory anabolism at 15 DAA. Furthermore, down-regulated DEGs were enriched in GO terms related to transcription and translation (GO:0010467, GO:0006396, GO:0003723, GO:0003735, GO:0006412, GO:0005840 and GO:00529), while up-regulated genes were enriched in GO terms related to protein modification

Table 4 Differentially expressed genes associated with nucleotide metabolic pathways

| No. | Gene id | KEGG Number and description | FPKM | | Log$_2$ |
			6 DAA	15 DAA	
1	Os10g0556700	K12604 CNOT1; CCR4-NOT transcription complex subunit 1	2.20	7.94	1.85
2	Os10g0556600	K12604 CNOT1; CCR4-NOT transcription complex subunit 1	6.46	19.83	1.62
3	Os03g0743900	K13811 PAPSS; 3'-phosphoadenosine 5'-phosphosulfate synthase [EC:2.7.7.4 2.7.1.25]	22.79	67.60	1.57
4	Os07g0661000	K01488 add; adenosine deaminase [EC:3.5.4.4]	7.10	19.48	1.46
5	Os05g0151000	K03006 RPB1; DNA-directed RNA polymerase II subunit RPB1 [EC:2.7.7.6]	0.40	1.03	1.37
6	Os04g0680400	K01466 allB; allantoinase [EC:3.5.2.5]	20.50	51.62	1.33
7	Os04g0111200	K13811 PAPSS; 3'-phosphoadenosine 5'-phosphosulfate synthase [EC:2.7.7.4 2.7.1.25]	35.63	87.25	1.29
8	Os06g0151900	K00850 pfkA; 6-phosphofructokinase 1 [EC:2.7.1.11]	16.82	32.47	0.95
9	Os07g0249600	K12611 DCP1B; mRNA-decapping enzyme 1B [EC:3.-.-.-]	10.47	19.87	0.92
10	Os08g0175600	K02335 DPO1; DNA polymerase I [EC:2.7.7.7]	6.01	11.22	0.90
11	Os11g0148500	K00873 PK; pyruvate kinase [EC:2.7.1.40]	72.24	128.11	0.83
12	Os08g0109300	K00939 adk; adenylate kinase [EC:2.7.4.3]	115.98	197.90	0.77
13	Os03g0316900	K12608 CAF16; CCR4-NOT complex subunit CAF16	8.36	14.02	0.75
14	Os03g0748800	K14442 DHX36; ATP-dependent RNA helicase DHX36 [EC:3.6.4.13]	8.84	14.64	0.73
15	Os01g0256900	K12623 LSM4; U6 snRNA-associated Sm-like protein LSm4	28.87	43.89	0.60
16	Os04g0677500	K00873 PK; pyruvate kinase [EC:2.7.1.40]	56.13	82.28	0.55
17	Os03g0313600	K01756 purB; adenylosuccinate lyase [EC:4.3.2.2]	51.84	34.23	−0.60
18	Os12g0589100	K00759 APRT; adenine phosphoribosyltransferase [EC:2.4.2.7]	146.48	96.30	−0.61
19	Os04g0428950	K14525 RPP25; ribonucleases P/MRP protein subunit RPP25 [EC:3.1.26.5]	69.88	45.16	−0.63
20	Os03g0831500	K01933 purM; phosphoribosylformylglycinamidine cyclo-ligase [EC:6.3.3.1]	18.98	11.83	−0.68
21	Os05g0389300	K12624 LSM5; U6 snRNA-associated Sm-like protein LSm5	64.94	39.91	−0.70
22	Os05g0389500	K12603 CNOT6; CCR4-NOT transcription complex subunit 6 [EC:3.1.13.4]	53.51	32.55	−0.72
23	Os08g0270200	K12592 C1D; exosome complex protein LRP1	28.37	16.60	−0.77
24	Os07g0495000	K14977 ylbA; (S)-ureidoglycine aminohydrolase [EC:3.5.3.26]	46.29	26.69	−0.79
25	Os02g0168900	K10745 RNASEH2C; ribonuclease H2 subunit C	14.48	8.31	−0.80
26	Os04g0443900	K04077 groEL; chaperonin GroEL	434.11	239.14	−0.86
27	Os03g0320900	K00942 E2.7.4.8; guanylate kinase [EC:2.7.4.8]	25.64	13.99	−0.87
28	Os08g0206600	K00602 purH; phosphoribosylaminoimidazolecarboxamide formyltransferase/IMP cyclohydrolase [EC:2.1.2.3 3.5.4.10]	34.51	18.77	−0.88
29	Os07g0168000	K00962 pnp; polyribonucleotide nucleotidyltransferase [EC:2.7.7.8]	80.00	42.30	−0.92
30	Os03g0780500	K00088 guaB; IMP dehydrogenase [EC:1.1.1.205]	92.09	47.95	−0.94
31	Os01g0865100	K00365 uaZ; urate oxidase [EC:1.7.3.3]	37.21	19.36	−0.94
32	Os06g0168600	K10807 RRM1; ribonucleoside-diphosphate reductase subunit M1 [EC:1.17.4.1]	5.53	2.75	−1.01
33	Os03g0130400	K00939 adk; adenylate kinase [EC:2.7.4.3]	126.52	62.57	−1.02
34	Os05g0349200	K01490 AMPD; AMP deaminase [EC:3.5.4.6]	65.88	32.58	−1.02
35	Os05g0595400	K00940 ndk; nucleoside-diphosphate kinase [EC:2.7.4.6]	84.30	41.54	−1.02
36	Os07g0611600	K03024 RPC7; DNA-directed RNA polymerase III subunit RPC7	17.52	8.47	−1.05
37	Os06g0257450	K10808 RRM2; ribonucleoside-diphosphate reductase subunit M2 [EC:1.17.4.1]	10.51	5.08	−1.05
38	Os03g0844450	K12587 MTR3; exosome complex component MTR3	10.41	4.91	−1.08
39	Os05g0270800	K00601 E2.1.2.2; phosphoribosylglycinamide formyltransferase [EC:2.1.2.2]	9.31	4.25	−1.13
40	Os04g0388900	K12625 LSM6; U6 snRNA-associated Sm-like protein LSm6	17.50	7.50	−1.22
41	Os01g0888500	K01952 purL; phosphoribosylformylglycinamidine synthase [EC:6.3.5.3]	64.76	27.15	−1.25
42	Os05g0430800	K00764 purF; amidophosphoribosyltransferase [EC:2.4.2.14]	5.86	2.44	−1.26
43	Os04g0692500	K18213 PRORP; proteinaceous RNase P [EC:3.1.26.5]	35.63	14.75	−1.27
44	Os10g0457500	K01519 ITPA; inosine triphosphate pyrophosphatase [EC:3.6.1.19]	10.45	4.16	−1.33
45	Os04g0684900	K12581 CNOT7_8; CCR4-NOT transcription complex subunit 7/8	2.03	0.79	−1.36
46	Os03g0699300	K01939 purA; adenylosuccinate synthase [EC:6.3.4.4]	41.47	16.12	−1.36
47	Os09g0482680	K00784 rnz; ribonuclease Z [EC:3.1.26.11]	138.72	46.97	−1.56
48	Os08g0116800	K12587 MTR3; exosome complex component MTR3	3.64	0.93	−1.97
49	Os02g0273800	K18213 PRORP; proteinaceous RNase P [EC:3.1.26.5]	79.60	15.86	−2.33
50	Os12g0548300	K00940 ndk; nucleoside-diphosphate kinase [EC:2.7.4.6]	38.89	7.56	−2.36

(GO:0006464 and GO:0043687) and protein phosphorylation (GO:0006468, GO:0004672, GO:0004674, GO:0016301 and GO: 0004713). This suggests that at 15 DAA, regulation of metabolism was achieved through the modulation of existing proteins rather than synthesis of new proteins, which is consistent with a change from anabolism towards catabolism and the remobilization of nutrients at 15 DAA. Resources are recycled and removed rather than re-invested within the flag leaf, which is supported by the finding that up-regulated DEGs were enriched in GO terms related to transport (GO:0006810, GO:0022857, GO:0022804, GO:0015291, GO:0005215, GO:0015297). Collectively, this clearly indicates a shift towards senescence at 15 DAA.

Markedly enriched up-regulated DEGs were found in GO terms related to P metabolism (GO:0006796, GO:0006793, GO:0016310, GO:0016772 and GO:0016773), which corresponds to the observation of enhanced P remobilization from flag leaves at 15 DAA. However, none of the identified 115 PSR genes were found among the enriched genes within P metabolism-related GO terms. We therefore employed a targeted approach by individually comparing the expression levels of the PSR genes between 6 DAA and 15 DAA.

A number of PSR genes and other genes from PT, SPX and PAP families had expression profiles consistent with involvement in the remobilization of P from senescing leaves during grain filling. One of these genes, OsPAP26, is the rice orthologue of AtPAP26 which plays a role in P remobilization from senescing leaves to developing Arabidopsis seeds (Robinson et al., 2012). Interestingly, OsPAP26 has not been implicated in plant responses to P starvation at the vegetative stage, and therefore, its role in P remobilization may be specific to leaf senescence during grain filling (Stigter and Plaxton, 2015). In contrast, the other three OsPAP genes with highly up-regulated expression at 15 DAA in flag leaves, OsPAP3c, OsPAP9b and OsPAP10a, show increased expression in rice under P starvation at the vegetative stage (Secco et al., 2013; Zhang et al., 2011) and may have a broader role in P remobilization from leaves.

Two genes in the SPX-MFS subfamily OsSPX-MFS1 and OsSPX-MFS2 were highly up-regulated at 15 DAA. Wang et al. (2012) observed OsSPX-MSF1 mutant plants accumulated Pi in leaves which led them to suggest OsSPX-MFS1 is a P transporter in leaves during vegetative growth. Up-regulation of these same genes during senescence is consistent with them functioning as P transporters during remobilization of Pi from leaves during senescence. OsSPX1 and OsSPX2 were two of the 26 PSR genes that were up-regulated at 15 DAA, and although up-regulated expression may have arisen in response to the reduced levels of available Pi in the leaves, the similarity in structure and expression levels to OsSPX-MFS1 and OsSPX-MFS2 suggests these genes may have been in part responsible for the reduced levels of available Pi in the leaves. While both OsSPX-MFS1 and OsSPX-MFS2 are regulated by the transcription factor OsPHR2 under P starvation (Wang et al., 2012), up-regulation of OsSPX-MFS1 and OsSPX-MFS2 at 15 DAA in the present study did not correspond to increased expression of OsPHR2. This suggests that, as with the regulation of the Arabidopsis AtPAP26 (Stigter and Plaxton, 2015), key regulatory genes in the P starvation response are not necessarily involved in regulation of P remobilization from vegetative tissues during grain filling.

Three P transporters (OsPT5, OsPT19 and OsPT20) were up-regulated while the expression of another four (OsPT8, OsPT14, OsPT16 and OsPT22) was down-regulated at 15 DAA. RNAi evidence has suggested OsPT8 is associated with P translocation

from vegetative organs to reproductive tissues (Jia et al., 2011) which is contrary to the data presented here. A similar role has been assigned to AtPHT2; 1, an Arabidopsis orthologue of OsPT14 that was also down-regulated at 15 DAA. The expression level of two P transporters, OsPT21 and OsPT24, were similar at the two stages although their expression was much higher than all other P transporters at both 6 DAA and 15 DAA. It has been suggested that OsPT21–OsPT26 transport P between the cytosol and plastids in leaves and roots (Guo et al., 2008) so high expression of these genes may be indicative of the need to maintain Pi required for photosynthesis (Sivak and Walker, 1986). Despite a recent review of P remobilization during leaf senescence concluding that Pi transporters facilitate the translocation of P liberated from organic P sources to developing grain (Stigter and Plaxton, 2015), we found no evidence for up-regulation of any of the 26 known Pi transports in rice in transport of Pi from senescing flag leaves.

Several genes that participate in dephosphorylation such as OsPAP1d, OsPAP3c, OsPAP10c, OsPLDz2 and glycerophosphoryl diester phosphodiesterase and genes related to lipid metabolism, such as digalactosyldiacylglycerol (DGDG) and monogalactosyl-diacylglycerol (MGDG) synthases, were up-regulated at 15 DAA. Enhanced activity of lipid phosphatases may be related to the activity of plastid-located MGDG and DGDG synthases for maintenance of membrane lipids in these organelles, which has been observed under low temperature stress (Campos et al., 2003). In particular, OsPLDz2 has been implicated in dephosphorylation of plasma membranes and endoplasmic reticulum phospholipids, generating DAG and releasing phosphate upon P starvation (Cruz-Ramírez et al., 2006) with the phosphate becoming available for translocation to the developing grains. Moreover, three genes whose products are directly related to de novo synthesis of DAG, glycerol 3-phosphate dehydrogenase, dihydroxyacetone kinase and glycerol 3-phosphate O-acyltransferase were down-regulated, indicating that this process is reduced during senescence and highlighting the importance of phospholipid degradation for the release of DAG and Pi. Sulphoquinovosyl diacylglycerols substitute for phosphatidylglycerol to maintain the proportion of anionic lipids under P starvation (Yu and Benning, 2003). The replacement of phospholipids with sulpholipids and galactolipids is a key strategy used by plants from the Proteaceae family to maintain photosynthesis in P-limited environments (Lambers et al., 2012). These observations therefore support the notion that the genes involved in recycling P from phospholipids during P starvation in leaves during vegetative growth are also involved in recycling P from leaf phospholipids during leaf senescence which could be translocated to the filling grain. Whether the replacement of phospholipids with other lipids successfully enables flag leaves to maintain high levels of photosynthesis, as occurs when phospholipids are replaced in P-efficient Proteaceae (Lambers et al., 2012), or whether there is a resultant decline in photosynthesis, is not known.

RNA-P is a relatively large component of the metabolically active P pool in plants (Veneklaas et al., 2012), and it was hypothesized RNS2 is part of the P scavenging system under P starvation stress (Abel et al., 2002). Shane et al. (2014) demonstrated RNase activity is highly up-regulated during leaf senescence in Hakea prostrata, and transcriptome analyses indicated that the S-like RNase, AtRNS2, plays a role in the P starvation response during vegetative growth, but is also induced by senescence in Arabidopsis (Hillwig et al., 2011; Taylor et al., 1993). Our results demonstrate that two genes related to RNA degradation were significantly up-regulated at 15 DAA ($\log_2$ = 1.8 and 0.9; Table 4). The genes coded by Os10g0556700 and Os10g0556600 correspond to CCR4-NOT transcription complex subunit 1, a deadenylase participating in poly (A) shortening (Temme *et al.*, 2014), while Os07g0249600 codes for an mRNA-decapping enzyme 1B (Lykke-Andersen, 2002). These two enzymes act on two of the RNA protecting mechanisms, capping and polyadenylation, indicating that during rice leaf senescence, RNA degradation occurs leading to the release of P for its redistribution to the seeds. So far, neither of these two enzymes have been identified as induced by P starvation or leaf senescence; thus, our results widen our knowledge of alternative RNases participating in the liberation of P during leaf senescence. Interestingly, at 15 DAA, the expression of Os07g0661000 and Os04g0680400, that code for an adenosine deaminase and an allantoinase, respectively, were also up-regulated, opening the possibility to associate the activity of the corresponding enzymes with that of CCR4-NOT. Deadenylation by the latter would lead to an increase in cytoplasmic adenosine which would be eventually degraded by the consorted activity of adenosine deaminase and allantoinase, to finally generate CO_2 and NH_3, which could be reassimilated via the glutamine oxoglutarate aminotransferase (GOGAT) pathway (Moffatt and Ashihara, 2002), indicating that nucleotide degradation also serves to recirculate N during leaf senescence.

Conclusions

Several genes involved in the phosphate starvation response were found to have gene expression profiles consistent with a role in the remobilization of P from senescing flag leaves. While some of these genes have previously been implicated in the P starvation response, it was evident that some of these genes (e.g. *OsSPX-MFS1* and *OsSPX-MFS2*) are under different regulatory control during senescence. The fact that a number of genes that have not been implicated in the P starvation response were identified provides potential novel targets for manipulation to reduce the quantity of P loaded into cereal grains, thus reducing the amount of P removed from farmers' fields at harvest. Further studies are evidently needed to confirm the involvement of genes identified in the present study and to explore whether further novel genes or signalling molecules that have not been implicated in the P starvation response play a role in P remobilization from senescing leaves to developing grains.

Experimental procedures

Soil and plant material

Soil from the 0- to 10-cm horizon was collected from Southern Cross University's Brookside field site on the Lismore campus (28°49′3.74″ S, 153°18′21.49″ E). A detailed description of the soil is given in Rose *et al.* (2015). Briefly, the soil was pH 5.81 (1:5 H_2O); total carbon 2.21%; electrical conductivity 0.48 dS/m; Bray P 0.6 mg/kg; and effective cation exchange capacity 13.6 cmol$^+$/kg. The soil was air-dried and passed through a 2 mm sieve before being added at 8 kg/pot to 10 L, nondraining, black plastic pots. Immediately prior to transplanting rice seedlings (see below), nitrogen (N, as urea) and P (as superphosphate) were mixed into the soil for each pot at rates equivalent to 100 kg N/ha and 50 kg P/ha, respectively, on a pot soil surface area basis, before soils were flooded with tap water.

Seeds of the widely grown rice (mega) variety IR64 were sterilized using $HClO_3$ for 2 min before germination in Petri dishes in the dark at 30 °C for 2 days. Germinated seeds were transferred to a mesh floating on a solution containing 0.1 mm calcium (Ca) and 36 μm iron (Fe). After 10 days, the solution was changed to half strength Yoshida solution (Yoshida *et al.*, 1976) and plants were grown for another 2 weeks, with nutrient solution replaced every week. Seedlings were then transplanted into soil with three evenly sized seedlings planted in each pot.

Plants were grown under controlled glasshouse conditions at Southern Cross University (Lismore, NSW, Australia) with a mean day/night air temperature of 29 °C/21 °C and relative humidity (RH) of 75%. Additional N (50 kg/ha as urea) was top-dressed at tillering stage (60 days after transplanting) to ensure low soil N did not induce premature leaf senescence.

Individual panicles in each pot were tagged at anthesis, with anthesis defined as when 50% of florets on a panicle had flowered. This occurred at around 85 days after sowing (DAS). At anthesis, and every 3 DAA until maturity, panicles of the same age (excluding panicles of the main tiller) from three separate pots were harvested and separated into grain, husk, rachis, stem, flag leaf and other leaves. This occurred until 33 DAA, when plants reached physiological maturity, with no plant sampled more than once. After weighing, fresh tissue samples were divided into half: one subsample was immediately frozen in liquid nitrogen and kept at −80 °C for later RNA extraction and RNA-seq analysis, and the second subsample was dried for 7 days at 40 °C in a drying room.

Tissue biomass and P measurements

Phosphorus concentration in flag leaf tissue was measured by digesting 0.2 g tissue with 5 mL of nitric acid (HNO_3) in a MARS Xpress microwave oven (CEM Corporation, NC, Charlotte). After digestion, each sample was diluted by addition of purified water (Milli-Q, Millipore, Billerica, MA, US) to a final volume of 25 mL and P concentration measured by inductively coupled plasma mass spectrometry (ICP-MS) (Perkin Elmer, Melbourne, Victoria, Australia).

RNA extraction and library preparation

RNA extraction was undertaken on flag leaf tissue samples from two time points selected from the P audit (above). Total RNA was extracted from rice flag leaf tissue using the RNeasy Mini Kit (Qiagen, Victoria, Australia, Chadstone) according to the manufacturer's instructions. After extraction, RNA was quantified using the Nanodrop (ND1000; Labtech, Paris, France) and RNA quality was then examined using a 2100 Bioanalyzer (Agilent Technologies, CA, Santa Clara). High-quality RNA samples for library construction were selected based on 260/280 nm ratio and RNA integrity number (RIN) above 2.0 and 8.0, respectively. Samples that did not meet these quality criteria were re-extracted. The TruSeq mRNA stranded kit (Illumina, Scoresby, Vicotria, Australia) was used to prepare Illumina RNA-seq libraries for each sample from 1.3 μg of total RNA. The concentration of each library was measured by qPCR using the KAPA library quantification kit (KAPA Biosystems, MA, Wilmington) to determine the required number of PCR cycles for library amplification. The final concentration of the amplified library was measured using Qubit® Fluorometric Quantitation (Life Technologies, CA, Carlsbad). Library sequencing was undertaken with the Illumina Hi-Seq 2500 system (Illumina) at the Biomolecular Facility (BRF) at the John Curtin School for Medical Research (JCSMR), Australian

National University (ANU), ACT, Australia. The raw sequencing data have been uploaded to ENA (European Nucleotide Archive) database (accession number: PRJEB11899).

RNA-seq data analysis

Raw sequencing reads in FASTQ format were first filtered for quality using FASTQC (Andrews, 2010) followed by removal of adapter sequences, poly-N stretches and low-quality reads using the BBDuk module of the BBMap software package (http://sourceforge.net/projects/bbmap/). All subsequent analyses were based on high-quality sequencing reads. Bowtie v2.2.4 (Langmead and Salzberg, 2012) was used to index the genome. Retained high-quality paired-end reads were mapped against the rice genome IRGSP 1.0 using TopHat (Trapnell et al., 2009), which is a splice aware aligner of RNA-Seq reads. The Ensembl Plants (http://plants.ensembl.org/Oryza_sativa/Info/Annotation) O. sativa cv. Nipponbare (ssp. japonica) reference genome annotation was utilized. TopHat identified the exon–exon junctions and produced the read vs genome alignment in BAM (Binary Alignment Map) format. Cufflinks (Trapnell et al., 2012) then used the TopHat-generated alignment to assemble a set of reference-based transcripts. Finally, the CuffDiff module of Cufflinks was used to identify differentially expressed genes between samples and CummeRbund (http://compbio.mit.edu/cummeRbund/) R package was used for subsequent analyses.

Involvement of PSR genes in phosphorus remobilization from flag leaves during grain filling

To investigate whether any PSR genes were involved in remobilization of P from leaves during grain filling, we identified a total of 115 PSR genes, including 26 OsPT genes and eight RNS genes, based on recent literature (Lin et al., 2009; Liu et al., 2011; MacIntosh et al., 2010; Secco et al., 2013) (Table S1). First, we assessed the absolute FPKM value of the known 115 PSR genes at the two selected time points. We also compared their expression to the expression of the 30 most highly expressed genes to assess their relative level of expression. Secondly, we assessed the relative change in their expression between the two time points when the flag leaf had a stable P concentration compared with when the leaf acted as P source tissue.

Gene expression was evaluated based on FPKM values (i.e. FPKM > 0). Significant DEGs were defined based on P and FDR (i.e. <0.05).

GO analysis of DEGs using RNA-seq data

To analyse the GO enrichment analysis, firstly, significantly differential expressed genes were retrieved by cutting off based on P and FDR values <0.05, respectively, and $\log_2$ fold change >1.5. Total of 1643 DEGs were retrieved of 5160 DEGs. GO terms for each rice gene were obtained using PAGE analysis in AgriGO public web tool (http://bioinfo.cau.edu.cn/agriGO/analysis.php?method=PAGE) which can be accepted an arbitrarily large input list with fold change (Du et al., 2010).

Expression of OsPTs, OsPAPs and OsSPX domains in the flag leaf at 6 DAA and 15 DAA

Several OsPTs, OsPAPs and OsSPXs were among the 115 PSR genes with differential expression between the two time points. We therefore investigated the expression of all known members of these gene families at 6 DAA and 15 DAA, including those that were not among the 115 PSR genes. This included the OsSPX-MFS genes that are classified as SPX domain-possessing

proteins that are regulated by Pi starvation in rice and preferentially expressed in the leaves (Lin et al., 2010; Wang et al., 2012).

Biological pathway analysis

We examined whether metabolic pathways that include RNase or phospholipase genes were more active when the flag leaf was acting as a P source (15 DAA) using the KEGG (Kyoto Encyclopedia of Genes and Genomes) pathway tool. We assumed that genes differentially expressed between 6 DAA and 15 DAA, particularly with elevated expression 15 DAA, might be involved in P remobilization. We therefore retrieved from the KEGG pathway database all the activating enzymes/genes related with lipid and nucleotide metabolism, such as glycerophospholipid metabolism, glycerolipid metabolism, sphingolipid metabolism, glycosphingolipid metabolism and RNA transport/degradation. Those retrieved genes were then mapped to the RNA-seq data to verify their involvement based on $\log_2$ fold change and significance.

Acknowledgements

We thank Dr. Norman Warthmann and Peter Crisp for their help with the library construction. This research work was supported by funding from the Global Rice Science Partnership (GRiSP) New Frontiers Research Project. O.P. was supported by a Sabbatical Fellowship by DGAPA-UNAM.

References

Abel, S., Ticconi, C.A. and Delatorre, C.A. (2002) Phosphate sensing in higher plants. Physiol. Plant. 115, 1–8.

Andrews, S. (2010) FastQC: a quality control tool for high throughput sequence data. online available http://www.bioinformatics.babraham.ac.uk/projects/fastqc (8 march 2016, date last accessed).

Araújo, A.P. and Teixeira, M.G. (2003) Nitrogen and phosphorus harvest indices of common bean cultivars: implications for yield quantity and quality. Plant Soil, 257, 425–433.

Batten, G.D. (1992) A review of phosphorus efficiency in wheat. Plant Soil, 146, 163–168.

Breeze, E., Harrison, E., McHattie, S., Hughes, L., Hickman, R., Hill, C., Kiddle, S. et al. (2011) High-resolution temporal profiling of transcripts during Arabidopsis leaf senescence reveals a distinct chronology of processes and regulation. Plant Cell, 23, 873–894.

Buchanan-Wollaston, V. (1997) The molecular biology of leaf senescence. J. Exp. Bot. 48, 181–199.

Buchanan-Wollaston, V., Earl, S., Harrison, E., Mathas, E., Navabpour, S., Page, T. and Pink, D. (2003) The molecular analysis of leaf senescence – a genomics approach. Plant Biotechnol. J. 1, 3–22.

Campos, P.S., Quartin, V.n., Ramalho, J.c. and Nunes, M.A. (2003) Electrolyte leakage and lipid degradation account for cold sensitivity in leaves of Coffea sp. plants. J. Plant Physiol. 160, 283–292.

Cruz-Ramírez, A., Oropeza-Aburto, A., Razo-Hernández, F., Ramírez-Chávez, E. and Herrera-Estrella, L. (2006) Phospholipase DZ2 plays an important role in extraplastidic galactolipid biosynthesis and phosphate recycling in Arabidopsis roots. Proc. Natl Acad. Sci. USA, 103, 6765–6770.

Dong, H., Li, W., Eneji, A.E. and Zhang, D. (2012) Nitrogen rate and plant density effects on yield and late-season leaf senescence of cotton raised on a saline field. Field. Crop. Res. 126, 137–144.

Du, Z., Zhou, X., Ling, Y., Zhang, Z. and Su, Z.. (2010) agriGO: a GO analysis toolkit for the agricultural community. Nucleic Acids Res. 38, W64–W70.

Duan, M. and Sun, S.S. (2005) Profiling the expression of genes controlling rice grain quality. Plant Mol. Biol. 59, 165–178.

Guo, Y., Cai, Z. and Gan, S. (2004) Transcriptome of Arabidopsis leaf senescence. Plant, Cell Environ. 27, 521–549.

Guo, B., Jin, Y., Wussler, C., Blancaflor, E.B., Motes, C.M. and Versaw, W.K. (2008) Functional analysis of the *Arabidopsis* PHT4 family of intracellular phosphate transporters. *New Phytol.* **177**, 889–898.

Hillwig, M.S., Contento, A.L., Meyer, A., Ebany, D., Bassham, D.C. and MacIntosh, G.C. (2011) RNS2, a conserved member of the RNase T2 family, is necessary for ribosomal RNA decay in plants. *Proc. Natl Acad. Sci. USA*, **108**, 1093–1098.

Jia, H., Ren, H., Gu, M., Zhao, J., Sun, S., Zhang, X., Chen, J. *et al.* (2011) The phosphate transporter gene OsPht1;8 is involved in phosphate homeostasis in rice. *Plant Physiol.* **156**, 1164–1175.

Kong, X., Luo, Z., Dong, H., Eneji, A.E., Li, W. and Lu, H. (2013) Gene expression profiles deciphering leaf senescence variation between early-and late-senescence cotton lines. *PLoS ONE*, **8**, e69847.

Kong, L., Guo, H. and Sun, M. (2015) Signal transduction during wheat grain development. *Planta*, **241**, 789–801.

Lambers, H., Cawthray, G.R., Giavalisco, P., Kuo, J., Laliberté, E., Pearse, S.J., Scheible, W.R. *et al.* (2012) Proteaceae from severely phosphorus-impoverished soils extensively replace phospholipids with galactolipids and sulfolipids during leaf development to achieve a high photosynthetic phosphorus-use-efficiency. *New Phytol.* **196**, 1098–1108.

Langmead, B. and Salzberg, S.L. (2012) Fast gapped-read alignment with Bowtie 2. *Nat. Methods*, **9**, 357–359.

Lin, W.Y., Lin, S.I. and Chiou, T.J. (2009) Molecular regulators of phosphate homeostasis in plants. *J. Exp. Bot.* **60**, 1427–1438.

Lin, S.-I., Santi, C., Jobet, E., Lacut, E., El Kholti, N., Karlowski, W.M., Verdeil, J.-L. *et al.* (2010) Complex regulation of two target genes encoding SPX-MFS proteins by rice miR827 in response to phosphate starvation. *Plant Cell Physiol.* **51**, 2119–2131.

Lin, M., Pang, C., Fan, S., Song, M., Wei, H. and Yu, S. (2015) Global analysis of the *Gossypium hirsutum* L. Transcriptome during leaf senescence by RNA-Seq. *BMC Plant Biol.* **15**, 43.

Liu, F., Chang, X.J., Ye, Y., Xie, W.B., Wu, P. and Lian, X.M. (2011) Comprehensive sequence and whole-life-cycle expression profile analysis of the phosphate transporter gene family in rice. *Mol. Plant*, **4**, 1105–1122.

Lott, J.N.A., Ockenden, I., Raboy, V. and Batten, G.D. (2000) Phytic acid and phosphorus in crop seeds and fruits: a global estimate. *Seed Sci. Res.* **10**, 11–33.

Lykke-Andersen, J. (2002) Identification of a human decapping complex associated with hUpf proteins in nonsense-mediated decay. *Mol. Cell. Biol.* **22**, 8114–8121.

MacIntosh, G., Hillwig, M., Meyer, A. and Flagel, L. (2010) RNase T2 genes from rice and the evolution of secretory ribonucleases in plants. *Mol. Genet. Genomics*, **283**, 381–396.

Moffatt, B.A. and Ashihara, H. (2002) Purine and pyrimidine nucleotide synthesis and metabolism. In: Somerville CR, Meyerowitz EM (eds) *The Arabidopsis Book*. American Society of Plant Biologists, 1–20.

Nagy, R., Grob, H., Weder, B., Green, P., Klein, M., Frelet-Barrand, A., Schjoerring, J.K. *et al.* (2009) The *Arabidopsis* ATP-binding cassette protein AtMRP5/AtABCC5 is a high affinity inositol hexakisphosphate transporter involved in guard cell signaling and phytate storage. *J. Biol. Chem.* **284**, 33614–33622.

Pariasca-Tanaka, J., Vandamme, E., Mori, A., Segda, Z., Saito, K., Rose, T.J. and Wissuwa, M. (2015) Does reducing seed-P concentrations affect seedling vigor and grain yield of rice? *Plant Soil*, **392**, 253–266.

Raboy, V. (2009) Approaches and challenges to engineering seed phytate and total phosphorus. *Plant Sci.* **177**, 281–296.

Raboy, V., Cichy, K., Peterson, K., Reichman, S., Sompong, U., Srinives, P. and Saneoka, H. (2014) Barley (*Hordeum vulgare* L.) low phytic acid 1-1: an endosperm-specific, filial determinant of seed total phosphorus. *J. Hered.* **105**, 656–665.

Reimmann, C. and Dudler, R. (1993) Circadian rhythmicity in the expression of a novel light-regulated rice gene. *Plant Mol. Biol.* **22**, 165–170.

Robinson, W.D., Carson, I., Ying, S., Ellis, K. and Plaxton, W.C. (2012) Eliminating the purple acid phosphatase AtPAP26 in *Arabidopsis thaliana* delays leaf senescence and impairs phosphorus remobilization. *New Phytol.* **196**, 1024–1029.

Rose, T.J. and Wissuwa, M. (2012) Chapter five – Rethinking internal phosphorus utilization efficiency: a new approach is needed to improve PUE in grain crops. In *Advances in Agronomy*, **116**, 185–217.

Rose, T.J., Rengel, Z., Ma, Q. and Bowden, J.W. (2007) Differential accumulation patterns of phosphorus and potassium by canola cultivars compared to wheat. *J. Plant Nutr. Soil Sci.* **170**, 404–411.

Rose, T.J., Pariasca-Tanaka, J., Rose, M.T., Fukuta, Y. and Wissuwa, M. (2010) Genotypic variation in grain phosphorus concentration, and opportunities to improve P-use efficiency in rice. *Field. Crop. Res.* **119**, 154–160.

Rose, T., Liu, L. and Wissuwa, M. (2013) Improving phosphorus efficiency in cereal crops: Is breeding for reduced grain phosphorus concentration part of the solution? *Front. Plant Sci.* **4**, 1–6.

Rose, T.J., Raymond, C.A., Bloomfield, C. and King, G.J. (2015) Perturbations of nutrient source-sink relationships by post-anthesis stresses results in differential accumulation of nutrients in wheat grain. *J. Plant Nutr. Soil Sci.* **178**, 89–98.

Secco, D., Wang, C., Arpat, B.A., Wang, Z., Poirier, Y., Tyerman, S.D., Wu, P. *et al.* (2012) The emerging importance of the SPX domain-containing proteins in phosphate homeostasis. *New Phytol.* **193**, 842–851.

Secco, D., Jabnoune, M., Walker, H., Shou, H., Wu, P., Poirier, Y. and Whelan, J. (2013) Spatio-temporal transcript profiling of rice roots and shoots in response to phosphate starvation and recovery. *Plant Cell*, **25**, 4285–4304.

Shane, M.W., Stigter, K., Fedosejevs, E.T. and Plaxton, W.C. (2014) Senescence-inducible cell wall and intracellular purple acid phosphatases: implications for phosphorus remobilization in *Hakea prostrata* (Proteaceae) and *Arabidopsis thaliana* (Brassicaceae). *J. Exp. Bot.* **65**, 6097–6106.

Shi, J., Wang, H., Schellin, K., Li, B., Faller, M., Stoop, J.M., Meeley, R.B. *et al.* (2007) Embryo-specific silencing of a transporter reduces phytic acid content of maize and soybean seeds. *Nat. Biotechnol.* **25**, 930–937.

Sivak, M.N. and Walker, D.A. (1986) Photosynthesis in vivo can be limited by phosphate supply. *New Phytol.* **102**, 499–512.

Smith, A.P., Fontenot, E.B., Zahraeifard, S. and DiTusa, S.F. (2015) Molecular components that drive phosphorus-remobilisation during leaf senescence. *Annu. Plant Rev.* **48**, 159–186.

Stigter, K.A. and Plaxton, W.C. (2015) Molecular mechanisms of phosphorus metabolism and transport during leaf senescence. *Plants*, **4**, 773–798.

Subramanian, A., Tamayo, P., Mootha, V.K., Mukherjee, S., Ebert, B.L., Gillette, M.A., Paulovich, A. *et al.* (2005) Gene set enrichment analysis: a knowledge-based approach for interpreting genome-wide expression profiles. *Proc. Natl Acad. Sci. USA*, **102**, 15545–15550.

Sugano, S., Jiang, C.-J., Miyazawa, S.-I., Masumoto, C., Yazawa, K., Hayashi, N., Shimono, M. *et al.* (2010) Role of OsNPR1 in rice defense program as revealed by genome-wide expression analysis. *Plant Mol. Biol.* **74**, 549–562.

Taylor, C.B., Bariola, P.A., delCardayré, S.B., Raines, R.T. and Green, P.J. (1993) Rxml: a senescence-associated RNase of *Arabidopsis* that diverged from the S-RNases before speciation. *Proc. Natl Acad. Sci. USA*, **90**, 5118–5122.

Temme, C., Simonelig, M. and Wahle, E. (2014) Deadenylation of mRNA by the CCR4–NOT complex in *Drosophila*: molecular and developmental aspects. *Front Genet.* **5**, 1–11.

Trapnell, C., Pachter, L. and Salzberg, S.L. (2009) TopHat: discovering splice junctions with RNA-Seq. *Bioinformatics*, **25**, 1105–1111.

Trapnell, C., Roberts, A., Goff, L., Pertea, G., Kim, D., Kelley, D.R., Pimentel, H. *et al.* (2012) Differential gene and transcript expression analysis of RNA-seq experiments with TopHat and Cufflinks. *Nat. Protoc.* **7**, 562–578.

Uauy, C., Distelfeld, A., Fahima, T., Blechl, A. and Dubcovsky, J. (2006) A NAC gene regulating senescence improves grain protein, zinc, and iron content in wheat. *Science*, **314**, 1298–1301.

Veneklaas, E.J., Lambers, H., Bragg, J., Finnegan, P.M., Lovelock, C.E., Plaxton, W.C., Price, C.A. *et al.* (2012) Opportunities for improving phosphorus-use efficiency in crop plants. *New Phytol.* **195**, 306–320.

Wang, Z., Hu, H., Huang, H., Duan, K., Wu, Z. and Wu, P. (2009) Regulation of OsSPX1 and OsSPX3 on expression of OsSPX domain genes and Pi-starvation signaling in rice. *J. Integr. Plant Biol.* **51**, 663–674.

Wang, C., Huang, W., Ying, Y.H., Li, S., Secco, D., Tyerman, S., Whelan, J. *et al.* (2012) Functional characterization of the rice SPX-MFS family reveals a key role of OsSPX-MFS1 in controlling phosphate homeostasis in leaves. *New Phytol.* **196**, 139–148.

Wang, F., Rose, T., Jeong, K., Kretzschmar, T. and Wissuwa, M. (2016) The knowns and unknowns of P loading into grains, and implications for P efficiency in cropping systems. *J. Exp. Bot.* **67**, 1221–1229.

White, P.J. and Veneklaas, E.J. (2012) Nature and nurture: the importance of seed phosphorus content. *Plant Soil*, **357**, 1–8.

Wu, P., Shou, H., Xu, G. and Lian, X. (2013) Improvement of phosphorus efficiency in rice on the basis of understanding phosphate signaling and homeostasis. *Curr. Opin. Plant Biol.* **16**, 205–212.

Ye, H.X., Zhang, X.Q., Broughton, S., Westcott, S., Wu, D.X., Lance, R. and Li, C.D. (2011) A nonsense mutation in a putative sulphate transporter gene results in low phytic acid in barley. *Funct. Integr. Genomics*, **11**, 103–110.

Yoshida, S., Forno, D.A., Cock, J.H. and Gomez, K.A. (1976) *Laboratory Manual for Physiological Studies of Rice*, 1–82. Los Banos (Philippines): International Rice Research Institute.

Yu, B. and Benning, C. (2003) Anionic lipids are required for chloroplast structure and function in *Arabidopsis. Plant J.* **36**, 762–770.

Zhang, Q., Wang, C., Tian, J., Li, K. and Shou, H. (2011) Identification of rice purple acid phosphatases related to phosphate starvation signalling. *Plant Biol.* **13**, 7–15.

Zhang, W.Y., Xu, Y.C., Li, W.L., Yang, L., Yue, X., Zhang, X.S. and Zhao, X.Y. (2014) Transcriptional analyses of natural leaf senescence in maize. *PLoS ONE*, **9**, e115617.

Zhu, T., Budworth, P., Chen, W.Q., Provart, N., Chang, H.S., Guimil, S., Su, W.P. *et al.* (2003) Transcriptional control of nutrient partitioning during rice grain filling. *Plant Biotechnol. J.* **1**, 59–70.

4

A high-density, SNP-based consensus map of tetraploid wheat as a bridge to integrate durum and bread wheat genomics and breeding

Marco Maccaferri[1,*], Andrea Ricci[1], Silvio Salvi[1], Sara Giulia Milner[1], Enrico Noli[1], Pier Luigi Martelli[2], Rita Casadio[2], Eduard Akhunov[3], Simone Scalabrin[4,5], Vera Vendramin[4,5], Karim Ammar[6], Antonio Blanco[7], Francesca Desiderio[8], Assaf Distelfeld[9], Jorge Dubcovsky[10,11], Tzion Fahima[12], Justin Faris[13], Abraham Korol[12], Andrea Massi[14], Anna Maria Mastrangelo[15], Michele Morgante[4,5], Curtis Pozniak[16], Amidou N'Diaye[16], Steven Xu[13] and Roberto Tuberosa[1]

[1]Department of Agricultural Sciences (DipSA), University of Bologna, Bologna, Italy

[2]Biocomputing Group, University of Bologna, Bologna, Italy

[3]Department of Plant Pathology, Kansas State University, Manhattan, KS, USA

[4]Istituto di Genomica Applicata, Udine, Italy

[5]Dipartimento di Scienze Agrarie e Ambientali, University of Udine, Udine, Italy

[6]CIMMYT Carretera Mexico, Texcoco, Mexico

[7]Dipartimento di Biologia e Chimica Agro-forestale ed ambientale, Università di Bari, Aldo Moro, Bari, Italy

[8]Consiglio per la ricerca e la sperimentazione in agricoltura, Genomics Research Centre, Fiorenzuola d'Arda, Italy

[9]Faculty of Life Sciences, Department of Molecular Biology and Ecology of Plants, Tel Aviv University, Tel Aviv, Israel

[10]Department of Plant Sciences, University of California, Davis, CA, USA

[11]Howard Hughes Medical Institute, Chevy Chase, MD, USA

[12]Department of Evolutionary and Environmental Biology, Institute of Evolution, Faculty of Science and Science Education, University of Haifa, Haifa, Israel

[13]USDA-ARS Cereal Crops Research Unit, Fargo, ND, USA

[14]Società Produttori Sementi Bologna (PSB), Argelato, Italy

[15]Consiglio per la ricerca e la sperimentazione in agricoltura, Cereal Research Centre, Foggia, Italy

[16]Crop Development Centre and Department of Plant Sciences, University of Saskatchewan, Saskatoon, SK, Canada

*Correspondence

e-mail marco.maccaferri@unibo.it

Keywords: durum wheat, single nucleotide polymorphism, consensus map, genomics-assisted breeding, homeologous loci, chromosome translocation events, anchor markers.

Summary

Consensus linkage maps are important tools in crop genomics. We have assembled a high-density tetraploid wheat consensus map by integrating 13 data sets from independent biparental populations involving durum wheat cultivars (*Triticum turgidum* ssp. *durum*), cultivated emmer (*T. turgidum* ssp. *dicoccum*) and their ancestor (wild emmer, *T. turgidum* ssp. *dicoccoides*). The consensus map harboured 30 144 markers (including 26 626 SNPs and 791 SSRs) half of which were present in at least two component maps. The final map spanned 2631 cM of all 14 durum wheat chromosomes and, differently from the individual component maps, all markers fell within the 14 linkage groups. Marker density per genetic distance unit peaked at centromeric regions, likely due to a combination of low recombination rate in the centromeric regions and even gene distribution along the chromosomes. Comparisons with bread wheat indicated fewer regions with recombination suppression, making this consensus map valuable for mapping in the A and B genomes of both durum and bread wheat. Sequence similarity analysis allowed us to relate mapped gene-derived SNPs to chromosome-specific transcripts. Dense patterns of homeologous relationships have been established between the A- and B-genome maps and between nonsyntenic homeologous chromosome regions as well, the latter tracing to ancient translocation events. The gene-based homeologous relationships are valuable to infer the map location of homeologs of target loci/QTLs. Because most SNP and SSR markers were previously mapped in bread wheat, this consensus map will facilitate a more effective integration and exploitation of genes and QTL for wheat breeding purposes.

Introduction

Durum wheat [*Triticum turgidum* ssp. *durum* (Desf.) Husn.] is grown worldwide on 17 M ha for a production of 35.6 Mt in 2013 (International Grain Council, IGC Grain Market reports). The main durum wheat product, semolina, is at the base of the pasta industrial chain (13.5 MT per year, IPO 2013).

Cultivated durum wheat is tetraploid and its direct ancestor cultivated emmer (*T. turgidum* ssp. *dicoccum* Schrank) was derived from the domestication of a wild tetraploid species (wild emmer, *T. turgidum* ssp. *dicoccoides* Körn. ex Asch. & Graebner, Schweinf.). The tetraploidization event involved a cross between *T. urartu* (genome AA) with an unknown close relative of *Aegilops speltoides* (genome BB) and occurred <0.8 Mya (Marcussen *et al.*, 2014; Mori, 2003; Ozkan *et al.*, 2005; Salamini *et al.*, 2002). A further hybridization with *Ae. tauschii* (D genome) gave rise to hexaploid wheat (bread wheat, *Triticum aestivum* L., Kihara, 1944; McFadden and Sears, 1946; Dvorak *et al.*, 1998), The genetic constitution of and relationships among tetraploid and hexaploid wheat species were then shaped by domestication, selection bottlenecks, migrations and extensive gene flow (mainly from tetraploid to hexaploid wheat, Nesbitt and Samuel, 1996; Salamini *et al.*, 2002; Ozkan *et al.*, 2005; Dvorak *et al.*, 1998, 2006; Caldwell *et al.*, 2004; Giles and Brown, 2006). Hence, any useful information accrued in durum wheat genetics is likely to be valuable also to bread wheat breeders.

Until now, durum wheat molecular genetics and genomics have largely relied on hexaploid wheat tools (Feuillet *et al.*, 2012; Gupta *et al.*, 2008; Röder *et al.*, 1998; Varshney *et al.*, 2007). Molecular markers (RFLP, SSR, AFLP, DArT®, etc.) developed for hexaploid wheat are generally useful for genetic mapping experiments in tetraploid wheat based on the sharing of the two A and B genomes (Blanco *et al.*, 1998; Korzun *et al.*, 1999; Lotti *et al.* 2000; Mantovani *et al.*, 2008). However, markers' polymorphism information content in durum wheat germplasm differs from that observed in hexaploid wheat (Maccaferri *et al.*, 2003, 2005). Single nucleotide polymorphisms (SNPs) have become the most widely utilized markers and are being extensively developed in crops, including wheat (Allen *et al.*, 2013; Cavanagh *et al.*, 2013; Ganal *et al.*, 2014; Lai *et al.*, 2012; van Poecke *et al.*, 2013; Tiwari *et al.*, 2014; Trebbi *et al.*, 2011). A high-density (90K) wheat SNP array has been recently developed (Wang *et al.*, 2014), mostly based on bread wheat SNPs, but including approx. 8000 SNPs from durum cultivars.

Linkage maps are crucial tools for the identification and cloning of genes and quantitative trait loci (QTL), marker-assisted selection and genome structure analysis. The inherent constraints of linkage maps based on single populations can be overcome by the production of consensus maps of several maps and independent crosses. Consensus maps provide large number of markers, extend the mapped genome portions (including regions identical by descent), overcome local loss of genetic resolution, validate marker order, identify chromosome rearrangements and increase the effectiveness of genome-wide association studies and QTL meta-analyses (Salvi *et al.*, 2010; Swamy *et al.*, 2011).

Integration of mapping information from individual mapping populations has been performed using various approaches (Ronin *et al.*, 2012). Currently, the *genetic merging* method has been used for medium-density (up to 5000 markers) projects (Wang *et al.*, 2011). The *graph*-based method has been used for high-density (of more than 10 000 loci) consensus maps (Gautami *et al.*, 2012; Wang *et al.*, 2014). The *regression* method is also suitable for high-density maps but requires high-quality initial maps (Blenda *et al.*, 2012; Hudson *et al.*, 2012).

In hexaploid wheat, a SSR-based consensus map by Somers *et al.* (2004) has been widely used. In durum wheat, two consensus maps mainly based on SSR and DArT® markers have been developed (Marone *et al.*, 2012; Maccaferri *et al.*,). SNP-based consensus maps are quickly replacing or updating former maps (Bowers *et al.*, 2012; Hyten *et al.*, 2010; Muñoz-Amatriaín *et al.* 2012; Goretti *et al.*, 2014; van Poecke *et al.*, 2013). In bread wheat, six and eight biparental cross-populations were used to build SNP-based consensus maps (Cavanagh *et al.*, 2013; Wang *et al.*, 2014).

In this study, 13 independent tetraploid wheat mapping populations were utilized to assemble a durum wheat consensus map with the highest number of markers (>30 000) to date. These populations included elite × elite, elite × cultivated emmer and elite × wild emmer. Ten of these populations were genotyped with the recently developed Illumina 9K and 90K wheat SNP arrays (Cavanagh *et al.*, 2013; Wang *et al.*, 2014). Informativeness and potential applications of this consensus map were analysed and discussed in terms of genome coverage and marker density, diversity and distribution.

Results

Linkage maps from the individual populations (component maps)

The genotypic data of 13 tetraploid wheat (AABB; $2n = 28$) mapping populations were used for assembling the consensus map. In total, the populations included 1928 lines (1773 RILs and 155 DHs, Table 1). The elite × elite populations that were genotyped with the Illumina 90K SNP array (six in total) allowed us to map between 3676 and 6163 SNPs with a mean of 5175 SNPs and a mean map length of 2012 cM. The two durum × cultivated emmer populations allowed us to map 8952 and 10 811 SNPs (3028 and 2363 cM, respectively) while the Svevo × Zavitan (durum × wild emmer) population included 10 911 SNPs (2258 cM).

Despite the high density of polymorphic markers detected using the 90K SNP assay, none of the six elite × elite populations produced component maps with a number of linkage groups equal to the nominal chromosome pair number (14). Rather, linkage groups ranged from 20 to 33 (28 on average) with an average density of 2.2–3.0 SNPs/cM and an average SNP intermarker distance of 0.39 cM/SNP (from 0.33 in Cl × Ld to 0.46 cM/SNP in G9586). In contrast, the component maps derived from the three cultivated emmer and wild emmer-derived maps genotyped with the Illumina 90K SNPs showed 14–15 linkage groups with a more even genome coverage as compared to any map from cultivated durum wheat. However, the increase in marker density of the SNPs mapped in the polymorphic regions (0.23–0.34 cM/SNP for the emmer-derived maps and 0.20 cM/SNP for the wild emmer-derived map) was only slightly higher as compared to the elite × elite populations.

Consensus map features

The genotypic data of the 1773 RILs were used to construct a framework consensus map comprised of 16 384 common markers (hereafter referred to as anchor markers, 54.3% of the total). In particular, anchors included 15 827 Illumina wheat SNPs, 297 SSRs, 242 DArT® markers and 18 SNPs/STS markers of diverse origin. The anchor markers were grouped into 14 linkage groups, corresponding to the tetraploid wheat chromosomes (Table 2), and were mapped at a mean LOD score = 61.9. The framework map was used to infer the position of the 13 760 markers uniquely mapped in single component populations, leading to a final consensus map including 30 144 markers. This map was based on 5279 unique recombination events (377 per chromosome) and spanned 2631 cM for an average intermarker distance of 0.087 cM, corresponding to 11 markers per cM. The consensus map was thus 30.7 and 17.5% longer than the mean map length of the six component elite × elite durum populations genotyped with the Illumina 90K SNP array and the longest one (Mr × Cd; 2239 cM), respectively. However, the consensus map was very similar in length to those of the three durum × cultivated emmer maps (2631 *vs.* 2639 cM) and slightly longer (14.2%) than the durum × wild emmer Svevo × Zavitan map.

Table 1 Details of the 13 mapping populations and of the corresponding single component maps

Mapping populations					Molecular markers					Linkage group		
		Type	Size		Genomic SSR	DArT®	Illumina SNP[†]	Others[‡]	Total	Linkage group	Total length	Intermarker distance[§]
Parents	Acronym	DH/RIL	no.		no.	no.	no.	no.	no.	no.	cM	cM/marker
T. durum × *T. durum*												
Colosseo × Lloyd[¶]	Cl × Ld	RIL	176		184	372	6163	1227	7946	20	2063.9	0.33
Meridiano × Claudio[¶]	Mr × Cd	RIL	181		178	608	5097	87	5970	27	2238.8	0.43
Simeto × Levante[¶]	Sm × Lv	RIL	180		142	335	5315	6	5798	30	2184.7	0.40
Mohawk × Cocorit69	Mh × Cr	RIL	81		–	–	5554	–	5554	31	2012.7	0.36
Svevo × Ciccio[¶]	Sv × Cc	RIL	103		16	213	5246	12	5487	26	1887.6	0.36
W9292-260D3 × Kofa	G9586	DH	155		34	–	3676	2	3712	33	1685.0	0.46
Kofa × Svevo[¶]	Kf × Sv	RIL	249		205	–	–	38	243	18	1256.2	–
Kofa × UC1113[¶]	Kf × UC	RIL	93		172	–	–	31	203	24	755.1	–
T. durum × *T. dicoccum*												
Ben × PI41025	Bn × PI_41025	RIL	200		111	–	2456	–	2567	14	2526.9	–
Simeto × Molise Colli	Sm × Ml	RIL	136		26	–	8926	–	8952	15	3028.4	0.34
Latino × MG5323	Lt × MG_5323	RIL	82		216	–	10 572	23	10 811	14	2363.4	0.23
T. durum × *T. dicoccoides*												
Langdon × G18-16[¶]	Ln × G18-16	RIL	152		120	148	–	–	268	20	1577.3	–
Svevo × Zavitan	Sv × Zv	RIL	140		–	–	10 911	–	10 911	14	2258.0	0.20

RIL, recombinant inbred line; DH, double haploid; SSR, simple sequence repeat; DArT®, Diversity Array Technology; SNP, single nucleotide polymorphism.

*UNIBO, University of Bologna; UNIUD, University of Udine; PSB, Produttori Sementi Bologna; CIMMYT, International Maize and Wheat Improvement Center; USASK, University of Saskatchewan; UNIBA, University of Bari; AAFC, Agriculture and Agri-Food Canada; USDA-ARS, Cereal Crop Research Unit, Fargo; CRA, Consiglio per la Ricerca e la Sperimentazione in Agricoltura; UHAIFA, University of Haifa; UTELAVIV, Tel Aviv University.

[†] Illumina iSelect 90K wheat SNP array used for 9 populations; Illumina iSelect 9K wheat SNP array used for Bn × PI41025.

[‡] Include sequence tagged sites, morphological and biochemical markers, and 1065 sequence-based genotyping SNPs for Cl × Ld.

[§] Referred to the Illumina iSelect 90K wheat SNPs only.

[¶] Details of mapping populations in Materials and Methods.

Table 2 Features of the consensus framework map summarized by chromosome

Chromosome	Illumina SNP no.	SSR no.	DArT no.	Others (SNP, STS) no.	Total no.	Common markers no.	Singleton no.	Informative maps no.	Recombination bins no.	LOD LOD units	Length cM	Intermarker distance cM/marker
1A	1705	44	78	35	1862	975	887	3.0 ± 1.0	336	78.1 ± 45.3	171.6	0.09
1B	2450	69	105	180	2804	1481	1323	2.9 ± 1.0	398	45.1 ± 32.6	178.0	0.06
2A	1860	63	72	106	2101	1123	978	3.2 ± 1.3	320	67.4 ± 39.1	215.5	0.10
2B	2978	85	121	140	3324	1789	1535	3.1 ± 1.1	493	60.1 ± 38.5	199.8	0.06
3A	1472	60	74	92	1698	964	734	3.3 ± 1.3	326	68.6 ± 47.6	184.7	0.10
3B	2091	60	150	122	2423	1183	1240	2.9 ± 1.1	470	55.2 ± 36.1	212.5	0.09
4A	1346	44	108	65	1563	813	750	3.0 ± 1.2	320	61.7 ± 40.4	176.5	0.11
4B	1254	39	56	77	1426	924	502	3.8 ± 1.3	307	71.7 ± 41.9	136.5	0.10
5A	1442	70	18	78	1608	860	748	2.9 ± 1.1	377	58.1 ± 30.4	218.6	0.14
5B	2140	65	65	53	2323	1279	1044	3.4 ± 1.3	435	72.9 ± 40.7	206.2	0.09
6A	1635	27	92	106	1860	978	882	3.0 ± 1.1	303	63.7 ± 36.1	135.1	0.07
6B	2093	46	132	133	2404	1337	1067	3.2 ± 1.2	367	50.7 ± 42.9	156.0	0.07
7A	2084	49	108	92	2333	1314	1019	2.3 ± 1.4	418	69.8 ± 38.9	214.0	0.09
7B	2076	70	145	124	2415	1364	1051	4.0 ± 1.2	409	43.5 ± 32.9	226.3	0.09
Mean	1901	57	95	100	2153	1170	982	3.1 ± 1.2	377	61.9 ± 38.8	187.9	0.09
Total	26 626	791	1324	1403	30 144	16 384	13 760	–	5279	–	2631.3	–
Mean A genome	1649	51	79	82	1861	1004	857	–	343	–	188.0	0.10
Mean B genome	2155	62	111	118	2446	1337	1109	–	411	–	187.9	0.08
Total A genome	11 544	357	550	574	13 025	7027	5998	–	2400	–	1316	–
Total B genome	15 082	434	774	829	17 119	9357	7762	–	2879	–	1315	–

SSR, simple sequence repeat; DArT®, Diversity Array Technology; SNP, single nucleotide polymorphism; LOD, logarithm of the odds.

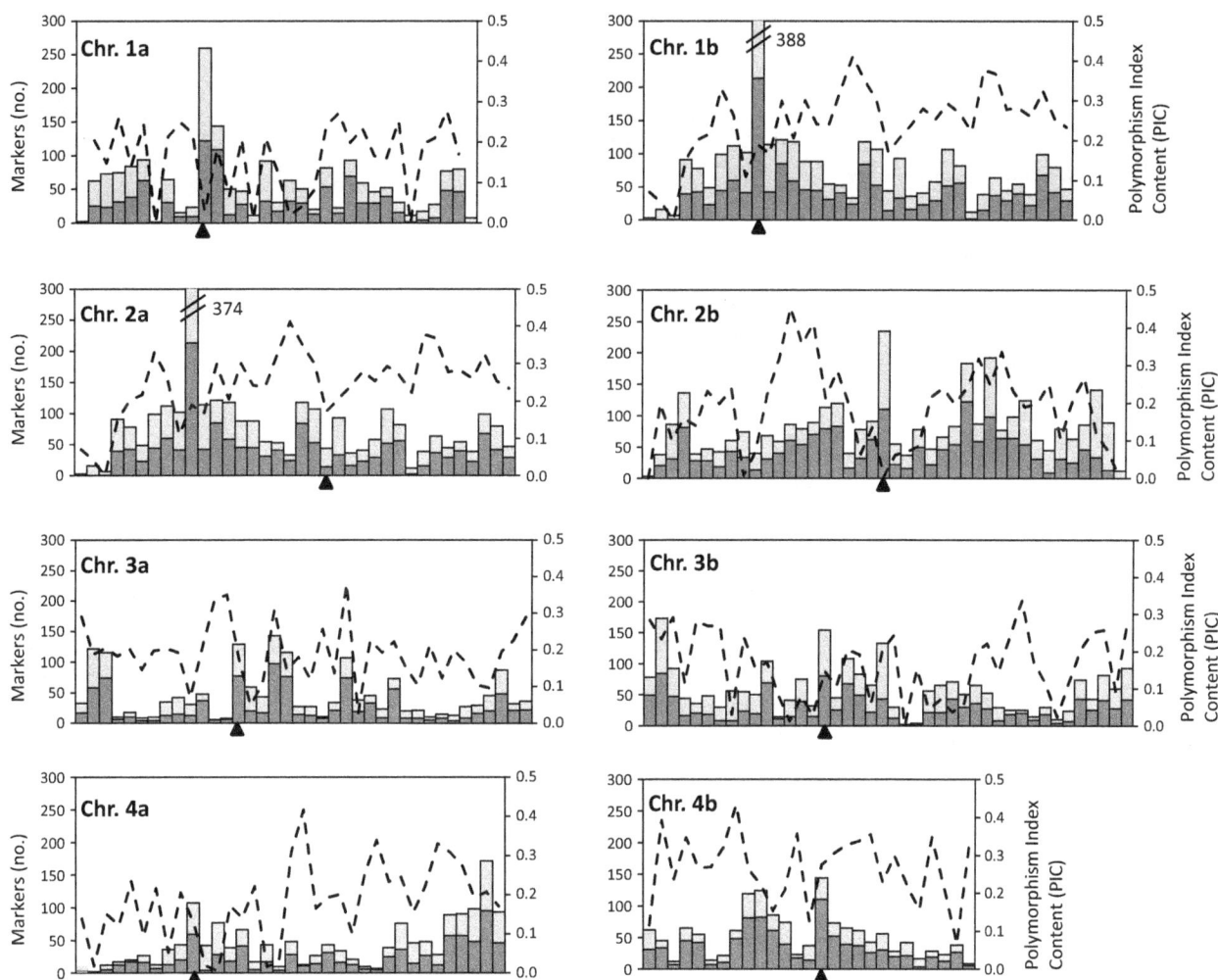

Figure 1 Distribution of markers on the consensus map. Markers were counted based on 5-cM bins. Anchor and singleton markers are reported separately. Mean Polymorphism Index Content (PIC) is from 55 of elite durum accessions from worldwide. Putative centromere positions are reported.

Considering only materials mapped with the Illumina 90K SNP array, the elite × elite populations allowed us to map 5055 anchor markers and 1064 singletons, whereas the cultivated and wild emmer-derived populations were effective in mapping an additional 7932 anchor SNPs and 2289 singletons. On average, each anchor marker was mapped approximately across 3.0 mapping populations (2.9–3.4). The number of anchor markers and total map length for each chromosome are reported in Table 2. Anchor markers mapped in the elite × elite populations were more informative in terms of the overall number of segregating biparental populations per SNP as compared to the anchors mapped to the interspecific crosses, as expected and most probably due to ascertainment bias effects and divergence of allele frequency between durum and emmer germplasm. Distribution of markers for all maps is reported in details in Tables S1 and S2.

In the consensus map, the A and B genomes had nearly identical map length (1316 and 1315 cM, corresponding to 188.0 and 187.9 cM per chromosome, respectively). However, a greater number of markers mapped to the B genome (17 119) vs. the A genome (13 025). This proportion was conserved across all marker classes, and average intermarker distance was equal to 0.079 cM/marker for the B genome and 0.101 cM/marker for the A genome.

For most chromosomes, previously indicated centromere positions (Somers et al., 2004; Sourdille et al., 2004) were coincident with clear spikes in marker density located within a single 5-cM bin (Figure 1), with the exception of chromosome 6A. Conversely, low-marker density regions spanning 10–30 cM were detected in chromosome arms 1AS, 2AS, 3AS, 4AS (proximal regions) and the distal regions of 7AS and 7BS. Although the patterns of PIC value distribution fluctuated widely (Figure 1), the diversity of the Illumina SNPs in the elite durum germplasm could be related to marker distribution and/or chromosome features. A decrease in diversity was observed in the centromeric and pericentromeric regions of several chromosomes (e.g. 1A, 1B, 2A, 2B, 3B, 5B, 6A, 7B) despite the relatively high-marker density of these regions. Conversely, extended regions characterized by low-marker density and diversity were also observed (e.g. in chromosome arms 1BS, 2AS, 4AS, 5AS).

The consensus map contained both transcript-derived SNP markers and nongenic SSR and DArT markers (considered together with SNPs obtained from reduced-representation genomic libraries). The relative distribution of these three marker classes along chromosomes is represented in Figure 2 as standardized relative marker ratio (proportion) per 5-cM bin. Transcript-derived SNP markers showed a tendency towards enrichment in the pericentromeric and centromeric regions (e.g.

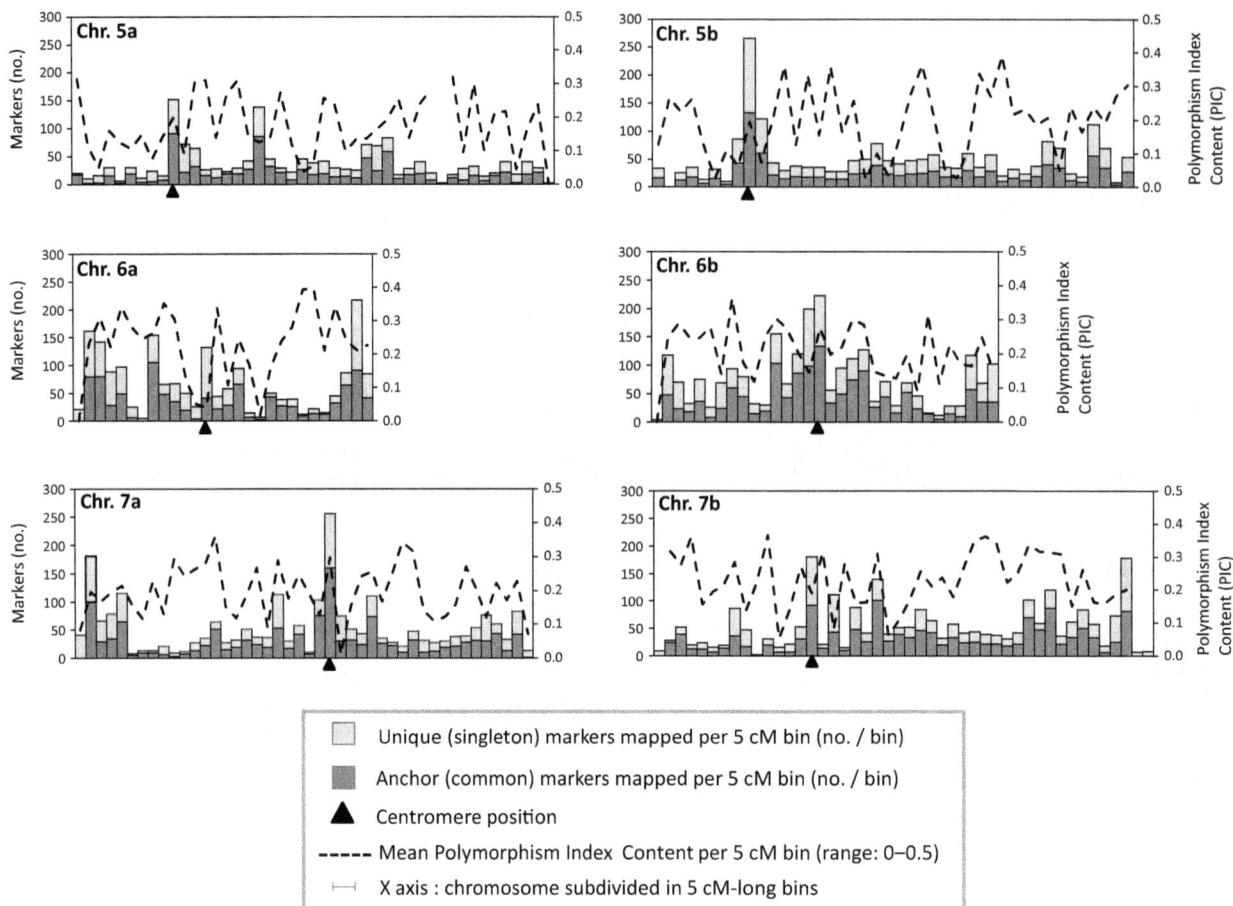

Figure 1 Continued.

in chromosomes 1A, 2A, 3A, 3B, 4A, 6B, 7A). Additionally, their distribution was not regular along chromosomes, with local peaks of high and low density. DArT markers and genomic SNPs obtained from reduced-representation libraries derived from methylation-sensitive restriction enzymes (*Pst*I: KBO and SBG markers) were clearly over-represented in the distal euchromatic regions and under-represented in the centromeric and pericentromeric chromosome regions (enriched in heterochromatin) for most of the chromosomes. The mapped SSR markers, usually chosen based on the previous knowledge on their mapping position, showed a tendency towards even distribution along the chromosomes.

In the elite × elite maps, the marker density differed markedly along the linkage groups, with extensive presence of low-marker density regions and gaps of no marker coverage scattered along chromosomes (Figures S1 and 3, which is magnified in Figure S2). This feature was attenuated in maps involving emmer and wild emmer crosses.

As to the degree of marker order correspondence between the consensus and the component maps, of the 182 pairwise LG comparisons, 147 (80.7%) showed ρ values >0.95 and 56 (30.8%) showed ρ values >0.99 (Figure S1). Conflicts in marker order were mainly limited to a single map among those sharing common markers. Most conflicts involved small distances and the two alternative orders presented nearly equal LOD scores; therefore, the consensus marker order was retained, and the conflicts were not declared. Declared conflicts were detected in 7.4% of markers per chromosome, and the conflict order

information was retained in the consensus map by marker renaming. A small fraction of conflicts involved large structural rearrangements of several linked markers (e.g. on chromosome 3B, data not reported) while the majority was limited to distances <10 cM.

Genetic-to-physical correspondence between tetraploid and hexaploid chromosome 3B

A total of 1886 gene-derived SNPs genetically mapped on the chromosome 3B tetraploid consensus had significant matches to the 774 Mb pseudomolecule sequence of hexaploid wheat (Choulet *et al.*, 2014). The majority of markers anchored to the pseudomolecule, following a close-to sigmoidal pattern (Figure 4). However, 193 SNPs (10.2%) had significant matches but anchored to noncollinear regions and their positions were scattered throughout the pseudomolecule without any clear pattern. Only 40% of the 193 noncollinear SNPs were univocally positioned on the pseudomolecule, while the remaining 60% showed multiple BLAST hits (from 2 to 10, based on e^{-10} filtering), possibly due to occurrence in duplicated/multiple gene family members. The localization of noncollinear SNPs that were univocally positioned on the hexaploid pseudomolecule supports the occurrence of rearrangements of gene order between the tetraploid wheat parental genotypes and hexaploid wheat.

The genetic-to-physical relationship for chromosome 3B can be described according to subpatterns at five main segmented regions (two distal regions, two proximal regions and the centromeric/pericentromeric region). In the two distal regions

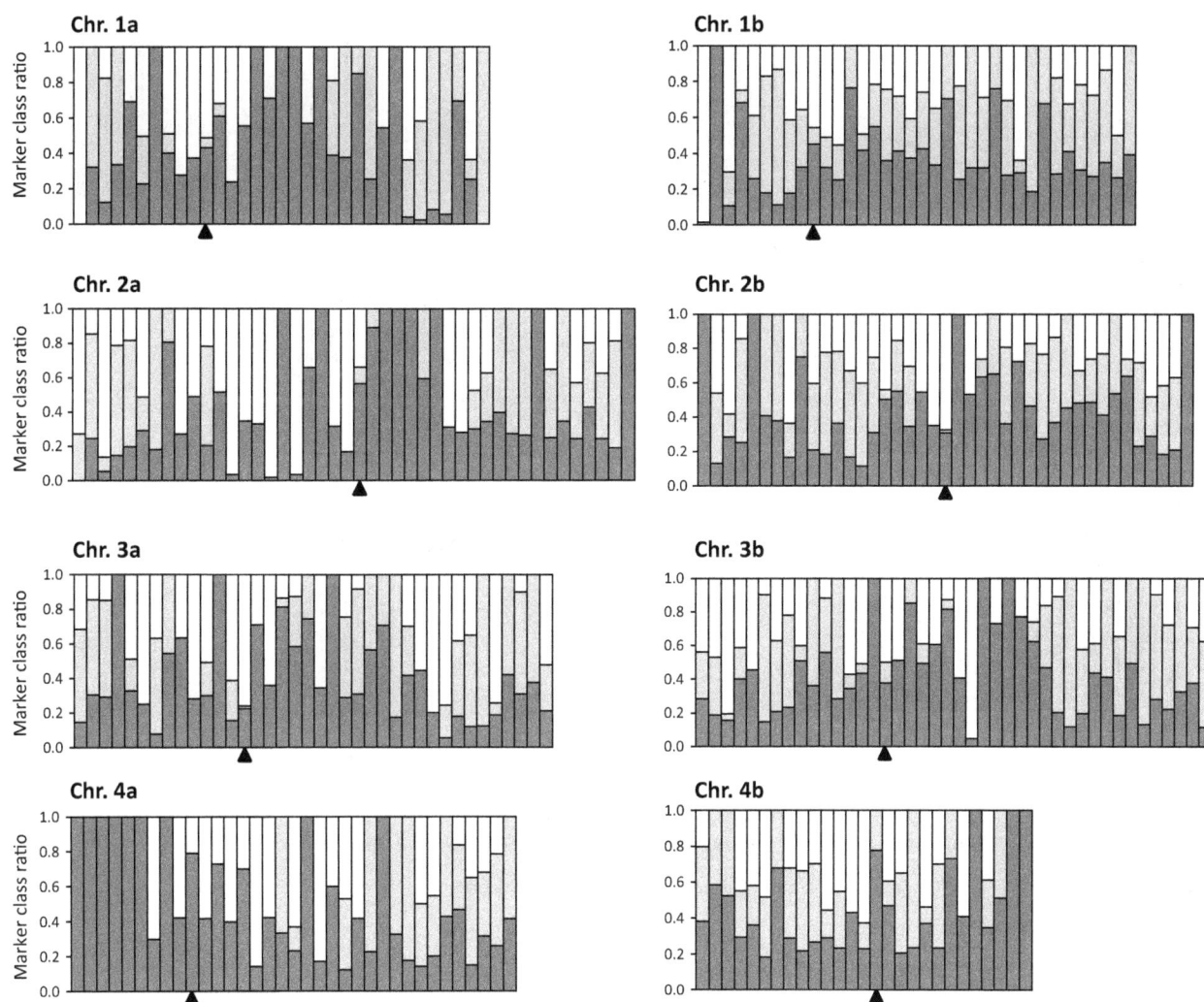

Figure 2 Relative frequency distribution of genic and nongenic markers on the consensus map. Transcript-associated SNPs, nongenic DArT and SNP markers, nongenic SSR and STS were considered as the main three marker classes. Relative frequencies are reported along chromosomes based on 5-cM bins.

corresponding to 0–61.6 cM in 3BS and 133.4–210.8 cM in 3BL, the physical-to-genetic distance relationship was mostly linear as from the regression results (Figure 4a). These two distal regions extended over ca. 95 (distal 3BS) and 127 Mb (distal 3BL), that accounted for 12.2 and 16.4% of the whole chromosome, and were characterized by an overall recombination rate of 0.64 and 0.61 cM/Mb, respectively. In the two proximal regions, the meiotic recombination rate dropped to 0.15 cM/Mb. The centromeric region was positioned in a 1.4 cM interval (75.2–76.8 cM) extending over 127 Mb (16.4% of the chromosome) and was characterized by strong suppression of recombination. SNP PIC values plotted using a 30-markers sliding window showed higher polymorphism level in the two distal regions than in the rest of the chromosome, apart from localized regions (Figure 4a).

The comparison between the hexaploid (Wang *et al.*, 2014) and the tetraploid consensus maps for chromosome 3B using the SNPs shared between the two maps is reported in Figure 4b. Spearman rank correlation between the two maps (ρ value) was equal to 0.95, thus showing a good collinearity conservation between the two subspecies. The tetraploid wheat map showed a higher homogeneity in recombination event distribution as compared to the hexaploid map, with the latter showing a

markedly lower recombination rate throughout the centromeric–pericentromeric region (as shown in Figure 4b). Despite the overall good correspondence, the collinearity at a finer level of resolution (down to 5 cM) was compromised, mostly due to the low resolution of both maps. When comparing the collinearity (marker order) of the two consensus maps to the 3B pseudo-molecule, the two genetic maps performed similarly based on rank correlation ρ values = 0.89 and 0.84 for the tetraploid and hexaploid map, respectively.

Homeologous relationships between the A and B genome explored by gene-derived SNPs

To precisely define the homeologous (and paralogous) relationships between the coding sequences tagged by mapped SNPs, the latter were related to the *T. aestivum* genome-specific transcripts. Based on a conserved-sequence similarity search (BLASTN, 1e^{-10}), it was possible to relate 21 547 gene-associated SNPs to 29 520 A-, B- and D-genome transcripts with 78 311 hits in total, restricted to 20 045 unique transcripts for the A and B genomes only (53 316 hits). High similarity matches were thus found for 2.66 SNPs per transcript, on average. Unique transcript-SNP relationships (hits) were observed for 9276 cases (46.3% of

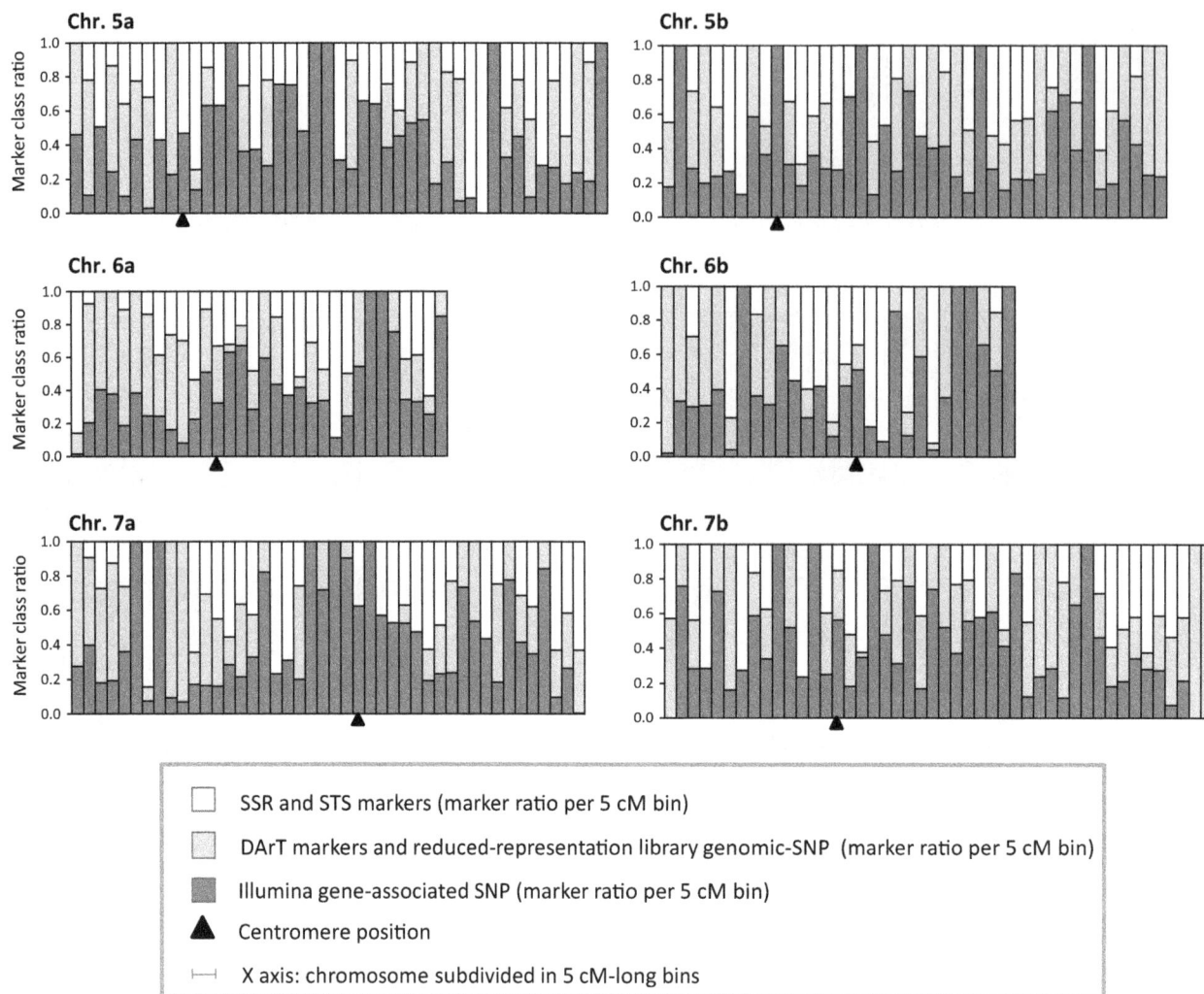

Figure 2 Continued.

coding sequences) while 8680 (43.3%) transcripts were tagged by 2 to 5 SNP hits (including SNPs from homeologous chromosome regions); the remaining 8.04 and 2.37% were tagged by multiple SNPs (up to 10 and up to 26 tagged SNP, respectively).

The conserved-sequence similarity analysis highlighted the homeologous and paralogous relationships between gene-associated SNPs and transcripts mapped to syntenic regions of homeologous pairs, as well as to other chromosomes. Genome-specific transcripts-SNP hits are reported as Tables S3 and S4. As detailed in Table S3, among the 53 316 genome-specific transcript-SNP hits, 22 579 (42.3%) and 18 372 (34.4%) were traced to direct homologous and indirect homologous relationships within the seven homeologous groups, respectively (76.7% of the total number of matches). SNP-transcript matches between putative homeologous chromosomes are reported in Table S5. A mean of 558 nonredundant reference Chinese Spring transcripts per homeologous group was tagged by SNP clusters whose components were mapped to homeologous/paralogous positions, for a mean of 1254 SNPs in total (2.25 SNPs per transcript). In homeologous groups 1, 2 and 3, the homeologous relationships were abundant, with 627, 818 and 813 nonredundant transcripts and 1516, 1937 and 1330 nonredundant SNPs, respectively. Chromosome groups 4, 5 and 7 showed the lowest

number of nonredundant transcripts tagged by SNPs in putative homeologous positions, with 130, 426 and 519 reference transcripts and 370, 933 and 1408 SNPs, respectively. This was expected based on the multiple translocation events occurred among these chromosomes. Complex, nonsyntenic homeologous relationships among chromosomes 4A, 5A and 7B, tagging the multiple known translocations involving chromosome arms 4AL, 4AS, 5AL and 7BS (T4A/5A and T7B/4A), were identified. In addition, preliminary evidences of additional putative translocations were obtained. Relationships are detailed in Text S1 and Table S6.

Comparison between the tetraploid and hexaploid wheat consensus maps

The tetraploid consensus map was compared with the hexaploid Illumina 90K SNP consensus map reported by Wang *et al.* (2014). Some 21 145 SNPs were in common between the two maps (79.4% of the SNP mapped in tetraploid wheat), with 5481 SNPs mapped exclusively in the tetraploid consensus map. Among the common SNPs, 19 202 (90.8%) showed agreement in chromosome assignment between the two maps, whereas 1268 (6.0%) were discordant in assignment to the nominal homeologs and the remaining 675 (3.2%) were assigned to different chromosome

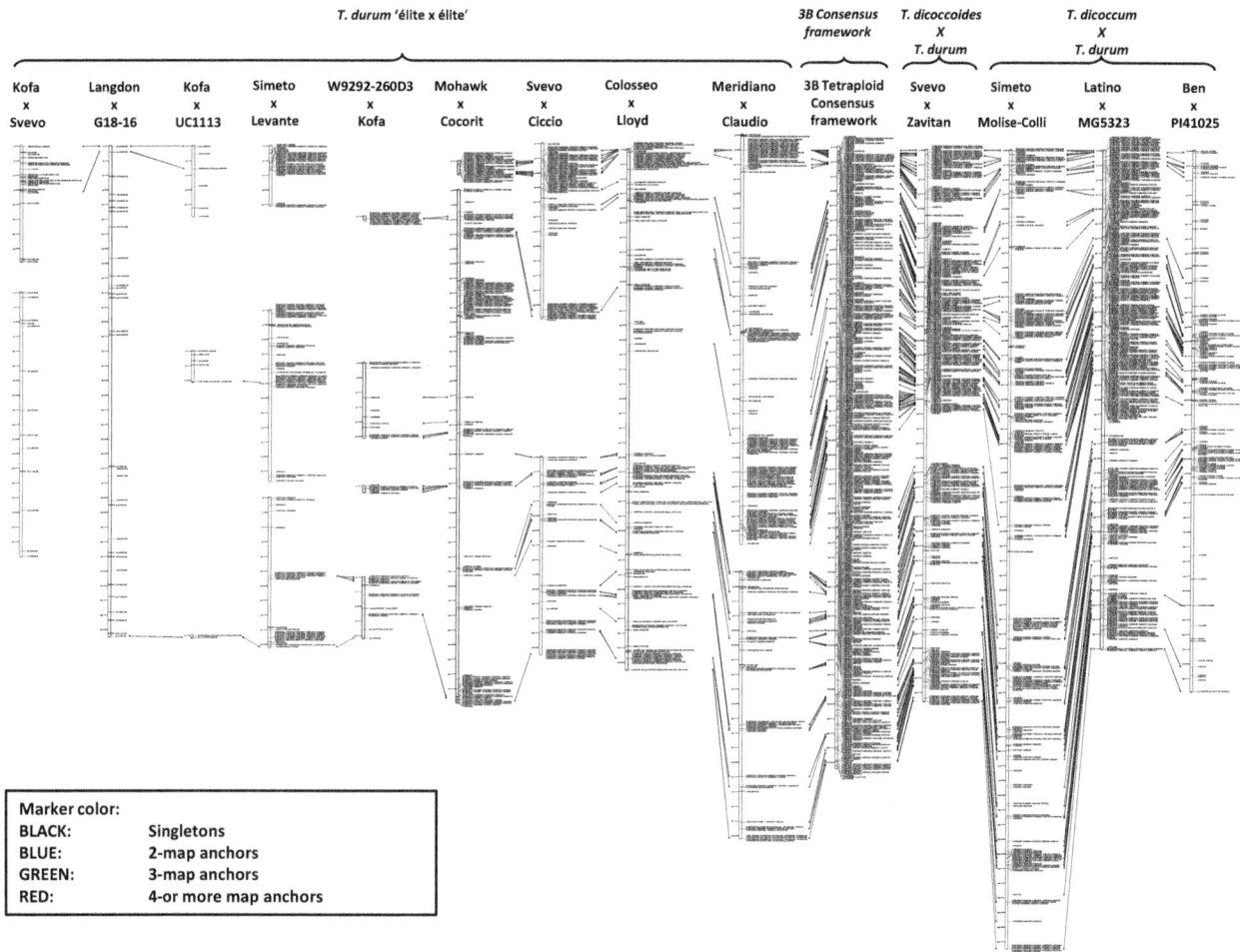

Figure 3 Single component and consensus map representations for chromosome 3B. The consensus framework map is central. Elite × elite durum maps are reported on the left while durum × cultivated and wild emmer maps on the right. Singleton markers are in black and anchor markers for two, three and four or more maps are reported in blue, green and red, respectively. Connectors join highly informative marker positions. For sake of clarity, unique singleton markers are not reported in the tetraploid consensus.

groups, in balanced frequencies across chromosome groups with the notable exceptions of the few main translocations. One hundred and sixteen SNPs were differentially located to the homeologous 4A-7A and 4A-7D positions in the two maps, thus tagging alternate homeologous sites (tracing to the T4AL/7BS) depending on the maps. Additionally, 65 SNPs showed discordant location to groups 4 and 5 (tracing to T4AL/5AL, with 18 SNPs mapped in the distal 5BL (170.7–206.2 cM interval) on the tetraploid consensus and in the 4AL distal in the hexaploid map, and 13 SNPs mapped in opposite relationship. Additionally, traces of a putative T5A/2B translocation were detected by 15 SNPs differentially mapped in chromosome 2B (tetraploid map, in 75.4–158.3 cM interval)—5B (hexaploid map) or in opposite relationship.

Overall, 9.2% of Illumina SNP assays tagged different homeologous or highly similar loci between durum and bread wheat, a finding already observed by Wang *et al.* (2014) that poses the problem of uniqueness (i.e. adherence to Mendelian inheritance) of the SNP assays. Wang demonstrated that most of these assays detected at least two polymorphisms (target SNPs plus interfering SNPs at the genotyping oligo-annealing site). The precise causes behind this finding remain to be identified. As to the single component populations, it appeared

that most of them held the ancient translocations, with the only notable exception for the 7A-1 linkage group in W9292-260D3 × Kofa, where a notable portion (10.8%) of linked SNPs was mapped in chromosome 4A in other populations and in the consensus map.

The Spearman rank correlations between homologous chromosomes of the two consensus maps ranged from 0.90 to 0.99. Rank-order plots of data pairs from the two consensus (see Figure S3) evidenced extensive regions with completely suppressed recombination in the hexaploid wheat consensus, particularly in the pericentromeric regions of chrs. 1A, 2A, 3A, 4A, 4B, 5A, 6A, 7A, similarly to what we observed for the 3B pericentromeric chromosome region where only the recombination rate of the pericentromeric region was lower as compared to the tetraploid wheat map.

Discussion

A tetraploid-specific consensus map as a reference for the A and B genomes of wheat

The consensus mapping data set reported herein builds on a sizeable amount of genotyping data obtained for tetraploid wheat in the last decade. The consensus map integrates data

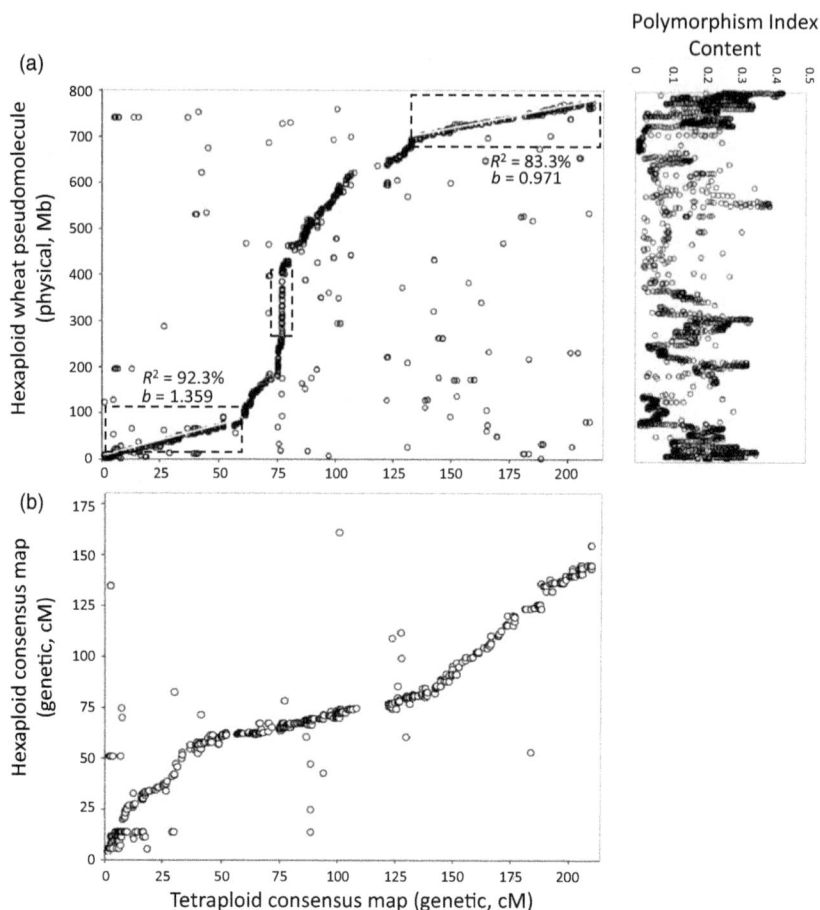

Figure 4 Physical-to-genetic and genetic-to-genetic relationships between the hexaploid physical map, the tetraploid and the hexaploid consensus genetic maps for chromosome 3B. (a) Plot of the hexaploid physical map (Y axis, Mb) on the tetraploid consensus genetic map (X axis, cM). Correspondence has been established by conserved-sequence similarity search (BLASTN). The two distal chromosome regions with linear relationships between physical and genetic distances and the centromeric/pericentromeric region are indicated by dashed-lined rectangular boxes. Polymorphism Index Content (PIC) of mapped markers from 55 elite durum wheat cultivars is plotted along the physical maps. (b) Plot of the hexaploid consensus genetic map (Y axis, cM) on the tetraploid consensus genetic map (X axis, cM).

from mapping populations obtained from crosses between the cultivated durum varieties and crosses involving the closest ancestors domesticated emmer and wild emmer. This study integrates in a unique framework different types of markers (mostly SSR, DArT® and SNP), hence allowing one to trace the results of genes and QTL mapping studies and genetic diversity analysis in both tetraploid and hexaploid wheat for the A and B genomes. Additionally, our consensus map integrates a large number (26 626) of SNP markers from the recently released wheat 90K SNP array that includes ca. 90 000 gene-associated assays (Wang et al., 2014). This provides a unique opportunity to switch from neutral markers to a large set of functional, physically anchored markers and cross-reference gene/QTL map information and other chromosome and genomic features between bread and durum wheat, thus allows for a highly accurate comparative genotyping in both cereals.

Our results show the utility of the 90K SNP array for map construction providing nearly complete genome coverage in T. dicoccum- and T. dicoccoides-derived populations compared to elite × elite populations. Gaps in the linkage maps from the elite × elite durum wheat crosses suggest that the genomes of the parental durum wheat genotypes share extensive regions of identity by descent (IBD), difficult to populate with informative markers. The presence of shared long-range haplotypes has been previously demonstrated in elite durum wheat using mapped multi-allelic SSR markers (Maccaferri et al., 2005, 2007). As the IBD regions are randomly scattered across the individual maps, the consensus map approach provided the opportunity to fill linkage gaps and to reconstruct a map with nearly complete

genome coverage. It should be noted, however, that the adoption of an open genotyping technology (such as genotyping by sequencing, Saintenac et al., 2013) could probably partially mitigate the low polymorphism observed in the elite × elite crosses.

Consensus map construction and map features

To produce a robust consensus map, we used a combination of *genetic merging* for framework mapping, based on common (=anchor) markers, and *interpolation* for singleton markers. Advantages from both mapping methods were thus retained, such as the ability of (i) producing a robust map of anchor markers with nearly complete genome coverage and recombination rate estimates considered as representative for the tetraploid germplasm, and (ii) adding unique markers to a reliable framework using a fast *regression*-based method (Cone et al., 2002; Sosnowski et al., 2012).

Graph-based methods are faster alternatives. However, they do not rely on the original set of marker distance matrices (Ronin et al., 2012). A further approach based on the search of best multilocus ordering from consensus analysis of the component data sets was provided by Ronin et al. (2012). A more efficient use of anchor loci is expected to speed up such analysis (A. Korol, personal communication).

The *genetic merging* method was chosen based on the availability of genotypic data from 1773 RIL, with 54% of anchor markers, a number sufficiently large to provide sufficiently large to provide genetic distances representative of the distribution of recombination rates in tetraploid wheat. Second, the mapping populations compared to hexaploid wheat showed a remarkable homogeneity in terms of marker order and recombination rate. In bread wheat, the genomes of modern lines frequently include alien chromosomes transferred from wild relatives that result in extensive suppressions of recombination or aberrations of recombination frequency in mapping populations (Neu et al., 2002; Jayatilake et al., 2013; Xie et al., 2012 supporting materials in Wang et al., 2014). Marker order conservation across maps and relative to the consensus map demonstrated that the marker order was generally well conserved in the tetraploid data set, making it a distinct feature of the tetraploid germplasm as compared to T. aestivum, which shows both higher frequency of local rearrangements as well as frequent suppressed recombination events as shown for chromosome 3B and for several other chromosomes as well (Wang et al., 2014). One possible explanation is that, compared to hexaploid wheat, tetraploid wheat germplasm has been less subjected to improvement through introgressions and chromosome engineering from the secondary and tertiary gene pools (Ceoloni et al., 2005). In summary, this consensus map provides an important resource for its resolving power at pericentromeric regions in both durum and bread wheat. On the other hand, the extensive IBD among parents of durum germplasm shown by our study constrain mapping in populations derived from elite × elite crosses. This trend could result from extensive genetic diversity reduction in the elite durum wheat as a whole (Haudry et al., 2007) and consequent hindrance of breeders' effort, or ascertainment bias caused by inadequate sampling of genetic diversity in durum wheat.

Marker density in the consensus map varied widely along chromosomes, even within intervals of 20 cM or less, likely due to both variation in recombination frequency (Erayman et al., 2004; Saintenac et al., 2011) and/or in genetic diversity (ancient genetic drift and selection sweep effects and/or more recent breeder-driven selection). As compared to the hexaploid wheat consensus map and the maps derived from either wild or cultivated emmer (Marone et al., 2012; Maccaferri et al., 2014), at least three large chromosome regions with reduced marker density have been consistently found in chromosomal arms 1AS, 2AS and 3AS of durum wheat across mapping populations and germplasm diversity studies.

Based on the relatively high number of component maps used to produce the consensus and their representativeness of the durum germplasm, these low-marker density regions were considered as a constitutive feature of the cultivated durum wheat genome. The local drop in marker density in these regions was found in all the durum maps but not in the emmer- and wild emmer-derived maps. The projection of the Illumina SNP polymorphism level in elite worldwide durum germplasm (expressed as PIC) to their genetic locations using the consensus map evidenced a marked decline in diversity and recombination rate in the centromeric/pericentromeric regions as well as strong fluctuation in diversity along the chromosomes, as previously shown in wheat and other species (Akhunov et al., 2010; Wang et al., 2011). The distribution of transcript-associated SNPs, with a slight enrichment in the pericentromeric and centromeric regions most probably caused by the lower recombination rates in such regions, supports the hypothesis of relatively even gene distribution along chromosomes, a result that matches the direct sequence observations reported for chromosome 3B (Choulet et al., 2014). Irregular clustering of transcript-associated SNPs may be indicative of presence of gene-rich regions. Conversely,

the skewedness towards distal euchromatic regions observed for DArT markers and genomic SNPs from reduced-representation libraries based on the use of methylation-sensitive restriction enzymes was expected based on previous reports (Akbari et al., 2006; Mantovani et al., 2008; Wenzl et al., 2006), including the absence of DArT coverage in the 4AS chromosome arm (Mantovani et al., 2008).

Hexaploid physical-to-tetraploid genetic map comparison for chromosome 3B

The graphical representation of the genetic vs. physical distances along the chromosome 3B showed a classic sigmoidal pattern that could be split into five regions showing more homogeneous relationships with the physical sequence. The two distal regions showing linear physical-to-genetic distance relationship corresponded closely to the two distal subregions identified by Choulet et al. (2014) based on segmentation analysis of the recombination rate and characterized by an elevated recombination rate (of 0.60 and 0.96 cM/Mb) clearly distinct from the low recombination rate (of 0.05 cM/Mb) observed for the large proximal region. These recombination rate values were confirmed in our chromosome 3B tetraploid consensus map. The localization of centromere based on suppressed recombination was also consistent with the physical centromere position in hexaploid wheat, namely 127 Mb in our case and 122 Mb in Choulet et al. (2014).

In spite of heterogeneity in the distribution of PIC values along the physical map in the durum wheat germplasm, a strong decline in polymorphism content outside of the two small distal regions—in agreement with observations by Akhunov et al. (2010) and Wang et al. (2014)—was evident. Meiotic recombination is the likely cause of the variability.

A database of homeologous relationships for the mapped gene-derived SNPs

In wheat, differential transcriptional regulation of homeologs and duplicated gene copies is a mechanism for fine tuning the phenotypic response (Dubcovsky and Dvorak, 2007), therefore the possibility to easily cross-refer homologous genes/QTLs (and their corresponding tagging SNPs) between their A and B genomes facilitates genetic dissection of traits (Quarrie et al., 2005; Diaz et al. 2005; Wilhelm et al. 2013; Zheng et al. 2013; Avni et al. 2014; Nitcher et al. 2014; Hurni et al. 2014). To date, homeologous relationships were mainly investigated by RFLPs (Qi et al., 2004) and SSRs (Röder et al. 2008) that recognize homeologous copies as distinct patterns in the same genetic profile. Conversely, SNP markers are generally single Mendelian homeologous-specific assays. However, the 90K wheat SNP array has been mostly assembled from wheat transcripts (Wang et al., 2014) and the SNP array includes multiple SNPs per single wheat transcript. These two features were used to establish robust relationships between SNPs mapped to homeologous/paralogous positions.

The mapped SNP sequences were related to the database of chromosome-specific hexaploid transcripts [The International Wheat Sequencing Consortium (IWGSC) (2014)] by conserved-sequence similarity analysis while data sets of SNPs were related to homeologous and paralogous transcript copies have been produced for each homeologous group. These data sets provide homeologous relationship patterns at relatively high density (up to 1937 SNPs per homeologous group for 818 nonredundant transcript sets). Homeologous relationships between chromosomes harbouring ancestral translocation events (involving chromosome groups 4, 5 and 7; Ma et al., 2013) were also clearly tagged and defined in their extent. In this case, the results were also supported by the absence of the homeologous transcripts in the nominal chromosomes of Chinese Spring (e.g. chrs. 4A-4B, 5A-5B, 7A-7B). Additionally, SNP-transcript relationships between non-nominal homeologous groups were established that could be ascribed to smaller and more fragmented translocation events already noticed based on SSR markers (Marone et al., 2012). In some cases, connections were established between transcript sets of different groups (e.g. chromosome groups 1–3) for which the presence of the nominal homeologous transcripts in the corresponding chromosome-specific database deserves further investigation, possibly ascribed to ancestral genome duplications (Pont et al., 2013).

Conclusions

By bringing the number of marker loci specifically mapped in tetraploid wheat to a number close to that reached for hexaploid wheat, the present consensus map provides a significant advancement for durum wheat genetics. Most importantly, the majority of mapped SNPs are gene-derived markers that provide valuable anchor points for locus dissection and for bridging durum and bread wheat genetics. The proposed homeologous SNP and transcript database will be useful to define the homeologous relationships between chromosome regions harbouring target genes and QTLs. Additionally, the consensus map provides a framework of unprecedented marker density and genome coverage for association mapping, meta-QTL analysis and positional cloning, thus facilitating the application of molecular breeding approaches to enhance durum wheat productivity and quality. It is worth mentioning that the tetraploid consensus map includes mostly markers common to hexaploid wheat and also tetraploid-specific markers as well, and for the first time, the map integrates SSR, DArT and SNP markers in a single framework. Moreover, the map showed an improved marker order resolution in centromeric and pericentromeric regions as compared to hexaploid wheat map. The tetraploid consensus map will also have potential applications for several aspects of hexaploid wheat research and breeding.

Experimental procedures

Mapping populations used for consensus map assembling

Twelve recombinant inbred line (RIL) and one double haploid (DH) mapping populations were derived as follows: (i) eight populations from elite × elite durum wheat crosses including Colosseo × Lloyd (Maccaferri et al., 2008b; Mantovani et al., 2008), Meridiano × Claudio (Maccaferri et al., 2011), Simeto × Levante (Maccaferri et al., 2012), Mohawk × Cocorit69 (K. Ammar, unpublished), Svevo × Ciccio (Gadaleta et al., 2011), W9292-260D3 × Kofa (C. Pozniak, unpublished), Kofa × Svevo (Maccaferri et al., 2008a) and Kofa × UC1113 (Zhang et al., 2008); (ii) three populations from durum × cultivated emmer including Ben × PI_41025 (Faris et al., 2014), Simeto × Molise Colli (A. Mastrangelo, unpublished) and Latino × MG_5323 (F. Desiderio, unpublished); (iii) two populations from durum × wild emmer crosses including Langdon × G18-16 (Peleg et al., 2008) and Svevo × Zavitan (A. Distelfeld, unpublished). Additional details are reported in Table 1.

Single map analysis

Genotyping was carried out at each contributing institution following standard methods (Mantovani et al., 2008; Peleg et al., 2008; Wang et al., 2014). Segregation data were obtained either as raw or postprocessed data. Low-quality data filtering was carried out according to rules: (i) missing data allowed to a maximum frequency of 0.10 (markers and lines), (ii) data points showing residual heterozygosity were converted to missing data, (iii) marker segregation distortion (departure from the expected 1:1 segregation ratio) allowed up to a threshold $P = 1E^{-04}$, corresponding to a segregation distortion not exceeding 0.7:0.3, above which the markers were eliminated due to residual heterogeneity in the parental lines or excessive distortion. SNPs segregating as dominant markers were retained only upon confirmation in the parental lines, provided that their segregation distortion did not exceed a stringent 0.60:0.40 ratio. After a round of provisional mapping, the distribution in recombination events in each mapping population was tested for departure from the expected Poisson' distribution (Minitab v16, Minitab inc. 2010). This step was critical for uncovering unreliable portions of genotypic data, most likely caused by erroneous DNA/genotyping data handling.

Filtered genotypic data were used to recalculate the linkage maps using Carthagene v. 1.2.3 (De Givry et al., 2005). For the single component maps, nonsegregating SNPs were collapsed (mrkdouble function). Robust initial linkage groups (LG) were obtained using the group function at recombination frequency = 0.3 and LOD = 6.0. Assignment of LG to wheat chromosomes and their orientation were carried out by cross-checking the position of common SSR loci in the hexaploid wheat consensus map (Ta-SSR-2004, http://wheat.pw.usda.gov/GG2/index.shtml and/or http://www.triticarte.com.au). Once groups were attributed to chromosomes, their genotypic data were pooled on a chromosome-by-chromosome basis and regrouped at a more relaxed threshold of recombination frequency = 0.3 and LOD = 3 or progressively lowered. Mapping was carried using the multipoint maximum likelihood algorithm (build command) and the Haldane mapping function based on iterative cycles of: (i) mapping with marker order refinement (greedy, flips and polish commands), (ii) investigation of graphical genotypes for suspicious data points, for example singletons that recombined within ≤5 cM, not supported by other markers and mostly derived from unlikely events, particularly in case of SSR and DArT® markers; (iii) raw data checking and, in case, replacement with missing data. The reconstructed maps are thereafter reported as component maps.

Construction of the consensus framework map and marker interpolation

The consensus framework map was constructed using the markers common to two or more populations (anchor) by collapsing the genotype data sets and using the same procedure described for the single populations. Prior to framework mapping, interpopulation recombination rate heterogeneity in the common chromosome regions was assessed. In the case of strong heterogeneity, a preselection of LGs from populations with relatively homogeneous recombination rates was carried out. Anchors led to obtain a robust framework map with intermarker genetic map distances estimated from mean recombination frequencies across populations. Loci mapped in single populations only were projected by interpolation from the positions of the two closest flanking common markers (Cone et al., 2002) in Biomercator v.3 (Sosnowski et al., 2012). Conflicts in marker order conservation in single maps compared to the framework or consensus maps were detected in Biomercator (control of monotony step). Re-inspection of the segregation data for single maps carrying conflicts was carried out to assess the multipoint likelihood of the map order previous and after resolving the conflicts. For conflicts that were not alternative to the consensus order, the markers were considered as tagging two potentially different loci and thus renamed using a suffix. As the local conflicting orders across maps were usually two, of which one in low and the other in high frequency, only two potentially conflicting 'marker versions' ('a' and 'b') were used. After marker renaming, one or two cycles of single population and consensus mapping were carried out as necessary. The degree of correspondence and monotony between the single components and consensus maps was inspected by calculating pairwise rho (ρ) Spearman rank marker order correlations, a nonparametric measure of monotonic correlation, that is considering only the direction of marker concatenation.

Assessment of SNP polymorphism information content in the durum wheat germplasm

A panel of 55 durum wheat accessions (mainly cultivars) including parents of mapping populations and reference cultivars originating from Mediterranean countries (Italy, Morocco, Spain, Syria and Tunisia), North and Southwestern USA, Canada and from CIMMYT and ICARDA breeding programs was used to evaluate the polymorphism information content (PIC) of mapped SNPs and to relate it to the chromosome map location. PIC was computed as follows:

$$PIC = 1 - \sum p_j^2$$

where p_j is the frequency of each of the two alleles across accessions (Powell et al., 1996).

Investigation of the relationships between the hexaploid physical and tetraploid genetic map of chromosome 3B

The hexaploid wheat chromosome 3B assembled physical sequence (pseudomolecule) of Chinese Spring was downloaded from the URGI repository (http://wheat-urgi.versailles.inra.fr/Seq-Repository/Reference-sequence, traes3bPseudomoleculeV1) (Choulet et al., 2014). The sequences of chromosome 3B-mapped SNPs from the Illumina manifest file (http://triticeaetoolbox.org/wheat/display_genotype.php?trial_code=TCAP90K_SWWpanel) were used to conduct a conserved-sequence search on the chromosome 3B pseudomolecule by BLASTN with a filtering threshold e-value $<1e^{-10}$.

Identification of homeologous relationships between gene-derived SNPs

The mapped Illumina iSelect wheat SNP sequences were used as queries to search the chromosome-specific transcript database of the hexaploid wheat Sequence Survey database (ftp://ftp.ensemblgenomes.org/pub/plants/release22/fasta/triticum_aestivum/cdna/Triticum_aestivum.IWGSP1.22.cdna.all.fa.gz). Based on the recovered hits, SNPs and transcripts were interrelated based on the SNP chromosome and mapping positions and the chromosome-specific origin of transcripts in Excel.

Acknowledgements

We wish to thank Maria C. Sanguineti for her help and suggestions in the statistical analysis of the data sets. Research supported by the AGER Agroalimentare e Ricerca—Project 'From seed to pasta—Multidisciplinary approaches for a more sustainable and high-quality durum wheat production. CJP acknowledges the financial support of Agriculture and Agri-Food Canada, Genome Canada, Genome Prairie, The Saskatchewan Ministry of Agriculture and the Western Grains Research Foundation. EA and JD acknowledge the financial support of the National Research Initiative Competitive Grants 2011-68002-30029 (Triticeae-CAP). AMM acknowledges the financial support of the special grant ISCOCEM.

References

Akbari, M., Wenzl, P., Vanessa, C., Carling, J., Xia, L., Yang, S., Uszynski, G., Mohler, V., Lehmensiek, A., Kuchel, H., Hayden, M.J., Howes, N., Sharp, P., Rathmell, B., Vaughan, P., Huttner, E. and Kilian, A. (2006) Diversity Arrays Technology (DArT) for high-throughput profiling of the hexaploid wheat genome. Theor. Appl. Genet. **113**, 1409–1420.

Akhunov, E.D., Akhunova, A.R., Anderson, O.D., Anderson, J.A., Blake, N., Clegg, M.T., Coleman-Derr, D., Conley, E.J., Crossman, C.C., Deal, K.R., Dubcovsky, J., Gill, B.S., Gu, Y.Q., Hadam, J., Heo, H., Huo, N.X., Lazo, G.R., Luo, M.C., Ma, Y.Q., Matthews, D.E., McGuire, P.E., Morrell, P.L., Qualset, C.O., Renfro, J., Tabanao, D., Talbert, L.E., Tian, C., Toleno, D.M., Warburton, M.L., You, F.M., Zhang, W.J. and Dvorak, J. (2010) Nucleotide diversity maps reveal variation in diversity among wheat genomes and chromosomes. BMC Genom. **11**, 702.

Allen, A.M., Barker, G.L.A., Wilkinson, P., Burridge, A., Winfield, M., Coghill, J., Uauy, C., Griffiths, S., Jack, P., Berry, S., Werner, P., Melichar, J.P.E., McDougall, J., Gwilliam, R., Robinson, P. and Edwards, K.J. (2013) Discovery and development of exome-based, co-dominant single nucleotide polymorphism markers in hexaploid wheat (Triticum aestivum L.). Plant Biotechnol. J. **11**, 279–295.

Avni, R., Zhao, R.R., Pearce, S., Jun, Y., Uauy, C., Tabbita, F., Fahima, T., Slade, A., Dubcovsky, J. and Distelfeld, A. (2014) Functional characterization of GPC-1 genes in hexaploid wheat. Planta, **239**, 313–324.

Blanco, A., Bellomo, M.P., Cenci, A., De Giovanni, C., D'Ovidio, R., Iacono, E., Laddomada, B., Pagnotta, M.A., Porceddu, E., Sciancalepore, A., Simeone, R. and Tanzarella, O.A. (1998) A genetic linkage map of durum wheat. Theor. Appl. Genet. **97**, 721–728.

Blenda, A., Fang, D.D., Rami, J.F., Garsmeur, O., Luo, F. and Lacape, J.M. (2012) A high density consensus genetic map of tetraploid cotton that integrates multiple component maps through molecular marker redundancy check. PLoS One, **7**, e45739.

Bowers, J.E., Nambeesan, S., Corbi, J., Barker, M.S., Rieseberg, L.H., Knapp, S.J. and Burke, J.M. (2012) Development of an ultra-dense genetic map of the sunflower genome based on single-feature polymorphisms. PLoS One, **7**, e51360.

Caldwell, K.S., Dvorak, J., Lagudah, E.S., Akhunov, E., Luo, M.C., Wolters, P. and Powell, W. (2004) Sequence polymorphism in polyploidy wheat and their D-genome diploid ancestor. Genetics, **167**, 941–947.

Cavanagh, C.R., Chao, S.M., Wang, S.C., Huang, B.E., Stephen, S., Kiani, S., Forrest, K., Saintenac, C., Brown-Guedira, G.L., Akhunova, A., See, D., Bai, G.H., Pumphrey, M., Tomar, L., Wong, D.B., Kong, S., Reynolds, M., da

Silva, M.L., Bockelman, H., Talbert, L., Anderson, J.A., Dreisigacker, S., Baenziger, S., Carter, A., Korzun, V., Morrell, P.L., Dubcovsky, J., Morell, M.K., Sorrells, M.E., Hayden, M.J. and Akhunov, E. (2013) Genome-wide comparative diversity uncovers multiple targets of selection for improvement in hexaploid wheat landraces and cultivars. *Proc. Natl Acad. Sci. USA*, **110**, 8057–8062.

Ceoloni, C., Forte, P., Gennaro, A., Micali, S., Carozza, R. and Bitti, A. (2005) Recent developments in durum wheat chromosome engineering. *Cytogenet. Genome Res.* **109**, 328–334.

Choulet, F., Alberti, A., Theil, S., Glover, N., Barbe, V., Daron, J., Pingault, L., Sourdille, P., Couloux, A., Paux, E., Leroy, P., Mangenot, S., Guilhot, N., Le Gouis, J., Balfourier, F., Alaux, M., Jamilloux, V., Poulain, J., Durand, C., Bellec, A., Gaspin, C., Safar, J., Dolezel, J., Rogers, J., Vandepoele, K., Aury, J.-M., Mayer, K., Berges, H., Quesneville, H., Wincker, P. and Feuillet, C. (2014) Structural and functional partitioning of bread wheat chromosome 3B. *Science*, **345**, 1249721.

Cone, K.C., McMullen, M.D., Bi, I.V., Davis, G.L., Yim, Y.S., Gardiner, J.M., Polacco, M.L., Sanchez-Villeda, H., Fang, Z.W., Schroeder, S.G., Havermann, S.A., Bowers, J.E., Paterson, A.H., Soderlund, C.A., Engler, F.W., Wing, R.A. and Coe, E.H. (2002) Genetic, physical, and informatics resources for maize on the road to an integrated map. *Plant Physiol.* **130**, 1598–1605.

De Givry, S., Bouchez, M., Chabrier, P., Milan, D. and Schiex, T. (2005) Carthagene: multipopulation integrated genetic and radiated hybrid mapping. *Bioinformatics*, **21**, 1703–1704.

Diaz, A., Zikhali, M., Turner, A.S., Isaac, P. and Laurie, D.A. (2012) Copy number variation affecting the *photoperiod-B1* and *Vernalization-A1* genes is associated with altered flowering time in Wheat (*Triticum aestivum*). *PLoS One*, **7**, e33234.

Dubcovsky, J. and Dvorak, J. (2007) Genome plasticity a key factor in the success of polyploid wheat under domestication. *Science*, **316**, 1862–1866.

Dvorak, J., Luo, M.C., Yang, Z.L. and Zhang, H.B. (1998) The structure of the *Aegilops tauschii* genepool and the evolution of hexaploid wheat. *Theor. Appl. Genet.* **97**, 657–670.

Dvorak, J., Akhunov, E.D., Akhunov, A.R., Deal, K.R. and Luo, M.C. (2006) Molecular characterization of a diagnostic DNA marker for domesticated tetraploid wheat provides evidence for gene flow from wild tetraploid wheat to hexaploid wheat. *Mol. Biol. Evol.* **23**, 1386–1396.

Erayman, M., Sandhu, D., Sidhu, D., Dilbirligi, M., Baenziger, P.S. and Gill, K.S. (2004) Demarcating the gene-rich regions of the wheat genome. *Nucleic Acids Res.* **32**, 3546–3565.

Faris, J.D., Zhang, Q., Chao, S., Zhang, Z. and Xu, S.S. (2014) Analysis of agronomic and domestication traits in a durum × cultivated emmer wheat population using a high-density single nucleotide polymorphism-based linkage map. *Theor. Appl. Genet.* **127**, 2333–2348.

Feuillet, C., Stein, N., Rossini, L., Praud, S., Mayer, K., Schulman, A., Eversole, K. and Appels, R. (2012) Integrating cereal genomics to support innovation in the Triticeae. *Funct. Integr. Genomics*, **12**, 573–583.

Gadaleta, A., Nigro, D., Giancaspro, A. and Blanco, A. (2011) The glutamine synthetase (*GS2*) genes in relation to grain protein content of durum wheat. *Funct. Integr. Genomics*, **11**, 665–670.

Ganal, M., Wieseke, R., Luerssen, H., Durstewitz, G., Graner, E.-M., Plieske, J. and Polley, A. (2014) High-throughput SNP Profiling of Genetic Resources in Crop Plants Using Genotyping Arrays. In *Genomics of Plant Genetic Resources*(Tuberosa, R., Graner, A. and Frison, E., eds), pp. 113–130. Netherlands: Springer.

Gautami, B., Fonceka, D., Pandey, M.K., Moretzsohn, M.C., Sujay, V., Qin, H.D., Hong, Y.B., Faye, I., Chen, X.P., BhanuPrakash, A., Shah, T.M., Gowda, M.V.C., Nigam, S.N., Liang, X.Q., Hoisington, D.A., Guo, B.Z., Bertioli, D.J., Rami, J.F. and Varshney, R.K. (2012) An international reference consensus genetic map with 897 marker loci based on 11 mapping populations for tetraploid groundnut (*Arachis hypogaea* L.). *PLoS One*, **7**, e41213.

Giles, R.J. and Brown, T.A. (2006) GluDy allele variations in *Aegilops tauschii* and *Triticum aestivum*: implications for the origins of hexaploid wheats. *Theor. Appl. Genet.* **112**, 1563–1572.

Goretti, D., Bitocchi, E., Bellucci, E., Rodriguez, M., Rau, D., Gioia, T., Attene, G., McClean, P., Nanni, L. and Papa, R. (2014) Development of single nucleotide polymorphisms in *Phaseolus vulgaris* and related *Phaseolus* spp. *Mol. Breeding*, **33**, 531–544.

Gupta, P.K., Mir, R.R., Mohan, A. and Kumar, J. (2008) Wheat genomics: present status and future prospects. *Int. J. Plant Genomics*, **2008**, 896451.

Haudry, A., Cenci, A., Ravel, C., Bataillon, T., Brunel, D., Poncet, C., Hochu, I., Poirier, S., Santoni, S., Glemin, S. and David, J. (2007) Grinding up wheat: a massive loss of nucleotide diversity since domestication. *Mol. Biol. Evol.* **24**, 1506–1517.

Hudson, C.J., Freeman, J.S., Kullan, A.R.K., Petroli, C.D., Sansaloni, C.P., Kilian, A., Detering, F., Grattapaglia, D., Potts, B.M., Myburg, A.A. and Vaillancourt, R.E. (2012) A reference linkage map for *Eucalyptus*. *BMC Genom.* **13**, 240.

Hurni, S., Brunner, S., Stirnweis, D., Herren, G., Peditto, D., McIntosh, R.A. and Keller, B. (2014) The powdery mildew resistance gene *Pm8* derived from rye is suppressed by its wheat ortholog *Pm3*. *Plant J.* **79**, 904–913. doi:10.1111/tpj. 12593.

Hyten, D.L., Song, Q., Fickus, E.W., Quigley, C.V., Lim, J.-S., Choi, I.-Y., Hwang, E.-Y., Pastor-Corrales, M. and Cregan, P.B. (2010) High-throughput SNP discovery and assay development in common bean. *BMC Genom.* **11**, 475.

IPO International Pasta Organization (2013) *The World Pasta Industry Status Report*. www.internationalpasta.org p. 44.

Jayatilake, D.V., Tucker, E.J., Bariana, H., Kuchel, H., Edwards, J., McKay, A.C., Chalmers, K. and Mather, D.E. (2013) Genetic mapping and marker development for resistance of wheat against the root lesion nematode *Pratylenchus neglectus*. *BMC Plant Biol.* **13**, 230.

Kihara, H. (1944) Discovery of the DD-analyser, one of the ancestors of *Triticum vulgare*. *Agric. Hortic. (Tokyo)*, **19**, 13–14.

Korzun, V., Röder, M.S., Wendehake, K., Pasqualone, A., Lotti, C., Ganal, M.W. and Blanco, A. (1999) Integration of dinucleotide microsatellites from hexaploid bread wheat into a genetic linkage map of durum wheat. *Theor. Appl. Genet.* **98**, 1202–1207.

Lotti, C., Salvi, S., Pasqualone, A., Tuberosa, R. and Blanco, A. (2000) Integration of AFLP markers into an RFLP-based map of durum wheat. *Plant Breeding*, **119**, 393–401.

Lai, K., Duran, C., Berkman, P.J., Lorenc, M.T., Stiller, J., Manoli, S., Hayden, M.J., Forrest, K.L., Fleury, D., Baumann, U., Zander, M., Mason, A.S., Batley, J. and Edwards, D. (2012) Single nucleotide polymorphism discovery from wheat next-generation sequence data. *Plant Biotechnol. J.* **10**, 743–749.

Ma, J., Stiller, J., Berkman, P.J., Wei, Y., Rogers, J., Feuillet, C., Dolezel, J., Mayer, K.F., Eversole, K., Zheng, Y.-L. and Liu, C. (2013) Sequence-based analysis of translocations and inversions in bread wheat (*Triticum aestivum* L.). *PLoS One*, **8**, e79329.

Maccaferri, M., Sanguineti, M.C., Donini, P. and Tuberosa, R. (2003) Microsatellite analysis reveals a progressive widening of the genetic basis in the elite durum wheat germplasm. *Theor. Appl. Genet.* **107**, 783–797.

Maccaferri, M., Sanguineti, M.C., Noli, E. and Tuberosa, R. (2005) Population structure and long-range linkage disequilibrium in a durum wheat elite collection. *Mol. Breeding*, **15**, 271–289.

Maccaferri, M., Stefanelli, S., Rotondo, F., Tuberosa, R. and Sanguineti, M.C. (2007) Relationships among durum wheat accessions. I. Comparative analysis of SSR, AFLP, and phenotypic data. *Genome*, **50**, 373–384.

Maccaferri, M., Sanguineti, M.C., Corneti, S., Ortega, J.L.A., Ben Salem, M., Bort, J., DeAmbrogio, E., Fernando Garcia del Moral, L., Demontis, A., El-Ahmed, A., Maalouf, F., Machlab, H., Martos, V., Moragues, M., Motawaj, J., Nachit, M., Nserallah, N., Ouabbou, H., Royo, C., Slama, A. and Tuberosa, R. (2008a) Quantitative trait loci for grain yield and adaptation of durum wheat (*Triticum durum* Desf.) across a wide range of water availability. *Genetics*, **178**, 489–511.

Maccaferri, M., Mantovani, P., Tuberosa, R., DeAmbrogio, E., Giuliani, S., Demontis, A., Massi, A. and Sanguineti, M.C. (2008b) A major QTL for durable leaf rust resistance widely exploited in durum wheat breeding programs maps on the distal region of chromosome arm 7BL. *Theor. Appl. Genet.* **117**, 1225–1240.

Maccaferri, M., Ratti, C., Rubies-Autonell, C., Vallega, V., Demontis, A., Stefanelli, S., Tuberosa, R. and Sanguineti, M.C. (2011) Resistance to Soil-borne cereal mosaic virus in durum wheat is controlled by a major QTL on chromosome arm 2BS and minor loci. *Theor. Appl. Genet.* **123**, 527–544.

Maccaferri, M., Francia, R., Ratti, C., Rubies-Autonell, C., Colalongo, C., Ferrazzano, G., Tuberosa, R. and Sanguineti, M.C. (2012) Genetic analysis of *Soil-Borne Cereal Mosaic Virus* response in durum wheat: evidence for the role of the major quantitative trait locus QSbm.ubo-2BS and of minor quantitative trait loci. *Mol. Breeding*, **29**, 973–988.

Maccaferri, M., Cane', M.A., Colalongo, C., Massi, A., Clarke, F., Pozniak, C., Korol, A., Fahima, T., Dubcovsky, J., Xu, S., Karsai, I., Knox, R., Clarke, J.M., Salvi, S., Sanguineti, M.C. and Tuberosa, R. (2014) A consensus framework map of durum wheat (Triticum durum Desf.) suitable for linkage disequilibrium and genome-wide association mapping. *BMC Genom.* **15**, 873.

Mantovani, P., Maccaferri, M., Sanguineti, M.C., Tuberosa, R., Catizone, I., Wenzl, P., Thomson, B., Carling, J., Huttner, E., DeAmbrogio, E. and Kilian, A. (2008) An integrated DArT-SSR linkage map of durum wheat. *Mol. Breeding*, **22**, 629–648.

Marcussen, T., Sandve, S.R., Heier, L., Spannagl, M., Pfeifer, M.; The International Wheat Genome Sequencing Consortium, Jakobsen, K.S., Wulff, B.B.H., Steuernagel, B., Mayer, K.F.X. and Olsen, O.A. (2014) Ancient hybridizations among the ancestral genomes of bread wheat. *Science*, **345**, 1250092.

Marone, D., Laido, G., Gadaleta, A., Colasuonno, P., Ficco, D.B.M., Giancaspro, A., Giove, S., Panio, G., Russo, M.A., De Vita, P., Cattivelli, L., Papa, R., Blanco, A. and Mastrangelo, A.M. (2012) A high-density consensus map of A and B wheat genomes. *Theor. Appl. Genet.* **125**, 1619–1638.

McFadden, E.S. and Sears, E.R. (1946) The origin of *Triticum spelta* and its free-threshing hexaploid relatives. *J. Hered.* **37**, 81–89.

Mori, N. (2003) Wheat domestication: when, where and how? *Plant Cell Physiol.* **44**, S2–S2.

Muñoz-Amatriaín, M., Moscou, M.J., Bhat, P.R., Svensson, J.T., Bartos, J., Suchankova, P., Simkova, H., Endo, T.R., Fenton, R.D., Lonardi, S., Castillo, A.M., Chao, S., Cistue, L., Cuesta-Marcos, A., Forrest, K.L., Hayden, M.J., Hayes, P.M., Horsley, R.D., Makoto, K., Moody, D., Sato, K., Valles, M.P., Wulff, B.B.H., Muehlbauer, G.J., Dolezel, J. and Close, T.J. (2011) An improved consensus linkage map of barley based on flow-sorted chromosomes and single nucleotide polymorphism markers. *Plant Genome*, **4**, 238–249.

Nesbitt, M. and Samuel, D. (1996) From stable crop to extinction? The archaeology and history of hulled wheats. In: S. Padulosi, K. Hammer, J. Hellers (Eds) *Hulled wheat*. International Plant Genetic Resources Institute, Rome, pp. 41–100.

Neu, C., Stein, N. and Keller, B. (2002) Genetic mapping of the *Lr20-Pm1* resistance locus reveals suppressed recombination on chromosome arm 7AL in hexaploid wheat. *Genome*, **45**, 737–744.

Nitcher, R., Distelfeld, A., Tan, C., Yan, L. and Dubcovsky, J. (2013) Increased copy number at the *HvFT1* locus is associated with accelerated flowering time in barley. *Mol. Genet. Genomics*, **288**, 261–275.

Ozkan, H., Brandolini, A., Pozzi, C., Effgen, S., Wunder, J. and Salamini, F. (2005) A reconsideration of the domestication geography of tetraploid wheats. *Theor. Appl. Genet.* **110**, 1052–1060.

Peleg, Z., Saranga, Y., Suprunova, T., Ronin, Y., Roeder, M.S., Kilian, A., Korol, A.B. and Fahima, T. (2008) High-density genetic map of durum wheat x wild emmer wheat based on SSR and DArT markers. *Theor. Appl. Genet.* **117**, 103–115.

van Poecke, R.M.P., Maccaferri, M., Tang, J., Truong, H.T., Janssen, A., van Orsouw, N.J., Salvi, S., Sanguineti, M.C., Tuberosa, R. and van der Vossen, E.A.G. (2013) Sequence-based SNP genotyping in durum wheat. *Plant Biotechnol. J.* **11**, 809–817.

Pont, C., Murat, F., Guizard, S., Flores, R., Foucrier, S., Bidet, Y., Quraishi, U.M., Alaux, M., Dolezel, J., Fahima, T., Budak, H., Keller, B., Salvi, S., Maccaferri, M., Steinbach, D., Feuillet, C., Quesneville, H. and Salse, J. (2013) Wheat syntenome unveils new evidences of contrasted evolutionary plasticity between paleo- and neoduplicated subgenomes. *Plant J.* **76**, 1030–1044.

Powell, W., Morgante, M., Andre, C., Hanafey, M., Vogel, J., Tingey, S. and Rafalski, A. (1996) The comparison of RFLP, RAPD, AFLP and SSR (microsatellite) markers for germplasm analysis. *Mol. Breeding*, **2**, 225–238.

Qi, L.L., Echalier, B., Chao, S., Lazo, G.R., Butler, G.E., Anderson, O.D., Akhunov, E.D., Dvorak, J., Linkiewicz, A.M., Ratnasiri, A., Dubcovsky, J.,

Bermudez-Kandianis, C.E., Greene, R.A., Kantety, R., La Rota, C.M., Munkvold, J.D., Sorrells, S.F., Sorrells, M.E., Dilbirligi, M., Sidhu, D., Erayman, M., Randhawa, H.S., Sandhu, D., Bondareva, S.N., Gill, K.S., Mahmoud, A.A., Ma, X.F., Miftahudin, Gustafson, J.P., Conley, E.J., Nduati, V., Gonzalez-Hernandez, J.L., Anderson, J.A., Peng, J.H., Lapitan, N.L.V., Hossain, K.G., Kalavacharla, V., Kianian, S.F., Pathan, M.S., Zhang, D.S., Nguyen, H.T., Choi, D.W., Fenton, R.D., Close, T.J., McGuire, P.E., Qualset, C.O. and Gill, B.S. (2004) A chromosome bin map of 16,000 expressed sequence tag loci and distribution of genes among the three genomes of polyploid wheat. *Genetics*, **168**, 701–712.

Quarrie, S.A., Steed, A., Calestani, C., Semikhodskii, A., Lebreton, C., Chinoy, C., Steele, N., Pljevljakusic, D., Waterman, E., Weyen, J., Schondelmaier, J., Habash, D.Z., Farmer, P., Saker, L., Clarkson, D.T., Abugalieva, A., Yessimbekova, M., Turuspekov, Y., Abugalieva, S., Tuberosa, R., Sanguineti, M.C., Hollington, P.A., Aragues, R., Royo, A. and Dodig, D. (2005) A high-density genetic map of hexaploid wheat (*Triticum aestivum* L.) from the cross Chinese Spring X SQ1 and its use to compare QTLs for grain yield across a range of environments. *Theor. Appl. Genet.* **110**, 865–880.

Röder, M.S., Korzun, V., Wendehake, K., Plaschke, J., Tixier, M.H., Leroy, P. and Ganal, M.W. (1998) A microsatellite map of wheat. *Genetics*, **149**, 2007–2023.

Ronin, Y., Mester, D., Minkov, D., Belotserkovski, R., Jackson, B.N., Schnable, P.S., Aluru, S. and Korol, A. (2012) Two-phase analysis in consensus genetic mapping. *G3-Genes Genomes Genet.* **2**, 537–549.

Saintenac, C., Faure, S., Remay, A., Choulet, F., Ravel, C., Paux, E., Balfourier, F., Feuillet, C. and Sourdille, P. (2011) Variation in crossover rates across a 3-Mb contig of bread wheat (*Triticum aestivum* L.) reveals the presence of a meiotic recombination hotspot. *Chromosoma*, **120**, 185–198.

Saintenac, C., Jiang, D., Wang, S. and Akhunov, E. (2013) Sequence-based mapping of the polyploid wheat genome. *G3-Genes Genomes Genet.* **3**, 1105–1114.

Salamini, F., Ozkan, H., Brandolini, A., Schafer-Pregl, R. and Martin, W. (2002) Genetics and geography of wild cereal domestication in the Near East. *Nat. Rev. Genet.* **3**, 429–441.

Salvi, S., Castelletti, S. and Tuberosa, R. (2010) An updated consensus map for flowering time QTLs in maize. *Maydica*, **54**, 501–512.

Somers, D.J., Isaac, P. and Edwards, K. (2004) A high-density microsatellite consensus map for bread wheat (*Triticum aestivum* L.). *Theor. Appl. Genet.* **109**, 1105–1114.

Sosnowski, O., Charcosset, A. and Joets, J. (2012) BioMercator V3: an upgrade of genetic map compilation and quantitative trait loci meta-analysis algorithms. *Bioinformatics*, **28**, 2082–2083.

Sourdille, P., Singh, S., Cadalen, T., Brown-Guedira, G.L., Gay, G., Qi, L., Gill, B.S., Dufour, P., Murigneux, A. and Bernard, M. (2004) Microsatellite-based deletion bin system for the establishment of genetic-physical map relationships in wheat (*Triticum aestivum* L.). *Funct. Integr. Genomics*, **4**, 12–25.

Swamy, B.P., Vikram, P., Dixit, S., Ahmed, H.U. and Kumar, A. (2011) Meta-analysis of grain yield QTL identified during agricultural drought in grasses showed consensus. *BMC Genom.* **12**, 319.

The International Wheat Genome Sequencing Consortium (IWGSC). (2014) A chromosome-based draft sequence of the hexaploid bread wheat (*Triticum aestivum*) genome. *Science*, **345**, 1251788. doi:10.1126/science.

Tiwari, V.K., Wang, S., Sehgal, S., Vrana, J., Friebe, B., Kubalakova, M., Chhuneja, P., Dolezel, J., Akhunov, E., Kalia, B., Sabir, J. and Gill, B.S. (2014) SNP Discovery for mapping alien introgressions in wheat. *BMC Genom.* **15**, 273.

Trebbi, D., Maccaferri, M., de Heer, P., Sorensen, A., Giuliani, S., Salvi, S., Sanguineti, M.C., Massi, A., van der Vossen, E.A.G. and Tuberosa, R. (2011) High-throughput SNP discovery and genotyping in durum wheat (*Triticum durum* Desf.). *Theor. Appl. Genet.* **123**, 555–569.

Varshney, R.K., Langridge, P. and Graner, A. (2007) Application of genomics to molecular breeding of wheat and barley. *Adv. Genet.* **58**, 121–155.

Wang, J., Lydiate, D.J., Parkin, I.A.P., Falentin, C., Delourme, R., Carion, P.W.C. and King, G.J. (2011) Integration of linkage maps for amphiploid Brassica napus and comparative mapping with Arabidopsis and Brassica rapa. *BMC Genom.* **12**, 101.

Wang, S., Wong, D., Forrest, K., Allen, A., Chao, S., Huang, B.E., Maccaferri, M., Salvi, S., Milner, S.G., Cattivelli, L., Mastrangelo, A.M., Whan, A.,

Stephen, S., Barker, G., Wieseke, R., Plieske, J.; International Wheat Genome Sequencing C, Lillemo, M., Mather, D., Appels, R., Dolferus, R., Brown-Guedira, G., Korol, A., Akhunova, A.R., Feuillet, C., Salse, J., Morgante, M., Pozniak, C., Luo, M.-C., Dvorak, J., Morell, M., Dubcovsky, J., Ganal, M., Tuberosa, R., Lawley, C., Mikoulitch, I., Cavanagh, C., Edwards, K.J., Hayden, M. and Akhunov, E. (2014) Characterization of polyploid wheat genomic diversity using a high-density 90 00 single nucleotide polymorphism array. *Plant Biotechnol. J.* **12**, 787–796.

Wenzl, P., Li, H., Carling, J., Zhou, M., Raman, H., Paul, E., Hearnden, P., Maier, C., Xia, L., Caig, V., Ovesná, J., Cakir, M., Poulsen, D., Wang, J., Raman, R., Smith, K.P., Muehlbauer, G.J., Chalmers, K.J., Kleinhofs, A., Huttner, E. and Kilian, A. (2006) A high-density consensus map of barley linking DArT markers to SSR, RFLP and STS loci and agricultural traits. *BMC Genom.* **7**, 206.

Wilhelm, E.P., Boulton, M.I., Al-Kaff, N., Balfourier, F., Bordes, J., Greenland, A.J., Powell, W. and Mackay, I.J. (2013) *Rht-1* and *Ppd-D1* associations with height, GA sensitivity, and days to heading in a worldwide bread wheat collection. *Theor. Appl. Genet.* **126**, 2233–2243.

Xie, W., Ben-David, R., Zeng, B., Dinoor, A., Xie, C., Sun, Q., Roeder, M.S., Fahoum, A. and Fahima, T. (2012) Suppressed recombination rate in 6VS/6AL translocation region carrying the *Pm21* locus introgressed from *Haynaldia villosa* into hexaploid wheat. *Mol. Breeding*, **29**, 399–412.

Zhang, W., Chao, S., Manthey, F., Chicaiza, O., Brevis, J.C., Echenique, V. and Dubcovsky, J. (2008) QTL analysis of pasta quality using a composite microsatellite and SNP map of durum wheat. *Theor. Appl. Genet.* **117**, 1361–1377.

Zheng, B., Biddulph, B., Li, D., Kuchel, H. and Chapman, S. (2013) Quantification of the effects of *VRN1* and *Ppd-D1* to predict spring wheat (*Triticum aestivum*) heading time across diverse environments. *J. Exp. Bot.* **64**, 3747–3761.

5

Fine-tuning levels of heterologous gene expression in plants by orthogonal variation of the untranslated regions of a nonreplicating transient expression system

Yulia A. Meshcheriakova, Pooja Saxena[†] and George P. Lomonossoff*

Department of Biological Chemistry, John Innes Centre, Norwich, UK

*Correspondence

email george.lomonossoff@jic.ac.uk
†Present address: Department of Chemistry, Indiana University, Bloomington, Indiana 47405, USA.

Keywords: Translational efficiency, viral 3'UTR, CPMV-HT, Y-shaped secondary structure, pEAQ vectors.

Summary

A transient expression system based on a deleted version of Cowpea mosaic virus (CPMV) RNA-2, termed CPMV-HT, in which the sequence to be expressed is positioned between a modified 5′ UTR and the 3′ UTR has been successfully used for the plant-based expression of a wide range of proteins, including heteromultimeric complexes. While previous work has demonstrated that alterations to the sequence of the 5′ UTR can dramatically influence expression levels, the role of the 3′ UTR in enhancing expression has not been determined. In this work, we have examined the effect of different mutations in the 3′UTR of CPMV RNA-2 on expression levels using the reporter protein GFP encoded by the expression vector, pEAQexpress-HT-GFP. The results showed that the presence of a 3′ UTR in the CPMV-HT system is important for achieving maximal expression levels. Removal of the entire 3′ UTR reduced expression to approximately 30% of that obtained in its presence. It was found that the Y-shaped secondary structure formed by nucleotides 125–165 of the 3′ UTR plays a key role in its function; mutations that disrupt this Y-shaped structure have an effect equivalent to the deletion of the entire 3′ UTR. Our results suggest that the Y-shaped secondary structure acts by enhancing mRNA accumulation rather than by having a direct effect on RNA translation. The work described in this paper shows that the 5′ and 3′ UTRs in CPMV-HT act orthogonally and that mutations introduced into them allow fine modulation of protein expression levels.

Introduction

The Cowpea mosaic virus hypertranslatable (CPMV-HT) system and its associated pEAQ vectors have become a popular system for high-level transient expression of heterologous proteins in plants via agro-infiltration (for a recent review, see Peyret and Lomonossoff, 2013). This system involves positioning the target gene between a modified 5′UTR, from which upstream AUG codons have been deleted, and a wild-type 3′UTR from CPMV RNA-2, to create a 'hypertranslatable' (HT) cassette within the T-DNA region of a binary vector. Transcription from the HT cassette is driven by a Cauliflower mosaic virus (CaMV) 35S promoter and is terminated by a nopaline synthase (nos) terminator. This results in the production of an mRNA in which the inserted gene is flanked at its 5′ end by the modified CPMV 5′ UTR and at its 3′ end by the 3′ UTR from CPMV RNA-2, followed by a linker sequence and the 3′ UTR from the nos gene (Figure 1b).

While the importance of the modified 5′ UTR in achieving high-level expression using the CPMV-HT system is well documented (Peyret and Lomonossoff, 2013; Sainsbury and Lomonossoff, 2008; Sainsbury et al., 2010; Thuenemann et al., 2013), the role of the 3′ UTR, if any, is unknown. The sequence of the 3′ UTR was originally included in CPMV-based expression plasmids to allow replication of RNA transcripts by the CPMV RNA-1-encoded polymerase (Cañizares et al., 2006), an ability not required in the CPMV-HT system. However, the sequence was retained on the grounds that many plant viral mRNAs contain a cap-independent translation element (3′-CITE) within the 3′UTR (for a review, see

Miller et al., 2007). These elements can be transplanted onto heterologous open reading frames and have been shown to promote their translation. Thus, it was rationalized that the sequence of the 3′ UTR of CPMV RNA-2 may assist high-level expression from RNAs transcribed from HT cassettes. In this paper, we show that this is, indeed, the case and that the enhancing effect operates by increasing mRNA accumulation. Furthermore, we show that the contributions of the 5′ and 3′ UTRs towards enhancing expression act orthogonally, and constructs directing varying levels of expression can be obtained by mixing and matching 5′ and 3′ UTRs.

Results

Comparison of GFP expression levels in the presence and in the absence of the 3′ UTR of CPMV RNA-2

The potential role of the 3′ UTR of CPMV RNA-2 in enhancing expression levels from pEAQ plasmids (Sainsbury et al., 2009) was investigated using the construct pEAQexpress-HT-GFP (Figure 1a). This construct encodes the sequence of GFP within a CPMV-HT cassette and gives high-level expression of GFP when agro-infiltrated into 3-week-old Nicotiana benthamiana plants (Sainsbury et al., 2009). In common with other pEAQ-based constructs, transcripts from the CPMV-HT cassette are anticipated to consist of the modified CPMV RNA-2 5′ UTR, followed by the structural gene (in this case gfp) and a chimaeric 3′ UTR (Figure 1b). The latter consists of 183 nucleotides corresponding to the 3′ UTR of CPMV RNA-2, a 51-nucleotide linker consisting

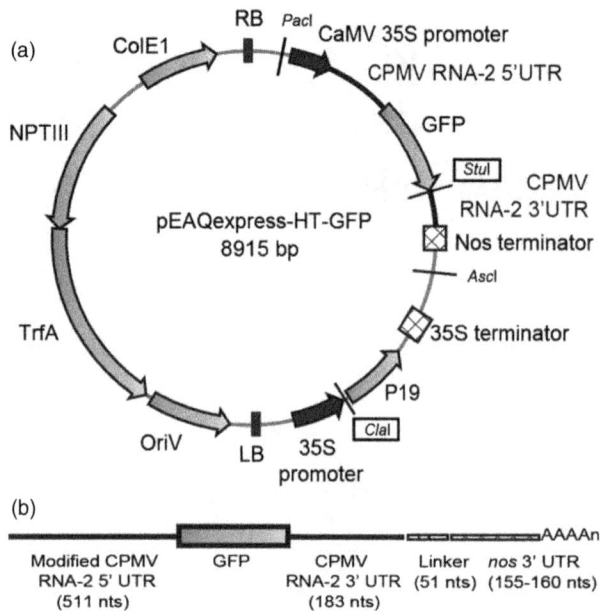

Figure 1 Schematic representation of construct used for GFP expression (a) Plasmid diagram of pEAQexpress-HT-GFP. Dark arrows represent promoters, light arrows represent open reading frames, checked boxes represent terminators, and solid lines represent the UTRs from CPMV RNA-2 system. (b) Schematic representation of the GFP-specific mRNA transcribed from construct pEAQexpress-HT-GFP. Solid lines denote UTRs from CPMV RNA-2, the lined box denotes the 51 nucleotide linker, and the checked box denotes the 3′ UTR from the *nos* gene.

of 18As derived from the poly(A) tail of RNA-2 plus a polylinker and, finally, 155-160 nucleotides (depending on the precise site of termination; Bevan *et al.*, 1983) from the 3′ UTR of the *nos* gene from pBINPLUS (van Engelen *et al.*, 1995; Figure S1). The transcripts are also expected to be capped at their 5′ end and polyadenylated at their 3′ end.

To determine whether the sequences from the 3′ UTR of CPMV RNA-2 and the linker have any role in enhancing expression from the pEAQ vectors, either a 234-nucleotide sequence encompassing the entire 183-nucleotide RNA-2-derived 3′ UTR sequence plus the 51 nucleotide linker or just the RNA-2-specific 183 nucleotides were deleted from pEAQexpress-HT-GFP to give plasmids pEAQexpress-HT-GFP-Del-234 and pEAQexpress-HT-GFP-Del-183, respectively (Figure 2a). In pEAQexpress-HT-GFP-Del-234, the 3′ UTR from the *nos* gene is immediately adjacent to the sequence encoding GFP, while in pEAQexpressGFP-Del-183, it is separated by the 51-nucleotide linker. As controls, the 234 nucleotides deleted from pEAQexpress-HT-GFP-Del-234 were re-inserted in either the correct (pEAQexpress-HT-GFP-Del-234-corr 234) or opposite orientation (pEAQexpress-HT-GFP-Del-234-opp 234), and the 183-nucleotide sequence corresponding to the CPMV RNA-2 3′ UTR was inserted in the correct orientation (pEAQexpress-HT-GFP-Del-234-corr 183) using the unique *StuI* site of pEAQexpress-HT-GFP-Del-234 (Figure 2a). Following transformation into *Agrobacterium tumefaciens* and infiltration of *N. benthamiana* leaves, the level of GFP expression was assessed visually under UV light (Figure 2b), and the accumulation of GFP was quantified by spectrofluorometric analysis of the protein extracts of agro-infiltrated leaves. Elimination of the 234-nucleotide sequence encompassing both the CPMV RNA-2 3′ UTR and the linker or just the 183 nucleotides corresponding to

the 3′ UTR resulted in a decrease in GFP expression levels to 43.64% and 33.94%, respectively, of that obtained with the original pEAQexpress-HT-GFP construct (Figure 2c). Re-insertion of the 234-nucleotide fragment in the correct but not the reverse orientation fully restored GFP expression; re-insertion of just the 183-nucleotide CPMV-specific region in the correct orientation was equally effective. The relative levels of fluorescence were precisely correlated with the levels of GFP detected on Coomassie blue-stained SDS-polyacrylamide gels (data not shown) consistent with previous results (Sainsbury and Lomonossoff, 2008; Sainsbury *et al.*, 2009). These results indicate that the presence of the 3′ UTR of CPMV RNA-2 in the correct orientation boosts expression of GFP up to threefold compared with the levels obtained when just the *nos* 3′ UTR is present.

A secondary structure motif within the 3′ UTR of CPMV RNA-2 is important for enhancing expression

To define more precisely the region(s) of the 3′ UTR involved in enhancing expression levels, a series of deletion mutants based on pEAQexpress-HT-GFP was constructed (Figure 3b). Deletion of 65 and 107 nucleotides from the region of 3′ UTR immediately downstream of the GFP stop codon, resulting in plasmids pEAQexpress-HT-GFP-3′UTR(66-183) and pEAQexpress-HT-GFP-3′UTR(108-183), had only a limited effect on reducing expression levels, whereas deletion of 141 nucleotides in pEAQexpress-HT-GFP-3′UTR(142-183) reduced the expression level to that of pEAQexpress-HT-GFP-Del-234, which lacks the entire 3′ UTR (Figure 3c). This suggests that there is a region downstream of nucleotide 107 of the 3′ UTR of CPMV RNA-2 that is critical for maintaining high levels of expression from the pEAQ vectors, although upstream sequences may also play some role. To verify this, 93 nucleotides, including the 3′ terminal 42 nucleotides of the 3′ UTR plus the 51 nucleotides downstream, were deleted to give pEAQexpress-HT-GFP-3′UTR (1-141) (Figure 3b). Again, expression levels from this construct were equivalent to those obtained with pEAQexpress-HT-GFP-Del-234 (Figure 3c).

Secondary structure predictions of the 3′ UTR of CPMV RNA-2 have indicated that the region between nucleotides 125 and 165 can be folded into a stable Y-shaped stem-loop structure (Eggen *et al.*, 1989; Rohll *et al.*, 1993); (Figure 3a). This structure was subsequently shown to be important for efficient accumulation of RNA-2 during replication in protoplasts (Rohll *et al.*, 1993). As the deletions in pEAQexpress-HT-GFP-3′UTR (142-183) and pEAQexpress-HT-GFP-3′UTR(1-141) extend into this Y-shaped structure while those in pEAQexpress-HT-GFP-3′UTR(66-183) and pEAQexpress-HT-GFP-3′UTR(108-183) do not, it is plausible that the differential expression levels obtained relate to the presence or absence of the intact Y-shaped structure. To investigate this possibility, point mutations were introduced into pEAQexpress-HT-GFP, which were predicted either to disrupt the Y-shaped structure (C132U) or to leave it intact (U136G) (Figure 4a). Assessment of the GFP expression levels obtained with the plasmids harbouring these two mutations (pEAQexpress-HT-GFP-MutC132U and pEAQexpress-HT-GFP-MutU136G, respectively) showed that the mutation that disrupted the Y-shaped structure (C132U) reduced GFP expression levels to that found with pEAQexpress-HT-GFP-Del-234, while the mutation which maintained it (U136G) had little effect (Figure 4b). These results indicate that an intact Y-shaped secondary structure is, indeed, critical for maintaining high GFP expression levels.

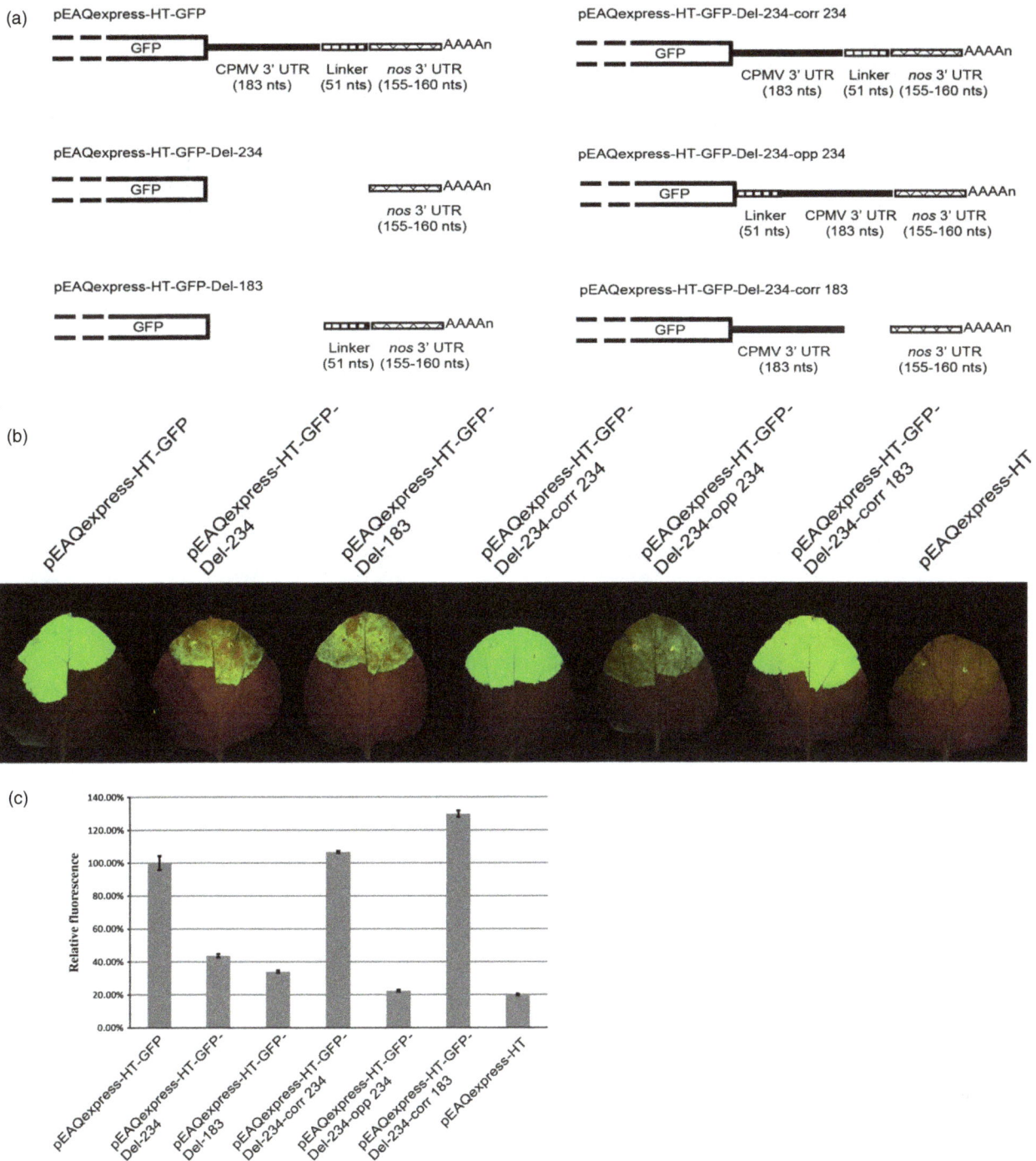

Figure 2 Analysis of the role of the 3′ UTR in achieving high-level expression. (a) Schematic representation of the series of mutants made upon deletion of different sections within the 3′ UTR of pEAQexpress-HT-GFP. (b) Leaves infiltrated with the various constructs (as indicated), photographed at 6 dpi under UV light. (c) Relative GFP expression levels based on spectrofluorometric analysis of extracts from leaves harvested at 6 dpi. Values indicate relative expression levels averaged from three technical replicates ± standard error.

Complementarity between the 5′ and 3′ UTRs of CPMV RNA-2 is not required for enhanced expression

Nucleotides 134–140 (GUUUAGC), in the minor loop of the predicted Y-shaped structure of the 3′UTR, are complementary to nucleotides 136–142 (CAAAUCG) of the 5′UTR. As regions of complementarity between the 5′ and 3′ UTRs have been shown to be important in controlling expression levels in several plant

viral RNAs (Fabian and White, 2004; Guo *et al.*, 2001), the importance of maintaining this complementarity to achieve high levels of expression from the CPMV-*HT* system was investigated. To this end, the seven-nucleotide sequence in the 3′ UTR was either deleted (pEAQexpress-HT-GFP-Del7) or three point mutations were introduced to change the sequence to G**G**UG**A**G**A** (changes shown in bold), thereby destroying its complementarity with the corresponding sequence in the 5′ UTR (pEAQexpress-HT-

(a)

(b)

pEAQexpress-HT-GFP-3'UTR (1-183)

pEAQexpress-HT-GFP-3'UTR (66-183)

pEAQexpress-HT-GFP-3'UTR (108-183)

pEAQexpress-HT-GFP-3'UTR (142-183)

pEAQexpress-HT-GFP-3'UTR (1-141)

(c)

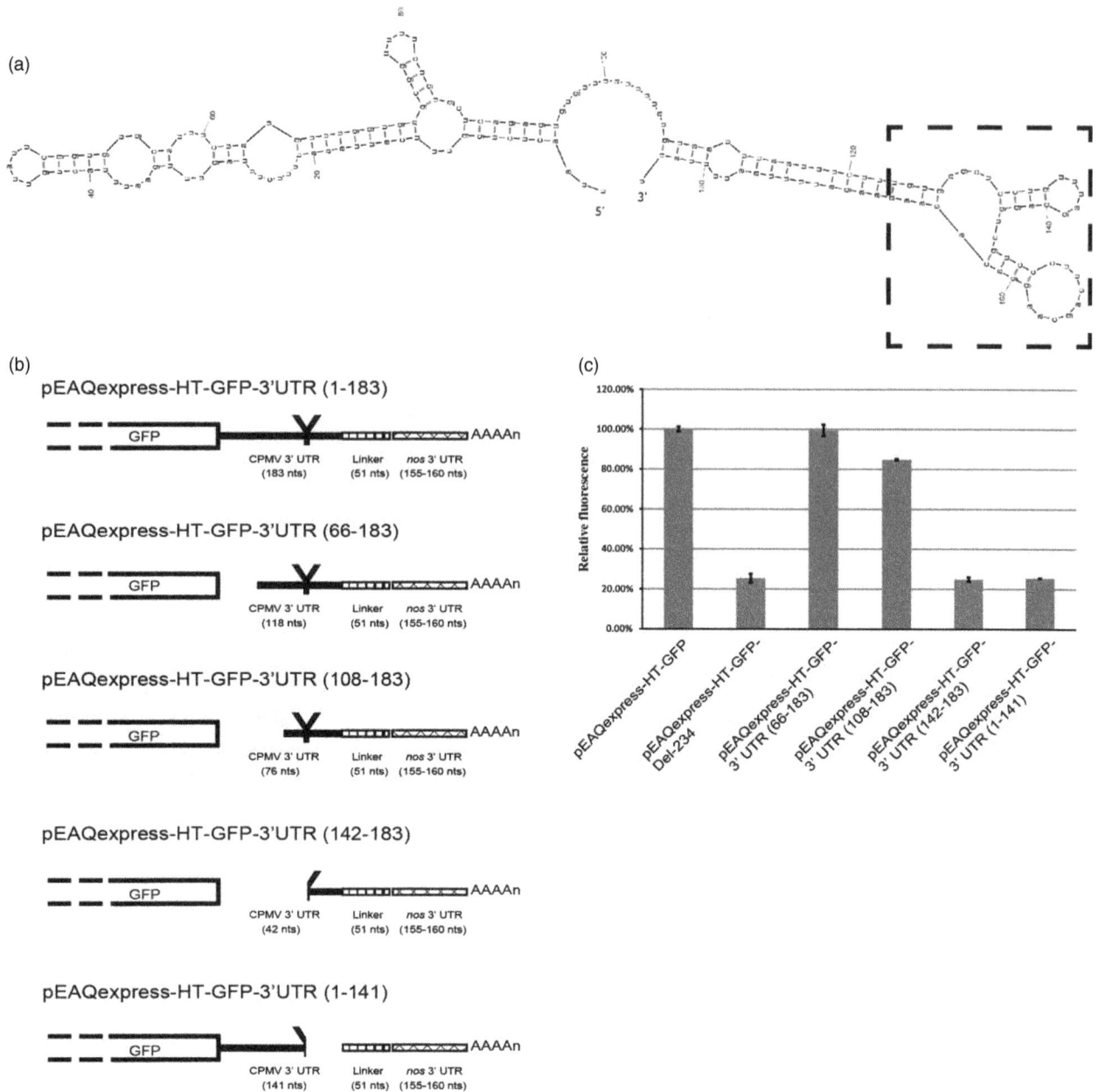

Figure 3 Deletion mutagenesis of the 3' UTR. (a) Predicted secondary structure of the 3'UTR of CPMV RNA-2. The Y-shaped loop has been highlighted. (b) Schematic representation of the series of mutants constructed upon deletion of different sections within the 3' UTR of CPMV RNA-2. (c) Relative GFP expression levels based on spectrofluorometric analysis of extracts from leaves harvested at 6 dpi. Values indicate relative expression levels averaged from three technical replicates ± standard error.

GFP-3Mut). In both cases, this resulted in at least a threefold decrease in the GFP expression level (Figure 4b). However, the introduction of compensatory mutations in the 5' UTR of pEAQexpress-HT-GFP-3Mut (to give the sequence CCACUCU, thereby restoring complementarity between the UTRs in pEAQexpress-HT-GFP-3Mut-Restored) did not restore GFP expression (Figure 4b). Moreover, introduction of the same three point mutations into the 5'UTR of pEAQexpress-HT-GFP (encoding the native 3'UTR) to generate pEAQexpress-HT-GFP-3Mut5'UTR did not reduce GFP expression levels (Figure 4b). Taken together, these results indicate that although nucleotides 134–140 in the 3' UTR play an important role in maintaining expression levels, this

does not operate through direct base pairing between 5' and 3' UTRs. It is more likely that these nucleotides are critical for maintaining the correct functioning of the Y-shaped structure as both of the above mutations are predicted to destabilize it.

Replacement of the 3' UTR of CPMV RNA-2 with 3' UTRs from other viruses

If it is the Y-structure rather than the linear sequence of the 3' UTR of CPMV RNA-2 that is important for enhancing expression, other sequences forming such a structure should give a similar effect. To investigate this possibility, the CPMV RNA-2 3' UTR was replaced with the either (i) the 3' UTR of RNA-2 of another

(a)

(b)

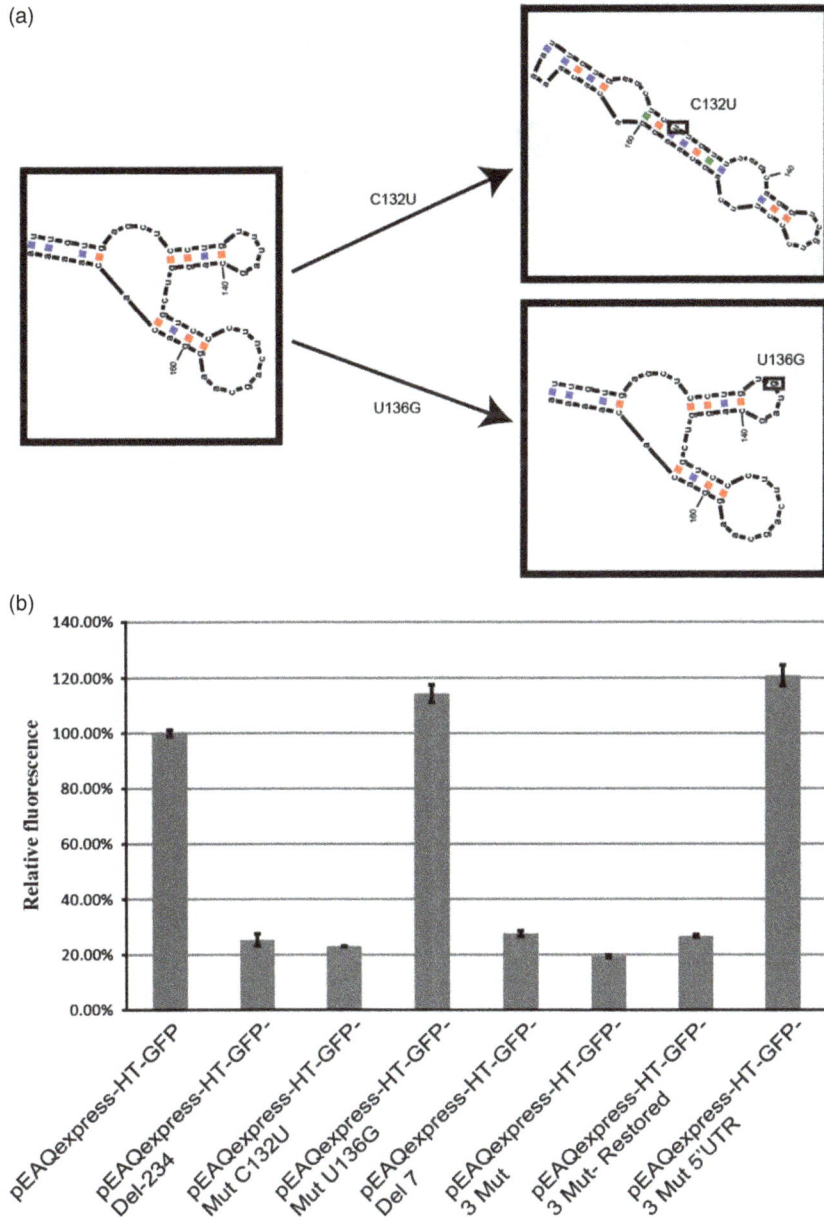

Figure 4 Importance of secondary structure in the 3′ UTR. (a) Secondary structures predicted for 3′ UTR point mutants C132U and U136G. (b) Relative GFP expression levels based on spectrofluorometric analysis of extracts from leaves harvested at 6 dpi. Values indicate relative expression levels averaged from three technical replicates ± standard error.

comovirus, Red clover mottle virus (RCMV); (ii) the 3′ UTR of CPMV RNA-1 or (iii) the 3′ UTR of an unrelated virus, Tobacco mosaic virus (TMV) strain IM (Figure 5a). The 3′ UTR from RNA-2 of RCMV is longer than that of CPMV RNA-2 (262 versus 183 nucleotides) and has an overall sequence identity of only 32%; however, it contains a region in which 44 of 48 nucleotides are identical to nucleotides 127–175 of the CPMV RNA-2 sequence and which can be folded into a similar Y-shaped structure, with the four differences making compensating changes in one of the stems. The 3′ UTR of CPMV RNA-1 is considerably shorter than its RNA-2 equivalent (82 versus 183 nucleotides) but has a similar Y-shaped structure near its 3′ end (Figure 5a). By contrast, the TMV-IM 3′ UTR is comparable in length (203 versus 183 nucleotides) but is predicted to form a different secondary structure, devoid of any Y-shaped structures similar to those predicted for CPMV RNA-2 but containing a series of pseudo-knots (Figure 5a; van Belkum et al., 1985). To analyse the effects of these alternative 3′UTRs on GFP expression levels, synthetic sequences corresponding to them were inserted into StuI-

digested pEAQexpress-HT-GFP-Del-234 to create pEAQexpress-HT-GFP-RCMV 3′UTR, pEAQexpress-HT-GFP- RNA-1 3′UTR and pEAQexpress-HT-GFP- TMV 3′UTR, and the levels of GFP expression obtained from these vectors were determined. The presence of the 3′UTR from RNA-2 of RCMV conferred expression levels approximately 75% achieved using the native 3′UTR from CPMV RNA-2 (Figure 5b), while the shorter 3′UTR of CPMV RNA-1gave only half the level. The 3′UTR from TMV was the least effective out of those tested in this experiment, giving expression levels even lower than that seen in the absence of any virus-specific 3′ UTR (Figure 5b). These results are consistent with the requirement for the Y-shaped structure positioned a minimum distance from the end of the structural gene, as suggested by deletion analysis, to achieve maximum enhancement of expression levels.

The Y-shaped structure enhances mRNA accumulation

The difference in expression levels of GFP observed for vectors with altered 3′ UTRs could stem either from different translational

(a)

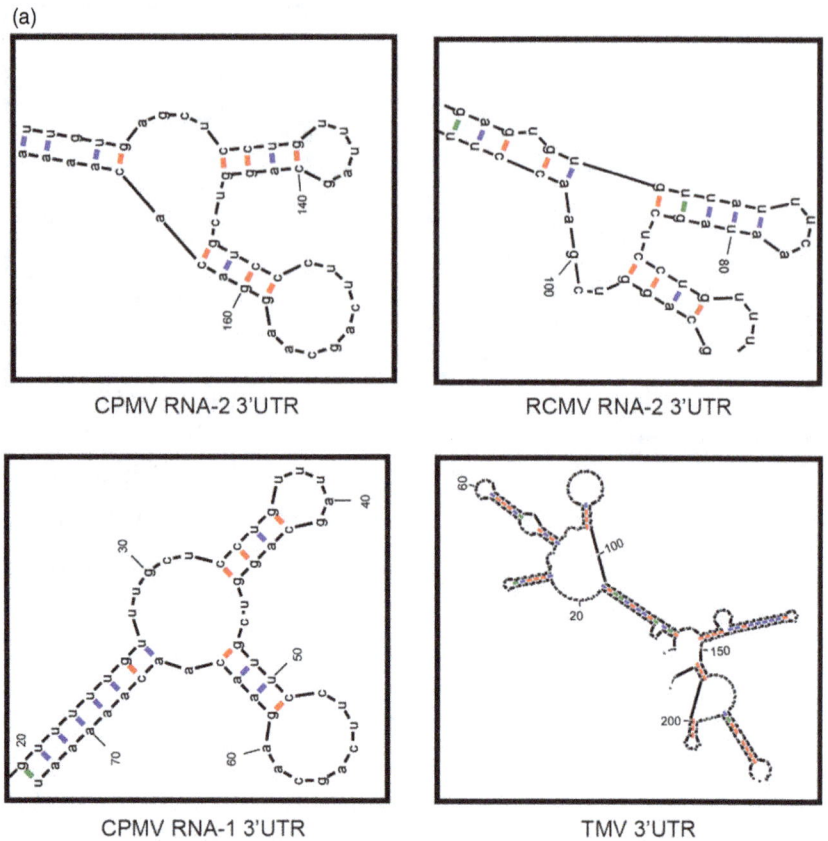

CPMV RNA-2 3'UTR

RCMV RNA-2 3'UTR

CPMV RNA-1 3'UTR

TMV 3'UTR

(b)

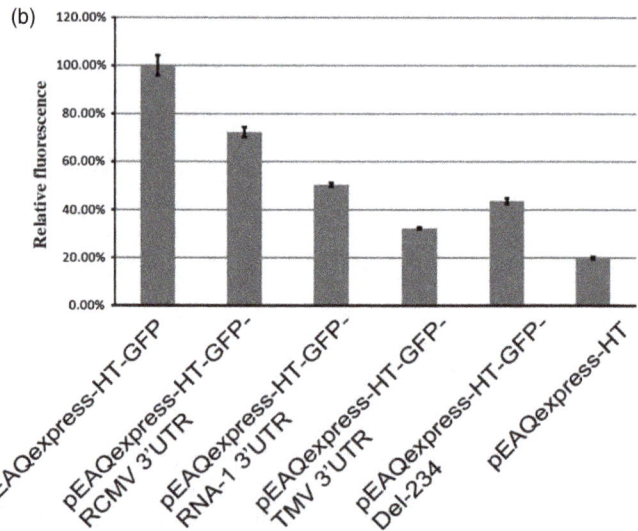

Figure 5 Effect of 3′ UTR substitutions. (a) Secondary structures predicted for 3′ UTRs of RCMV RNA-2, CPMV RNA-1 and TMV and RCMV RNA-2. (b) Relative GFP expression levels based on spectrofluorometric analysis of extracts from leaves harvested at 6 dpi. Values indicate relative expression levels averaged from three technical replicates ± standard error.

efficiencies of the GFP mRNAs or, alternatively, from differences in the relative accumulation of the mRNAs. To distinguish between these two possibilities, real-time quantitative reverse transcription (qRT)-PCR was performed to compare relative levels of GFP mRNA in leaf tissue infiltrated with different constructs, two of which [pEAQexpress-HT-GFP and pEAQexpress-HT-GFP-3′UTR(66-183)] were known to give high levels of GFP expression and possess the Y-shaped structure in their 3′ UTR and two which gave low levels of GFP expression and lacked the Y-shaped structure (pEAQexpress-HT-GFP-Del-234 and pEAQexpress-HT-GFP-3Mut). The results show a strong correlation between mRNA levels and GFP expression (Figure 6), suggesting that the higher levels of expression are due to the ability of the Y-shaped

structure to enhance mRNA accumulation, rather than any direct effect on translational efficiency. This contrasts with the situation as previously found to be the case for the 5′ UTR of CPMV RNA-2 where enhancement of expression does not correlated with mRNA levels (Sainsbury and Lomonossoff, 2008).

Creation of expression cassettes of varying strength

The above data indicated that the 5′ and 3′ UTRs of CPMV RNA-2 enhance expression independently through different mechanisms. Hence, it was envisaged that it should be possible to vary the effects of the 5′ and 3′ UTRs orthogonally to create a series of cassettes of varying translational strengths. To investigate this possibility, an expression system similar in design to CPMV-*HT* but

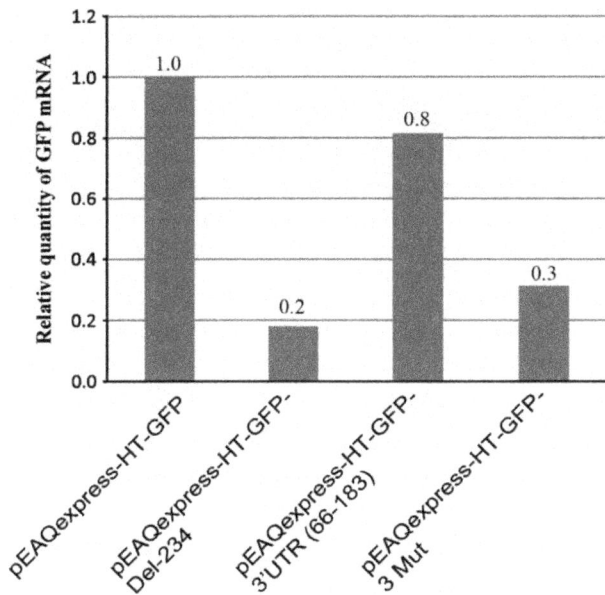

Figure 6 Quantitation of GFP mRNA in plant tissue. Relative levels of mRNA of GFP in leaf tissue infiltrated with four different constructs, as determined by real-time qRT-PCR.

utilizing the UTRs from CPMV RNA-1 was created. Briefly, a synthetic cassette consisting of the CaMV 35S promoter, the 5′ UTR of CPMV RNA-1, a polylinker, the 3′ UTR of CPMV RNA-1 and the *nos* terminator was inserted into the T-DNA region of pEAQexpress (Sainsbury et al., 2009) to generate pEAQexpress-RT (Figure S2). The gene encoding GFP was inserted into the polylinker to generate plasmid pEAQexpress-RT-GFP, anticipated to be transcribed to give a mRNA in which the sequence encoding GFP is flanked by the 5′UTR of RNA-1 (206 nucleotides) and a 3′ UTR consisting of the 3′UTR of RNA-1 (82 nucleotides) and the *nos* 3′ UTR (155–160 nucleotides; Figure 7a).

The kinetics of GFP expression obtained with pEAQexpress-RT-GFP were compared with those obtained with pEAQexpress-HT-GFP by preparing protein extracts at various times after infiltration and calculating the amount of GFP per Kg fresh weight of leaf tissue. The results showed that expression from HT vectors was always higher than that obtained from RT vectors and persisted for longer; however, this difference was at its lowest at the earliest time point examined (3 dpi; Figure 7b). The different kinetics of expression are suggestive of differences in mRNA accumulation, rather than translatability, a factor shown to be controlled by the 3′UTR in the case of CPMV-HT. To investigate this possibility, the 3′UTR in pEAQexpress-RT-GFP was either deleted (to give pEAQexpress-RT-GFP-del 3′ UTR) or replaced with

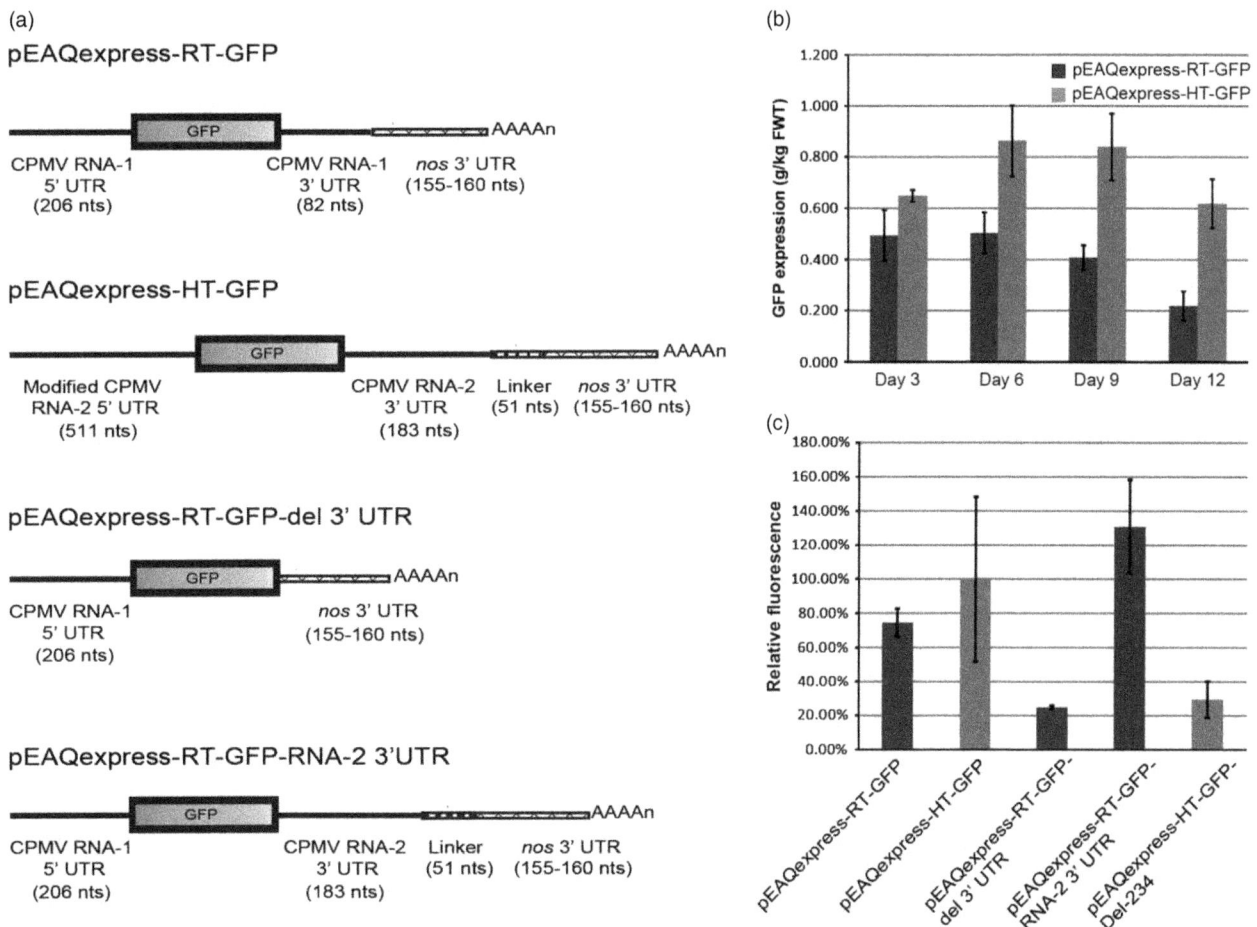

Figure 7 Orthogonal effects of 5′ and 3′ UTRs. (a) Schematic representation of the expression cassettes of the CPMV-RT and CPMV-HT vectors. (b) Expression levels of GFP in the *RT* system (black bars) and the *HT* system (grey bars) based on spectrofluorometric analysis of extracts from leaves harvested on 3, 6, 9 and 12 dpi. (c) Relative GFP expression levels based on spectrofluorometric analysis of extracts from leaves harvested on 6 dpi. Values indicate average expression levels from three biological replicates ± standard error.

the 3′ UTR of RNA-2 (to generate pEAQexpress-RT-GFP-RNA-2 3′UTR), and the GFP expression levels of these constructs were compared with those obtained with pEAQexpress-RT-GFP, pEAQexpress-HT-GFP and pEAQexpress-HT-GFP-Del-234.

As found previously with the HT system, deletion of the entire 3′ UTR from pEAQexpress-RT-GFP resulted in at least a threefold reduction in GFP expression (Figure 7c). On the other hand, insertion of the 3′ UTR of RNA-2 in place of that of RNA-1 resulted in a level of GFP expression higher than that obtained with pEAQexpress-RT-GFP (Figure 7c), a result consistent with the earlier observation that the 3′ UTR of RNA-2 is more effective than that of RNA-1 at enhancing expression. These results show that the levels and duration of foreign gene expression can be modulated by switching 5′ and 3′ UTRs.

Discussion

The results presented here show, for the first time, that the 3′UTR of RNA-2 of CPMV plays an important role in ensuring high-level expression from the CPMV-*HT* system and its associated pEAQ plasmids through stabilizing the transcribed mRNA. The most important region of the RNA-2 3′UTR in terms of this function is the Y-shaped structure formed by nucleotides 125–165 of the 3′ UTR, and any mutations that delete or disrupt this structure are equivalent to deletion of the entire 3′ UTR. Deletions between the GFP termination codon and the Y-shaped structure have less drastic effects, although removal of 107 nucleotides from this region, as seen in case of pEAQexpress-HT-GFP-3′UTR(108-184), reduces expression to 80% of that achieved with the native 3′ UTR.

The CPMV RNA-2 3′UTR can be partially substituted with that of RCMV RNA-2, a sequence which is predicted to fold into an identical Y-shaped structure. The much shorter 3′ UTR of CPMV RNA-1, despite possessing the Y-shaped structure, is considerably less effective, indicating that the Y-shaped structure must be positioned a minimum distance from the termination codon to operate most efficiently, a result supported by the deletion analysis.

Site-directed mutagenesis experiments provided no evidence that direct interactions between the 5′ and 3′ UTRs are required for high-level expression from CPMV-HT cassettes but, rather, suggested that the two UTRs exert their effects independently. In contrast to the 5′UTR where translation can be modulated without affecting mRNA levels (Sainsbury and Lomonossoff, 2008), deletion of the 3′ UTR dramatically reduces mRNA accumulation. This ability to enhance accumulation suggests that the RNA-2 3′ UTR, and the Y-shaped structure in particular, either improves mRNA stability or enhances transcription. Previous experiments have shown that mutants that disrupt the Y-shaped structure adversely affect RNA-2 accumulation during CPMV replication in protoplasts (Rohll *et al.*, 1993); however, whether this was due to reduced replication, reduced RNA stability or a combination of the two was not addressed. The ability of a plant viral 3′ UTR to stabilize an RNA sequence has previously been reported in the case of TMV where the UTR was able to stabilize RNA molecules in protoplasts, although only when present at the very 3′ end (Gallie and Walbot, 1990). However, when positioned between the sequence of GFP and the *nos* 3′UTR in construct pEAQexpress-HT-GFP-TMV 3′UTR, it proved ineffective, suggesting that the 3′UTR of CPMV RNA-2 and that of TMV stabilize RNA molecules through different mechanisms. Our current data do not permit us to distinguish whether the 3′ UTR enhances mRNA accumulation via increased mRNA stability or enhanced transcription.

A question arising from the current study is its relevance to the mechanism of translation of native CPMV RNA molecules. The transcripts generated during *Agrobacterium*-mediated transient expression from the pEAQ vectors will be capped at the 5′ termini and will contain the sequence of the *nos* 3′UTR, including a poly (A) tract, downstream of the CPMV-specific 3′ UTR. By contrast, natural CPMV RNAs have a small protein (VPg) at their 5′ termini, which is not removed prior to translation (de Varennes *et al.*, 1986) and terminate in a poly(A) tract immediately after the 3′ UTR. Thus, for example, the failure to find evidence for interactions between regions of complementarity in the 5′ and 3′ UTRs may simply reflect the fact that an artificial, cap-dependent mechanism for translation is being used. On the other hand, it should be noted that agro-infiltration of full-length cDNA copies of the CPMV RNAs inserted between the CaMV 35S promoter and the *nos* terminator is a highly efficient means of initiating infection (Liu and Lomonossoff, 2002), implying that the transcribed RNA is biologically active.

Given the independent activities of the 5′ and 3′UTRs in promoting expression, it is now possible to create a range of translational cassettes with different expression properties by 'mixing and matching' different UTRs. By varying the efficiency of translation (by mutations within the 5′ UTR) and the level of an mRNA (by altering the 3′ UTR) orthogonally, it is possible to vary both the amount and duration of expression of a heterologous gene. The ability to control the expression levels of different genes expressed within the same cell has immense potential in the field of synthetic biology, for instance in cases where differential enzymes levels are required in a metabolic pathway.

Experimental procedures

Design and construction of vectors

pEAQexpress-HT-GFP (Figure 1a; Sainsbury *et al.*, 2009) was used in experiments to investigate the role of the native 3′UTR of CPMV RNA-2. To create constructs lacking either the 3′UTR of RNA-2 plus the 51-nt linker or just the 3′UTR, the region between the unique *Cla*I and *Stu*I sites in the T-DNA region of pEAQexpress-HT-GFP was amplified by PCR using primers P1 and either P2 or P3 (Table S1). This resulted in the generation of fragments encompassing the sequence from the *Cla*I at the start of the *p19* gene to either the beginning of the *nos* terminator or the beginning of the 51-nt linker with *Stu*I sites immediately upstream (Figure S2). Substitution of these fragments, after digestion with *Stu*I and *Cla*I, into similarly digested pEAQexpress-HT-GFP gave constructs pEAQexpress-HT-GFP-Del-234 and pEAQexpress-HT-GFP-Del-183, respectively.

To re-insert the 234-nucleotide sequence encompassing the CPMV RNA-2 3′UTR plus the 51-nucleotide linker sequence into pEAQexpress-HT-GFP-Del-234, a PCR fragment containing this region was amplified from pEAQexpress-HT-GFP using primers P3 and P4, which introduced *Stu*I sites at each end of the amplified fragment. The amplified fragment was digested with *Stu*I and cloned in both orientations into the *Stu*I site of pEAQexpress-HT-GFP-Del-234 to give pEAQexpress-HT-GFP-Del-234-corr 234 and pEAQexpress-HT-GFP-Del-234-opp 234. To re-insert just the 183 nucleotides corresponding to the CPMV RNA-2 3′UTR, this region was amplified using primers P6 and P4 and cloned to generate pEAQexpress-HT-GFP-Del-234-corr 183 (Figure 2a).

To create deletions within the CPMV RNA-2 3'UTR, the region between the *Cla*I and *Stu*I sites of pEAQexpress-HT-GFP was amplified using primer combinations P1 and P7 to obtain construct pEAQexpress-HT-GFP-3'UTR(66-183), P1 and P8 to obtain pEAQexpress-HT-GFP-3'UTR(108-183), and P1 and P9 to obtain pEAQexpress-HT-GFP-3'UTR(142-183). These fragments were ligated into *Cla*I-*Stu*I-digested pEAQexpress-HT-GFP. To create pEAQexpress-HT-GFP-3'UTR(1-141), a fragment of the vector pEAQexpress-HT-GFP was amplified by PCR using primers P11 and P10, digested with *Stu*I and ligated into *Stu*I-digested pEAQexpress-HT-GFP-Del-234.

The cDNA sequences of 3'UTRs of RCMV RNA-2 (accession number NC_003738.1), CPMV RNA-1 (accession number NC_003549) and TMV (accession number AB369276.1) were obtained from GenBank®. DNA was synthesized by GENEART® (Life Technologies, Grand Island, NY) and cloned into *Stu*I-digested construct pEAQexpress-HT-GFP-Del-234.

A cDNA comprising the 5' UTR of RNA-1 (206 bp), a polylinker (52 bp) and the 3' UTR of RNA-1 (82 bp) (Figure S2) was cloned into pEAQexpress (Sainsbury et al., 2009) to create construct pEAQexpress-*RT*. The gene encoding GFP was cloned into the *Xho*I and *Xma*I sites of the polylinker to generate pEAQexpress-*RT*-GFP, in which *gfp* is flanked by the 5' UTR and the 3' UTR of CPMV RNA-1 and is under the control of a 35S promoter and a *nos* terminator.

pEAQexpress-RT-GFP-del3'UTR was created by amplifying the 1.5-kb region immediately downstream of the 3'UTR of pEAQexpress-RT-GFP using primers P26 and P27 and inserting the amplified fragment upstream of the 3'UTR using sites *Xma*I and *Bam*HI, thereby deleting the entire 3'UTR. Similarly, pEAQexpress-RT-GFP-RNA-2 3'UTR was created by amplifying a 1.8-kb region from pEAQ-HT (Sainsbury et al., 2009) consisting of the 3'UTR of RNA-2 and DNA downstream of it (using P26 and P28) and inserting the amplified product upstream of the 3'UTR of pEAQexpress-RT-GFP, thereby replacing the 3'UTR of RNA-1 with that of RNA-2.

Site-directed mutagenesis

Point mutations were introduced into 5' and 3' UTR regions of pEAQexpress-HT-GFP using the QUIKCHANGE® (Agilent Technologies Inc. Santa Clara, CA) method according to the instructions provided by the manufacturer (Agilent technologies). Site-directed primers were designed using the Web program provided by the manufacturer.

RNA secondary structure prediction

RNA secondary structures mentioned in this study were predicted using Mfold Web Server, available from http://mfold.rna.albany.edu/?q=mfold/RNA-Folding-Form (Zuker, 2003).

Agro-infiltration

Binary plasmids were maintained in *Agrobacterium tumefaciens* strain LBA4404, which was transformed by electroporation and grown on LB containing 50 µg/mL rifampicin and 50 µg/mL kanamycin. Stationary-phase liquid cultures were pelleted by centrifugation at 2000 *g* for 20 min and resuspended in MMA {10 mM MES [2-(N-morpholino)ethanesulfonic acid] at pH 5.6, 10 mM MgCl$_2$, 100 µM acetosyringone (4-hydroxy-3,5-dimethoxyacetophenone)} to make a solution of final optical density at 600 nm (OD$_{600}$) = 0.4. The suspensions were left at room temperature for 0.5–3 h prior to infiltration to allow acetosyringone in the buffer to induce virulence of agrobacteria. The agro-

suspension was pressure-infiltrated into young fully expanded leaves of 3- to 4-week-old *N. benthamiana* plants with the help of a needle-less syringe. Infiltrated leaves were harvested at various time points, up to 12 days postinfiltration.

Preparation of total protein extract

For each construct used, 4–6 leaf discs were generated from infiltrated leaves using a sterile cork borer. The leaf discs were homogenized in three volumes of extraction buffer [50 mM Tris–HCl pH 7.25, 150 mM NaCl, 2 mM ethylenediaminetetraacetic acid (EDTA)] using ceramic beads in THERMO® Savant FastPrep FP120 Homogenizer (Thermo Fischer Scientific, Loughborough, UK). The homogenate was clarified by centrifugation for 15 min at 16 000 *g* using a cold bench-top centrifuge. The supernatant represented the total soluble protein extract.

GFP fluorescence assay

GFP fluorescence measurements were made using a protocol modified from Richards et al. (2003). Soluble protein extracts were diluted in the ratio 1:100 in 0.1 M Na$_2$CO$_3$ and loaded in triplicate onto a fluorescently neutral black COSTAR® 96-well plate (Sigma-Aldrich, Suffolk, UK). Recombinant GFP (Clontech, Saint-Germain-en-Laye, France), which is the same variant of GFP as expressed by the pEAQ vectors, was diluted in plant extract to generate standard curves. Excitation (wavelength of 395 nm) and emission (509 nm) maxima were matched to Clontech's GFP and read using a SPECTRAmax spectrofluorometer (Molecular Devices, Sunnyvale, CA). Following subtraction of the signal from a control extract of non-infiltrated leaves, values representing GFP fluorescence were entered into the linear regression for the standard curve to calculate the amount of GFP present in each sample. To calculate relative expression levels for all constructs, expression achieved from construct pEAQexpress-HT-GFP was set to 100%, and expression from other constructs was calculated as a percentage of the expression from pEAQexpression-HT-GFP. Measurements were made in triplicate to account for experimental variation and averaged to give a final value for each construct. In the case of time-course experiments, the average from three biological replicates is reported. Error bars denote standard error of the mean of the three (technical or biological) replicates with a 95% confidence interval. This was calculated using the formula: $1.96 \times (\text{standard deviation})/\sqrt{\text{sample size}}$ (where sample size = 3).

Real-time quantitative RT-PCR

Total RNA from agro-infiltrated *N. benthamiana* leaves was isolated using RNeasy® Plant Mini Kit (QIAGEN®, Manchester, UK). After treatment with DNaseI, 2 µg of RNA was used for reverse transcription with oligo(dT) primers using ProtoScript® m-MuLV First Strand cDNA Synthesis Kit (New England BioLabs, Hertfordshire, UK). Quantitative RT-PCR was performed using SYBR®Green JumpStart™ Taq ReadyMix™ (Sigma-Aldrich) with primers P22 and P23 (to detect GFP transcripts) and primers P24 and P25 (to detect β-actin transcripts) in the Chromo4 Q-PCR Opticon cycler (Bio-Rad, Hercules, CA). Relative RNA amounts were determined by the ΔΔCt method using β-actin mRNA for normalization (Livak and Schmittgen, 2001).

Visualization and Photography

GFP expression in infiltrated *N. benthamiana* plants was monitored with a 100 AP handheld UV lamp (Blak Ray®, Upland, CA). A Nikon® D700 (Nikon, Tokyo, Japan) digital camera with a

60-mm macro lens was used for image acquisition under visible light or, for the detection of GFP, under UV illumination. Images were edited using Photoshop CS4 and Illustrator (Adobe, San Jose, CA).

Acknowledgements

The authors would like to thank Andrew Davis for photography and Dr. Thomas Girin for help with quantitative PCR experiments. The work described in this paper was supported by grants BB/G024197/1 and BB/J004561/1 from the Biotechnology and Biological Science Research Council (BBSRC) U.K. and the John Innes Foundation.

References

van Belkum, A., Abrahams, J.P., Pleij, C.W.A. and Bosch, L. (1985) Five pseudoknots are present at the 204 nucleotide long 3′ noncoding region of tobacco mosaic virus RNA. *Nucleic Acids Res.* **13**, 7673–7686.

Bevan, M., Barnes, W.M. and Chilton, M.D. (1983) Structure and transcription of the nopaline synthase gene region of T-DNA. *Nucleic Acids Res.* **11**, 369–385.

Cañizares, M.C., Liu, L., Perrin, Y., Tsakiris, E. and Lomonossoff, G.P. (2006) A bipartite system for the constitutive and inducible expression of high levels of foreign proteins in plants. *Plant Biotechnol. J.* **4**, 183–193.

Eggen, R., Verver, J., Wellink, J., Pleij, K., van Kammen, A. and Goldbach, R. (1989) Analysis of sequences involved in cowpea mosaic virus RNA replication using site-specific mutants. *Virology* **173**, 456–464.

van Engelen, F.A., Molthoff, J.W., Conner, A.J., Nap, J.P., Pereira, A. and Stiekema, W.J. (1995) pBINPLUS – an improved plant transformation vector based on pBIN19. *Transgenic Res.* **4**, 288–290.

Fabian, M.R. and White, K.A. (2004) 5′-3′ RNA-RNA interaction facilitates cap- and poly (A) tail-independent translation of Tomato bushy stunt virus mRNA. *J. Biol. Chem.* **279**, 28862–28872.

Gallie, D.R. and Walbot, V. (1990) RNA pseudoknot domain of tobacco mosaic virus can functionally substitute for a poly(A) tail in plant and animal cells. *Genes Dev.* **4**, 1149–1157.

Guo, L., Allen, E.M. and Miller, W.A. (2001) Base-pairing between untranslated regions facilitates translation of uncapped, nonpolyadenylated viral RNA. *Mol. Cell* **7**, 1103–1109.

Liu, L. and Lomonossoff, G.P. (2002) Agroinfection as a rapid method for propagating Cowpea mosaic virus-based constructs. *J. Virol. Methods*, **105**, 343–348.

Livak, K.J. and Schmittgen, T.D. (2001) Analysis of relative gene expression data using real- time quantitative PCR and the $2^{-\Delta\Delta CT}$ Method. *Methods*, **25**, 402–408.

Miller, W.A., Wang, Z. and Trader, K. (2007) The amazing diversity of cap – independent translation elements in the 3′-untranslated regions of plant viral RNAs. *Biochem. Soc. Trans.* **35**, 1629–1633.

Peyret, H. and Lomonossoff, G.P. (2013) The pEAQ vector series: the easy and quick way to produce recombinant proteins in plants. *Plant Mol. Biol.* **83**, 51–58.

Richards, H.A., Halfhill, M.D., Millwood, R.J. and Stewart, C.N., Jr. (2003) Quantitative GFP fluorescence as an indicator of recombinant protein synthesis in transgenic plants. *Plant Cell Rep.* **22**, 117–121.

Rohll, J.B., Holness, C.L., Lomonossoff, G.P. and Maule, A.J. (1993) 3′-terminal nucleotide sequences important for the accumulation of cowpea mosaic virus M-RNA. *Virology*, **193**, 672–679.

Sainsbury, F. and Lomonossoff, G.P. (2008) Extremely high-level and rapid transient protein production in plants without the use of viral replication. *Plant Physiol.* **148**, 1212–1218.

Sainsbury, F., Thuenemann, E.C. and Lomonossoff, G.P. (2009) pEAQ: versatile expression vectors for easy and quick transient expression of heterologous proteins in plants. *Plant Biotechnol. J.* **7**, 682–693.

Sainsbury, F., Sack, M., Stadlmann, J., Quendler, H., Fisher, R. and Lomonossoff, G.P. (2010) Rapid transient production in plants by replicating and non-replicating vectors yields high quality functional anti-HIV antibody. *PLoS ONE* **5**, e13976.

Thuenemann, E.C., Meyers, A.E., Verwey, J., Rybicki, E.P. and Lomonossoff, G.P. (2013) A method for rapid production of heteromultimeric protein complexes in plants: assembly of protective bluetongue virus-like particles. *Plant Biotechnol. J.* **11**, 839–846.

de Varennes, A., Lomonossoff, G.P., Shanks, M. and Maule, A.J. (1986) The stability of cowpea mosaic virus VPg in reticulocyte lysates. *J. Gen. Virol.* **67**, 2347–2354.

Zuker, M. (2003) Mfold web server for nucleic acid folding and hybridization prediction. *Nucleic Acids Res.* **31**, 3406–3415.

6

Potential of a tomato MAGIC population to decipher the genetic control of quantitative traits and detect causal variants in the resequencing era

Laura Pascual[1,†], Nelly Desplat[1,‡], Bevan E. Huang[2], Aurore Desgroux[1,§], Laure Bruguier[3], Jean-Paul Bouchet[1], Quang H. Le[4], Betty Chauchard[3], Philippe Verschave[3] and Mathilde Causse[1,*]

[1]INRA, UR1052, Génétique et Amélioration des Fruits et Légumes, Montfavet, France

[2]Computational Informatics and Food Futures Flagship, CSIRO, Dutton Park, Qld, Australia

[3]Vilmorin, Centre de La Costière, Ledenon, France

[4]Vilmorin & Cie, Route d'Ennezat, Chappes, France

*Correspondence
email mathilde.
causse@avignon.inra.fr

[†]Present address: Centre for Research in Agricultural Genomics (CRAG), CSIC-IRTA-UAB-UA, Universidad de Barcelona, Barcelona 08193, Spain.

[‡]Present address: BIOGEMMA, Centre de Recherche de Chappes, CS 90126, Chappes, 63720, France

[§]Present address: IGEPP, Domaine de la Motte, BP 35327, Le Rheu Cedex 35653, France.

Keywords: multiparental population, *Solanum lycopersicum*, resequencing, QTL, SNP.

Summary

Identification of the polymorphisms controlling quantitative traits remains a challenge for plant geneticists. Multiparent advanced generation intercross (MAGIC) populations offer an alternative to traditional linkage or association mapping populations by increasing the precision of quantitative trait loci (QTL) mapping. Here, we present the first tomato MAGIC population and highlight its potential for the valorization of intraspecific variation, QTL mapping and causal polymorphism identification. The population was developed by crossing eight founder lines, selected to include a wide range of genetic diversity, whose genomes have been previously resequenced. We selected 1536 SNPs among the 4 million available to enhance haplotype prediction and recombination detection in the population. The linkage map obtained showed an 87% increase in recombination frequencies compared to biparental populations. The prediction of the haplotype origin was possible for 89% of the MAGIC line genomes, allowing QTL detection at the haplotype level. We grew the population in two greenhouse trials and detected QTLs for fruit weight. We mapped three stable QTLs and six specific of a location. Finally, we showed the potential of the MAGIC population when coupled with whole genome sequencing of founder lines to detect candidate SNPs underlying the QTLs. For a previously cloned QTL on chromosome 3, we used the predicted allelic effect of each founder and their genome sequences to select putative causal polymorphisms in the supporting interval. The number of candidate polymorphisms was reduced from 12 284 (in 800 genes) to 96 (in 54 genes), including the actual causal polymorphism. This population represents a new permanent resource for the tomato genetics community.

Introduction

Identifying the genes responsible for the variation of adaptation and agronomic traits is a main goal for plant geneticists. The frequent polygenic control of these traits complicates the identification of the causal molecular variants (Morell *et al.*, 2012). Two main approaches, family-based quantitative trait loci (QTLs) mapping and genome-wide association studies (GWAS), have been employed to elucidate their genetic architecture (Mitchell-Olds, 2010).

Traditionally, family-based QTL mapping relies on populations derived from biparental crosses. These populations, like F2 and backcrosses, can be directly analysed (Clarke *et al.*, 1995; Grandillo and Tanksley, 1996), or studied after reaching homozygosity (Keurentjes *et al.*, 2007). This approach has led to the identification and positional cloning of several major genes underlying QTLs (Price, 2006). However, such populations allow only the analysis of alleles differing between two lines, and the resolution is limited to 10–30 cM, as the analysis mainly relies on recombination events taking place during the F1 meiosis (Hall *et al.*, 2010). This is true even in recombinant inbred lines (RILs),

as the number of efficient recombination decreases in advanced generations. GWAS overcome the limitations of biparental crosses. Based on collections of unrelated individuals, GWAS screen a wide range of diversity and take advantage of the historical recombination events that have accumulated over thousands of generations (Korte and Farlow, 2013). This approach has been useful to identify genetic associations with complex agronomic traits (Huang *et al.*, 2012a; Sauvage *et al.*, 2014). However, GWAS are limited by linkage disequilibrium (LD), which may vary greatly from one region to the other in the genome, and by population substructure, which can result in false positive or false negative results (Mitchell-Olds, 2010; Visscher *et al.*, 2012). Moreover, some interesting phenotypes might be caused by rare alleles, which are difficult to identify by GWAS (Kover and Mott, 2012). More complex experimental populations offer alternatives to these designs and address their limitations. To increase the number of recombinations, Darvasi and Soller (1995) proposed the development of advanced intercross lines (AILs). After the development of an F2 progeny, successive generations of random mating allow the accumulation of recombination break points. To increase the genetic variation analysed, nested association mapping (NAM) populations were developed in maize from a diverse set of parental lines crossed with a reference line (Yu *et al.*, 2008). However, the effect of genetic background and epistasis may affect QTL detection and are not taken into account in these populations (Rakshit *et al.*, 2012). To overcome these limitations, the AIL methodology was extended to multiple parent populations. This approach was first used to develop the mice heterogeneous stock (Yalchin *et al.*, 2005). Since then, it has been described by several acronyms, and many breeding designs (Rockman and Kruglyak, 2008; Valdar *et al.*, 2006). To avoid confusion, we will refer to them as multiparent advanced generation intercross (MAGIC) populations (Cavanagh *et al.*, 2008). MAGIC populations have been set up in the model species *Arabidopsis* (Kover *et al.*, 2009) and several cereal crops (Bandillo *et al.*, 2013; Huang *et al.*, 2012b) demonstrating the power of such resource to detect QTLs underlying quantitative traits.

Tomato (*Solanum lycopersicum*) is one of the most important vegetables consumed worldwide, but also the model species for studying fleshy fruit development (Giovannoni, 2004). During its domestication, the diversification of fruit aspect, as well as the adaptation to a wide range of environmental conditions, was simultaneous to a strong reduction of molecular diversity (Blanca *et al.*, 2012; Miller and Tanksley, 1990). This lack of genetic variation in the cultivated species limited the exploitation of intraspecific variation and thus led geneticists to study trait variation mostly in progenies of distant crosses involving wild species (Zamir, 2001). Recent association studies including cherry tomato accessions (*Solanum lycopersicum* var. *cerasiforme*) that have an intermediate position between cultivated tomato and its closest wild relative species (Ranc *et al.*, 2008) have shown the potential of this material to detect QTLs by GWAS (Ranc *et al.*, 2012; Xu *et al.*, 2013). The potential of MAGIC populations to include a wide range of variation and the recent publication of the tomato genome (Tomato Genome Consortium, 2012) open new avenues for the exploitation of this variation.

Here, we present the first tomato MAGIC population and describe its potential for (i) intraspecific variation exploitation, (ii) QTL mapping and (iii) causal polymorphism identification. We

have constructed the MAGIC population crossing eight tomato lines, selected to include a wide range of the genetic diversity of *S. lycopersicum* species. These eight founder lines have been deeply characterized following a systems biology approach (Pascual *et al.*, 2013) and their whole genomes resequenced allowing the identification of more than 4 million SNPs (Causse *et al.*, 2013). We used this information to develop a subset of markers especially designed to analyse the MAGIC population. The selected markers were employed to develop the first intraspecific saturated map in tomato. We phenotyped the population, discovered a wide range of variation through new allelic combinations and mapped QTLs. Finally, a strategy to fine map QTLs and identify the causal polymorphisms is proposed. We demonstrate the power of the MAGIC population when coupled with available genome sequence to restrict the number of putative causal polymorphisms underlying the QTLs.

Results

A tomato MAGIC population composed of 397 MAGIC lines was constructed as described in Figure 1, following four generations of crosses and three of selfing.

The custom-made genotyping platform is highly efficient to predict founder haplotypes

To genotype the MAGIC population, we designed a specific SNP platform to enhance the haplotype prediction and recombination detection in the MAGIC population. From more than four million SNPs detected in the founder lines when compared to the tomato reference genome (Causse *et al.*, 2013), we selected a subset of 1536 markers using a filtering pipeline in three steps (Table 1). First, based on general quality score criteria, we retained 408 795 SNPs. Second, we removed SNPs providing successive similar profiles over the eight founders, as these SNPs would provide redundant information, and reduced the number of SNPs to 149 808. Finally, we selected 1536 SNPs taking into account their physical and genetic position and the profile of the adjacent SNPs to enhance the founder imputation power. A total of 1486 SNPs were finally used for genotyping the MAGIC population (Table

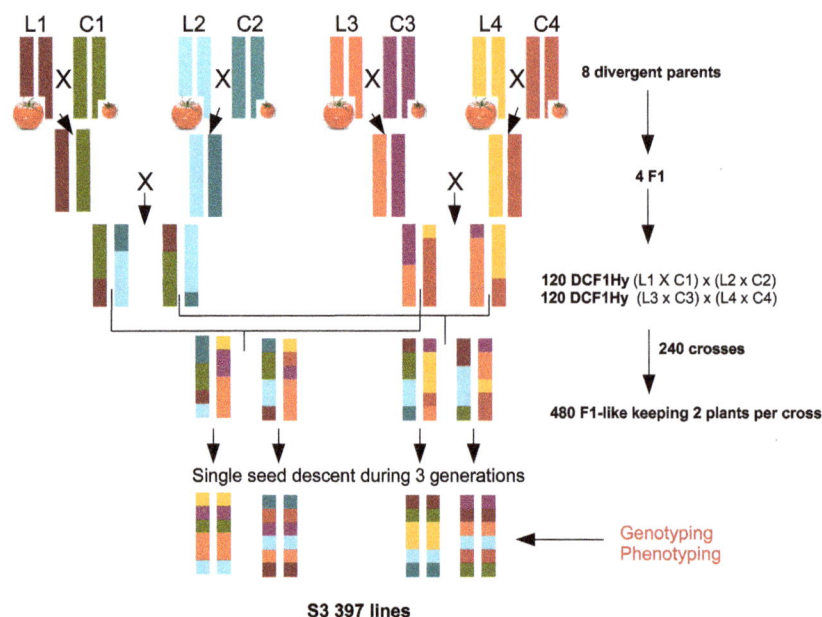

Figure 1 Construction of a tomato 8-way MAGIC population. Large fruited founders noted as L1 Levovil, L2 Stupicke PR, L3 LA0147, L4 Ferum. Small fruited founders noted as C1 Cervil, C2 Criollo, C3 Plovdiv24A, C4 LA1420. DCF1Hy: double cross F1 hybrid.

S1). The selection process allowed improving the haplotype prediction from 67% if the markers were randomly selected to 93.4% with the set of 1536 markers selected, as shown in Figure S1.

A genetic map twice as long as biparental maps

The MAGIC tomato population was then used to construct a genetic map. The final map included 1345 markers (Table S2), representing 524 unique map positions (genetic bins) with average intervals of 1.68 cM and 0.6 Mb (Table 2). The total map measured 2156 cM and covered 758 Mb (84% of the 900 Mb tomato genome size), and almost all the 760 Mb assembled genome (Tomato Genome Consortium, 2012). The 12 chromosomes were covered by 28 to 64 genetic bins. We did not find any clear correlation between genetic and physical map length, as, for example, chromosome 3 (215.2 cM), the longest, and chromosome 10 (120.96 cM), the smallest, covered 64.77 Mb and 64.8 Mb, respectively. When we compared physical and genetic positions, high recombination rates were found on the distal regions, while recombination was almost suppressed in large centromeric regions that comprised around 70% of the chromosomes (Figure 2, Figure S2).

We compared the MAGIC genetic map and the biparental tomato high-density genetic maps constructed by Sim et al. (2012a) (Table 2, Figure S2). The EXPEN 2012 map includes 3687 markers and was based on 160 F2 from a cross between Moneymaker (S. lycopersicum) and LA0716 (S. pennellii). The EXPIM 2012 map includes 4792 markers mapped with 183 F2 individuals from a cross between Moneymaker and LA0121 (S. pimpinellifolium). The MAGIC tomato map was 87% and 105% longer than the EXPEN 2012 and EXPIM 2012 map, respectively. This increase was not the same for all chromosomes, ranging from 43% to 155% with respect to EXPEN 2012 and 60% to 185% with respect to EXPIM 2012 (Table S3). Figure 2 illustrates the relationship between physical and genetic positions for the first three chromosomes. The recombination increase is limited to distal parts of the chromosomes as recombination in the centromeric regions is almost suppressed in all three maps (Figure S2). Genetic recombination also increased when compared to intraspecific tomato biparental populations. The MAGIC tomato map was 69% larger than the map constructed from a RIL population developed from a cross between Cervil and Levovil, the two most distant founders of the MAGIC population (data not shown).

No clear structure remained in the MAGIC population

The structure and LD in the MAGIC population will determine the power to detect genetic associations. The population structure was assessed using the 1345 SNP markers included in the genetic map. According to the Evanno et al. (2005) test, the most probable number of groups was one, indicating the absence of subgroups in the MAGIC population (Figure S3). LD was analysed between pairs of markers within each chromosome, and pairwise r^2 was plotted against genetic and physical distances between loci (Figure S4). With respect to genetic distances, LD within chromosomes fell to < 0.7 within 5 cM, and < 0.3 within 25 cM, intersecting with LD baseline value only at 90 cM. With respect to physical distance, LD decayed quickly from an average of 0.47 at 1 kb to < 0.2 at 2 Mb, reaching a minimum of 0.08 at 20 Mb. However, for more distant markers (40 Mb), LD increased again (higher than 0.13) to fall again to previous values at distances around 50 Mb. The kinship tended to be small with a third quartile of the values lower than 0.042, even though this value reached 0.8 for some pairs of lines (0.01%) (Table S4).

Finally, the haplotype structure of each line was analysed by identifying their founder allele at each marker (Figure 3, Figure S5). We predicted, on average, the marker origin for 89% of the genome (Figure 3a). This value was higher for all the chromosomes but 5 and 11, where three parental lines (Stupicke PR, LA0147 and Levovil) were difficult to distinguish. Along the chromosomes, haplotypes were predicted with lower accuracy at the end of the centromeric regions, probably due to the augmentation in recombination rates (Figure S5). Founder contribution was close to the expected value (0.125) along all the chromosomes, showing an increase in the contribution of LA1420 and Criollo coupled with a decrease of Cervil at the end of chromosome 4 (Figure 3b).

We then estimated the haplotype (segment inherited from a single founder) size and the number of haplotypes per chromosome for each line. With respect to genetic distances, haplotypes tended to be small (Figure 4a), with 25% of them being smaller than 11 cM and a median of 30 cM when all the chromosomes are considered together. The physical size distribution showed a bimodal distribution (Figure 4b), where most of the haplotypes were smaller than 5 Mb. An average size of 16 Mb was due to the large centromeric regions. The median number of haplotypes per chromosome was 4 or 3 depending of the chromosome. When all the chromosomes were considered together, the average number of recombination break points was 29.3, with all the tomato lines carrying variable parts of the eight founder lines.

Fruit weight QTL detection in the MAGIC population

Fruit weight (FW) is a key trait selected since domestication which has been widely studied in tomato. Several FW QTLs and associations have been already described (Grandillo et al., 1999; Xu et al., 2013) and two of them positionally cloned

Table 1 Summary of the SNP selection procedure for the construction of the genetic map

Chromosome	Total (1)	Quality filtering (2)	Successive profile filtering (3)	Final (4)
Ch1	140 192	22 790	9724	172
Ch2	274 273	37 292	9795	149
Ch3	357 900	47 148	5388	181
Ch4	505 272	45 926	24 068	166
Ch5	616 803	44 364	21 093	124
Ch6	109 945	14 480	4533	104
Ch7	385 516	48 566	16 409	112
Ch8	540 631	44 049	15 141	87
Ch9	411 088	18 348	7072	122
Ch10	63 524	10 232	3888	91
Ch11	499 546	29 416	13 984	106
Ch12	293 639	46 184	18 713	122
Total	4 198 329	40 8795	149 808	1536

(1) Total number of SNPs detected in the eight founders compared to the reference genome, (2) number of SNPs after filtering based on quality criteria, (3) number of SNPs after filtering against successive identical profiles and (4) final selection based on physical and genetic distances.

Table 2 Characteristics of the genetic map. Number of SNP markers, coverage in cM and Mb for each chromosome. Comparison of map length with tomato biparental maps

Chr.	Number of markers	Unique bins	Coverage (cM)	Marker interval (cM)		Coverage (Mb)	Marker interval (Mb)		Expansion in MAGIC (cM)	
				Max.	Average		Max	Average	EXPEN 2012	EXPIM 2012
1	156	58	214.4	13.20	1.38	90.16	7.13	0.58	83%	68%
2	131	47	200.2	14.53	1.54	49.57	4.22	0.38	82%	150%
3	161	64	215.2	25.58	1.34	64.77	9.52	0.40	104%	99%
4	147	64	200.1	13.31	1.37	63.37	6.08	0.43	85%	115%
5	111	40	168.2	22.93	1.53	64.99	6.36	0.59	76%	89%
6	85	36	140.9	20.94	1.68	45.96	7.28	0.55	61%	111%
7	99	35	191.0	21.02	1.95	65.11	7.68	0.66	155%	130%
8	79	33	129.4	17.68	1.66	62.97	7.69	0.81	68%	67%
9	109	40	206.1	16.69	1.91	67.64	7.57	0.63	113%	185%
10	73	28	120.9	13.83	1.68	64.80	10.21	0.90	43%	60%
11	90	40	183.3	13.50	2.06	53.18	10.23	0.60	86%	99%
12	104	39	186.2	19.55	2.00	65.47	13.80	0.64	88%	121%
Total	1345	524	2156	–	–	758.00	–	–	–	–

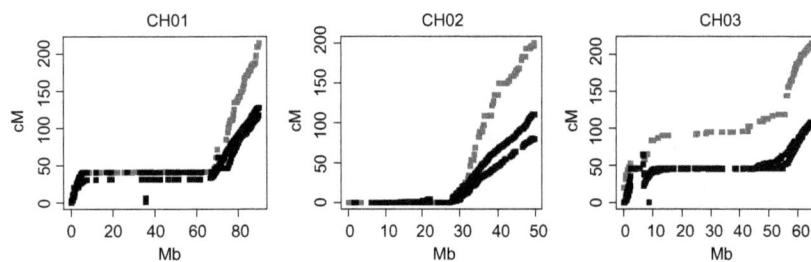

Figure 2 Relationship between the genetic and physical positions for the first three chromosomes. Positions are indicated in green for the MAGIC map, blue for the EXPIM 2012 and red for the EXPEN 2012 (adapted from Sim et al., 2012a,b).

(Chakrabarti et al., 2013; Frary et al., 2000). We thus chose FW as an example to analyse the power and precision of the MAGIC population for QTL mapping. The phenotypic characterization was performed at two locations in the south of France. In each location, the complete set of 397 RIL MAGIC lines (one plant per line) and five replicates of each founder were characterized (Table S5). FW distributions (Figure 5) illustrated the large range of phenotypic variation in the population, including transgressive lines, as well as a difference in average FW among locations. We thus analysed the data separately and then compared the QTLs obtained in each location. To detect QTLs and genetic associations, two approaches were tested, interval mapping adapted to MAGIC populations and GWAS.

To map QTLs by simple interval mapping (IM), we performed a joint Wald test for the significance of all founder effects at putative QTL positions along the genome. At location A, nine QTLs on chromosomes 2, 3 5, 7, 8 and 11 were detected (Table 3, Figure 6b). Support intervals (SI) ranged from 6 to 78 cM. The 78 cM interval actually corresponded to the second QTL peak detected on chromosome 11, including the first QTL peak inside the SI. At physical scale, SI ranged from 0.94 Mb to 5.61 Mb, except for the QTL located on chromosome 5 (57 Mb) that covered the centromeric region. Finally, after fitting all the QTLs in a unique model, we determined that the nine QTLs together explained 51% of the trait variation. At location B, three QTLs were detected on chromosomes 2, 3 and 11 (Table 3, Figure 6d). All these QTLs colocalized with those of location A QTLs (Table 3). For the QTLs on chromosomes 2 and 3, SI were 3

and 4 cM larger than for location A, but this difference is buffered when translated to physical distance. For the QTL on chromosome 11, SI was 14 cM (0.48 Mb) larger. The three QTLs explained 34% of the trait variation when fitted all together in a unique model.

According to the population structure analysis, the MAGIC population did not present any clear structure in subgroups; thus, we performed GWAS taking into account the kinship (Table S4) in a mixed linear model (MLM). To reduce the false positive associations, p-values were corrected to account for multiple testing, and only associations with corrected p-value lower than 0.05 were considered significant. At location A, 35 significant associations were detected (Figure 6a, Table S6), located on chromosomes 1, 2, 3, 5, 11 and 12 (Figure S6). When we compared QTL SI and the associations detected by MLM, no significant markers were detected for the QTL SI on chromosomes 7 and 8. By contrast, associations were detected by MLM on chromosomes 1 and 12 where no QTLs were called by IM (Table S4, Figure 6a,b).

At location B, we identified 30 associations on chromosomes 2, 3, 11 and 12 (Table S6, Figure 6c, Figure S6), 60% also identified at location A. When we compared the IM results with GWAS analysis, we detected associated markers along all the QTLs SI and associations were also found on chromosome 12 (Table S6, Figure 6c,d).

QTL effects and causal polymorphisms detection

The effect of each founder allele was calculated by IM for each QTL with respect to a reference founder, LA0147 (S. lycopersi-

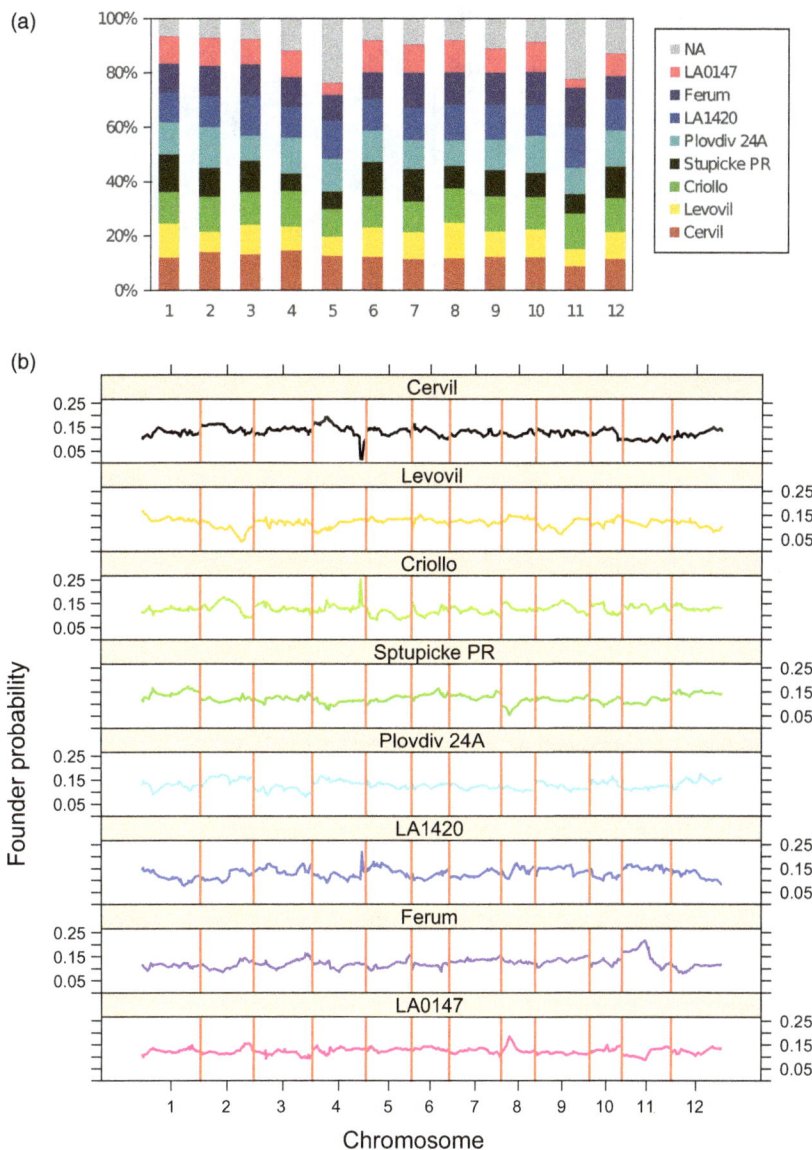

Figure 3 (a) Proportion of founder allele predicted in each chromosome. (b) Percentage of genome-wide founder assignment along the chromosomes.

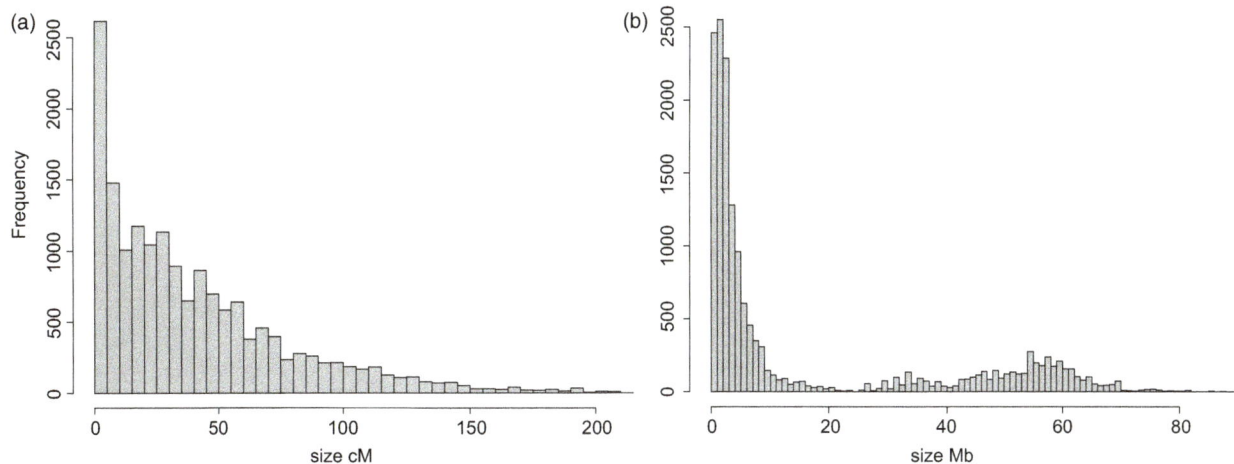

Figure 4 Size of haplotype blocks for all the MAGIC lines and chromosomes, relative to (a) genetic distances (cM), (b) physical distances (Mb).

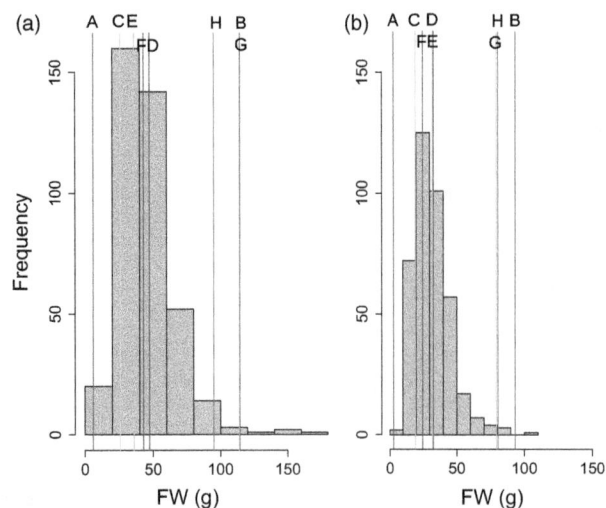

Figure 5 Distribution of fruit weight (gr) in the MAGIC lines grown in (a). Avignon. (b). La Costière. Founder trait values are indicated with vertical lines (A Cervil, B Levovil, C Criollo, D Stupicke PR, E Plovdiv24A, F LA1420, G Ferum, H LA0147).

cum). This line was chosen as reference because it was the closest to Heinz1706, the line used to sequence the tomato genome and then to detect polymorphism in the founder sequences. Effects were variable among QTLs (Figure 7). Cervil (the founder with the smallest fruits) alleles decreased or did not change fruit weight for all the QTLs. The other founder alleles decreased or increased fruit size depending on the QTL considered.

We hypothesized that coupling the QTL founder allele effects with the polymorphisms detected along the QTL SI should facilitate the identification of putative causal variants underlying the QTLs. We tested this hypothesis with two FW QTLs already cloned that colocalized with QTLs detected at both locations. For one of them, *fw3.2*, the polymorphism responsible for the phenotype has been recently identified in a cytochrome P450 gene, (Solyc03 g114940 at position 58 852 276; Chakrabarti

et al., 2013). The support interval for the corresponding QTL (detected at both locations) on chromosome 3 ranged between markers X03_58386463 and X03_60392846 and could correspond to the same gene. According to genomic annotations this interval comprised 800 genes (Tomato Genome Consortium, 2012). Between the two flanking markers, 12 284 SNPs and INDELs have been identified in the eight founders. We filtered out the polymorphisms according to expected allelic effects (Figure 7). Based on QTL effects, we supposed that Cervil, Criollo and LA1420 should have the same allele at the QTL, while Plovdiv24A, Stupicke PR and LA0147 should have the opposite allele. This procedure allowed us to reduce the number of polymorphisms to 96 SNPs and 3 INDELs covering 54 genes (Table S7). The list included the SNP corresponding to the one identified by Chakrabarti *et al.* (2013) as responsible for the *fw3.2* QTL.

For the other cloned QTL, *fw2.2* (Solyc02g090730, position 46 832 171), the causal polymorphism has not yet been identified (Frary *et al.*, 2000). The confidence interval for the QTL that colocalized with the gene ranged between markers X02_46353818 and X02_47498009. In this interval, 6510 polymorphisms have been identified. According to the founder effects, the alleles from the two lines, Cervil and Stupicke PR, clearly reduced fruit weight with respect to the reference. Thus, we first screened for polymorphisms common to these lines and different from the rest of the founders. We ended up with 18 SNPs and 1 INDEL. However, none of them colocalized or was near to the *fw2.2* gene (Table S8). When we analysed all the polymorphisms located inside or close to the *fw2.2* gene (Table S9), we identified a total of 43 SNPs and 3 INDELs, all of them being polymorphic only in one founder line (Cervil). This suggested the presence of two linked QTLs in the region, *fw2.2* being specific to Cervil.

Discussion

We have constructed and analysed a MAGIC population in tomato, one of the most important vegetables consumed worldwide and the model plant for fleshy fruit. Quantitative trait

Table 3 Characteristics of QTLs detected for fruit weight by interval mapping in the MAGIC population. (A) Phenotyping data from location A (Avignon). (B) Phenotyping data form location B (La Costiere)

Chr.	Pos.	LeftMrk	RightMrk	SI (cM)	SI (Mb)	*P*-value
A						
2	186	X02_47433596	X02_47498009	179–192	46.35–47.49	0
2	152	X02_42399961	X02_42773566	150–156	42.39–43.47	2.47×10^{-11}
3	178	X03_58754293	X03_58846611	170–190	57.98–60.24	8.62×10^{-7}
3	202	X03_62140362	X03_62287203	168–207	57.98–63.29	1.41×10^{-5}
5	50	X05_05638011	X05_05886227	16–72	2.8–60	1.12×10^{-4}
7	148	X07_60966290	X07_61091852	102–156	57.73–61.53	9.84×10^{-5}
8	70	X08_56902554	X08_57091589	58–86	56.26–58.33	8.63×10^{-6}
11	154	X11_51548415	X11_51631459	152–161	51.35–52.66	2.49×10^{-8}
11	118	X11_48934628	X11_49059536	105–183	47.57–53.18	5.41×10^{-6}
B						
2	184	X02_47433596	X02_47498009	179–192	46.35–47.49	4.44×10^{-16}
3	176	X03_58754293	X03_58846611	170–190	57.98–60.24	1.47×10^{-5}
11	148	X11_51176762	X11_51308212	137–160	50.5–52.29	1.16×10^{-9}

Including chromosome (Chr), position of maximum *P*-value peak (Pos in cM), left and right markers flanking the peak position, 1-LOD support interval (SI) in cM and Mb, *P*-value at the peak position.

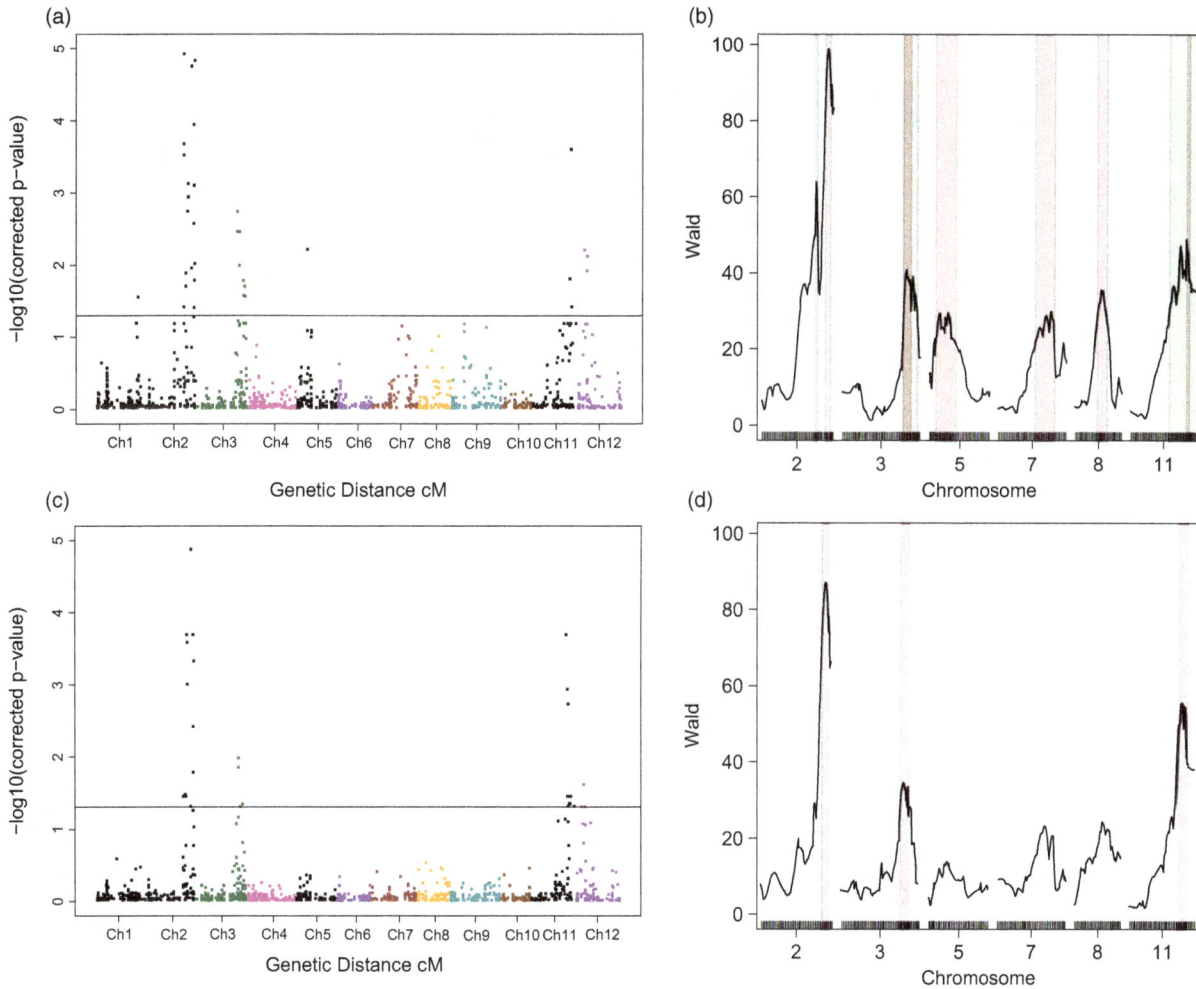

Figure 6 QTL detection in the MAGIC population for FW. (a and c) Manhattan plot (MLM) showing corrected p-values, red line indicates significance threshold. (b and d) Wald test profile on chromosomes with significant QTL. Red regions indicate 1-LOD support intervals, green regions indicate 1-LOD support intervals for QTLs detected from the original QTL profile with the function *findqtl2*. (a and b): location A (Avignon). (c and d): location B (La Costière).

dissection has been especially challenging for species like tomato, with low genetic diversity and strong population structure. The lack of intraspecific genetic variation has led to the study of interspecific progenies involving related species, but the potential of intraspecific variation remains poorly explored. The MAGIC population enabled the construction of the first high-density intraspecific map in tomato. We assessed the power of such population to generate new variation through new allelic combinations and to map QTLs. Finally, we showed its potential to identify putative causal variants and closely linked QTLs.

The population presented here together with the whole genome sequences of the founder lines constitute a useful resource for the scientific community, which overcomes the main disadvantages linked to collections of accessions employed for GWAS or to the interspecific biparental populations developed until now (Table 4). The main issues when using natural populations to detect genetic associations are (i) the presence of population substructure, (ii) variable LD block length along chromosomes and (iii) unbalanced allele frequencies (Visscher et al., 2012). In tomato, the structure is especially strong (Blanca et al., 2012; Ranc et al., 2008). Indeed, the species can be

divided in two major groups: *S. lycopersicum* and *S. lycopersicum var. cerasiforme*, and FW is highly correlated to the classification (Ranc et al., 2012). We used four accessions of each group as founders of the population. The population showed a wide range of phenotypes increasing the phenotypic range of the founders, without any remaining subgroups. The design used to develop the MAGIC population effectively mixed the genomes of all the founder genotypes, leading to balanced allele frequencies along the genome. In contrast to other MAGIC populations (Huang et al., 2012b), < 1.5% of the markers showed segregation distortion, even though some founder lines carried introgressions from distant wild species (Causse et al., 2013).

LD extent, as well as the specific size and genomic structure of the centromeres, impacts the number of markers needed to detect QTLs and should be taken into account when designing genotyping strategies. The tomato genome structure character-ized by very low recombination rates in the centromeric regions that comprise around 70% of the chromosomes (Sim et al., 2012a) greatly affected the LD among markers. LD decay is slower in the MAGIC population than in natural populations, where the LD baseline is reached before 50 cM (Sauvage et al.,

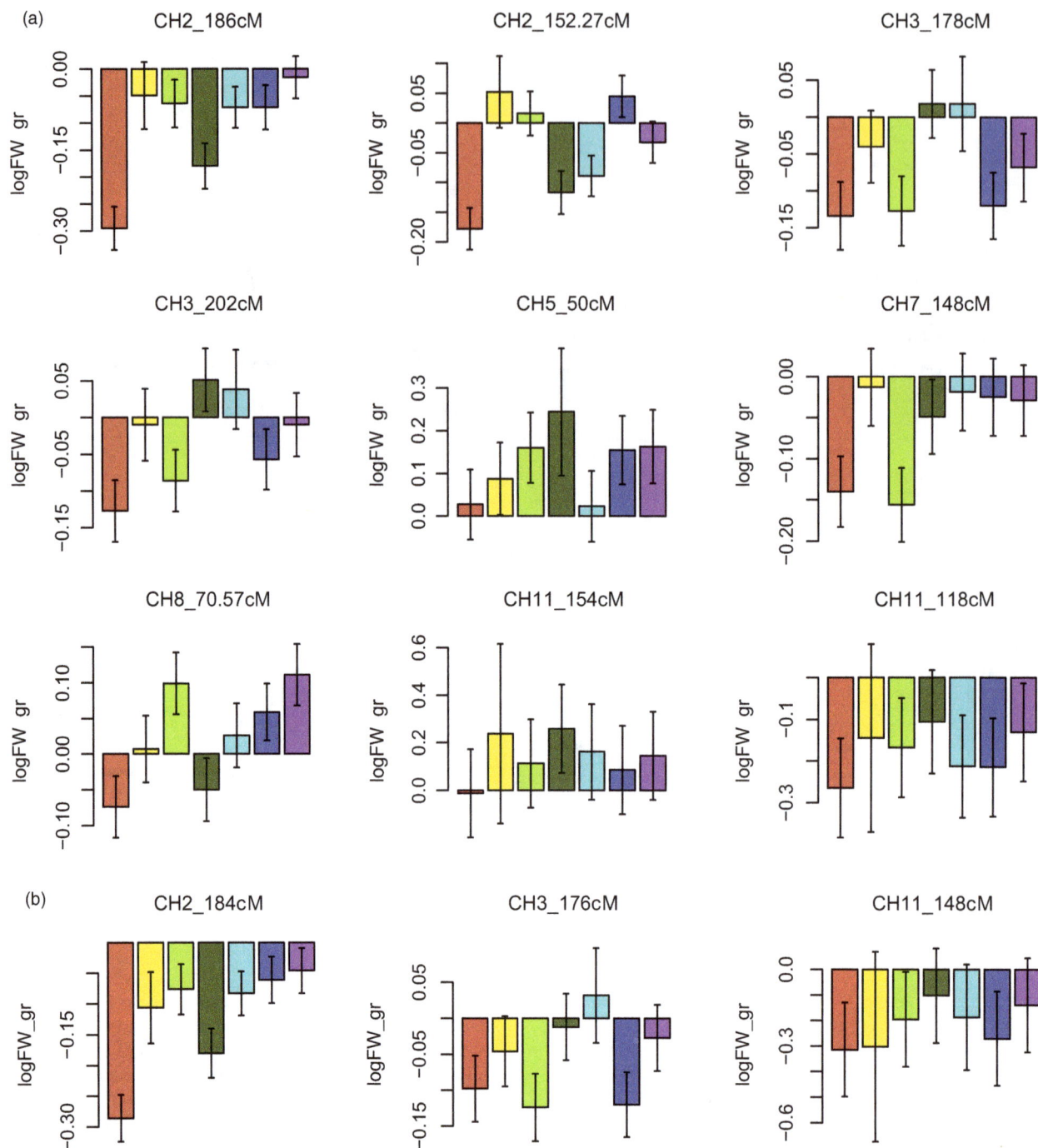

Figure 7 Founder allelic effects at the fruit weight QTL. Effects and standard errors for the alleles of Cervil, Levovil, Criollo, Stupicke PR, Plovdiv24A, LA1420 and Ferum (from left to right), relative to LA0147. QTLs detected with data from location A (Avignon) and B. (La Costière).

2014; Sim *et al.*, 2012b). Compared to MAGIC populations developed in other species like *Arabidopsis,* where the LD baseline (0.05) is reached at 15 Mb (Kover *et al.*, 2009), our population reached the baseline at longer physical distances (40–60 Mb) due to the size of centromere regions.

The selection of the SNPs greatly impacted the power of haplotype prediction along the MAGIC line genomes and thus the chances to detect genetic associations through the haplotype reconstruction (Huang and George, 2011; Mott *et al.*, 2000). The availability of the founders' sequences allowed haplotype impu-

tation in most of the genomic regions. To characterize the remaining problematic regions, it may be necessary to character-ize the population by genotyping by sequencing (GBS) as it was performed in rice (Bandillo *et al.*, 2013), as long as the founder lines are not identical by descent. In such a case, clustering similar founder haplotypes may increase statistical power to detect QTLs, as shown by Bardol *et al.* (2013).

The MAGIC map size was 87% larger than biparental maps (Sim *et al.*, 2012a), showing the effectiveness of MAGIC popu-lation to enhance recombination and admixture. This increase

Table 4 Comparison of advantages and limits of biparental, MAGIC and association populations

	Advantages	Limits
Biparental progeny	Rapid to set up Useful for mapping rare alleles Easy analysis	Limited to two contrasting alleles Few recombination generations Large QTL support interval
MAGIC population	Several alleles and QTLs segregating Higher precision than biparental population Rapid fine mapping Useful for candidate SNP screening No population structure Suitable for selection	Time to establish Require more markers and larger populations than biparental population
Association panel	Existing collections, high diversity Natural recombination When LD limited, recise mapping	Require many markers Population structure When high LD, coarse mapping Rare alleles poorly identified

subsequently reduces the QTL support intervals when translated to physical size, facilitating QTL fine mapping and candidate gene selection. However, recombination was not increased in the centromere regions, comprising around 70% of the chromosomes but <40% of the genes (Sim et al., 2012a). Therefore, any QTL detected in/over the centromere regions will still have a very large support interval in physical distance, and the absence of recombination will make positional cloning impossible. Luckily, the detection of QTLs in these regions should be an exception, as most of the genes are located in the chromosome extremities (Tomato Genome Consortium, 2012).

To test the power of this resource to map QTLs, we analysed the fruit weight distribution in the population grown in two locations. Genetic associations were detected by IM and GWAS (accounting for population kinship). Using both methods and locations, we detected associations with markers close to the already cloned FW QTLs (Chakrabarti et al., 2013; Frary et al., 2000), showing the precision of QTL mapping. By IM, we detected 9 QTLs on six chromosomes at location A, among which three were also detected at location B. Five of the QTLs detected at location A (one at location B) were not detected in a biparental population derived from the cross between the two most distant founders of the MAGIC population (Saliba-Colombani et al., 2001), but colocalized with QTLs already detected in interspecific populations (reviewed by Grandillo et al., 1999). This showed that increasing the variability in the founders allows the discovery of new QTL.

GWAS analysis by MLM method taking into account the kinship may avoid the detection of false positives (Visscher et al., 2012) but failed to identify two of the QTLs identified by IM at location A (on chromosomes 7 and 8) that colocalized with already known QTLs (Causse et al., 2004; Grandillo et al., 1999; van der Knaap and Tanksley, 2003). On the other hand, we detected new associations on chromosome 1 at location A and on chromosome 12 at both locations. A QTL located at the top of chromosome 12 has been detected in the biparental progeny derived from two of the founders (Saliba-Colombani et al.,

2001), suggesting, that at least in this case, MLM method was more powerful than IM. This might occur in genomic regions where it is not possible to distinguish among several founders, as MLM relies on a biallelic model, while IM is comparing eight different haplotypes. For QTLs where the markers have an allele shared by founders with the same phenotypic value, MLM might be more powerful. On the contrary, if the allele is not shared among founders with the same phenotype, it will be impossible to detect association by MLM. Working with haplotypes, IM avoids this problem and permits the calculation of each founder allele effect for each QTL.

Allelic effects varied among QTLs, lines and locations. The same allelic effect was never shared by all the small or all the large fruited founders. The final fruit size is thus obtained from a specific combination of founder alleles. The knowledge of QTL allelic effects, coupled with the availability of the MAGIC population, constitutes a highly valuable resource to develop strategies for breeding and develop models to conduct genomic selection. On one hand, it is composed by a set of highly admixed lines from which we can obtain the genomic breeding values and develop models that encompass most of the species diversity (Morell et al., 2012). On the other hand, it allows the validation of model predictions, as it contains most of the possible allelic combinations. Additionally, the discovery of a large number of QTLs on different chromosomes allows the development of interesting breeding schemes as alternative allelic combinations could be used to create specific phenotypes (Rosyara et al., 2013).

Using a MAGIC population derived from resequenced founders allows to design strategies to identify causal polymorphisms. We showed that the list of candidate genes for a QTL interval can be strongly reduced and putative causal variants identified by coupling QTL effects with the founders' genome sequences. We analysed the genomic sequences underlying the two QTLs that colocalized with already cloned genes (fw3.2, Solyc03 g114940, Chakrabarti et al., 2013 and fw2.2, Solyc02 g090730, Frary et al., 2000). For fw3.2, we discarded 99.21% of the SNPs and INDELs located in the QTL SI by selecting variants differing between the three founders that equally decreased fruit weight and the three with opposite effect. The residual variants included the causal SNP identified by Chakrabarti et al. (2013). For the region around fw2.2, the causal polymorphism has not yet been identified. However, Frary et al. (2000) indicated that the phenotype was probably caused by one or more changes upstream in the promoter region. When we selected the variants that differed in the SI between the two founders that decreased fruit weight and the other founders, we did not find any variant linked to Solyc02 g090730. The analysis of all the variants surrounding the gene revealed that only one founder was different from the other lines in the region. However, there was another founder whose allele reduced fruit size but to a smaller extent. This might be caused by another gene located in close proximity. The presence of two linked QTLs might bias the estimation of founder effects and should be taken into account when looking for causal variants.

However, even if MAGIC populations allow QTL mapping at a subcentimorgan scale, this range might correspond to hundreds of kilobases. Thus, to identify the polymorphism underlying a QTL, it is still necessary to produce new recombinant plants and conduct positional cloning. The tomato MAGIC population was characterized after three selfing generations. The lines still carry residual heterozygosity that can be directly used to that end.

In conclusion, we have created the first intraspecific population of highly recombinant lines in tomato. This population segregates for many traits and can be analysed in different environments, providing a permanent resource to analyse the basis of phenotypic traits. We have illustrated its power for future fine mapping experiments. Our study has also highlighted the potential of the availability of the founder genome sequences. On the one hand, it enabled the efficient selection of a subset of markers especially designed to analyse the MAGIC population. On the other hand, it permitted to drastically restrict the number of putative causal polymorphisms underlying the QTLs.

Experimental procedures

Founder lines selection and population construction

Eight tomato lines, thoroughly characterized at different molecular and physiological levels (Pascual et al., 2013), were selected as founders of an eight-way MAGIC population. Founders included four S. lycopersicum (Levovil, Stupicke PR, LA0147 and Ferum) and four S. lycopersicum var. cerasiforme (Cervil, Criollo, Plovdiv24A and LA1420) lines. These lines were chosen to maximize the genetic diversity, based on a previous molecular characterization of 360 tomato accessions (Ranc et al., 2008).

Four crosses were performed between one S. lycopersicum line and one S. lycopersicum var. cerasiforme (Figure 1). F1 hybrids Levovil × Cervil and Stupicke PR × Criollo were crossed to obtain 120 DCF1Hy (Double Cross F1 hybrid) plants, while LA0147 × Plovdiv24A was crossed with Ferum × LA1420 to obtain another set of 120 DCF1Hy plants. The two subsets of 120 plants were intercrossed via 240 independent crosses using each DCF1Hy plant once as father and once as mother. Two offsprings per cross were then kept, producing 480 individuals (F1-like), each bearing parts of the eight founder genomes. These plants were propagated by single seed descent during three selfing generations, to create the set of 397 MAGIC lines (F4-like) employed in this study (Figure 1).

Genetic marker selection and construction of a MAGIC custom-made SNP platform

More than four million SNPs were detected by resequencing the genomes of the eight founder lines (Causse et al., 2013; Raw sequences deposited in ENA, accession numbers ERR327646 to ERR327656; SNPs and INDELs identified deposited in ENA, accession numbers ERZ015686 to ERZ015701). Among all these polymorphisms, 1536 SNPs were selected to construct the genetic map. The selection of the SNPs was performed with custom Python scripts (available upon request to the corresponding author) using a three steps filtering strategy. First, we selected 'the best quality SNPs' based on nucleotide prediction reliability, whether the SNPs were biallelic and the quality of the flanking sequences (60 bp). In the second step, we kept only one position for each set of successive positions with the same allelic profile over the eight lines, as they would provide redundant information. Finally, SNPs were selected from successive intervals with a maximum genetic distance of 8 cM. Genetic distances were assessed for each physical position by a linear regression using only the five nearest markers from EXPEN 2000 map (solgenomics.net) and giving to each marker a weight proportional to the physical distance from the candidate position. For this step, we performed a recursive search to select a combination of physical positions that had between them a minimal genetic distance (1–0.5 cM). Each combination should have balanced allele frequencies, enhancing founders with less SNPs. Besides, the combination of marker alleles should be specific for each founder, enhancing haplotype prediction.

DNA isolation and molecular marker genotyping

DNA was isolated from young leaves of each founder line and the 397 lines of the MAGIC population using DNeasy 96 Plant kit supplied by Qiagen (Hilden, Germany). From the 1536 SNPs selected to develop the genotype platform, 1486 passed the KASPar manufacturing quality control (K-Biosciences/LGC Genomics, Molsheim, France) and were finally used for genotyping the MAGIC population (Table S1). Genotyping was performed using the Fluidigm 96.96 Dynamic Arrays according to the manufacturer's protocol using the genotyping EP1 System (San Francisco, CA). Fluorescence intensity was measured with the EP1 reader (Fluidigm Corp, San Francisco, CA), and genotypic calls were made using the Fluidigm SNP Genotyping Analysis program (Fluidigm, 2011). All genotype calls were manually checked for accuracy and ambiguous data points that failed to cluster were scored as missing data. Heterozygous markers (caused by the residual heterozygosity in the MAGIC lines) were also scored as missing data.

Genetic map construction, recombination event prediction and LD analysis

The genetic map was constructed using the R package mpMap (Huang and George, 2011) version 1.24.3 and the available information of the physical location of the SNPs from the tomato reference genome version SL2.40 (Tomato Genome Consortium, 2012). First, we filtered out the markers with missing data in the founders, markers with more than 20% missing data in the population and markers with a segregation distortion P-value $< 5 \times 10^{-9}$. MAGIC lines with more than 20% missing markers were also removed. These lines were later included for QTL analysis.

Second, recombination fraction between each pair of markers was estimated with the 'mpestrf' mpMap function. This function maximizes the likelihood of observing data from a pair of markers over discrete values of recombination fractions (default values). Then, markers were grouped with the 'mpgroup' mpMap function. Markers in each groups were checked against the tomato reference genome and groups assigned to the different tomato chromosomes. Markers that were not in their expected chromosome were discarded. Then, markers were ordered within linkage groups, based on their physical position in the tomato reference genome.

Third, genetic distances were estimated with the 'computemap' mpMap function, using a 15-marker window and Haldane distances computed. The genetic distances among markers were plotted against the physical distances for each chromosome, and when inconsistency was found, problematic markers were removed and genetic distances re-estimated.

For each line, we calculated the multipoint probability that the genotype at a marker location (and positions spaced every 2 cM) was inherited from each of the eight founders ('mpprob' function from mpMap). Recombination events were imputed at locations where the founder allele changed along the chromosome. To assess the differences between MAGIC and biparental populations, we compared our genetic map with the tomato high-density genetic maps, constructed by Sim et al. (2012a) using interspecific biparental F2 tomato populations.

Multiallelic linkage disequilibrium (LD) r^2 was estimated between each pair of markers using the multipoint probabilities with the '*mpcalcld*' function as described by Huang et al. (2012b). We compared the values for r^2 between markers along the chromosomes and analysed the LD decay over genetic and physical distances, plotting r^2 values against the distances by chromosomes. Values were fitted by nonlinear regression.

Phenotypic data

The population was grown in two locations in the south of France in Avignon (location A) and La Costière (location B). In each location, the 397 lines (one plant per line) and five replicates of each founder were grown in greenhouses during spring–summer 2012, as described in Pascual et al. (2013). Fruit weight (FW) was evaluated from a minimum of 10 ripe fruits per genotype, harvested from truss two to six. Before detecting the FW QTLs, a log_{10} transformation was carried out to normalize the phenotype.

Population structure and association analysis

Population structure was inferred with Structure v2.1 software (Pritchard et al., 2000) in the complete MAGIC population using the 1345 mapped SNP markers. We used the admixture model for the ancestry of individuals and linkage to correct for the effect of nearby markers. The structure was modelled with a burn-in of 2.5×10^5 cycles followed by 10^6 Markov chain Monte Carlo repeats. Probabilities were estimated for population number (k) between 1 and 20, computing 10 replicates for each k. The Evanno transformation method (Evanno et al., 2005) was then used to detect the number of subgroups in the MAGIC population. Kinship matrix was calculated with the SPAGeDi software (Hardy and Vekemans, 2002). According to Yu et al. (2006), the diagonal matrix was set to two and the negative values were set to zero. Association analysis was performed independently for FW measured in each location. Analyses were conducted with the TASSEL version 3.0 software (Bradbury et al., 2007) employing mixed linear models (MLM) incorporating the kinship among individuals. The p-values were adjusted with the Benjamini and Hochberg (2000) procedure using the R package '*multtest*' (Pollard et al., 2012). Associations with an adjusted P-value < 0.05 were considered significant.

QTL detection and identification of putative causal variants

QTLs were called by simple interval mapping (IM) with the '*mpIM*' function from the R package *mpMap* (Huang and George, 2011). This function computes founder effects at a step size of 2 cM with a regression approach, based on the multipoint probabilities computed with '*mpprob*' function. Then, it performs a joint Wald test for the significance of all founder effects at each putative QTL position along the genome. QTLs were called when p-values were smaller than the empirical threshold p-value (1.72×10^{-4}) derived using the function '*sim.sigthr*' after computing 1000 permutations, to reflect a genome-wide significance threshold of 0.05. This approach called a single QTL by chromosome, so when the QTL profile showed more than one QTL peak by chromosome, multiple QTLs were considered significant when peaks were separated by more than 20 cM and the LOD score dropped by more than one. QTL support intervals were determined with a 1-LOD drop support. After QTL detection, QTL effects were simultaneously fitted in a single model with the function '*fit*' from R *mpMap* package to estimate the percentage of phenotypic variation explained by the QTLs. In order to test the potential of

the MAGIC population to detect causal variants, we analysed two QTLs that colocalized with already cloned genes. All the polymorphisms (Causse et al., 2013) present in the QTL support intervals were analysed and filtered according to the estimated founder effects.

Acknowledgements

We thank Yolande Carretero, Justine Gricourt, Frédérique Bitton, Esther Pelpoir and Renaud Duboscq for their help in phenotyping. We also thank the experimental team and Yolande Carretero for taking care of the plants in the greenhouse. This work was supported by the ANR MAGIC-Tom SNP project 09-GENM-109G. LP was supported by a postdoctoral INRA fellowship.

References

Bandillo, N., Raghavan, C., Muyco, P.A., Sevilla, M.A.L., Lobina, I.T., Dilla-Ermita, C.J., Tung, C.W., McCouch, S., Thomson, M., Mauleon, R., Singh, R.K., Gregorio, G., Redoña, E. and Leung, H. (2013) Multi-parent advanced generation inter-cross (MAGIC) populations in rice: progress and potential for genetics research and breeding. *Rice*, **6**, 11.

Bardol, N., Ventelon, M., Mangin, B., Jasson, S., Loywick, V., Couton, F., Derue, C., Blanchard, P., Charcosset, A. and Moreau, L. (2013) Combined linkage and linkage disequilibrium QTL mapping in multiple families of maize (*Zea mays* L.) line crosses highlights complementary between models based on parental haplotype and single locus polymorphism. *Theor. Appl. Genet.* **126**, 2717–2136.

Benjamini, Y. and Hochberg, Y. (2000) On the adaptive control of the false discovery rate in multiple testing with independent statistics. *J. Educ. Behav. Stat.* **25**, 60–83.

Blanca, J., Canizares, J., Cordero, L., Pascual, L., Diez, M.J. and Nuez, F. (2012) Variation revealed by SNP genotyping and morphology provides insight into the origin of the tomato. *PLoS ONE*, **7**, e48198.

Bradbury, P.J., Zhang, D., Kroon, D.E., Casstevens, T.M., Ramdoss, Y. and Buckler, E.S. (2007) TASSEL: software for association mapping of complex traits in diverse samples. *Bioinformatics*, **23**, 2633–2635.

Causse, M., Duffe, P., Gomez, M.C., Buret, M., Damidaux, R., Zamir, D., Gur, A., Chevalier, C., Lemaire-Chamley, M. and Rothan, C. (2004) A genetic map of candidate genes and QTLs involved in tomato fruit size and composition. *J. Exp. Bot.* **55**, 1671–1685.

Causse, M., Desplat, N., Pascual, L., Le Paslier, M.C., Sauvage, C., Bauchet, G., Bérard, A., RémiBounon, R., Tchoumakov, M., Brunel, D. and Bouchet, J.P. (2013) Whole genome profiles of nucleotide variations reveal breeding and introgression events in 8 lines of tomato. *BMC Genomics*, **14**, 791.

Cavanagh, C., Morell, M., Mackay, I. and Powell, W. (2008) From mutations to MAGIC: resources for gene discovery, validation and delivery in crop plants. *Curr. Opin. Plant Biol.* **11**, 215–221.

Chakrabarti, M., Zhang, N., Sauvage, C., Muños, S., Blanca, J., Cañizares, J., Diez, M.J., Schneider, R., Mazourek, M., McClead, J., Causse, M. and van der Knaap, E. (2013) A cytochrome P450 regulates a domestication trait in cultivated tomato. *Proc. Natl Acad. Sci. USA*, **110**, 17125–17130.

Clarke, J.H., Mithen, R., Brown, J.K. and Dean, C. (1995) QTL analysis of flowering time in *Arabidopsis thaliana*. *Mol. Gen. Genet.* **248**, 278–286.

Darvasi, A. and Soller, M. (1995) Advanced intercross lines, an experimental population for fine genetic mapping. *Genetics*, **141**, 199–1207.

Evanno, G., Regnaut, S. and Goudet, J. (2005) Detecting the number of clusters of individuals using the software structure: a simulation study. *Mol. Ecol.* **14**, 2611–2620.

Fluidigm (2011) *Fluidigm SNP Genotyping Analysis*. San Francisco: Fluidigm Corp.

Frary, A., Nesbitt, T.C., Grandillo, S., Knaap, E., Cong, B., Liu, J., Meller, J., Elber, R., Alpert, K.B. and Tanksley, S.D. (2000) fw2.2: a quantitative trait locus key to the evolution of tomato fruit size. *Science*, **289**, 547685–547688.

Giovannoni, J.J. (2004) Genetic regulation of fruit development and ripening. *Plant Cell*, **16**, S170–S180.

Grandillo, S. and Tanksley, S.D. (1996) QTL analysis of horticultural traits differentiating the cultivated tomato from the closely related species *Lycopersicon pimpinellifolium*. *Theor. Appl. Genet.* **92**, 935–951.

Grandillo, S., Ku, H.M. and Tanksley, S.D. (1999) Identifying the loci responsible for natural variation in fruit size and shape in tomato. *Theor. Appl. Genet.* **99**, 978–987.

Hall, D., Tegstrom, C. and Ingvarsson, P.K. (2010) Using association mapping to dissect the genetic basis of complex traits in plants. *Brief. Funct. Genomics*, **9**, 157–165.

Hardy, O.J. and Vekemans, X. (2002) SPAGeDi: a versatile computer program to analyze spatial genetic structure at the individual or population levels. *Mol. Ecol. Notes*, **2**, 618–620.

Huang, B.E. and George, A.W. (2011) R/mpMap: a computational platform for the genetic analysis of multiparent recombinant inbred lines. *Bioinformatics*, **27**, 727–729.

Huang, X., Zhao, Y., Wei, X., Li, C., Wang, A., Zhao, Q., Li, W., Guo, Y., Deng, L., Zhu, C., Fan, D., Lu, Y., Weng, Q., Liu, K., Zhou, T., Jing, Y., Si, L., Dong, G., Huang, T., Lu, T., Feng, Q., Qian, Q., Li, J. and Han, B. (2012a) Genome-wide association study of flowering time and grain yield traits in a worldwide collection of rice germplasm. *Nat. Genet.* **44**, 32–39.

Huang, B.E., George, A.W., Forrest, K.L., Kilian, A., Hayden, M.J., Morell, M.K. and Cavanagh, C.R. (2012b) A multiparent advanced generation inter-cross population for genetic analysis in wheat. *Plant Biotechnol. J.* **10**, 826–839.

Keurentjes, J.J., Bentsink, L., Alonso-Blanco, C., Hanhart, C.J., Blankestijn-De Vries, H., Effgen, S., Vreugdenhil, D. and Koornneef, M. (2007) Development of a near-isogenic line population of *Arabidopsis thaliana* and comparison of mapping power with a recombinant inbred line population. *Genetics*, **175**, 891–905.

van der Knaap, E. and Tanksley, S.D. (2003) The making of a bell pepper-shaped tomato fruit: identification of loci controlling fruit morphology in Yellow Stuffer tomato. *Theor. Appl. Genet.* **107**, 139–147.

Korte, A. and Farlow, A. (2013) The advantages and limitations of trait analysis with GWAS: a review. *Plant Methods*, **9**, 29.

Kover, P.X. and Mott, R. (2012) Mapping the genetic basis of ecologically and evolutionarily relevant traits in *Arabidopsis thaliana*. *Curr. Opin. Plant Biol.* **15**, 212–217.

Kover, P.X., Valdar, W. and Trakalo, J. (2009) A multiparent advanced generation inter-cross to fine-map quantitative traits in *Arabidopsis thaliana*. *PLoS Genet.* **5**, 7.

Miller, J.C. and Tanksley, S.D. (1990) RFLP analysis of phylogenetic relationships and genetic variation in the genus *Lycopersicon*. *Theor. Appl. Genet.* **80**, 437–448.

Mitchell-Olds, T. (2010) Complex-traits analysis in plants. *Genome Biol.* **10**, 113.

Morell, P.L., Buckler, E.S. and Ross-Ibarra, J. (2012) Crop genomics: advances and applications. *Nat. Rev. Genet.* **13**, 85–96.

Mott, R., Talbot, C.J., Turri, M.G., Collins, A.C. and Flint, J. (2000) A method for fine mapping quantitative trait loci in outbred stocks. *Proc. Natl Acad. Sci. USA*, **97**, 12649–12654.

Pascual, L., Xu, J., Biais, B., Maucourt, M., Ballias, P., Bernillon, S., Deborde, C., Jacob, D., Desgroux, A., Faurobert, M., Bouchet, J.P., Gibon, Y., Moing, A. and Causse, M. (2013) Deciphering genetic diversity and inheritance of tomato fruit weight and composition through a systems biology approach. *J. Exp. Bot.* **64**, 5737–5752.

Pollard, K.S., Gilbert, N.H., Ge, Y., Taylor, S. and Dudoit, S. (2012) *multtest: Resampling-based multiple hypothesis testing*. R package version 2.14.0.

Price, A.H. (2006) Believe it or not, QTLs are accurate!. *Trends Plant Sci.* **11**, 213–216.

Pritchard, J.K., Stephens, M. and Donelly, P. (2000) Inference of population structure using multilocus genotype data. *Genetics*, **155**, 945–959.

Rakshit, S., Rakshit, A. and Patil, J.V. (2012) Multiparent intercross populations in analysis of quantitative traits. *J. Genet.* **91**, 111–117.

Ranc, N., Munos, S., Santoni, S. and Causse, M. (2008) A clarified position for *Solanum lycopersicum* var. *cerasiforme* in the evolutionary history of tomatoes (Solanaceae). *BMC Plant Biol.* **8**, 130.

Ranc, N., Muños, S., Xu, J., Le Paslier, M.C., Chauveau, A., Bounon, R., Rolland, S., Bouchet, J.P., Brunel, D. and Causse, M. (2012) Genome-wide association mapping in tomato (*Solanum lycopersicum*) is possible using genome admixture of *Solanum lycopersicum* var. *cerasiforme*. *G3*, **2**, 853–864.

Rockman, M.V. and Kruglyak, L. (2008) Breeding designs for recombinant inbred advanced intercross lines. *Genetics*, **179**, 1069–1078.

Rosyara, U.R., C.A.M. Bink, M., van de Weg, E., Zhang, G., Wang, D., Sebolt, A., Dirlewanger, E., Quero-Garcia, J., Schuster, M. and Iezzoni, A.M. (2013) Fruit size QTL identification and the prediction of parental QTL genotypes and breeding values in multiple pedigreed populations of sweet cherry. *Mol. Breed.* **32**, 875–887.

Saliba-Colombani, V., Causse, M., Langlois, D., Philouze, J. and Buret, M. (2001) Genetic analysis of organoleptic quality in fresh market tomato. 1. Mapping QTLs for physical and chemical traits. *Theor. Appl. Genet.* **102**, 259–272.

Sauvage, C., Segura, V., Bauchet, G., Stevens, R., Thi Do, P., Nikoloski, Z., Fernie, A.R. and Causse, M. (2014) Genome wide association in tomato reveals 44 candidate loci for fruit metabolic traits. *Plant Physiol.* **165**, 1120–1132.

Sim, S.C., Durstewitz, G., Plieske, J., Wieseke, R., Ganal, M.W., Van Deynze, A., Hamilton, J.P., Buell, C.R., Causse, M., Wijeratne, S. and Francis, D.M. (2012a) Development of a large SNP genotyping array and generation of high-density genetic maps in tomato. *PLoS ONE*, **7**, e45520.

Sim, S.C., Van Deynze, A., Stoffel, K., Douches, D.S., Zarka, D., Ganal, M.W., Chetelat, R.T., Hutton, S.F., Scott, J.W., Gardner, R.G., Panthee, D.R., Mutschler, M., Myers, J.R. and Francis, D.M. (2012b) High-density SNP genotyping of tomato (*Solanum lycopersicum* L.) reveals patterns of genetic variation due to breeding. *PLoS ONE*, **7**, e40563.

Tomato Genome Consortium (2012) The tomato genome sequence provides insights into fleshy fruit evolution. *Nature*, **485**, 635–641.

Valdar, W., Flint, J. and Mott, R. (2006) Simulating the collaborative cross: power of QTL detection and mapping resolution in large sets of recombinant inbred strains of mice. *Genetics*, **172**, 1783–1797.

Visscher, P.M., Brown, M.A., McCarthy, M.I. and Yang, J. (2012) Five years of GWAS Discovery. *Am. J. Hum. Genet.* **90**, 7–24.

Xu, J., Ranc, N., Muños, S., Rolland, S., Bouchet, J.P., Desplat, N., Le Paslier, M.C., Liang, Y., Brunel, D. and Causse, M. (2013) Phenotypic diversity and association mapping for fruit quality traits in cultivated tomato and related species. *Theor. Appl. Genet.* **126**, 567–581.

Yalchin, B., Flint, J. and Mott, R. (2005) Using progenitor strain information to identify quantitative trait nucleotides in out bred mice. *Genetics*, **171**, 673–681.

Yu, J., Pressoir, G., Briggs, W.H., Vroh Bi, I., Yamasaki, M., Doebley, J.F., McMullen, M.D., Gaut, B.S., Nielsen, D.M., Holland, J.B., Kresovich, S. and Buckler, E.S. (2006) A unified mixed-model method for association mapping that accounts for multiple levels of relatedness. *Nat. Genet.* **38**, 203–208.

Yu, J., Holland, J.B., McMullen, M.D. and Buckler, E.S. (2008) Genetic design and statistical power of nested association mapping in maize. *Genetics*, **178**, 539–551.

Zamir, D. (2001) Improving plant breeding with exotic genetic libraries. *Nat. Rev. Genet.* **2**, 983–989.

7

RNA-Seq bulked segregant analysis enables the identification of high-resolution genetic markers for breeding in hexaploid wheat

Ricardo H. Ramirez-Gonzalez[1], Vanesa Segovia[2,†], Nicholas Bird[2], Paul Fenwick[3], Sarah Holdgate[4,‡], Simon Berry[3], Peter Jack[4], Mario Caccamo[1] and Cristobal Uauy[2,5,*]

[1]The Genome Analysis Centre, Norwich, UK
[2]John Innes Centre, Norwich, UK
[3]Limagrain UK Ltd, Rothwell, UK
[4]RAGT Seeds, Saffron Walden, UK
[5]National Institute of Agricultural Botany, Cambridge, UK

*Correspondence
email cristobal.uauy@jic.ac.uk
†Present address: Regional Cereal Rust Research Center, International Center for Agricultural Research in the Dry Areas (ICARDA), PO Box 35661, Menemen, Izmir, Turkey.
‡Present address: National Institute of Agricultural Botany, Huntingdon Road, Cambridge, CB3 OLE, UK.

Keywords: bulk frequency ratio, Yr15, marker-assisted selection, haplotype, yellow rust, SRA project numbers, PolyMarker.

Summary

The identification of genetic markers linked to genes of agronomic importance is a major aim of crop research and breeding programmes. Here, we identify markers for Yr15, a major disease resistance gene for wheat yellow rust, using a segregating F_2 population. After phenotyping, we implemented RNA sequencing (RNA-Seq) of bulked pools to identify single-nucleotide polymorphisms (SNP) associated with Yr15. Over 27 000 genes with SNPs were identified between the parents, and then classified based on the results from the sequenced bulks. We calculated the bulk frequency ratio (BFR) of SNPs between resistant and susceptible bulks, selecting those showing sixfold enrichment/depletion in the corresponding bulks (BFR > 6). Using additional filtering criteria, we reduced the number of genes with a putative SNP to 175. The 35 SNPs with the highest BFR values were converted into genome-specific KASP assays using an automated bioinformatics pipeline (PolyMarker) which circumvents the limitations associated with the polyploid wheat genome. Twenty-eight assays were polymorphic of which 22 (63%) mapped in the same linkage group as Yr15. Using these markers, we mapped Yr15 to a 0.77-cM interval. The three most closely linked SNPs were tested across varieties and breeding lines representing UK elite germplasm. Two flanking markers were diagnostic in over 99% of lines tested, thus providing a reliable haplotype for marker-assisted selection in these breeding programmes. Our results demonstrate that the proposed methodology can be applied in polyploid F_2 populations to generate high-resolution genetic maps across target intervals.

Introduction

Wheat is a major crop providing over 20% of the world's calorie and 25% of its protein intake (FAO, 2012). With increased population growth and demand for cereal crops, breeders are under constant pressure to deliver high-performing varieties (Galushko and Gray, 2014). The use of new technologies, such as marker-assisted selection (MAS), has reduced the cost and development time of new elite varieties across major crops (Bernardo, 2008). More recently, the advent of next-generation sequencing (NGS) is revolutionizing molecular breeding either through genomic selection approaches (Collard and Mackill, 2008; Gupta et al., 2008) or through the identification of large numbers of SNPs which can be converted into MAS assays (Allen et al., 2013). Despite this potential, identifying SNPs in bread wheat (Triticum aestivum L.) is challenging due to its hexaploid nature ($6n = 42$; AABBDD) and large genome size ~17 Gbp (Gupta et al., 2008; Shewry, 2009), although large-scale efforts to identify intervarietal SNP are becoming more common (Lai et al., 2014; Wang et al., 2014). The three genomes (A, B and D) are referred to as homoeologues and are related, and yet distinct, sharing a complementary set of genes which have between 96% and 98% sequence identity across coding regions (Krasileva et al., 2013).

A high-priority trait in wheat breeding programmes is the introduction of major disease resistance (R) genes using MAS. Despite modern agricultural practices, low-level persistent plant diseases have been predicted to cause losses equal to enough calories to feed ~8.5% of the world (Fisher et al., 2012). MAS is especially relevant in disease resistance breeding strategies as it is the only quick and reliable way of pyramiding multiple R-genes within a single variety. In the past decade, new races of the wheat yellow rust pathogen (Puccinia striiformis f. sp. tritici) have emerged with expanded virulence profiles (Hovmøller et al., 2010; Milus et al., 2008), and more recently a series of new races (collectively termed 'Warrior') have appeared in Europe. The Warrior races have overcome the majority of the major resistance genes in European germplasm (GRRC, 2014). However, a few major R-genes, such as Yr15, still provide effective resistance against these and the major international isolates. Therefore, the availability of closely linked markers, which are diagnostic for R-genes such as Yr15, is of great interest to breeders.

The recently published high-density SNP arrays in wheat are providing an extremely valuable resource in diversity and genomewide association studies (Cavanagh et al., 2013; Wang et al., 2014). However, their potential is not fully realized in F_2 and similar large mapping populations because of their unit cost and the difficulty to reliably call heterozygous individuals, especially in a polyploid species such as wheat. In addition, the restriction to a predefined set of allelic variants limits their use when targeting genes introgressed from progenitor species, such as Yr15 originally from wild emmer (Triticum dicoccoides) (McIntosh et al., 1995).

Several strategies have been proposed to identify SNPs, which are targeted to specific chromosomal intervals (Hodges et al., 2007; Michelmore et al., 1991; Mortazavi et al., 2008; Paux et al., 2010; Trick et al., 2012). Among these, bulked segregant analysis (BSA) (Michelmore et al., 1991) can be readily combined with NGS to assist in the development of genetic markers for specific target loci. BSA consists of generating pools (bulks) of individuals with contrasting phenotypes for a specified trait and then identifying markers that are enriched for the corresponding parental allele in the relevant bulk. This approach is very flexible as different types of segregating populations can be used to develop the pools (Ehrenreich et al., 2010), and different NGS-enabled approaches can be used to identify SNPs or markers [exome capture (Hodges et al., 2007), RNA-Seq (Pickrell et al., 2010), whole-genome resequencing (Schneeberger et al., 2009), among others]. Several modifications of these core techniques have been recently implemented in model (reviewed in (James et al., 2013) and crop species (Abe et al., 2012; Liu et al., 2012; Mascher et al., 2014).

In this study, we propose a comprehensive methodology to identify high-resolution markers using NGS and high-throughput genotyping. We used near-isogenic lines (NILs) differing for the presence of Yr15 to develop a segregating F_2 population. After phenotyping, we implemented BSA and RNA-Seq to identify SNPs associated with Yr15. We developed an automated bioinformatics pipeline for the design of genome-specific assays, which circumvents the limitations associated with the polyploid wheat genome. These SNP markers were found to be closely linked to Yr15, suggesting that the proposed methodology can be applied to other agronomically important traits in polyploid species.

Results and discussion

Generation of bulks segregating for resistance to yellow rust

Near-isogenic lines segregating for the presence of Yr15 in the Avocet 'S' genetic background (AVS) (Wellings and McIntosh, 1998) were used to develop a segregating population (Figure 1a). F_2 seeds from three independent F_1 plants were sown in three 96-well trays, and their tissue sampled and stored. This sampling was conducted prior to fungal inoculation to reduce false associations in the expression data resulting from the resistance or susceptible response. F_2 plants were challenged with P. striiformis f. sp. tritici at 3-leaf stage where Yr15 is known to confer complete resistance. Within trays, segregation ratios were the expected 3 : 1 (resistant : susceptible) of a single dominant resistance gene such as Yr15 (χ^2 ranging from $P = 0.12$ to 0.42), although across all trays the segregation ratio deviates slightly from this expectation (χ^2 $P = 0.049$; 187 resistant and 45 susceptible F_2 plants). Previous studies using the same Yr15 donor germplasm have also shown segregation distortion (Randhawa

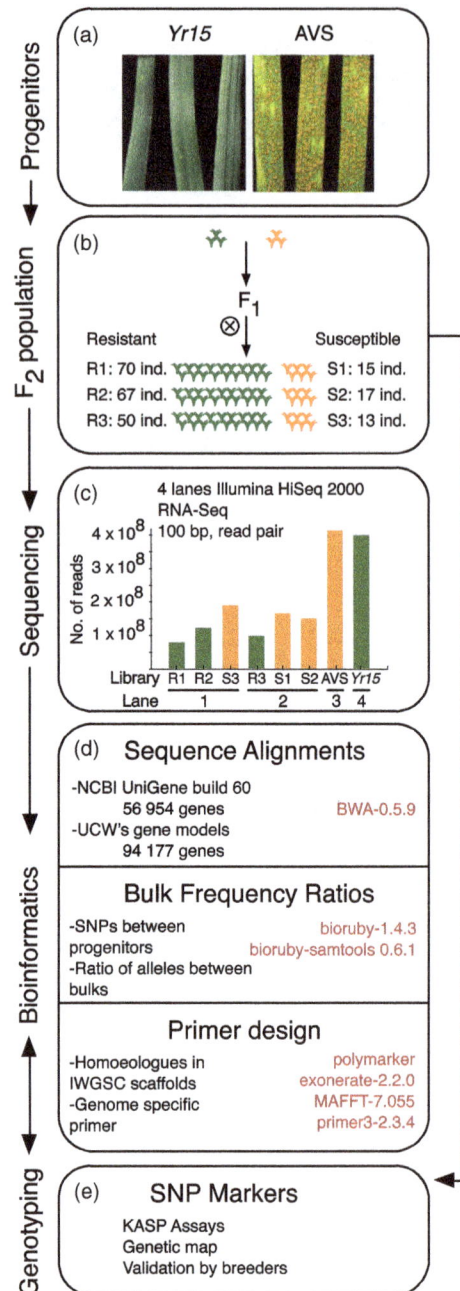

Figure 1 Overview of experimental methodology. (a) Phenotype of AVS + Yr15 (Yr15) and AVS seedling leaves 14 days after inoculation with Puccinia striiformis. (b) The F_2 population was developed by crossing AVS and Yr15, followed by self-pollination of the F_1 plants. The F_2 progeny was phenotyped and bulked according to the resistant or susceptible phenotype. (c) Number of reads generated by RNA-Seq of the parents and the bulks across 4 HiSeq 2000 lanes. (d) A bioinformatics pipeline was implemented to first align the transcriptome to the gene models, then call for SNPs and calculate the Bulk Frequency Ratios, and finally design the KASP assays. The resources used for each step of the pipeline are in black, whereas red text denotes the software and version used. (e) The SNP markers were used to genotype the segregating population and to develop a genetic map. The SNP markers were also validated in independent breeding material.

et al., 2009), although an alternative explanation for the increased number of resistant plants in the F_2 population could be a small number of escapes in the virulence assays (i.e. plants without the *Yr15* gene scored as resistant). Based on the phenotypic response, tissue of the 232 F_2 plants described above was independently pooled into three resistant (R1–R3) and three susceptible (S1–S3) bulks as outlined in Figure 1b and RNA-Seq libraries prepared for sequencing.

Sequencing and mapping

We performed RNA-Seq of each parent on a single Illumina HiSeq 2000 lane, whereas the susceptible and resistant bulks were multiplexed across two lanes to avoid lane-induced bias (Figure 1c) (Pickrell *et al.*, 2010). This design was implemented to look for consistency across bulks and to have an *in silico* mix of the resistant and susceptible bulks with the same expected coverage as the RNA-Seq from the parents.

There is no complete gene annotation for hexaploid wheat; therefore, we conducted two independent sequence alignments for the eight samples. First, we aligned the RNA-Seq reads to the 56 954 gene clusters in the NCBI UniGene build 60 database (NCBI, 2012) for hexaploid wheat (*T. aestivum*) (Figure 1d). The UniGene gene clusters provide a canonical representation of expressed sequence tags, high-throughput cloning and messenger RNA. To complement the UniGenes, we also aligned the reads to 94 177 gene models from tetraploid and hexaploid wheat that have been assembled and phased to distinguish between homoeologues, henceforth referred as UCW gene models (Krasileva *et al.*, 2013). We hypothesized that the presence of genome-specific gene models should improve the alignments to the targeted region compared to the canonical representation of the UniGenes.

The percentage of genes with an average coverage of at least 20× was 39% and 45% for AVS + *Yr15* (*Yr15*, hereafter) and AVS, respectively, across both sets of reference gene models (Figure 2a, Table S1). The performance of both reference sets was also similar across the resistant and susceptible bulks although the percentage of genes with coverage over 20× was lower in these samples (between 17% and 37%; Table S1) due to the lower number of sequencing and mapped reads (Table S2). To improve on these statistics and be able to score SNPs in genes with relatively low expression, two *in silico* mixes were prepared by merging the alignments of (i) susceptible bulks 1 with 2 (S1 + S2) and resistant bulks 1 with 2 (R1 + R2) and (ii)

all the susceptible (S1 + S2 + S3) and resistant bulks (R1 + R2 + R3). By merging all the bulks, we manage to increase the percentage of genes with coverage over 20× to 44% and 50% in the resistant and susceptible bulks (Table S1), respectively, which is similar to the values obtained from the parents.

SNP calling

Using a base coverage threshold of at least 20×, we identified SNPs between parental lines and the gene references. Roughly 3% more genes with polymorphisms were identified in AVS compared to *Yr15* (Figure 2b) for those genes longer than 500 bp and with at least 50% breadth coverage. This suggests that the lower coverage of *Yr15* is affecting the number of putative SNPs identified. Across both parental data sets, the use of the UCW gene models resulted in a higher number of monomorphic genes relative to the UniGene set, leading to a higher SNP frequency in the UniGenes (Figure 2b, Table S3). This is most likely due to the fact that multiple homoeologues are represented as a single sequence within the UniGene set as opposed to the homoeologue-specific UCW set in which a higher percentage of reads map to the correct homoeologues, thereby reducing the number of genes with SNPs.

As both references were generated from varieties different to AVS and are not complete, we called a base when at least 20% of the reads contained the nucleotide. When more than one putative consensus was identified, the corresponding International Union of Pure and Applied Chemistry (IUPAC) ambiguity code was used (Cornish-Bowden, 1985). We next compared SNPs between parental lines; those found in both parents are most likely homoeologous SNP or common varietal polymorphism between AVS and the reference varieties. These shared SNPs are not informative for this study and were thus eliminated from the analysis. On the other hand, SNPs that are unique to a single parent represent putative varietal SNPs and were therefore examined further. Focusing on these unique varietal SNPs, we identified 66 426 putative SNPs across 16 022 UCW genes (17% of UCW genes) and 52 262 putative SNPs across 11 056 UniGenes (19.4% of UniGene gene models) (Figure S1).

The relatively high number of genes with putative SNPs was not expected as BC_6 NILs had been used to develop the F_2

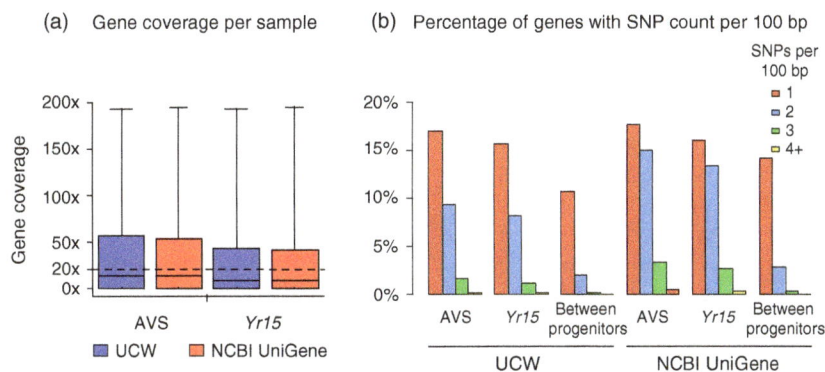

Figure 2 Gene coverage and SNP calling in parents. (a) Box plot distribution of the gene coverage of the parent reads (AVS and *Yr15*) across the UCW (blue) and the UniGene (red) gene models. The dashed line represents the 20× minimum coverage required for SNP calling. The full line represents the average coverage across all gene models. (b) Percentage of genes exhibiting SNPs across references. The number of SNPs between the parent reads and the corresponding references was calculated (per 100 bp, rounded). The 'between-parents' category corresponds to putative SNPs when comparing the consensus sequence between AVS and *Yr15*.

mapping population (<1% of genetic background is expected to be segregating between BC_6 NILs). We aligned both gene references to the Chinese Spring chromosome arm survey sequence scaffolds (CSS; IWGSC, 2014) with BLAT (Kent, 2002) (Data S1). A total of 80 031 UCW gene models (85.0%) and 41 118 UniGenes (72.2%) were assigned to the CSS assemblies (Table S4). Using the CSS as a common reference, we mapped the SNPs between parents using a recently published genetic map (Wang et al., 2014) as described below (in silico map). The SNPs distributed evenly across all chromosomes (Figure S2), suggesting that the AvocetS line used as the recurrent parent in the Yr15 NIL development (University of Sydney, Australia) was different to the AvocetS line used as the parent in the F_2 population (JIC, UK).

We confirmed the distinct nature of the AvocetS seed stocks by sourcing the original DNA from Sydney and comparing to our internal JIC stock through the iSelect 90k wheat SNP chip. When comparing two independent AvocetS seed stocks from JIC, only 58 of 71 972 valid assays (0.08%) were polymorphic. However, when comparing either JIC-AvocetS to the Sydney AvocetS stock, over 5000 assays (>7.5%) were polymorphic (Table S5). Although unexpected, this is not surprising considering that the AvocetS seed source was of different origin and the fact that commercial varieties with the same name have been released in both the UK and Australia. This exemplifies the importance of using genetic stocks of common origin between research groups.

Bulk frequency ratios

Our objective was to identify SNPs that were enriched for the corresponding parental allele in the appropriate bulk: AVS-derived SNPs for the susceptible bulks and Yr15-derived SNPs for the resistant bulks. The total number of SNPs scored from the individual bulks ranged from 15 261 to 31 891 across both reference sets (24.5%–48.0% of SNPs). When merging the three bulks, we could score over 95% of the SNPs from both reference sets (Table S6), suggesting that the coverage of the individual bulks is insufficient to score all the putative SNPs identified in the parents. To classify and prioritize the SNPs, we calculated the BFRs as previously described (Trick et al., 2012). Briefly, across each bulk, we calculated the frequency of the allele at each SNP position (SNP index; Takagi et al., 2013) and then determined the ratio between the bulks (BFR) for each SNP (Figure 1d). Thus, a high BFR in a Yr15-derived SNP is indicative of an allele that is both very frequent in the resistant bulk and depleted in the susceptible bulk.

We observed enrichment in SNPs from the short arm of group 1 chromosomes (1S) as we increased the minimum BFR threshold. Using the complete set of SNPs between parents (BFR = 0; no enrichment), the total called SNPs that map to group 1S is ~3.6% across the bulks, equivalent to the values observed across other chromosomes (Table S4). As we increase the threshold, the relative number of SNPs that map to group 1S increases until a BFR range of between 5 and 7. The relative enrichment only grows marginally after this, but the number of putative SNPs is reduced as the threshold is increased (Table S7, Figure S3). Therefore, we decided to set a threshold of BFR > 6 to select putative SNPs for further validation. We also catalogued presence/absence polymorphisms between bulks within the BFR > 6 category (BFR of infinity). This was an improvement over previous studies in polyploid species (Trick et al., 2012) that were unable to assess this type of variation as the SNP index had a base of 0 (i.e. no coverage in one of the bulks). In total, 1582 SNPs across 1173 genes had a BFR > 6.

In silico mapping

Across data sets, we found that ~60% of the mapped SNPs with BFR > 6 (872 of 1470 mapped SNPs) aligned to scaffolds in the group 1 chromosomes of hexaploid wheat, with no other chromosome group having over 4% (Table S8). Within this chromosomal group, the highest proportion of SNPs mapped to the B genome (54%, Figure 3a), with 255 assigned to the long arm (54%) and 214 to the short arm (46%). These results are consistent with previous studies that have located Yr15 in the short arm of chromosome 1B, near the centromere (Murphy et al., 2009; Peng et al., 2000; Sun et al., 1997), and the fact that the Yr15 introgression includes chromosomal regions from both the short arm and long arm of T. dicoccoides.

To further refine this location, we cross-referenced the SNPs with BFR > 6 (including infinity) with a recently published genetic map of 40 266 SNP markers (Wang et al., 2014) (Figure 3b). Using the CSS assemblies as the common reference across both data sets, we found that only 678 SNPs (across 474 genes) were successfully assigned to the genetic map. This represents only 43% of the 1582 SNP with a BFR > 6 (across 1173 genes). In this analysis, the unit of comparison was the CSS assemblies, not the SNPs or genes themselves. Therefore, this suggests that many of the CSS assemblies with SNPs in our parental lines did not have equivalent SNP in the 40 266 marker data set. Despite the fact that this genetic map was not sufficient to make best use of over 50% of the SNPs identified in the parental lines, it provided a more fine-scale resolution of the genetic positions compared to the wider chromosome arm information of the CSS. Of the 678 SNPs that were located on the genetic map, 325 mapped to chromosome 1B and 311 of them were located within a 30-cM interval (Figure 3c, Figure S4).

Although we do not find a smooth curve with a defined peak as those identified in sequenced diploid organism using QTL-seq (Takagi et al., 2013) or similar NGS-enabled genetic approaches (James et al., 2013), in practice, we observed a set of clusters with high BFR near the centromere of chromosome 1B (Figure 3c). This distribution would not be expected on a random set of genes as wheat chromosomes have relatively higher gene densities as distances increase from the centromere (Akhunov et al., 2003). Therefore, the relatively narrow distribution of SNPs in the centromeric region of chromosome 1B suggests that the approach was successful in enriching for SNPs that are closely linked to Yr15. The lack of a clear single peak is most likely due to a combination of factors including the bias induced by the difference in expression, the relative low resolution of the genetic map and the absence of a contiguous physical sequence with which to order a larger number of SNPs with higher confidence. This determines that multiple criteria need to be considered when prioritizing SNPs for marker development as genetic map position or high BFR alone is not sufficient.

Assay selection

We used three independent criteria to classify SNPs before marker development and validation. First, we used the BFR > 6 as the threshold for inclusion as detailed above. We used only SNPs that were present in at least two of the bulk replicates or in one of the in silico mixes to ensure consistency while still including SNPs with low coverage in a single bulk. Next, we selected SNPs in chromosome group 1S according to our preliminary in silico map information and previous studies (Murphy et al., 2009; Peng et al., 2000; Sun et al., 1997). Finally, we selected SNPs that

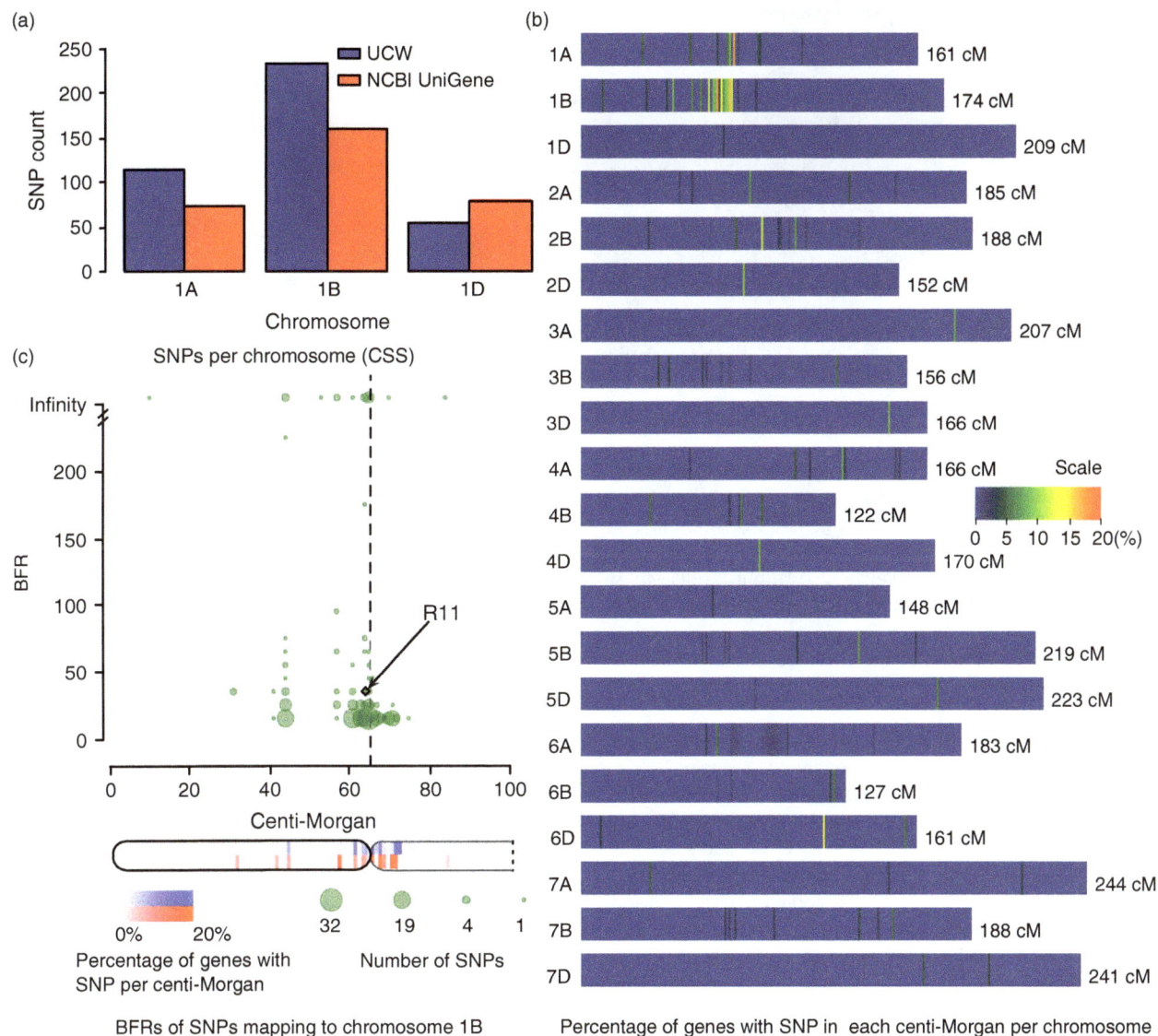

Figure 3 Physical and genetic location of SNPs with a bulk frequency ratio (BFR) > 6. (a) Location of SNPs according to the best alignment of the UniGene (red) and UCW (blue) gene models to the flow-sorted group 1 chromosomes from the Chinese Spring Survey sequence (CSS). (b) Percentage of genes with BFR > 6 SNPs per centi-Morgan (cM) across the 21 wheat chromosomes. The location of the genes was determined by the best alignment to the CSS scaffolds, and the location of these was determined by their position on a genetic map (Wang et al., 2014). (c) BFRs of mapped genes with SNPs on chromosome 1B. The area of the circle represents the number of SNPs clustered by location (windows size: 10 cM) and BFR (window size: 5). R11 is the only marker near the *Yr15* locus that had a corresponding position in the genetic map. The percentage of genes with SNPs per cM is also illustrated based on UCW (blue) and UniGene (red) gene models. The centromere is imputed by the centre of a window of 10 cM where the short arm switches to the long in the genetic map. BFRs correspond to those from the mixed *in silico* bulk S1 + S2 + S3/R1 + R2 + R3.

originate from the *Yr15* parent. This was done to ensure that the SNP would tag the *Yr15 T. dicoccoides* introgression as opposed to a unique SNP present only in the AVS genetic background which would be of less use in breeding programmes.

The multiple criteria reduced the number of genes with putative SNP from over 27 000 to just 175 (98 and 77 in the UCW and UniGene references, respectively) (Figure 4). As the two references come from independent sources, we aligned the selected genes to the opposite gene set and found that roughly half of the selected genes overlapped between references while each contained a similar proportion of unique genes (Figure 4). From the original 175 genes, we selected the 50 SNPs with highest BFR for marker development, of which 15 were redundant between gene sets, thereby resulting in 35 SNPs to test.

We next examined the behaviour of the SNPs across the individual bulks and *in silico* mixes to evaluate the usefulness of the multiple bulk strategy. We initially expected the number of SNPs with BFRs of infinity to drop in the *in silico* mixes as the additional coverage would reduce instances of calling for the absence of an allele. However, the opposite occurred as new genes with low expression and which were not represented in any single bulk comparison were rescued in the mix due to the higher combined coverage (Figure 5). Other SNPs were present consistently across all bulks and mixes, although their exact BFR values changed dependent on the bulk (Figure 5, marker R5). In other cases, individual SNPs would be missing from an individual bulk, but present in all others (marker R8). These results highlight the importance of adequate coverage across the bulks and argue

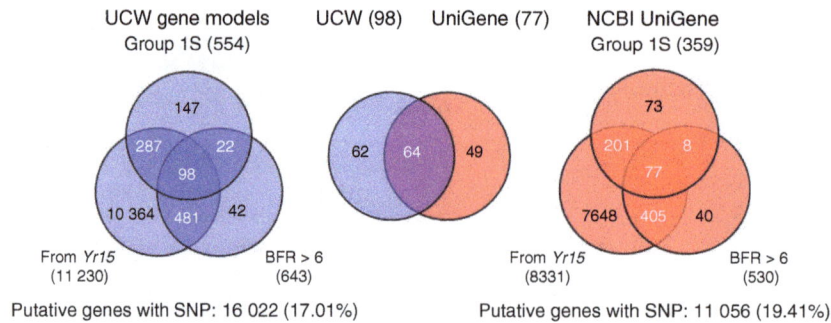

UCW gene models
Group 1S (554)   UCW (98)   UniGene (77)   NCBI UniGene
Group 1S (359)

Putative genes with SNP: 16 022 (17.01%)   Putative genes with SNP: 11 056 (19.41%)

Figure 4 Selection criteria for marker design. Venn diagrams based on the three selection criteria (SNP in the short arm of chromosome group 1; SNP has a bulk frequency ratio > 6; and SNP is from the *Yr15* parent) for the UCW (blue) and UniGene (red) gene models. The centre diagram shows the intersection between common genes matching all three criteria across both data sets. Note that the numbers are not directly additive as in cases, multiple models from one reference set will relate to a single gene model in the other.

for a strategy in which improved coverage is the main objective rather than having independent replications of the bulks themselves. Although previous studies using genomic DNA found that relatively low coverage (<5×) is sufficient to call for SNPs in model organisms (Schneeberger and Weigel, 2011), our results are in agreement with studies using populations for SNP calling (Abe *et al.*, 2012; Takagi *et al.*, 2013). The use of RNA-Seq data leads to a nonuniform distribution of the coverage across genes, and this is particularly evident on genes with low expression levels (Mortazavi *et al.*, 2008).

Approximately 60% of the reference gene models produced a unique hit with >99% sequence identity to a single IWGSC scaffold (Table S9). This suggests that in many cases, there is no unique homoeologue reference and hence the reads from at least two genomes will map to the same reference. We addressed this in the SNP calling by masking polymorphic positions within each parental line using the IUPAC ambiguity code. However, this also meant that a high coverage (20×) was required to maintain a high confidence for SNP calling. We had previously shown in tetraploid wheat that increasing coverage from 8× to 16× reduces the total number of putative SNPs identified by 60%, but increases the validation rate from 57% to 83% (Trick *et al.*, 2012). The balance between increased coverage, reduced number of SNPs and increased validation rates needs to be weighed in each situation according to the objectives of the study.

SNP validation

To validate the putative SNPs and generate a genetic map of the *Yr15* locus, we designed KASP assays for the 35 selected SNPs. We developed an automated bioinformatics pipeline, called PolyMarker, to increase the probability of designing homoeologue-specific assays. This feature is particularly relevant as most SNP platforms cannot easily distinguish between heterozygote and homozygote individuals in polyploid species. Briefly, Poly-Marker generates a multiple alignment between the target SNP sequence and the CSS scaffolds (IWGSC, 2014) for each of the three wheat genomes. Informative positions (varietal and homoeologous SNPs) are highlighted with respect to the target genome (1BS for *Yr15*). The common primer is selected to incorporate a homoeologue-specific or semi-specific base at the 3′ end, whereas the competing diagnostic primers are designed to incorporate the alternative varietal bases at their 3′ ends (Figure 6a). PolyMarker designed 17 specific and 9 semi-specific assays to chromosome arm 1BS, whereas 9 assays were not specific due to missing information from the CSS scaffolds

(Figures 1d and 6a). The mask generated by PolyMarker enabled the identification of putative homoeologous variants (between genomes as opposed to varietal SNPs) which were present in the data set, but that were not identified previously. By exploiting the latest genomic resources in wheat, PolyMarker increases the likelihood of generating codominant SNP assays that are required for commercial MAS programmes and genotyping of F_2 mapping populations like the one used in this study.

In the first instance, we tested the 35 SNPs in a test panel consisting of the parents (AVS and *Yr15*) plus six commercial wheat varieties, three of which contain *Yr15* and three without it. Two of the lines without *Yr15* include *T. dicoccoides* (the *Yr15* donor species; McIntosh *et al.*, 1995) in their pedigree and were used to assess the specificity of the *Yr15*-derived SNPs (Table S10). From the 35 SNPs tested, 28 were polymorphic across the parents (80%), with three diagnostic for *Yr15* in the panel (R5, R8 and R33, Figure 6b). We also identified five SNPs as homoeologous, three of which were monomorphic across the parents and two polymorphic SNPs that were not diagnostic across the panel. To determine the genetic position of these SNPs relative to *Yr15*, we developed a genetic map using the phenotyped F_2 individuals (Figure 1b,e).

Genetic map

To determine the genetic position of the 28 polymorphic SNP markers with respect to *Yr15*, we first genotyped a subset of the F_2 population (66 plants). Twenty-three of the 28 SNP markers were linked to *Yr15* (82%; Figure S5), with several markers mapping within a few cM. This suggests that the multiple criteria for assay selection performed effectively in selecting markers across the *Yr15* region. To improve the resolution of the genetic map, the complete F_2 population was then assessed with the seven most closely linked markers, which included the three diagnostic markers across the variety panel (R5, R8, R33). In addition, we mapped two microsatellite markers that are currently used by breeders to select for *Yr15* in UK germplasm (*Xbarc8* and *Xgwm413*), and we included one additional marker based on barley–wheat synteny (R43). From the 232 F_2 plants with phenotypic information, 196 individuals were genotyped reliably, with no more than 1 missing marker data point. Using these markers, we mapped *Yr15* to a 0.77-cM interval, with *Yr15* mapping 0.26-cM distal to R8/*Xgwm413* and 0.51 cM proximal to R5/R11 (Figure 7).

Although we did not find a completely linked marker to *Yr15*, the sub-cM interval is within the expected resolution of an F_2

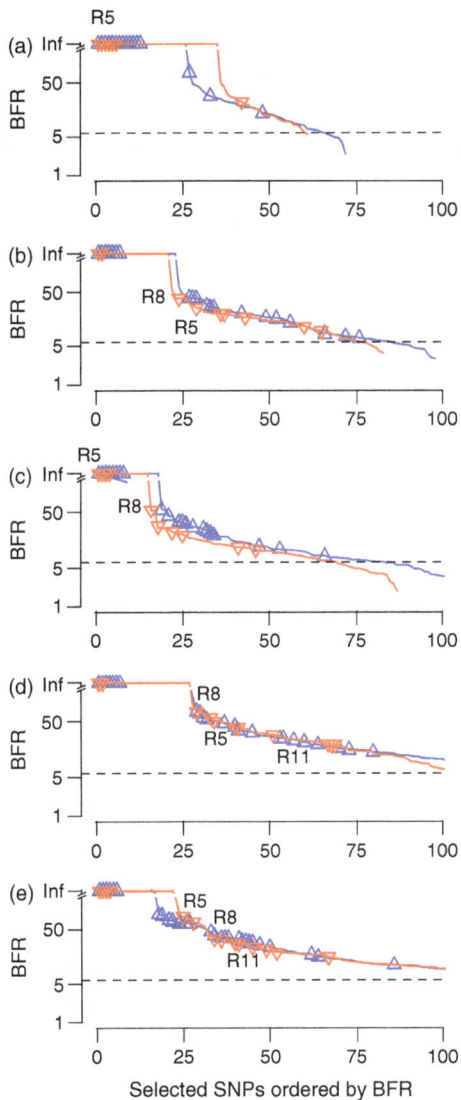

Figure 5 Bulk frequency ratio (BFR) of selected SNPs across the individual bulks (a–c) and *in silico* mixes (d, e). The BFR values of selected SNPs were sorted in descending order across the different bulks and according to their origin (UCW, red; UniGene, blue). The dotted line represents the BFR threshold of >6 (logarithmic scale). Validated SNPs are indicated by open triangles, and SNPs corresponding to markers R5, R8 and R11 are labelled across different bulks and mixes. Note that some SNPs are below the threshold in a specific bulk as they meet the BFR criteria across others.

population of 196 individuals (392 gametes should provide a 0.26-cM resolution). The shift from microsatellite to SNP markers is also of relevance as SNPs have become the most widely used marker system within MAS pipelines in a majority of breeding programmes (Bevan and Uauy, 2013).

Validation on breeding germplasm

To determine the usefulness of these markers in breeding programmes, we tested markers R11, R5 and R8 across 122 doubled haploid (DH) lines. These DH lines were derived from crosses between five different UK varieties/breeding lines to *Yr15* derivatives known to carry the resistance gene. The expected *Yr15* haplotype corresponded to T, G and C alleles at markers R11, R5 and R8, respectively (TGC haplotype). The DH lines were

tested at seedling stage for reaction to *P. striiformis*, with 84 showing complete resistance and 34 presenting an intermediate or completely susceptible reaction. The resistant lines all carried the complete *Yr15* haplotype (TGC, Figure 7c) across the three SNP markers with the exception of five lines which had a single missing data point, but were otherwise consistent. This compared favourably with the most diagnostic in-house SNP markers available within the breeding programmes. Using the three in-house markers, 79 resistant lines carried the expected haplotype, but five completely resistant DH lines were scored as false negative due to the presence of the non-*Yr15* haplotype. Within the intermediate and susceptible DH lines, all but one had a non-*Yr15* haplotype (CAT or TAT) across R11, R5 and R8 (Figure 7c). This single DH line was scored as a false positive as it carried the TGC *Yr15* haplotype, but was found to have an intermediate (chlorotic) reaction to *P. striiformis*. This line was also the only one scored as a false positive using the three in-house markers.

We further validated the markers in a larger set of 103 varieties and breeding lines representing UK elite germplasm. The *Yr15* status of these lines was known from pedigree analysis and characterization with SSR marker *Xbarc8* (8 resistant, 93 suscep-tible and 2 heterozygous lines). Marker R5 was completely diagnostic with 100% of called alleles agreeing with *Yr15* status, whereas marker R8 had a single false-positive result. Marker R11 was less diagnostic with a 12% false-positive error rate.

These results show that markers R5 and R8 are more diagnostic than the R11 SNP, which was found segregating among intermediate individuals and had a relatively high false-positive error in the UK elite germplasm. Markers R5 and R8 were linked in coupling and diagnostic in 112 of 113 DH lines tested (Figure 7c) and across 102 of the 103 varieties examined. Although flanking microsatellites markers have been available for some time and have proved useful in characterizing sources of resistance, they are not suited to large-scale MAS. Our results suggest that R5 and R8 can be used to replace the published microsatellites (and in-house marker) for *Yr15* MAS in the pedigree programmes being evaluated. These markers will enable effective pyramiding of *Yr15* with other resistance genes facili-tating the deployment of multiple *R*-gene combinations in wheat varieties. They also represent a diagnostic tool in predicting the *Yr15* status of lines of unknown performance or pedigree, although only lines with a nonrecombined haplotype should be unambiguously categorized. It will be important to now extend the haplotype analysis and evaluate the diagnostic nature of R5 and R8 across a wider set of world varieties and breeding lines.

Conclusion

This work exemplifies the use of BSA combined with RNA-Seq in an F_2 population to generate high-density genetic maps across target loci in polyploid wheat. Through the use of a bulk frequency ratio threshold and a series of complementary criteria, we were able to identify putative SNPs across the *Yr15* locus. We developed an automated bioinformatics pipeline, PolyMarker, to convert these SNPs into codominant genome-specific KASP assays, thereby overcoming a major barrier associated with the polyploid wheat genome. Using these SNPs, we mapped *Yr15* to a sub-cM interval, in line with the expected resolution of an F_2 population. The two most closely linked KASP assays were diagnostic across UK pedigree programmes, providing a reliable haplotype for MAS of *Yr15* in hexaploid wheat. The proposed methodology can be applied to generate high-density genetic

(a)

```
>gnl|UG|Ta#S58861868:A214G-AvocetS
cacagatatcacactgaacatactgatgaSacgattggttatgggccttgaAgatagMagcagatAacttcagtgtaatccaagttgactg

>gnl|UG|Ta#S58861868:A214G-Yr15
cacagatatcacactgaacatactgatgaSacgattggttatgggccttgaGgatagMagcagatGacttcagtgtaatccaagttgactg

>iwgsc_css_1AS_scaff_3271989
cacagatatcacactgaacatactgatgaCacgattggttatgggccttgaAgatagAagcagatAacttcagtgtaatccaagttgactg

>iwgsc_css_1BS_scaff_3474683
cacagatatcacactgaacatactgatgaGacgattggttatgggccttgaGgatagCagcagatAacttcagtgtaatccaagttgactg

>iwgsc_css_1DS_scaff_1913057
cacagatatcacactgaacatactgatgaCacgattggttatgggccttgaGgatagCagcagatAacttcagtgtaatccaagttgactg

MASK
------------------------------G--------------------:----c------&--------------------------
```

Genome specific Homoeologous Genome semi-specific Varietal

(b)

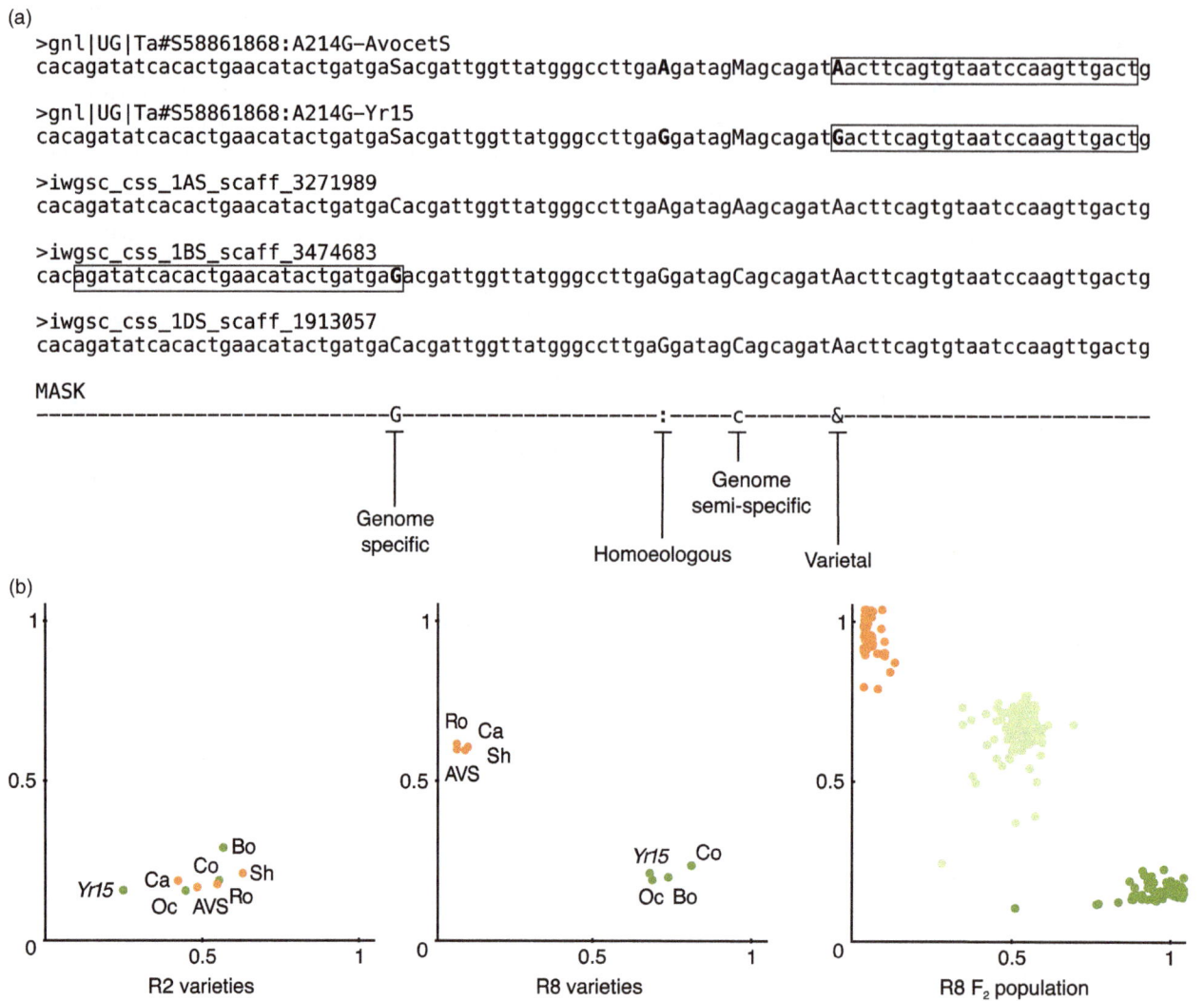

Figure 6 Genome-specific primer design and KASP assay. (a) Multiple alignment of the parent sequences and the Chinese Spring Survey scaffolds for each of the three wheat genomes. The mask highlights informative positions across the alignment. These include the target varietal polymorphism (&), homoeologous SNPs (:), and genome-specific and semi-specific positions (upper and lower case, respectively). The diagnostic primers incorporate the alternative varietal SNP at the 3′ end, but are otherwise similar. The third common primer is preferentially selected to include a genome-specific base (in this case, chromosome arm1BS), or a semi-specific base in the absence of an adequate genome-specific position. (b) KASP output from the wheat variety panel with (Ochre, Boston, Cortez) and without (Robigus, Cadenza and Shamrock) Yr15. Marker R2 is monomorphic while R8 is polymorphic between varieties know to carry the gene. Marker R8 results for the F_2 population showing three distinct clusters. The central cluster (light green) is comprised of heterozygous individuals, whereas clusters near the axes are homozygous for either AVS (VIC; orange) or Yr15 (FAM; dark green).

maps in F_2 populations and to provide closely linked markers for target traits in wheat and other polyploid species.

Experimental procedures

Plant material and pathology assays

Near-isogenic lines segregating for the presence of Yr15 in the Avocet 'S' (AVS) genetic background (Wellings and McIntosh, 1998) were crossed to generate an F_2 population. Prior to infection, samples were taken at seedling stage for DNA and RNA purification. A 2-cm section from the 1st leaf of each plant was used for DNA purification (Pallotta et al., 2003). Individual F_2 grains were sown into separate cells (2.5 cm × 2.5 cm) of single seed descent trays. The parents of the F_2 population (AVS and Yr15) were used as controls and replicated 6 times. Seedlings were inoculated at the three leaf stage with a mix of UK isolates

of P. striiformis f.sp. tritici known to be avirulent for Yr15 (08/21, 08/501, 03/7) (Segovia et al., 2014). Disease symptoms were recorded 2 weeks after inoculation and again after a further week although the scores remained the same. Plants were classified as resistant or susceptible based upon the presence of sporulation. Only two grain failed to germinate yielding data for 232 F_2 plants. Before RNA isolation, samples from the F_2 population were grouped into 6 bulks comprising either resistant [Bulk R1 (70 individuals), R2 (67) and R3 (50)] or susceptible individuals [Bulk S1 (15 individuals), S2 (17) and S3 (13)]. Five plants of each parental line (AVS and Yr15) were pooled for RNA and DNA extractions.

Sample preparation

For each bulk, total RNA was isolated using TRIzol reagent (Invitrogen, Paisley, UK) and treated with DNA-free DNase and

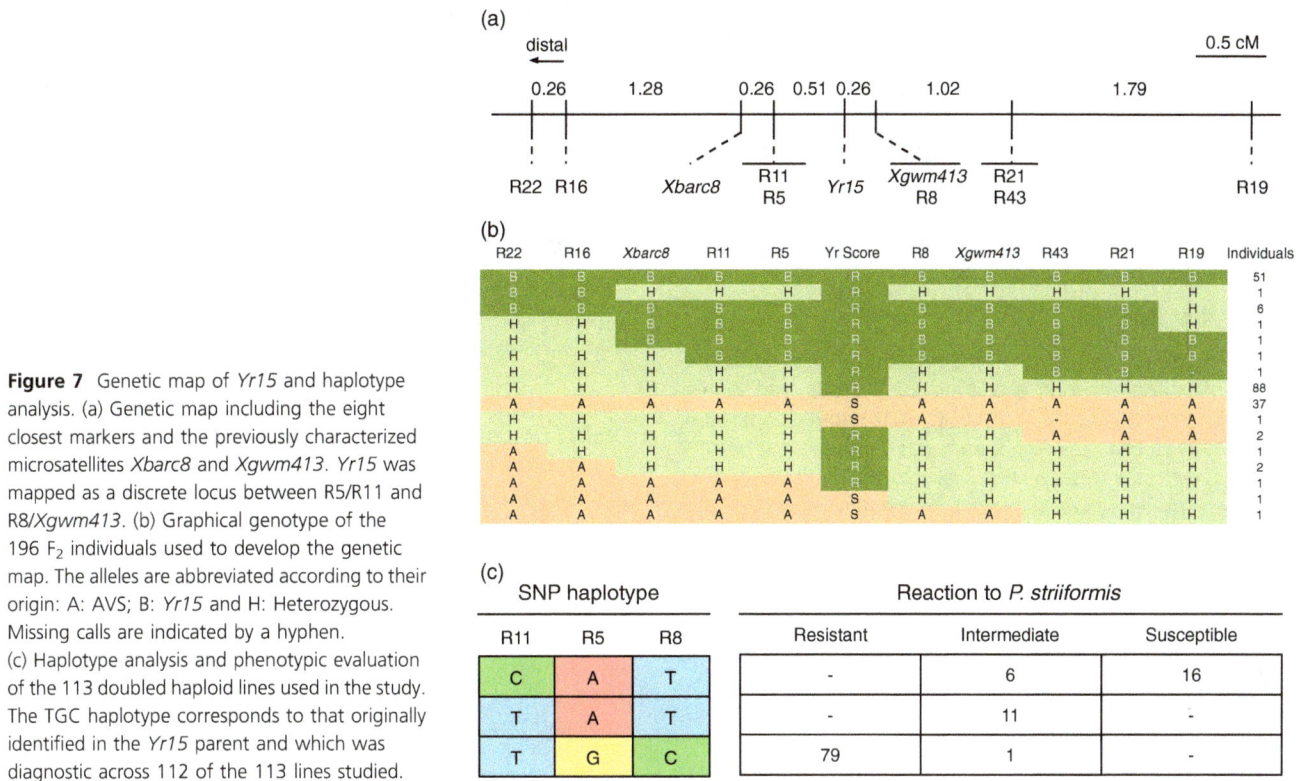

Figure 7 Genetic map of *Yr15* and haplotype analysis. (a) Genetic map including the eight closest markers and the previously characterized microsatellites *Xbarc8* and *Xgwm413*. *Yr15* was mapped as a discrete locus between R5/R11 and R8/*Xgwm413*. (b) Graphical genotype of the 196 F$_2$ individuals used to develop the genetic map. The alleles are abbreviated according to their origin: A: AVS; B: *Yr15* and H: Heterozygous. Missing calls are indicated by a hyphen. (c) Haplotype analysis and phenotypic evaluation of the 113 doubled haploid lines used in the study. The TGC haplotype corresponds to that originally identified in the *Yr15* parent and which was diagnostic across 112 of the 113 lines studied.

the Removal Reagents kit (Ambion, Paisley, UK) following the manufacturer's instructions. RNA was quantified by fluorometry using a Qubit fluorometer with QUANT-iT RNA assay (Invitrogen), and the quality was assessed by a Bioanalyser RNA chip [RNA integrity number (RIN) higher than 7]. The Illumina TruSeq version 2 library synthesis kit was used according to manufacturer's instructions to develop 8 sequencing libraries (AVS, *Yr15*, R1–3, S1–3). The average fragment size for all libraries was 300 bp. Sequencing was carried out on the Illumina HiSeq2000 instrument using sequencing chemistry v4 (FC-104-4001; Little Chesterford, Illumina) and software SCS 2.6 and RTA 1.6. Parental libraries (AVS and *Yr15*) were each run in a single lane for 100 cycles for each paired end, whereas the bulk libraries were assembled into two pools [(R1 + R2 + S3) and (R3 + S1 + S2); Figure 1] and each sequenced in a single lane for 100 cycles paired end.

Alignment of Illumina reads to gene models

The Illumina intensity files were converted to bases using Casava v1.8 (Illumina, 2011). The two lanes containing the bulks were deconvoluted with a tolerance of 1 mismatch in the barcode, whereas the parental lanes were not indexed. The FASTQ compressed files were left in chunks of 4 000 000 clusters, the default of the BCL conversion from Casava, to allow parallel processing in the downstream analysis. The quality of the sequencing was confirmed with FastQC v 0.10.1 (Babraham Bioinformatics, 2012). The RNA-Seq reads were aligned to the wheat UniGene build 60 (Pontius *et al.*, 2003) and to the UCW gene models (Krasileva *et al.*, 2013), including the *Triticum turgidum* ORFs and the complementary wheat ORFs (MAS Wheat, 2013), using BWA 0.5.9 (Li and Durbin, 2009) with the default parameters for read pair libraries. The alignments were stored as a single BAM file per each sample against a given reference, which

allow compression and random access to the alignments (Li and Durbin, 2009).

Bulk frequency ratios

To identify variant candidates, the methodology described in Trick *et al.* (2012) was used and extended to work with BAM files and to consider cases when a variant is completely absent from the parental sequences. The consensus from the two parents is called, allowing for ambiguities to take into account possible homoeologues when at least 20% of the bases differ from the reference. The consensus is stored with standard IUPAC codes (Cornish-Bowden, 1985). On the bases where the consensus is different between the parents, the bulk frequency ratio (BFR) was calculated as in Trick *et al.* (2012). The algorithm was implemented using Ruby on the top of BioRuby (Goto *et al.*, 2010) and extending the functionality of bio-samtools (Ramirez-Gonzalez *et al.*, 2012). The BFRs were calculated independently on each of the three comparisons (Bulk 1: S1–R1, Bulk 2: S2–R2 and Bulk 3: S3–R3) and with *in silico* mixes of bulks 1 and 2; and bulks 1, 2 and 3 produced by merging the BAM files using samtools (Li *et al.*, 2009). A list was produced with all the scores in a table format containing the BFR, coverage, ratio of the SNP base and the parent containing the SNP (Data S2–S4).

The UniGenes, UCW gene models and the contigs containing the 46 977 genetically mapped SNPs from Wang *et al.* (2014) were aligned with BLAT (Kent, 2002) to the CSS scaffolds (IWGSC, 2014). To assign the genes to a specific scaffold, we selected the best hit from the BLAT alignment, provided that at least 60% of the gene was covered in the scaffold with at least 90% of identity. This selection criterion reduces the number of spurious hits in repetitive motifs in the gene and allows the assignment to a homoeologous chromosome even when the correct homoeologue is missing from the CSS scaffolds. The origin of the scaffold was used to assign a putative chromosome arm. In addition, the gene models were also aligned to the cDNA of *Hordeum vulgare* (International Barley Genome Sequencing Consortium, 2012) from *Ensembl*! Plants, release 16 (Kersey *et al.*, 2012). The genetic position of the wheat contigs with SNPs was used to calculate the density of SNPs with a BFR > 6 and for all SNPs between AVS and *Yr15*. After selection using the criteria outlined in the results section, the selected genes were sorted by BFR in the mix between samples to select the top 50 SNPs for further validation via KASP markers.

Primer design

To design the primers for the 35 selected SNP candidates, the genes were aligned using exonerate 2.2.0 (Slater and Birney, 2005) to the IWGSC scaffolds (IWGSC, 2014) from chromosome arms 1AS, 1BS and 1DS. The search was limited to the best 10 matches with the 'est2genome' alignment model, which is optimized to recognize the exon–intron junctions. The best alignment for each chromosome was extracted, and 100 bases on either side of the SNP were kept. The consensus sequences from the RNA-Seq alignments containing a variation of each parent were extracted and trimmed to have at most 100 bases on either side of the variation. MAFFT v7.055 (Katoh and Standley, 2013) was used to produce a local alignment of the exons across the different genomes (mafft options: −maxiterate 1000, −localpair). From the local alignments, the positions of the polymorphisms between genomes were selected as putative positions for genome-specific primers. The forward oligos were designed to contain the polymorphism between the parents in the 3′ end. To increase the probability of genome-specific amplification, the reverse primers were selected to have a polymorphic base across the A, B and D genomes in the 3′ end (selecting the base observed in the B genome, Figure 6). The viability of the primers was assessed with primer3 v2.3.5 (Rozen and Skaletsky, 1999). Both the detailed code of the PolyMarker pipeline (https://github.com/TGAC/bioruby-polyploid-tools) and a web interface (http://polymarker.tgac.ac.uk) are available.

KASP assays

Oligos were ordered from Sigma-Aldrich (Gillingham, UK), with primers carrying standard FAM or HEX compatible tails (FAM tail: 5′ GAAGGTGACCAAGTTCATGCT 3′; HEX tail: 5′ GAAGGTCGGAGTCAACGGATT 3′) with the target SNP at the 3′ end. Primer mix was set up as recommended by LGC [46 µL dH$_2$O, 30 µL common primer (100 µM) and 12 µL of each tailed primer (100 µM)] (LGC Genomics, 2013). Assays were tested in 384-well format and set up as 4-µL reactions [2-µL template (10–20 ng of DNA), 1.944 µL of V4 2× Kaspar mix and 0.056 µL primer mix]. PCR cycling was performed on a Eppendorf Mastercycler pro 384 using the following protocol: hotstart at 95 °C for 15 min, followed by ten touchdown cycles (95 °C for 20 s; touchdown 65 °C, −1 °C per cycle, 25 s) then followed by 30 cycles of amplification (95 °C 10 s; 57 °C 60 s). As KASP amplicons are smaller than 120 bp, an extension step is unnecessary in the PCR protocol. 384-well optically clear plates (Cat. No. E10423000; Starlab Milton Keynes, UK) were read on a Tecan Safire plate reader. Fluorescence was detected at ambient temperature. If the signature genotyping clusters had not formed after the initial amplification, additional amplification cycles (usually 5–10) were conducted, and the samples were read again. Data analysis was performed manually using Klustercaller software (version 2.22.0.5; LGC Hoddesdon, UK).

Genetic map construction

JoinMap version 3 (van Ooijen and Voorrips, 2002) was used for linkage analysis and genetic map construction, using default settings. The linkage to *Yr15* was determined using a divergent log-of-odds (LOD) threshold of 3.0, and genetic distances were computed based on recombination frequency.

Acknowledgements

We would like to thank Martin Trick (JIC) for discussions on implementation of BFRs, Sarah Ayling and Jonathan Wright (TGAC) for assistance with the IWGSC CSS scaffolds, the IWGSC for prepublication access to the CSS scaffolds, and Prof. Bob McIntosh and Dr. Peng Zhang (University of Sydney) for providing DNA of AvocetS. This work was supported by grants BB/J004588/1, BB/J012017/1 and BB/J003557/1 from the UK Biotechnology and Biological Sciences Research Council (BBSRC) and a Norwich Research Park PhD Studentship and The Genome Analysis Centre Funding and Maintenance Grant to RHRG.

References

Abe, A., Kosugi, S., Yoshida, K., Natsume, S., Takagi, H., Kanzaki, H., Matsumura, H., Yoshida, K., Mitsuoka, C., Tamiru, M., Innan, H., Cano, L., Kamoun, S. and Terauchi, R. (2012) Genome sequencing reveals agronomically important loci in rice using MutMap. *Nat. Biotechnol.* **30**, 174–178.

Akhunov, E.D., Goodyear, A.W., Geng, S., Qi, L.-L., Echalier, B., Gill, B.S., Gustafson, J.P., Lazo, G., Chao, S. and Anderson, O.D. (2003) The organization and rate of evolution of wheat genomes are correlated with recombination rates along chromosome arms. *Genome Res.* **13**, 753–763.

Allen, A.M., Barker, G.L.A., Wilkinson, P., Burridge, A., Winfield, M., Coghill, J., Uauy, C., Griffiths, S., Jack, P., Berry, S., Werner, P., Melichar, J.P.E., McDougall, J., Gwilliam, R., Robinson, P. and Edwards, K.J. (2013) Discovery and development of exome-based, co-dominant single nucleotide polymorphism markers in hexaploid wheat (*Triticum aestivum* L.). *Plant Biotechnol. J.* **11**, 279–295.

Babraham Bioinformatics (2012) *FastQC A Quality Control tool for High Throughput Sequence Data.* http://www.bioinformatics.babraham.ac.uk/projects/fastqc/

Bernardo, R. (2008) Molecular markers and selection for complex traits in plants: learning from the last 20 years. *Crop Sci.* **48**, 1649–1664.

Bevan, M.W. and Uauy, C. (2013) Genomics reveals new landscapes for crop improvement. *Genome Biol.* **14**, 206.

Cavanagh, C.R., Chao, S., Wang, S., Huang, B.E., Stephen, S., Kiani, S., Forrest, K., Saintenac, C., Brown-Guedira, G.L., Akhunova, A., See, D., Bai, G., Pumphrey, M., Tomar, L., Wong, D., Kong, S., Reynolds, M., da Silva, M.L., Bockelman, H., Talbert, L., Anderson, J.A., Dreisigacker, S., Baenziger, S., Carter, A., Korzun, V., Morrell, P.L., Dubcovsky, J., Morell, M.K., Sorrells, M.E., Hayden, M.J. and Akhunov, E. (2013) Genome-wide comparative diversity uncovers multiple targets of selection for improvement in hexaploid wheat landraces and cultivars. *Proc. Natl Acad. Sci.* **110**, 8057–8062.

Collard, B.C.Y. and Mackill, D.J. (2008) Marker-assisted selection: an approach for precision plant breeding in the twenty-first century. *Philos. Trans. R. Soc. B Biol. Sci.* **363**, 557–572.

Cornish-Bowden, A. (1985) Nomenclature for incompletely specified bases in nucleic acid sequences: recommendations 1984. *Nucleic Acids Res.* **13**, 3021.

Ehrenreich, I.M., Torabi, N., Jia, Y., Kent, J., Martis, S., Shapiro, J.A., Gresham, D., Caudy, A.A. and Kruglyak, L. (2010) Dissection of genetically complex traits with extremely large pools of yeast segregants. *Nature*, **464**, 1039–1042.

Fisher, M.C., Henk, D.A., Briggs, C.J., Brownstein, J.S., Madoff, L.C., McCraw, S.L. and Gurr, S.J. (2012) Emerging fungal threats to animal, plant and ecosystem health. *Nature*, **484**, 186–194.

Food and Agriculture Organization of the United Nations (2012) *FAOSTAT.* http://faostat.fao.org/site/609/DesktopDefault.aspx?PageID=609

Galushko, V. and Gray, R. (2014) Twenty five years of private wheat breeding in the UK: lessons for other countries. *Sci. Public Policy*, doi: 10.1093/scipol/scu004.

Global Rust Reference Center (2014) *Pathotype by Country*. http://wheatrust.org/yellow-rust/pathotype-by-country/

Goto, N., Prins, P., Nakao, M., Bonnal, R., Aerts, J. and Katayama, T. (2010) BioRuby: bioinformatics software for the Ruby programming language. *Bioinformatics*, **26**, 2617–2619.

Gupta, P.K., Mir, R.R., Mohan, A. and Kumar, J. (2008) Wheat genomics: present status and future prospects. *Int. J. Plant Genomics*, **2008**, 36.

Hodges, E., Xuan, Z., Balija, V., Kramer, M., Molla, M.N., Smith, S.W., Middle, C.M., Rodesch, M.J., Albert, T.J., Hannon, G.J. and McCombie, W.R. (2007) Genome-wide in situ exon capture for selective resequencing. *Nat. Genet.* **39**, 1522–1527.

Hovmøller, M.S., Walter, S. and Justesen, A.F. (2010) Escalating threat of wheat rusts. *Science*, **329**, 369.

Illumina (2011) *CASAVA v1.8.2 User Guide*. pp. 19–46. San Diego, CA, USA: Illumina.

International Barley Genome Sequencing Consortium (2012) A physical, genetic and functional sequence assembly of the barley genome. *Nature*, **491**, 711–716.

IWGSC (2014) A chromosome-based draft sequence of the hexaploid bread wheat (*Triticum aestivum*) genome. *Science*, **345**, 1251788.

James, G., Patel, V., Nordstrom, K., Klasen, J., Salome, P., Weigel, D. and Schneeberger, K. (2013) User guide for mapping-by-sequencing in Arabidopsis. *Genome Biol.* **14**, R61.

Katoh, K. and Standley, D.M. (2013) MAFFT multiple sequence alignment software version 7: improvements in performance and usability. *Mol. Biol. Evol.* **30**, 772–780.

Kent, W.J. (2002) BLAT—the BLAST-like alignment tool. *Genome Res.* **12**, 656–664.

Kersey, P.J., Staines, D.M., Lawson, D., Kulesha, E., Derwent, P., Humphrey, J.C., Hughes, D.S., Keenan, S., Kerhornou, A. and Koscielny, G. (2012) Ensembl genomes: an integrative resource for genome-scale data from non-vertebrate species. *Nucleic Acids Res.* **40**, D91–D97.

Krasileva, K., Buffalo, V., Bailey, P., Pearce, S., Ayling, S., Tabbita, F., Soria, M., Wang, S., Consortium, I., Akhunov, E., Uauy, C. and Dubcovsky, J. (2013) Separating homeologs by phasing in the tetraploid wheat transcriptome. *Genome Biol.* **14**, R66.

Lai, K., Lorenc, M.T., Lee, H.C., Berkman, P.J., Bayer, P.E., Visendi, P., Ruperao, P., Fitzgerald, T.L., Zander, M., Chan, C.-K.K., Manoli, S., Stiller, J., Batley, J. and Edwards, D. (2014) Identification and characterization of more than 4 million intervarietal SNPs across the group 7 chromosomes of bread wheat. *Plant Biotechnol. J.* doi:10.1111/pbi.12240.

LGC Genomics (2013) http://www.lgcgroup.com/services/genotyping

Li, H. and Durbin, R. (2009) Fast and accurate short read alignment with Burrows-Wheeler transform. *Bioinformatics*, **25**, 1754–1760.

Li, H., Handsaker, B., Wysoker, A., Fennell, T., Ruan, J., Homer, N., Marth, G., Abecasis, G. and Durbin, R. (2009) The sequence alignment/map format and SAMtools. *Bioinformatics*, **25**, 2078–2079.

Liu, S., Yeh, C.-T., Tang, H.M., Nettleton, D. and Schnable, P.S. (2012) Gene mapping via bulked segregant RNA-seq (BSR-seq). *PLoS One*, **7**, e36406.

MAS Wheat (2013) *MAS Wheat Transcriptome*. Supplemental File 17. http://maswheat.ucdavis.edu/Transcriptome/index.htm.

Mascher, M., Jost, M., Kuon, J.-E., Himmelbach, A., Aßfalg, A., Beier, S., Scholz, U., Graner, A. and Stein, N. (2014) Mapping-by-sequencing accelerates forward genetics in barley. *Genome Biol.* **15**, R78.

McIntosh, R.A., Wellings, C.R. and Park, R.F. (1995) *Wheat Rusts: an Atlas of Resistance Genes*. East Melbourne, Victoria, Australia: CSIRO Publishing.

Michelmore, R.W., Paran, I. and Kesseli, R.V. (1991) Identification of markers linked to disease-resistance genes by bulked segregant analysis: a rapid method to detect markers in specific genomic regions by using segregating populations. *Proc. Natl Acad. Sci.* **88**, 9828–9832.

Milus, E.A., Kristensen, K. and Hovmøller, M.S. (2008) Evidence for increased aggressiveness in a recent widespread strain of *Puccinia striiformis* f. sp. *tritici* causing stripe rust of wheat. *Phytopathology*, **99**, 89–94.

Mortazavi, A., Williams, B.A., McCue, K., Schaeffer, L. and Wold, B. (2008) Mapping and quantifying mammalian transcriptomes by RNA-Seq. *Nat. Methods*, **5**, 621–628.

Murphy, L.R., Santra, D., Kidwell, K., Yan, G., Chen, X. and Campbell, K.G. (2009) Linkage maps of wheat stripe rust resistance genes and for use in marker-assisted selection. *Crop Sci.* **49**, 1786–1790.

NCBI (2012) *NCBI UniGene Build 60*. http://www.ncbi.nlm.nih.gov/UniGene/UGOrg.cgi?TAXID=4565.

van Ooijen, J.W. and Voorrips, R. (2002) *JoinMap: Version 3.0: Software for the Calculation of Genetic Linkage Maps*. Wageningen, the Netherlands: Wageningen University and Research Center.

Pallotta, M., Warner, P., Fox, R., Kuchel, H., Jefferies, S. and Langridge, P. (2003) Marker assisted wheat breeding in the southern region of Australia. In *Proc. 10th Int. Wheat Genet. Symp.* (Pogna, N.E., ed.), pp. 1–6. Paestum, Italy: Istituto Sperimentale per la Cerealicoltura.

Paux, E., Faure, S., Choulet, F., Roger, D., Gauthier, V., Martinant, J.-P., Sourdille, P., Balfourier, F., Le Paslier, M.-C., Chauveau, A., Cakir, M., Gandon, B. and Feuillet, C. (2010) Insertion site-based polymorphism markers open new perspectives for genome saturation and marker-assisted selection in wheat. *Plant Biotechnol. J.* **8**, 196–210.

Peng, J., Fahima, T., Röder, M., Huang, Q., Dahan, A., Li, Y., Grama, A. and Nevo, E. (2000) High-density molecular map of chromosome region harboring stripe-rust resistance genes YrH52 and Yr15 derived from wild emmer wheat, *Triticum dicoccoides*. *Genetica*, **109**, 199–210.

Pickrell, J.K., Marioni, J.C., Pai, A.A., Degner, J.F., Engelhardt, B.E., Nkadori, E., Veyrieras, J.-B., Stephens, M., Gilad, Y. and Pritchard, J.K. (2010) Understanding mechanisms underlying human gene expression variation with RNA sequencing. *Nature*, **464**, 768–772.

Pontius, J.U., Wagner, L. and Schuler, G.D. (2003) UniGene: a unified view of the transcriptome. In *The NCBI Handbook* (McEntyre, J. and Ostell, J. eds), Bethesda, MD: National Library of Medicine (US), NCBI.

Ramirez-Gonzalez, R., Bonnal, R.J., Caccamo, M. and MacLean, D. (2012) Bio-samtools: ruby bindings for SAMtools, a library for accessing BAM files containing high-throughput sequence alignments. *Source Code Biol. Med.* **7**, 6.

Randhawa, H.S., Mutti, J.S., Kidwell, K., Morris, C.F., Chen, X. and Gill, K.S. (2009) Rapid and targeted introgression of genes into popular wheat cultivars using marker-assisted background selection. *PLoS One*, **4**, e5752.

Rozen, S. and Skaletsky, H. (1999) Primer3 on the WWW for general users and for biologist programmers. In *Bioinformatics Methods and Protocols* (Misener, S and Krawetz, S.A., ed.), pp. 365–386. Clifton, NJ, USA: Springer.

Schneeberger, K. and Weigel, D. (2011) Fast-forward genetics enabled by new sequencing technologies. *Trends Plant Sci.* **16**, 282–288.

Schneeberger, K., Ossowski, S., Lanz, C., Juul, T., Petersen, A.H., Nielsen, K.L., Jorgensen, J.-E., Weigel, D. and Andersen, S.U. (2009) SHOREmap: simultaneous mapping and mutation identification by deep sequencing. *Nat. Methods*, **6**, 550–551.

Segovia, V., Hubbard, A., Craze, M., Bowden, S., Wallington, E., Bryant, R., Greenland, A., Bayles, R. and Uauy, C. (2014) Yr36 confers partial resistance at temperatures below 18°C to UK isolates of *Puccinia striiformis*. *Phytopathology*, **104**, 871–878.

Shewry, P.R. (2009) Wheat. *J. Exp. Bot.* **60**, 1537–1553.

Slater, G.S. and Birney, E. (2005) Automated generation of heuristics for biological sequence comparison. *BMC Bioinformatics*, **6**, 31.

Sun, G., Fahima, T., Korol, A., Turpeinen, T., Grama, A., Ronin, Y. and Nevo, E. (1997) Identification of molecular markers linked to the Yr15 stripe rust resistance gene of wheat originated in wild emmer wheat, *Triticum dicoccoides*. *Theor. Appl. Genet.* **95**, 622–628.

Takagi, H., Abe, A., Yoshida, K., Kosugi, S., Natsume, S., Mitsuoka, C., Uemura, A., Utsushi, H., Tamiru, M. and Takuno, S. (2013) QTL-seq: rapid mapping of quantitative trait loci in rice by whole genome resequencing of DNA from two bulked populations. *Plant J.* **74**, 174–183.

Trick, M., Adamski, N., Mugford, S., Jiang, C.-C., Febrer, M. and Uauy, C. (2012) Combining SNP discovery from next-generation sequencing data with bulked segregant analysis (BSA) to fine-map genes in polyploid wheat. *BMC Plant Biol.* **12**, 14.

8

McMYB10 regulates coloration via activating *McF3'H* and later structural genes in ever-red leaf crabapple

Ji Tian[†], Zhen Peng[†], Jie Zhang[†], Tingting Song[†], Huihua Wan, Meiling Zhang and Yuncong Yao*

Beijing Key Laboratory for Agricultural Application and New Technique, Department of Plant Science and Technology, Beijing University of Agriculture, Beijing, China

*Correspondence

email yaoyc_20@126.com
[†]Contributed equally to this work.

Summary

The ever-red leaf trait, which is important for breeding ornamental and higher anthocyanin plants, rarely appears in *Malus* families, but little is known about the regulation of anthocyanin biosynthesis involved in the red leaves. In our study, HPLC analysis showed that the anthocyanin concentration in ever-red leaves, especially cyanidin, was significantly higher than that in evergreen leaves. The transcript level of *McMYB10* was significantly correlated with anthocyanin synthesis between the 'Royalty' and evergreen leaf 'Flame' cultivars during leaf development. We also found the ever-red leaf colour cultivar 'Royalty' contained the known R_6: *McMYB10* sequence, but was not in the evergreen leaf colour cultivar 'Flame', which have been reported in apple fruit. The distinction in promoter region maybe is the main reason why higher expression level of *McMYB10* in red foliage crabapple cultivar. Furthermore, McMYB10 promoted anthocyanin biosynthesis in crabapple leaves and callus at low temperatures and during long-day treatments. Both heterologous expression in tobacco (*Nicotiana tabacum*) and Arabidopsis *pap1* mutant, and homologous expression in crabapple and apple suggested that *McMYB10* could promote anthocyanins synthesis and enhanced anthocyanin accumulation in plants. Interestingly, electrophoretic mobility shift assays, coupled with yeast one-hybrid analysis, revealed that McMYB10 positively regulates *McF3'H* via directly binding to AACCTAAC and TATCCAACC motifs in the promoter. To sum up, our results demonstrated that McMYB10 plays an important role in ever-red leaf coloration, by positively regulating *McF3'H* in crabapple. Therefore, our work provides new perspectives for ornamental fruit tree breeding.

Keywords: *Malus* crabapple, ever-red leaf, *McMYB10*, cyanidin, flavonoid 3'-hydroxylase, transformation.

Introduction

Red leaf colour is one of the most important factors determining ornamental landscape use, as it can provide a perfect visible sense and a great cultivation advantage due to its stress resistance. Plants with abundant anthocyanin accumulation are economic germplasm resources for ornamental breeding and pigment bioengineering.

Anthocyanins are the main typical pigments in *Malus* plants, and they are synthesized via the branched flavonoid biosynthetic pathway (Holton and Cornish, 1995; Winkel-Shirley, 2001). Anthocyanins contribute to the different colours of flowers, fruits, seeds and leaves and also perform several physiological and biochemical processes in plants including UV protection, insect attraction, herbivore defence and symbiosis (Gould and Lister, 2006; Koes *et al.*, 2005; Peters and Constabel, 2002). Moreover, the anthocyanins intake is linked to reduced risks for a range of human health problems, including heart disease, cancer and diabetes (Chalker-Scott, 1999; Dixon and Paiva, 1995; Hernández *et al.*, 2009; Sun *et al.*, 2013; Zhang *et al.*, 2011).

As described in *Arabidopsis* (Tohge *et al.*, 2005), grape (Deluc *et al.*, 2006), maize (Morohashi *et al.*, 2012) and pear (Feng *et al.*, 2010; Wang *et al.*, 2013), the anthocyanin biosynthesis pathway is controlled by structural genes encoding enzymes that synthesize these compounds, and by regulatory genes that control the expression of these enzymes (Jaakola, 2013). Most of these essential anthocyanin biosynthetic genes are expressed less in nonred organs/tissues than in red organs/tissues (Espley *et al.*,

2007, 2013). F3'H (flavonoid 3'-hydroxylase) is important for both flower and fruit coloration, and it is coordinately expressed with other genes in the anthocyanin biosynthetic pathway in apples (Han *et al.*, 2010). F3'5'H (flavanone 3', 5'-hydroxylase) has been successfully used to engineer the anthocyanin biosynthetic pathway to produce 3',5'-hydroxylated flavonoids to modulate petal colour in *petunia* (Brugliera *et al.*, 2013; Katsumoto *et al.*, 2007), but it has not yet been isolated from *Malus*. However, little pelargonin is synthesized by the late anthocyanin biosynthetic pathway, resulting in cyanin as the major anthocyanin component in apple fruit (Espley *et al.*, 2007).

MYB transcription factors (TFs), which belong to one of the largest plant TF families, play crucial roles in secondary metabolism, development, signal transduction and disease resistance. Some MYB TFs are involved in the regulation of anthocyanin biosynthesis. In apple, the anthocyanin pathway is controlled by the MYB transcription factors, *MdMYB1*, *MdMYB10*, and *MdMYBA*, which are allelic to each other; these genes have been isolated and characterized as key regulatory genes for anthocyanin accumulation and fruit coloration (Ban *et al.*, 2007; Espley *et al.*, 2007; Takos *et al.*, 2006). The transcription of *MdMYB1* and *MdMYBA* is activated when the fruit is exposed to sunlight after maintenance in the dark (Li *et al.*, 2012). *MdMYB10* is constitutively expressed in the whole plant via its direct binding to its own enhancer promoter in an auto-regulatory-loop manner, which leads to the production of red-fleshed apples (Espley *et al.*, 2009). Further analysis revealed that a structural difference in the promoter region is responsible for the expression difference of the *MdMYB10* gene between the white-fleshed and red-fleshed apples. That is, whereas the allele in the white-fleshed apples (*R1: MdMYB10*) carries a single MdMYB10 binding motif in the promoter, that in the red-fleshed apples (*R6:MdMYB10*) carries extra five tandem repeats of an MdMYB10 binding motif. This allelic rearrangement in the promoter of *R6:MdMYB10* allele results in the up-regulation of the anthocyanin pathway leading to red foliage, skin, and flesh via the autoregulation of the *MdMYB10* gene (Espley *et al.*, 2009). Recently, *MdMYB110a*, a paralog of *MdMYB10,* has been isolated from apple. The enhanced expression of this gene is associated with the red-fleshed cortex phenotype, and further analysis reveals that it up-regulates anthocyanin biosynthesis in transgenic tobacco (*Nicotiana tabacum*) (Chagné *et al.*, 2013).

TFs regulate the expression of their target genes by interacting with specific DNA *cis*-regulatory elements, which are usually localized upstream of the transcribed region (within the promoters). Therefore, the identification of these *cis*-elements is essential for a comprehensive understanding of transcriptional regulation. Gel-shift assays indicate that MdMYBA binds specifically to the *MdANS* promoter F2 fragment, which contains four types of MYB-binding sites (Ban *et al.*, 2007). Yeast one-hybrid screens show that NtMYBA1, NtMYBA2 and NtMYBB bind to the AACGG consensus sequence (Ito *et al.*, 1998). Many R2R3-MYB transcription factors recognize AC elements, DNA motifs that are enriched in adenosine and cytosine residues. In maritime pine (*Pinus pinaster* Ait.), MYB8 specifically binds a well-conserved eight-nucleotide-long AC-II element in the promoter regions of *PAT*, *PAL* and *GS1b*, thereby activating their expression (Craven-Bartle *et al.*, 2013). In loquat (*Eriobotrya japonica*), regulation of lignification by EjMYB1 and EjMYB2 is likely achieved via their competitive interaction with AC elements in the promoter region of lignin biosynthesis genes such as *Ej4CL1* (Xu *et al.*, 2014). Although there have been some progresses in the identification and functional analyses of MYB interactions with target genes in fruits, the potential interaction between MYB TFs and the *F3′H* promoter during red leaf development is relatively unexplored.

Previously, we found that *Malus* cv. 'Royalty' is an ever-red leaf cultivar with deep red colour flower and fruit; as such, so it is not only a favourable research material for exploring colour formation, but also meets the expectations of landscape architects and consumers (Tian *et al.*, 2011). We also isolated a MYB transcription factor gene named *McMYB10* from the petals of crabapples, which plays a vital role in regulating anthocyanins accumulation, and the massive expression of *McMYB10* develops a strong anthocyanin accumulation phenotype, suggesting that this regulator is able to increase anthocyanin content, presumably by the activation of endogenous anthocyanin biosynthetic genes (Jiang *et al.*, 2014).

To investigate the molecular mechanisms of ever-red leaf colour formation in *Malus* crabapple, we found the five tandem repeat sequences is present in *McMYB10* promoter in red foliage crabapple cultivar 'Royalty' but not in the green leaf cultivar 'Flame', which is correlated with an increase in *McMYB10* transcription in 'Royalty'. Transcription analysis after different environmental treatments showed that *McMYB10* expression is related to the expression of anthocyanin biosynthesis genes. Genetic transformation and DNA-binding assays demonstrated that the McMYB10 transcription factor is functionally active in driving anthocyanin accumulation by positively regulating *McF3′H* for crabapple red leaf coloration.

Results

HPLC analysis reveals polyphenolic diversity between the ever-red and evergreen leaf *Malus* crabapple cultivars

Two *Malus* crabapple cultivars, *Malus* cv. 'Royalty' and *Malus* cv. 'Flame', were chosen for anthocyanin content analysis based on their significant leaf colour phenotypes (Figure 1a). Developmental stages 1–5 within the *Malus* crabapple leaf-growing season were used to perform high-performance liquid chromatography (HPLC) analysis of flavonoid compounds composition and contents. The amount of flavonoids in ever-red leaf cultivar 'Royalty' was consistently higher than in the evergreen leaf 'Flame' cultivar in all leaf developmental stages (Figure 1c). HPLC analysis suggested that cyanidin was the main coloration anthocyanin component in *Malus* crabapple leaves, and the concentration of cyanidin was the most significant difference between 'Royalty' and 'Flame' in leaves (Figure 1b). 'Royalty' showed an abundant level of cyanidin at the first leaf developmental stage and the highest amount at stage 3, with approximately 118.08 μg/g flesh weight (FW). In contrast, cyanidin was almost undetectable in the leaves of the 'Flame' cultivar, and only trace amounts (7.41 μg/g FW) of cyanidin were detected at stage 1. The concentration of apigenin was also different in the 'Royalty' and 'Flame' leaves and was approximately 1.75- to 5.05-fold higher in red leaf cultivars than in the green leaf cultivar. Additionally, there were no apparent variations in amount of the flavonols (kaempferol, quercetin and rutin) (Figure 1c).

The RNA expression profile of anthocyanin biosynthetic enzyme genes suggests coordinating regulation in *Malus* crabapple leaf

We analysed expression profiles of eight anthocyanin biosynthetic genes from our crabapple microarray data (Our unpublished data, Table S1) between the ever-red leaf 'Royalty' and evergreen leaf 'Flame' cultivars during leaf developmental stages using qRT-PCR. The transcript levels of all the genes were higher at the early stages and then declined at the final stage in 'Royalty', but the evergreen leaf cultivar 'Flame' did not exhibit similar characteristics, except for *McUFGT* (Figure 2). For most transcripts, the abundance in 'Royalty' leaves at stages 1–4 was typically higher than in 'Flame' leaves at the same stages. *McF3′H*, *McDFR* and *McANS* showed the most significant increases of all the anthocyanin biosynthesis genes, with 4- to 30-fold increases throughout whole leaf developmental stages, except the last one, in 'Royalty' compared with 'Flame' leaves (Figure 2). The expression levels of *McPAL* and *McF3H* were 12- and 3-fold higher in 'Royalty' compared to 'Flame' at stage 1, respectively, and then gradually decreased to a lower level in 'Royalty' than in 'Flame' at stage 5. In contrast, transcriptions of the early genes in the anthocyanin synthesis pathway (*McCHS* and *McCHI*) were not associated with leaf colour in either cultivar. These results revealed differential transcript accumulation of the anthocyanin biosynthetic genes in differently pigmented leaf cultivars, and the expression results were consistent with the anthocyanin content of 'Royalty' and 'Flame' leaves.

Different sequence of *McMYB10* promoter caused different expression of McMYB10 in different crabapple cultivars

Previous studies showed that a 23-bp repeat motif in the upstream regulatory region of alleles of *MYB10* was only found

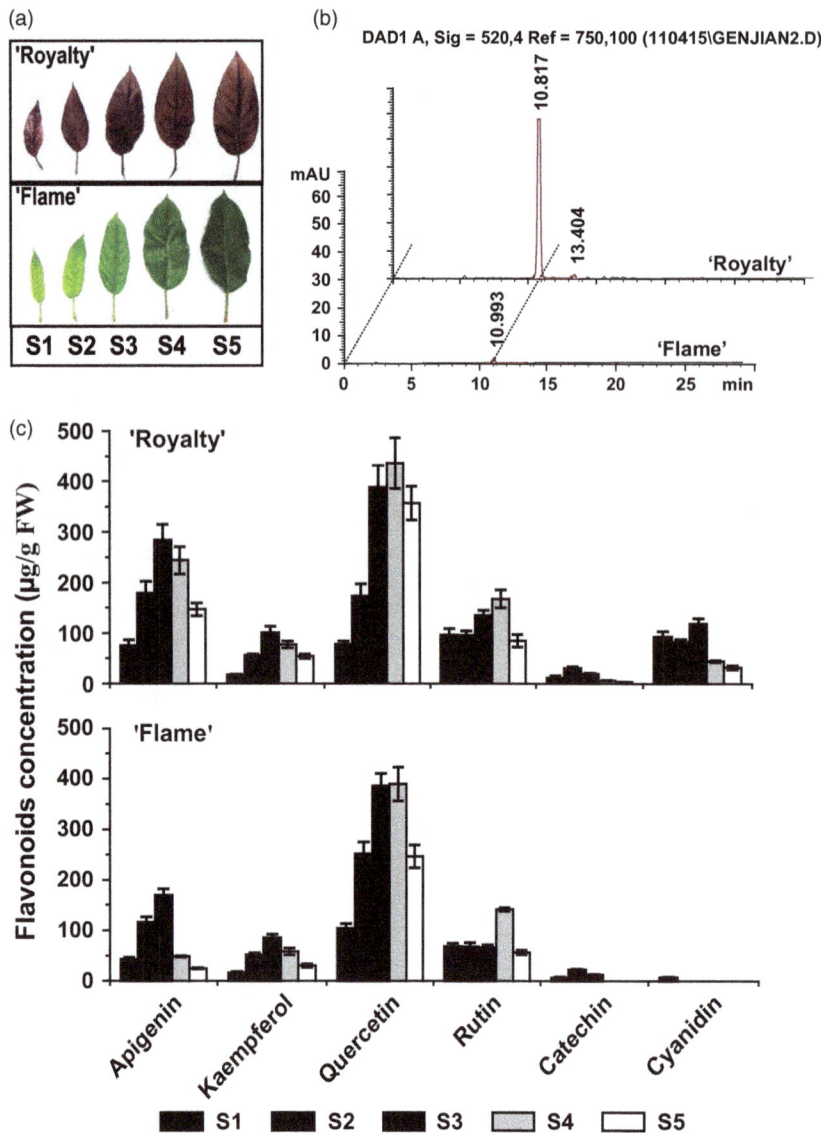

Figure 1 Analysis of flavonoid accumulation in 5 developmental stages of *Malus* crabapple leaves from the 'Royalty' and 'Flame' cultivars. (a) Five leaf development stages with different colour were used for analysis. (b) HPLC traces at 520 nm of ever-red and evergreen leaf cultivar 'Royalty' and 'Flame'. (c) HPLC analysis for flavonoid compounds in the two crabapple cultivars. S1, 3 days after budding; S2, 9 days after budding; S3, 15 days after budding; S4, 21 days after budding; S5, 30 days after budding. Error bars on each symbol indicate the mean ± SE of three replicate reactions.

in red-fleshed apples, and this promoter allele is responsible for the increased accumulation of anthocyanins (Espley *et al.*, 2009). To investigate the molecular basis of the ever-red leaf phenotype, we used homology-based cloning to isolate the upstream regulatory sequence of *McMYB10* from the ever-red leaf crabapple cultivar, *Malus* cv. 'Royalty', and the evergreen leaf cultivar, *Malus* cv. 'Flame'. The isolated DNA contained two sequences of different sizes in the same promoter region. One version was found in the promoter region of the ever-red leaf cultivar, *Malus* cv. 'Royalty', with six repeated minisatellites (GTTAGACTGGT AGCTATTAACAA) and a dinucleotide microsatellite (GTGT, Figure 3a). The promoter sequence in the evergreen leaf cultivar, *Malus* cv. 'Flame', was 105 bp smaller, with only one minisatellite and a dinucleotide microsatellite (Figure 3a). The same genetic polymorphism was discovered in red-fleshed apples and white cultivars (Espley *et al.*, 2009).

McMYB10 transcripts were measured using semiquantitative RT-PCR and qRT-PCR using gene-specific primers. The expression of *McMYB10* was noticeably different in the 'Royalty' and 'Flame' cultivars, with higher mRNA levels in the ever-red leaf than in the evergreen leaf in the whole leaf developmental stages (Figure 3b). qRT-PCR analysis quantified this distinction, with approximately 500- to 1000-fold higher levels in the ever-red cultivar than in the evergreen cultivar (Figure 3c). So we deduced the different promoter structure of *McMYB10* caused the different expression of *McMYB10* in the leaves of different crabapple cultivars.

In addition, *McMYB10* transcript levels were correlated with anthocyanin concentration and transcripts of anthocyanin biosynthesis genes, especially *McF3'H*, *McDFR* and *McANS*, that are required for anthocyanin biosynthesis in 'Royalty' and 'Flame' leaves.

Light and temperature alter the expression of *McMYB10* and anthocyanin biosynthesis genes

Light induces anthocyanin accumulation in many crops, such as apple (*Malus* × *domestica*) (Takos *et al.*, 2006), tobacco (*Nicotiana tabacum*) (Zhou *et al.*, 2008), *Arabidopsis thaliana* (Rowan *et al.*, 2009) and grape (*Vitis vinifera*) (Azuma *et al.*,

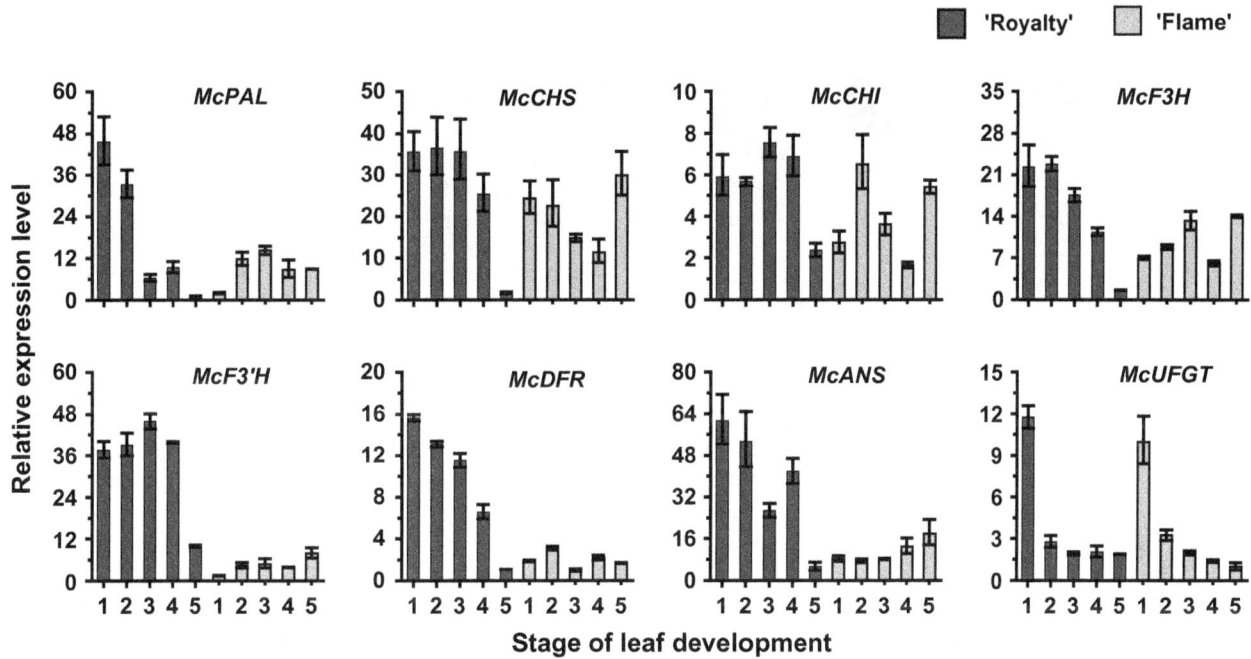

Figure 2 Expression profiles of *Malus* crabapples anthocyanin biosynthesis genes during the 5 leaf developmental stages. Real-time PCR was used to analyse *McPAL, McCHS, McCHI, McF3H, McF3'H, McDFR, McANS, McUFGT* expression patterns in leaves of the *Malus* cultivars 'Royalty' and 'Flame'. *18S* was used as the reference gene. 1–5 represents leaf developmental stages 1–5. Error bars on each symbol indicate the mean ± SE of three replicate reactions.

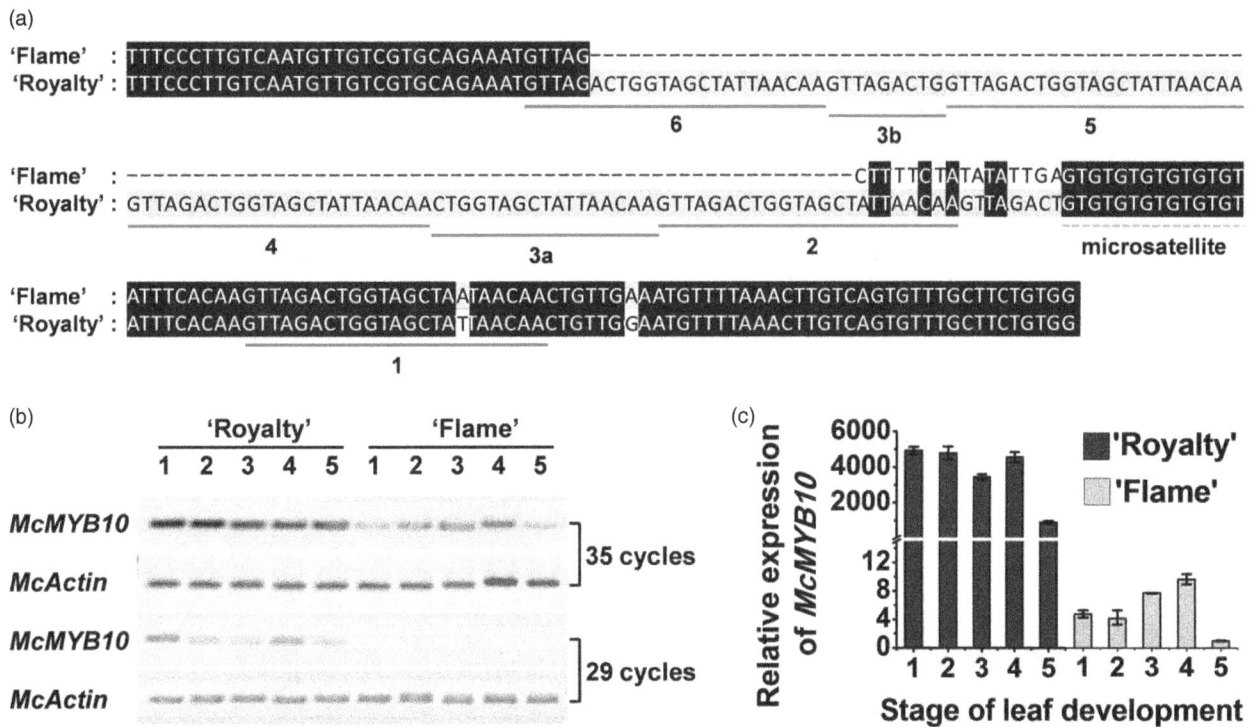

Figure 3 Sequence of *McMYB10* promoter and expression level of *McMYB10* in two crabapple cultivars. (a) Sequences alignment of *McMYB10* promoter regions in ever-red leaf 'Royalty' and evergreen leaf 'Flame' crabapple cultivars. Numbers and underlined regions represent the positions of the repeat units (GTTAGACTGGTAGCTATTAACAA); the dotted line indicates microsatellite regions. (b) *McMYB10* expression level was analysed using semiquantitative RT-PCR. *McActin* was used as the external control. (c) Quantitative RT-PCR analysis of *McMYB10* relative expression level. The numbers 1–5 represent leaf developmental stages 1–5. Error bars on each symbol indicate the mean ± SE of three replicate reactions

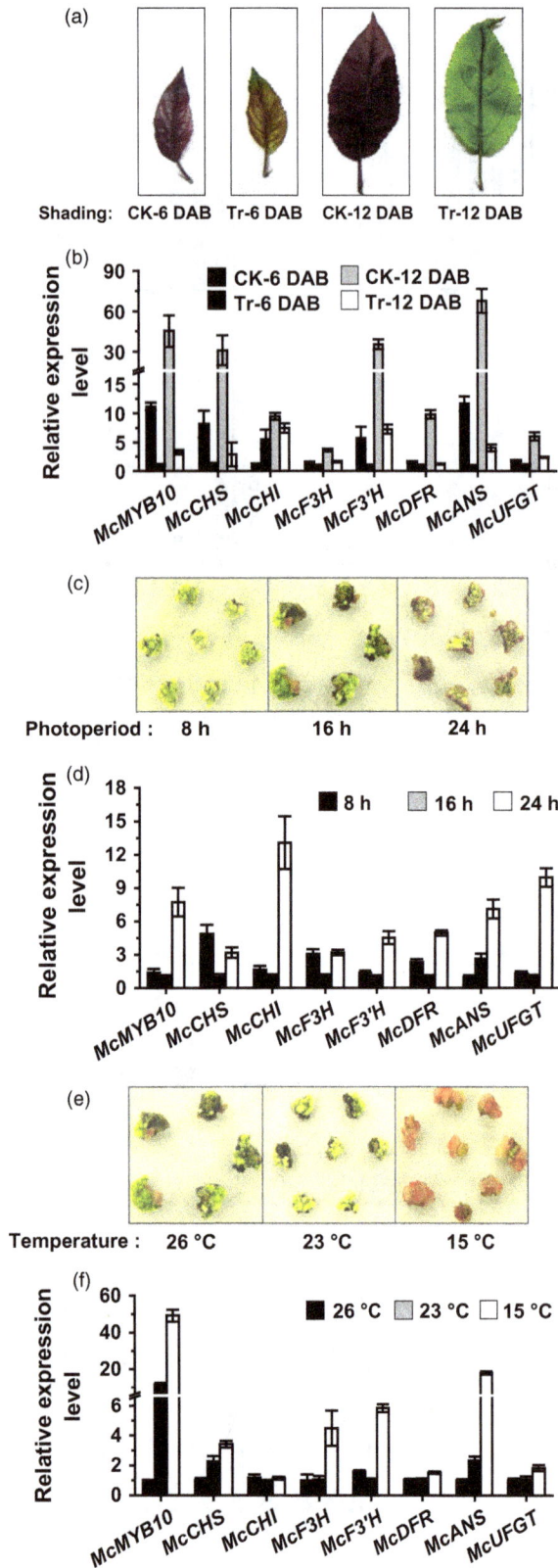

Figure 4 Different environmental conditions alter anthocyanin accumulation by changing the expression of *McMYB10* and anthocyanin biosynthesis genes. (a) Leaves harvested from the shading treatment plants (Tr-6 DAB, Tr-12 DAB) and control plants (CK-6 DAB, CK-12 DAB). DAB, days after budding. (b) The gene expression profiles under shading and control conditions. (c) The callus of 'Spring Snow' harvested from different photoperiod conditions. (d) The gene expression profile under different photoperiod conditions. (e) The callus of 'Spring Snow' harvested from different temperature conditions. (f) The gene expression profile under different temperature conditions. Error bars on each symbol indicate the mean ± SE of three replicate reactions.

yellow (at 6 DAB stage) or green (at 12 DAB stage), instead of the ever-red phenotype. Transcriptional analysis of *McMYB10* and anthocyanin pathway genes (*McCHS*, *McCHI*, *McF3H*, *McF3'H*, *McDFR*, *McANS* and *McUFGT*) via qRT-PCR indicated that all the genes were significantly down-regulated in the shaded conditions (Figure 4b).

Twenty-day-old 'Spring Snow' leaf callus was used to conduct 8-h, 16-h and 24-h light period lighting treatments. The callus displayed significant red coloration after 15 days of 24-h light, but the 8-h light period callus remained green (Figure 4c). *McMYB10* expression increased almost 5-fold in the callus exposed to 24-h light periods compared with 8-h light periods. Furthermore, all essential anthocyanin biosynthetic gene transcripts were increased (Figure 4d). Thus, longer periods enhanced anthocyanin biosynthesis and accumulation in leaf callus.

We also placed 20-day-old 'Spring Snow' callus at 26, 23 and 15 °C to test the effects of low temperature. After 15 days, the callus at 23 °C had some red coloration on its edge, and the whole callus turned into bright red in the 15 °C treatment (Figure 4e). At the reduced temperature, *McMYB10* transcripts increased, and the expression of anthocyanin biosynthesis genes (especially *McF3H*, *McF3'H* and *McANS*) also increased (Figure 4f).

These results indicated that long illumination time and low temperature promote anthocyanin accumulation as a result of up-regulated *McMYB10* expression. Additionally, because *McF3'H* expression was significantly changed by these treatments, it may play an important role in *McMYB10*-mediated anthocyanin variation.

Altered *McMYB10* expression changes anthocyanin accumulation in plants

To confirm whether the *McMYB10* gene regulates anthocyanins biosynthesis, we produced transgenic tobacco plants expressing *McMYB10* under the control of a constitutive CaMV35S promoter. Significant changes in flowers colour were observed in transgenic tobacco plants constitutively expressing *McMYB10*, which displayed deep pink coloration in transgenic but not wild-type petals. HPLC analysis demonstrated that anthocyanins increased almost 5-fold after the transformation of tobacco with *McMYB10*. Expression analysis was performed on RNA extracted from the petals of three independent transgenic lines of tobacco expressing *35S:McMYB10*. *McMYB10* activated the expression of several structural genes associated with the anthocyanin biosynthesis pathway (Figure 5a,b). The expression levels of *NtCHS*, *NtCHI*, *NtF3H*, *NtF3'H*, *NtDFR* and *NtANS* were slightly higher in the *35S:McMYB10* lines than in control flowers. In contrast, *NtF3H*, *NtF3'H* and *NtDFR* transcripts were greatly increased compared with other genes in the *35S:McMYB10* transgenic lines (Figure 5c).

2012). To understand the effect of light on coloration and the expression of anthocyanin-related genes in leaves, we covered trees with sunshade nets during the leaf developmental season to reduce sunlight. Anthocyanin accumulation was inhibited by covering 'Royalty' leaves (Figure 4a), which displayed bright

To further ascertain the role of *McMYB10* in a homologous system, transient gene overexpression assays were used. We injected *Agrobacterium* containing *35S:McMYB10* into the *Malus* crabapple evergreen leaf cultivar 'Donald'. Conspicuous anthocyanin accumulation was observed in the leaves and fruits of 'Donald' near the injection and scratch position, and control organs transformed with an empty pBI121 vector showed little or no pigmentation in the leaves and fruits (Figure 6a). HPLC measurement of the anthocyanin content confirmed that the transgenic organs contained more anthocyanin than those of the wild type. The cyanidin concentration in the nontransgenic leaf (3 µg/g FW) was much lower than in the *McMYB10*-overexpressed leaves (39 µg/g FW); the same phenotype was detected in nontransgenic fruit (8 µg/g FW) and *McMYB10*-overexpressed fruits (76 µg/g FW) (Figure 6b).

Further relative expression analysis in overexpressed tissues revealed the up-regulation of *McMYB10* accompanied by a proportional increase in the expression levels of some anthocyanin biosynthesis structural genes (*McCHS*, *McCHI*, *McF3H*, *McF3'H*, *McDFR*, *McANS* and *McUFGT*). All of these genes

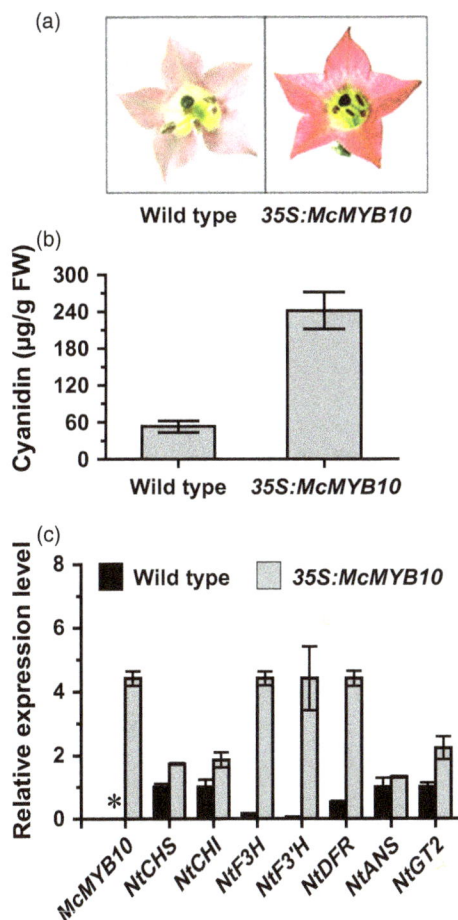

Figure 5 Overexpression of *McMYB10* in tobacco elevates anthocyanin production. (a) The phenotype of transgenic tobacco flowers overexpressing *McMYB10*. (b) The anthocyanin concentrations in samples. (c) qRT-PCR analysis of *McMYB10* and anthocyanin pathway genes in petals of transformed *35S:McMYB10* tobacco. The asterisks represent no detected gene expression. Error bars on each symbol indicate the mean ± SE of three replicate reactions.

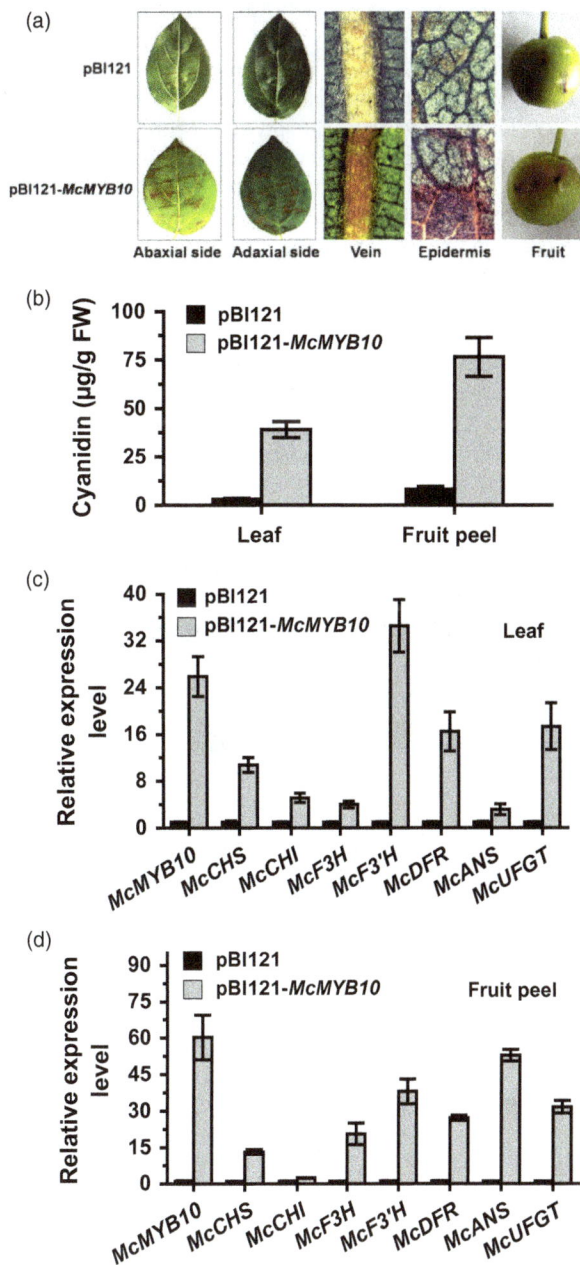

Figure 6 Transient overexpression analysis of *McMYB10* in 'Donald' leaves and fruits. (a) The phenotype of overexpressed *McMYB10* in 'Donald' leaves and fruits. Microscope images show anthocyanin accumulation in leaves infiltrated with *35S:McMYB10* and harvested at 5 days after transformation with magnifications of 20× (vein) and 40× (epidermis). *Scale bars* = 0.50 mm. (b) Anthocyanin concentrations in the leaves and peels of pBI121-*McMYB10* and pBI121 empty vector plants in µg/g flesh weight (FW). (c, d) Relative expression levels of *McMYB10* and anthocyanin pathway genes in leaves (c) and peels (d). Error bars on each symbol indicate the mean ± SE of three replicate reactions.

appeared up-regulated, ranging from 2- to 60-fold higher in pBI121-*McMYB10* leaves and fruit peels than in empty pBI121 vector plants (Figure 6c,d). Compared with the control, *McF3'H*, *McDFR*, *McANS* and *McUFGT*, members of late stage anthocyanin biosynthetic pathway were the most up-regulated in the *McMYB10*-overexpressed plants.

The viral recombinant plasmid pTRV2-McMYB10 containing a partial *McMYB10* ORF was used for infiltration, and the empty vector pTRV2 served as a control. Transient expression TRV-*McMYB10* suppressed the red coloration around the infiltration site compared with an empty vector control (Figure 7a,b). In contrast, the apples injected with the empty vector pTRV2 rapidly accumulated anthocyanins, resulting in red coloration at the injection sites (Figure 7a). qRT-PCR showed that *MYB10* transcripts in the *McMYB10*-silenced fruit skins decreased to approximately 10% of those in control fruits, and the expression of *MdF3H*, *MdF3'H* and *MdDFR* was strongly down-regulated by *MYB10* silencing through VIGS (Figure 7c). We also overexpressed *McMYB10* in apple skin to study the anthocyanin biosynthetic pathway. As expected, the apples of transient overexpressed *McMYB10* rapidly accumulated anthocyanins, resulting in red coloration at the injection sites compared with the controls and silenced apple skins (Figure 7a). Transcripts of *McMYB10* and anthocyanin biosynthetic genes (such as *MdCHS*, *MdF3H*, *MdF3'H*, *MdDFR* and *MdANS*) were up-regulated in these pBI121-*McMYB10* infiltrated tissues compared with control fruits injected with the empty vector (Figure 7c).

McMYB10 complements the Arabidopsis *pap1* mutant anthocyanin-deficient phenotype

The MYB transcription factor PAP1 is the predominant MYB expressed in seedlings. In this study, *pap1* insertion mutant seedlings (pst16228) in the No-0 ecotype germinated on 3% sucrose anthocyanin-deficient (Figure 8a) due to the loss of PAP1 function (Gonzalez *et al.*, 2008). To verify the function of *McMYB10*, the Arabidopsis *pap1* mutant (No-0 background) was selected for a functional complementation assay. The pBI121-*McMYB10* expression vector was genetically transformed

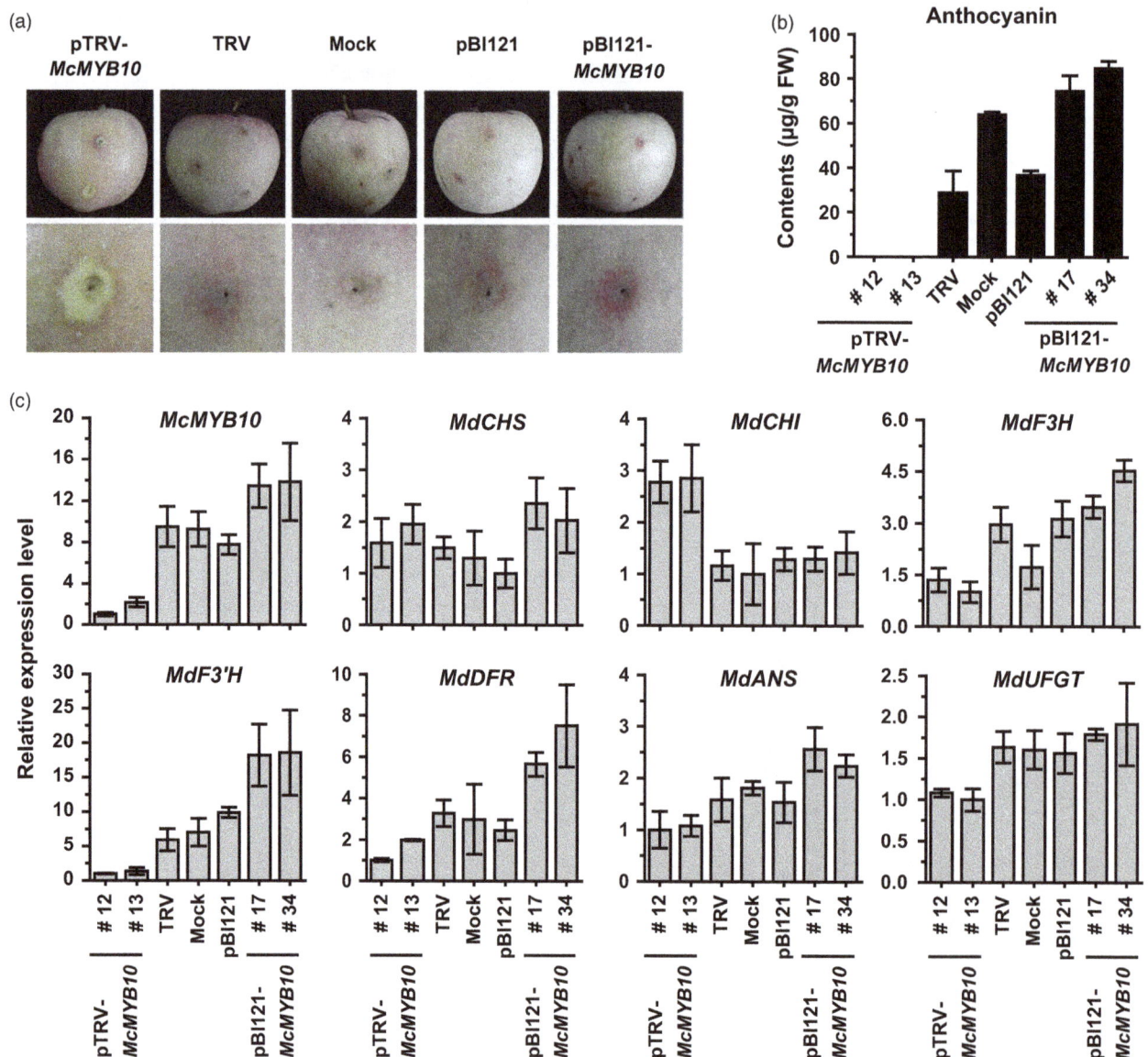

Figure 7 Transient expression of *McMYB10* in apple fruit. (a) Apple fruit peel coloration around the injection sites. *McMYB10* was used for suppression (pTRV2-*McMYB10*) or expression (pBI121-*McMYB10*). Apples injected with the empty TRV and pBI121 vectors and only infiltration buffer were used as controls. (b) Anthocyanin contents around the injection sites of apple peels in μg/g flesh weight (FW). (c) Relative expression levels were determined using qRT-PCR of apple fruit peel around the injection sites.

into the Arabidopsis *pap1* mutant. T2 transgenic lines 1 and 7 were confirmed using RT-PCR before phenotype and expression analysis. Purple coloration was observed in cotyledons after transgenic and wild-type plants were grown on 1/2 MS media (3% sucrose) with 3 days continuous light, whereas *pap1* mutants did not accumulate anthocyanin in the same conditions (Figure 8a).

When *McMYB10* was ectopically expressed in the *pap1* backgrounds, there was higher expression of *F3'H* in the transgenic seedlings than in the *pap1* seedlings grown in the sucrose-induced condition (Figure 8b). Thus, *McMYB10* complemented the function of the Arabidopsis anthocyanins regulator *PAP1,* and *McMYB10* may positively regulate the expression of *F3'H*.

Interaction of McMYB10 with the *McF3'H* promoter

We utilized both heterologous and homologous transcription data to profile dynamics of genes involved in anthocyanin biosynthetic pathway, the resulting data revealed *McF3'H* gene

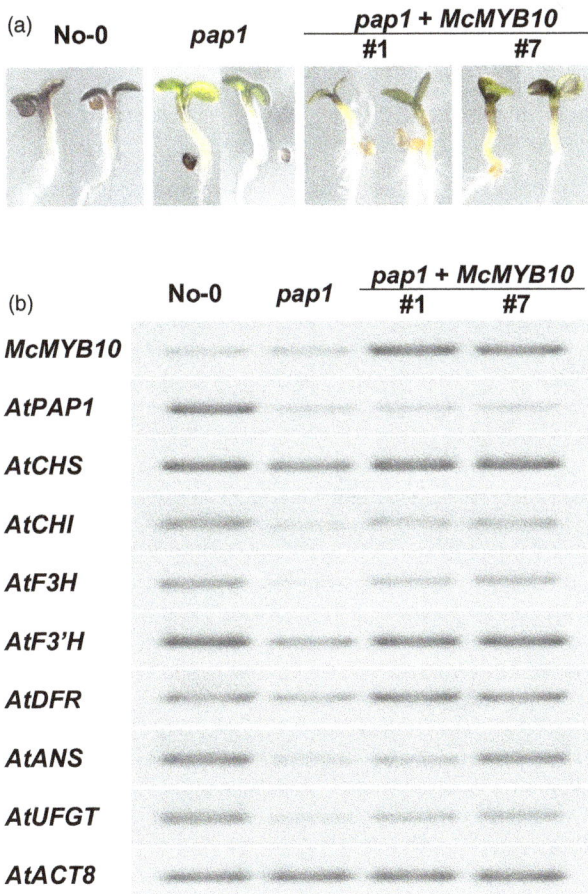

Figure 8 Complementation of the Arabidopsis *pap1* transposon mutant with crabapple *McMYB10*. (a) Comparison of the early seedling coloration phenotype in the wild type Nossen (No-0), *pap1* mutant, and the *pap1* mutant containing the *McMYB10* transgenic Arabidopsis. The seedlings were grown on one-half-strength MS medium with 3% sucrose for 3 d and photographed. (b) Transcript levels of *MYB10/PAP1*, *CHS*, *CHI*, *F3H*, *F3'H*, *DFR*, *ANS* and *UFGT* in the wild-type (No-0), *pap1* mutant, and the *McMYB10* transgenic Arabidopsis were measured using semiquantitative RT-PCR. Twenty seedlings from homozygous T2 lines were harvested for RNA extraction.

is the most upstream anthocyanin biosynthetic gene which response to McMYB10 regulation. So we deduced that the McMYB10 positively regulate the expression of *McF3'H* in crabapple. To test whether *McF3'H* is a direct downstream target gene of McMYB10, we isolated the promoter of *McF3'H* and performed DNA-binding assay.

To characterize the ability of McMYB10 to activate *McF3'H*, a yeast one-hybrid assay was employed. *LacZ* activity was detected in yeast harbouring AD-*McMYB10* together with BD-*proMcF3'H*, but not in control yeast harbouring BD-*proMcF3'H* together with AD lacking *McMYB10*, or in control yeast harbouring AD-*McMYB10* together with BD lacking *McF3'H* promoter (Figure 9a).

A promoter *cis*-element analysis with the PLACE database (http://www.dna.affrc.go.jp/PLACE/signalscan.html) showed that the *McF3'H* promoter contains several *cis*-regulatory elements, including six putative MYB-binding sequences (AAACCA, AACCTAAC, AACGG, TATCCAACC, CCAACC and TATCC) (Higo *et al.*, 1999). To determine whether McMYB10 binds to the MYB recognition sites in the *McF3'H* promoter, we expressed and purified recombinant His-McMYB10 fusion proteins. DNA fragments containing the six MYB-binding sites were used as probes to examine possible interactions with His-McMYB10 in an electrophoretic mobility shift assay (EMSA) (Figure 9b). Gel-shift assays showed that His-McMYB10 protein complexes were detected when the six *cis*-elements were used as labelled probes, especially probes Biop2 (AACCTAAC) and Biop4 (TATCCAACC, Figure 9c). The unlabelled probes competed with the labelled probes for McMYB10 binding, and mobility shifts were not observed in control reactions lacking McMYB10 protein. The specificity of competition confirmed the specific activated interaction between McMYB10 and the *McF3'H* promoter.

Discussion

The colour formation in ever-red leaf crabapple

Recently, more attention has been paid to the breeding of plants with coloured leaves (especially red/purple leaves) for ornamental use. Red leaf coloration is mainly due to the accumulation of flavonoids (anthocyanins) (Lightbourn *et al.*, 2008), pigments that impart a wide range of red and purple colours in plant leaves (Lebowitz, 1985; Nguyen and Cin, 2009), but the molecular mechanism of red-coloured-leaf formation is still unclear. 'Royalty' is an ever-red leaf cultivar of *Malus* crabapples; its deep red-coloured leaves endow it with valuable traits, such as disease and stress resistance, persistent colour, and aesthetics and favoured by consumers. In this study, we focused on the mechanism of ever-red leaf colouring and demonstrated that *McMYB10* and structural gene *McF3'H* are most likely the critical factors.

HPLC analysis and transcript quantifications suggested a possible mechanism controlling pigment patterns in crabapple leaves. We found that ever-red leaves were associated with higher cyanin accumulation (Figure 1c), which was consistent with the increased transcript levels of late anthocyanin pathway genes, including *McF3'H*, *McDFR* and *McANS*. The biosynthetic route to anthocyanins is proceeded by the action of *F3'H* on the precursor prior to their conversion by *DFR* and *ANS* to cyanidin (3'4'-hydroxylation) (Grotewold, 2006; Zhang *et al.*, 2010). Therefore, we hypothesize that the continuous high expression of *McF3'H* promotes cyanidin production, resulting in the ever-red leaf colour in 'Royalty'.

(a)

BD-*proMcF3'H* BD

AD-*McMYB10*

AD

SD/-Trp/-Ura

(b)

Biop1: 5' Biotin-gggtgattgt**AAACCA**gcttcttcagga-3'

Biop2: 5' Biotin-gtgaaaattc**AACCTAAC**ccacccaacaa-3'

Biop3: 5' Biotin-ggtagcatctaaag**AACGG**acacaatattc-3'

Biop4: 5' Biotin-cttcttaca**TATCCAACC**ttcaaaaaaatg-3'

Biop5: 5' Biotin-aagcacgatc**CCAACC**tacccagcgcacc-3'

Biop6: 5' Biotin-ctcctatatattg**TATCC**aacaccactat-3'

(c)

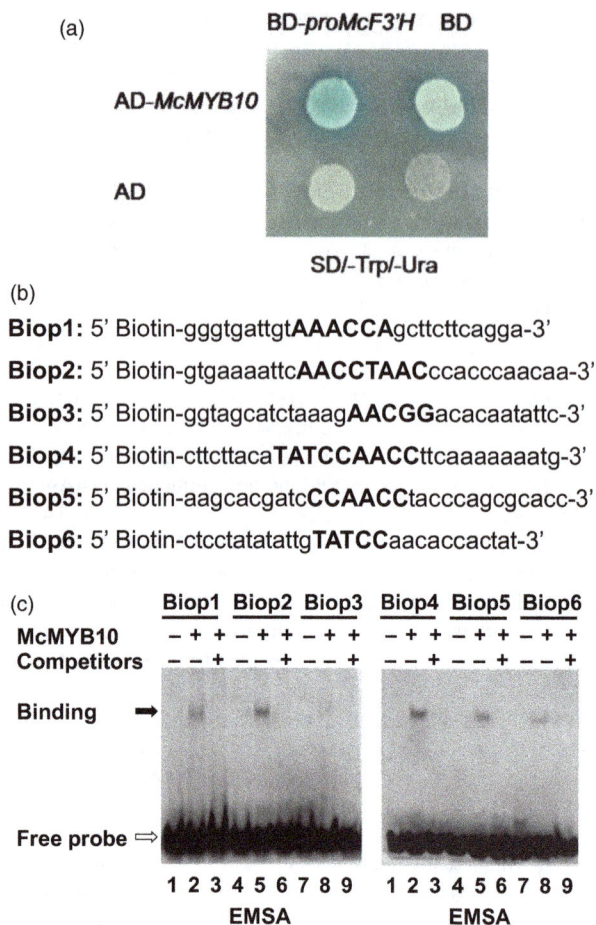

Figure 9 *Cis*-element binding ability and transcriptional activation assays of McMYB10 with *McF3'H*. (a) Interaction of McMYB10 with the *McF3'H* promoter, as determined by yeast one-hybrid assays. The panel shows yeast cells containing distinct effector and reporter constructs grown on an SD/-Trp/-Ura medium plate. The interaction of McMYB10 protein, fused to a GAL4 activation domain (AD-*McMYB10*) with *LacZ* driven by the *McF3'H* promoter (BD-*proMcF3'H*), is shown in the top left panel. Yeast transformed with AD-*McMYB10*/BD, AD/BD-*proMcF3'H* and AD/BD were used as controls. (b) The oligonucleotide sequences of six putative MYB-binding domains used as 5' biotin end-labelled probes for the electrophoretic mobility shift assays. (c) EMSA of six *McF3'H* promoter fragments and McMYB10. The black arrow indicates protein–DNA complexes, and the white arrow shows the positions of free probes. In lanes with competitor DNA, excess of unlabelled probe.

Transcriptional control of anthocyanin levels in crabapple leaves

It is likely that variations in promoter mutation and the activity of the R2R3 MYB TFs are responsible for controlling the spatial and temporal patterning of anthocyanin production in most plants (Espley *et al.*, 2009, 2013; Fraser *et al.*, 2013; Ravaglia *et al.*, 2013; Zhou *et al.*, 2008). In apple, three MYB genes mapping to the same locus, *MYB10*, *MYB1* and *MYBA*, have been independently isolated (Ban *et al.*, 2007; Espley *et al.*, 2007; Takos *et al.*, 2006). All of these genes contribute to anthocyanin variation and share a very high degree of sequence similarity, as they are alleles arising from the different varieties from which they were cloned (Lin-Wang *et al.*, 2010). Recently, *MdMYB110a*, a paralog of *MYB10*, was isolated from apple; the red-fleshed cortex pheno-

type is associated with enhanced *MYB110a* expression (Chagné *et al.*, 2013).

We isolated the promoter sequences of *McMYB10* from the leaves of ever-red cultivar 'Royalty' and evergreen cultivar 'Flame', our results showed that a 23-bp repeat motif in the upstream regulatory region of alleles of *McMYB10* in red foliage crabapple cultivars is responsible for the massive expression of *McMYB* during leaf development. Meanwhile, the expression pattern of *McMYB10* was very similar to the key anthocyanin biosynthesis genes which are important for cyanidin biosynthesis. A large increase in *McMYB10* expression in the red-coloured leaves suggested that it regulates the anthocyanin pathway. We demonstrated that *McMYB10* functions as an anthocyanin regulator by introducing the coding region of the cDNA into tobacco plants and crabapple leaves and fruits. Ectopic overexpression of *McMYB10* in tobacco plants generated a strong phenotype with highly pigmented tobacco petals (Figure 5) due to the induction of anthocyanin synthesis. Transient overexpression and silencing of *McMYB10* in crabapple leaves and fruits and apple fruits also altered anthocyanins accumulation. As a result, the infected parts appeared red instead of green or the opposite phenotype. Furthermore, *McMYB10* overexpression complemented the *pap1* mutant phenotype and induced ectopic anthocyanin accumulation in Arabidopsis. Therefore, our results suggest that *McMYB10* strongly regulates anthocyanins biosynthesis in *Malus* crabapple and other species.

Different environmental factors can impact anthocyanin content

Light and temperature also influence flavonoid production in a negative manner (Wang *et al.*, 2008). In grape skin, high temperature increases anthocyanin degradation and decreases the expression of flavonoid biosynthetic and *MYBA* genes. In contrast, low temperature increases anthocyanin production in grape (Mori *et al.*, 2005; Yamane *et al.*, 2006) and in maize, red orange and apple (Christie *et al.*, 1994; Lo Piero *et al.*, 2005; Ubi *et al.*, 2006). Light significantly increases the accumulation of flavonoids (Cortell and Kennedy, 2006) and the expression of their biosynthetic genes in grape (Jeong *et al.*, 2004). In apple, *MdMYB1* gene transcription increased within 1 days when fruit exposed skin to sunlight after removing bags. Light and temperature also influence flavonoid production in a negative manner (Wang *et al.*, 2008). In this study, we showed that shading inhibited the accumulation of anthocyanin and the expression of *McMYB10* and *McF3'H*. However, long light periods increased anthocyanin content and the transcription of *McMYB10* and *McF3'H*. In lower temperatures, the transcription of *McMYB10* and anthocyanin biosynthesis genes was gradually activated. The link between environmental conditions affecting anthocyanin accumulation and the transcription level of the corresponding genes indicated that *McMYB10* and *McF3'H* were pivotal in the anthocyanin pathway. In apple, *MdbHLH3* binds to the promoters of the anthocyanin biosynthesis genes *MdDFR* and *MdUFGT* and the regulatory gene *MdMYB1* to activate their expression in low temperatures (Xie *et al.*, 2012). Whether the same anthocyanin regulation mechanism acts in response to environmental variation in crabapple leaves is unclear, but future work will concentrate on understanding these regulatory pathways in crabapple.

The putative regulation mechanism of *McMYB10*

MdMYB10 transcription factors activate the *MdDFR* promoter (Espley *et al.*, 2007); *MdMYB1* activates endogenous *UFGT* expression (Takos et al., 2006), and MdMYBA binds specifically to an anthocyanidin synthase (MdANS) promoter region in apple fruit (Ban et al., 2007). Moreover MYB110a transactivates the CHS promoter (Chagné et al., 2013).

In our work, some anthocyanin structural genes were activated by McMYB10, and this activation was accompanied by the accumulation of anthocyanin in tobacco. The same results were observed in McMYB10-overexpressed crabapple tissues and those exposed to low temperature and long light periods. Interestingly, as the early anthocyanin biosynthetic gene, McF3'H is the most upstream anthocyanin biosynthetic gene which respond to McMYB10 regulation. The EMSA and Y1H assay provided direct evidences for interactions between McMYB10 and McF3'H promoter (Figure 9). Thus, McF3'H, an early response gene, plays an important role in cyanidin biosynthesis and is a candidate downstream gene of McMYB10, which contribute to red colour formation (Figure 10).

In summary, this functional characterization data indicates that McMYB10 is the potential regulator for the coloration of ever-red leaf, whose expression is high during red leaf development. The R_6:McMYB10 promoter is responsible for the massive expression of McMYB10 in red foliage crabapple cultivar, and genetic and biochemical analysis revealed that McMYB10 acts as a key regulator via positively regulating McF3'H in Malus crabapple. Undoubtedly, the work described in this report will trigger a series of exciting future research projects to elucidate the mechanisms governing plant coloration, and our results will improve the flavonoid pathway network and provide new perspectives for ornamental fruit trees breeding.

Experimental procedures

Plant materials and growth conditions

The plant materials used for the experiments included four Malus crabapple cultivars: (i) Malus cv. 'Royalty', an ever-red leaf cultivar, (ii) Malus cv. 'Flame', an evergreen leaf cultivar, (iii) Malus cv. 'Donald', a cultivar of red leaf in spring and (iv) Malus cv. 'Spring Snow', an evergreen leaf cultivar. These 5-year-old trees grafted on Malus hupehensis were planted in the Crabapple Germplasm Resources Nursery in the Beijing University of Agriculture. The 'Royalty' and 'Flame' leaves were collected at five different developmental stages (Figure 1a). The explants of 'Spring Snow' were harvested from one-year-old branches before spring buds germination and cultured on Murashige and Skoog medium supplemented with 0.1 mg/L 6-BA and 2 mg/L 2, 4-D at 23 °C for a 16-h light (200 µmol/s/m²)/8-h dark period, until environmental condition treatments.

Nicotiana tabacum cv. W38 was used for the gene transfer experiment and wild-type control. The plants were grown in a greenhouse at 27 °C under constant illumination. Transgenic plants were transformed with Agrobacterium tumefaciens strain GV3101 by suspension in the Agrobacterium solution.

Bagged apple fruits were harvested 145 days after full bloom (DAFB) from adult trees of the 'Red Fuji' cultivar (Malus pumila Milier). The skin of the fruits was peeled, obtaining <1 mm of cortical tissue, for anthocyanin measurements and other analyses (Xie et al., 2012).

The Arabidopsis (Arabidopsis thaliana) loss-of-function allele of PAP1 (pap1; pst16228), the result of a Ds insertion at the PAP1 locus, in the Nossen ecotype background (Kuromori et al., 2004), and the Nossen wild type (No-0; CS1394) were used in this study. Seeds were surface-sterilized using 3% commercial sodium hypochlorite and 75% ethyl alcohol, cold treated at 4 °C in the dark for 2 days, and plated on half-strength MS agar medium containing 2.16 g/L MS salts, 3% sucrose and 1% bacto-agar pH 5.7 adjusted with 1 M KOH at 21 °C in continuous white light (200 µmol/m²/s) for 3–5 days, to induce anthocyanin accumulation (Gonzalez et al., 2008; Teng et al., 2005). Half-strength MS agar medium with 1% sucrose plus 100 mg/L kanamycin was used for screening transgenic lines.

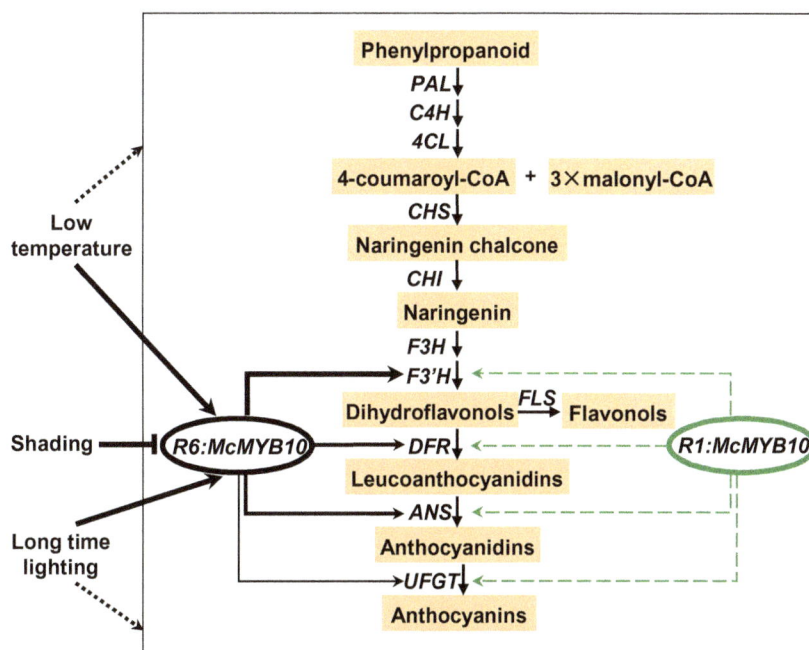

Figure 10 Schematic showing the effects low temperature, shading and long-day lighting on the anthocyanin pathway and McMYB10 expression, which significantly activating the expression of McDFR, McANS, McUFGT and especially McF3'H in ever-red crabapple leaves by known R_6: McMYB10 sequence, leading to massive anthocyanin accumulation. However, there was weaker effect from MYB TF, which was followed by lower transactivation of the late anthocyanin biosynthesis genes in evergreen crabapple leaves due to R_1: McMYB10 sequence.

Environmental condition treatments

Three 5-year-old 'Royalty' trees were wrapped in two layers of sunshade net shading (cutting off 80% of the ambient light), and the control plants were exposed to natural light. The leaves were harvested after 6 or 12 days after budding (DAB), frozen in liquid nitrogen and stored at −80 °C.

Temperature treatments were conducted on 20-old 'Spring Snow' callus, which was grown at 26, 23 and 15 °C. For photoperiod treatment, twenty-day-old 'Spring Snow' callus was grown at 23 °C for 8-h, 16-h and 24-h light periods. All samples were collected at 15 days after treatment, frozen in liquid nitrogen and stored at −80 °C.

HPLC analysis

The crabapple leaves of each samples (approximate 0.8–1.0 g flesh weight) were extracted with 10 mL extract solution (methanol: water: formic acid: trifluoroacetic acid = 70 : 27 : 2 : 1) (Hashimoto et al., 2000) in a test tube, at 4 °C in the dark for 72 h, with shaking every 6 h. The liquid was separated from the solid matrix by filtration through sheets of qualitative filter paper. The filtrate was further passed through 0.22-μm reinforced nylon membrane filters. Trifluoroacetic acid: formic acid: water (0.1 : 2 : 97.9) was selected as mobile phase A and trifluoroacetic acid: formic acid: acetonitrile: water (0.1 : 2: 48 : 49.9) was mobile phase B. The gradients used were as follows: 0 min, 30% B; 10 min, 40% B; 50 min, 55% B; 70 min, 60% B; 80 min, 30% B. Detection was performed at 520 nm for anthocyanin and 350 nm for flavonol (Revilla and Ryan, 2000). All samples were analysed in triplicate.

Cloning and sequence analysis of McMYB10 promoter region

To isolate the upstream promoter region, genomic DNA was isolated from Malus cv. 'Royalty' and Malus cv. 'Flame' using the Plant Genomic DNA Kit (TIANGEN BIOTECH CO., LTD, Beijing, China). The regions of upstream DNA immediately adjacent to the transcription start site were isolated from the genomic DNA of 'Royalty' and 'Flame' by homology-based cloning method. Primers P1 & P2 for cloning were designed according to Malus × domestica cultivar 'Royal Gala' MYB10 promoter region sequence (GenBank: EU518249) from NCBI database. PCR products were cloned into pMD-19T vector and sequenced.

Semiquantitative RT-PCR and qRT-PCR Analysis

Semiquantitative RT-PCR was carried out in 20-μL reactions with ~2 μL of 10× diluted cDNA template, and MdActin was used as the internal control. qRT-PCR was performed using the SYBR® Premix Ex TaqTM II (Perfect Real Time) (Takara, Ohtsu, Tokyo, Japan) on the CFX96TM Real Time System (Bio-rad, Hercules, CA). The PCR amplification conditions were described previously (Shen et al., 2012). The transcript levels were determined by relative quantification using the Malus 18S ribosomal RNA gene as the internal control and calculated using the $2^{(-\Delta\Delta Ct)}$ analysis method.

Specific primers (Table S2) for semiquantitative RT-PCR and qRT-PCR analysis were designed with primer 5 software.

Overexpression in crabapple and tobacco

The full length of McMYB10 open reading frame (ORF) was cloned into pBI121 vector through BamHI and SacI sites. The primers used for these constructs are shown in Table S2. Transient expression in Malus crabapple leaves was performed with the

'Donald' cultivar. The Agrobacterium strains containing the construct were harvested and resuspended in infiltration buffer (10 mм MES, 0.2 mм acetosyringone and 10 mм MgCl₂) to an ultimate concentration of $OD_{600} = 0.5$. Bacteria were used to infiltrate crabapple leaves and fruits attached to the plants, and an empty vector was used as the negative control. Seven days after infiltration, the infected leaves and fruits were collected to detect phenotype and expression differences. These recombinant strains were used to transform Nicotiana tabacum 'W38' by the leaf disc method (Horsch et al., 1989). Transgenic plants were selected by kanamycin resistance. Two independent lines from the T2 progeny of transgenic plants were used to compare colour changes with wild-type line.

Transient expression assays in apple fruit

Agrobacteria cells were grown, collected and resuspended in 10 mм MES, 10 mм MgCl₂ and 200 mм acetosyringone solution to a final optical density of 1.5 at 600 nm and then incubated at room temperature for 3–4 h without shaking. Before infiltration, A. tumefaciens cultures containing pTRV1 and pTRV2 or its derivatives were mixed in a 1 : 1 ratio.

The infiltration protocol for transient expression assays in apple fruit was adapted from Xie et al. (2012). Bagged apples that had been freshly harvested from trees were used for infiltration. The recipient apples were punctured with a hypodermic needle. The infiltrations were performed using a syringe without a needle, and approximately 100 μL inoculum (OD600 = 0.8–1.0) was pipetted into the tube until it was partially absorbed by the apples.

For coloration, the debagged apples were infiltrated with pTRV-McMYB10 and kept in the dark at 4 °C for 7 days (Tian et al., 2014). Subsequently, these apples were infiltrated and maintained for an additional 3–5 days under 24 h of continuous white light (200 μmol/m²/s) at 17 °C in a growth chamber. The debagged apples infiltrated with pBI121-McMYB10 were kept overnight in the dark at room temperature and were subsequently exposed to white light at 17 °C in a growth chamber for another 3–5 days for coloration. The inoculum with empty vector and only infiltration buffer were used as controls. All transient expression assays were repeated at least five times, and at least twenty fruits were used in each experiment.

Arabidopsis thaliana complementation assay

The McMYB10 gene construct was driven by the 35S promoter of Cauliflower mosaic virus in binary vectors, which were introduced into Agrobacterium tumefaciens GV3101 by freeze–thaw method. Subsequently, bacteria were used to transform Arabidopsis pap1 mutant (pst16228) with the floral-dip method (Clough and Bent, 1998). Wild-type Nossen (No-0) was used as a control where appropriate.

Electrophoretic mobility shift assays

McMYB10 was cloned into the pET-28a(+) expression vector and transformed into Rosetta (DE3) Escherichia coli competent cells. The His tag in pET-28a(+) was used to facilitate purification of the recombinant protein. Isopropyl β-D-1-thiogalactopyranoside (IPTG) was added to induce McMYB10 expression. The recombinant protein was purified with the Ni-Agarose His Protein Purification System (CWbio. Co. Ltd, Beijing, China). Electrophoretic mobility shift assay (EMSA) reactions were prepared according to the manufacturer's protocol (LightShift® Chemiluminescent EMSA Kit; Thermo Fisher Scientific, Waltha, MA) using binding buffer, 50 ng poly (dI: dC), proteins, cold oligos and biotin-labelled probes. The probes were labelled by annealing biotin-labelled oligonucleotides. The oligonucleotides used for EMSA are listed in Figure 10b. Approximately 10 µg of purified McMYB10 protein was used for each EMSA reaction.

Yeast one-hybrid assay

A yeast one-hybrid system was used to assay the transcriptional activation by the protein McMYB10. We prepared the effector construct AD-McMYB10, in which the *McMYB10* open reading frame was cloned into the *EcoRI* and *XhoI* sites of the pJG4–5 vector (Clontech, Palo Alto, CA) under the control of the galactokinase 1 (GAL1) promoter. The *McF3'H* promoter sequence was inserted upstream of the reporter gene *LacZ* in the vector pLacZi to prepare the reporter construct BD-pro*McF3'H*. The effector and reporter or control constructs were transformed into competent cells of the yeast strain EGY48, resulting in the following yeast cells: AD-McMYB10/BD-pro*McF3'H*, AD/BD-pro*McF3'H*, AD-McMYB10/BD, AD/BD. The cells were selected on synthetic dropout media lacking tryptophan and uracil, and positive colonies were spotted onto glucose plates containing X-gal at 28 °C for 2 days to confirm blue colour development.

Accession numbers

McMYB10 (JX162681), MdActin (AB638619), McPAL (JQ248934), McCHS (FJ599763), McCHI (FJ817485), McF3H (FJ817486), McDFR (FJ817487), McANS (FJ817488), 18S (DQ341382), NtActin (GQ339768), NtCHS (AF311783), NtCHI1 (AB213651), NtF3H (AF036169), NtF3'H (AB289449), NtDFR (AB289448), NtANS (AB723683), Nt3GT2 (AB723686), MdCHS (CN944824), MdCHI (CN946541), MdF3H (CN491664), MdF3'H (CN491664), MdDFR (AF117268), MdANS (AF117269), MdUFGT (AF117267), AtPAP1 (AF062908), AtCHS (NM121396), AtCHI (AJ287301), AtF3H (AF064064), AtF3'H (AH009204), AtDFR (AB033294), AtANS (JF681791), AtUFGT (AF196777), AtACT8 (NM103814).

Acknowledgements

We thank the Fruit Tree Key Laboratory at the Beijing University of Agriculture. We also thank the Beijing Key Laboratory for Agricultural Application and New Technique for providing experimental resources. We are grateful to all technicians in the BUA Crabapple Germplasm Resource Garden.

Financial support

Financial support was provided by the Project of Construction of Innovative Teams and Teacher Career Development for Universities and Colleges Under Beijing Municipality (IDHT20140509), the National High Technology Research and Development Program of China (863 program) (2011AA100204), Beijing Municipal Commission of Education Science and Technology Promotion Plan (PXM2014-014207-000081), scientific research improvement project of Beijing University of Agriculture (GL2012003) and National Modern Agricultural Science City Achievement for People Service Technology Demonstration Project (Z121100007412003). The funding agencies had no role in study design, data collection and analysis, decision to publish or preparation of the manuscript.

References

Azuma, A., Yakushiji, H., Koshita, Y. and Kobayashi, S. (2012) Flavonoid biosynthesis-related genes in grape skin are differentially regulated by temperature and light conditions. *Planta*, **236**, 1067–1080.

Ban, Y., Honda, C., Hatsuyama, Y., Igarashi, M., Bessho, H. and Moriguchi, T. (2007) Isolation and functional analysis of a MYB transcription factor gene that is a key regulator for the development of red coloration in apple skin. *Plant Cell Physiol.* **48**, 958–970.

Brugliera, F., Tao, G.Q., Tems, U., Kalc, G., Mouradova, E., Price, K., Stevenson, K., Nakamura, N., Stacey, I., Katsumoto, Y., Tanaka, Y. and Mason, J.G. (2013) Violet/blue chrysanthemums–metabolic engineering of the anthocyanin biosynthetic pathway results in novel petal colors. *Plant Cell Physiol.* **54**, 1696–1710.

Chagné, D., Lin-Wang, K., Espley, R.V., Volz, R.K., How, N.M., Rouse, S., Brendolise, C., Carlisle, C.M., Kumar, S., De Silva, N., Micheletti, D., McGhie, T., Crowhurst, R.N., Storey, R.D., Velasco, R., Hellens, R.P., Gardiner, S.E. and Allan, A.C. (2013) An ancient duplication of apple MYB transcription factors is responsible for novel red fruit-flesh phenotypes. *Plant Physiol.* **161**, 225–239.

Chalker-Scott, L. (1999) Environmental significance of anthocyanins in plant stress responses. *Photochem. Photobiol.* **70**, 1–9.

Christie, P.J., Alfenito, M.R. and Walbot, V. (1994) Impact of low-temperature stress on general phenylpropanoid and anthocyanin pathways: enhancement of transcript abundance and anthocyanin pigmentation in maize seedlings. *Planta*, **194**, 541–549.

Clough, S.J. and Bent, A.F. (1998) Floral dip: a simplified method for Agrobacterium-mediated transformation of *Arabidopsis thaliana*. *Plant J.* **16**, 735–743.

Cortell, J.M. and Kennedy, J.A. (2006) Effect of shading on accumulation of flavonoid compounds in (*Vitis vinifera* L.) Pinot noir fruit and extraction in a model system. *J. Agric. Food Chem.* **54**, 8510–8520.

Craven-Bartle, B., Pascual, M.B., Canovas, F.M. and Avila, C. (2013) A Myb transcription factor regulates genes of the phenylalanine pathway in maritime pine. *Plant J.* **74**, 755–766.

Deluc, L., Barrieu, F., Marchive, C., Lauvergeat, V., Decendit, A., Richard, T., Carde, J.-P., Mérillon, J.-M. and Hamdi, S. (2006) Characterization of a grapevine R2R3-MYB transcription factor that regulates the phenylpropanoid pathway. *Plant Physiol.* **140**, 499–511.

Dixon, R.A. and Paiva, N.L. (1995) Stress-induced phenylpropanoid metabolism. *Plant Cell*, **7**, 1085.

Espley, R.V., Hellens, R.P., Putterill, J., Stevenson, D.E., Kutty-Amma, S. and Allan, A.C. (2007) Red colouration in apple fruit is due to the activity of the MYB transcription factor, MdMYB10. *Plant J.* **49**, 414–427.

Espley, R.V., Brendolise, C., Chagne, D., Kutty-Amma, S., Green, S., Volz, R., Putterill, J., Schouten, H.J., Gardiner, S.E., Hellens, R.P. and Allan, A.C. (2009) Multiple repeats of a promoter segment causes transcription factor autoregulation in red apples. *Plant Cell*, **21**, 168–183.

Espley, R.V., Bovy, A., Bava, C., Jaeger, S.R., Tomes, S., Norling, C., Crawford, J., Rowan, D., McGhie, T.K., Brendolise, C., Putterill, J., Schouten, H.J., Hellens, R.P. and Allan, A.C. (2013) Analysis of genetically modified red-fleshed apples reveals effects on growth and consumer attributes. *Plant Biotechnol. J.* **11**, 408–419.

Feng, S., Wang, Y., Yang, S., Xu, Y. and Chen, X. (2010) Anthocyanin biosynthesis in pears is regulated by a R2R3-MYB transcription factor PyMYB10. *Planta*, **232**, 245–255.

Fraser, L.G., Seal, A.G., Montefiori, M., McGhie, T.K., Tsang, G.K., Datson, P.M., Hilario, E., Marsh, H.E., Dunn, J.K., Hellens, R.P., Davies, K.M., McNeilage, M.A., Silva, H.N. and Allan, A.C. (2013) An R2R3 MYB transcription factor determines red petal colour in an Actinidia (kiwifruit) hybrid population. *BMC Genom.* **14**, 28.

Gonzalez, A., Zhao, M., Leavitt, J.M. and Lloyd, A.M. (2008) Regulation of the anthocyanin biosynthetic pathway by the TTG1/bHLH/Myb transcriptional complex in Arabidopsis seedlings. *Plant J.* **53**, 814–827.

Gould, K.S. and Lister, C. (2006) Flavonoids: chemistry, biochemistry and applications. In *Flavonoid Function in Plants* (Andersen, Y.M. and Markham, K.R., eds), pp. 397–441. Boca Raton, FL: CRC Pressl Llc.

Grotewold, E. (2006) The genetics and biochemistry of floral pigments. *Annu. Rev. Plant Biol.* **57**, 761–780.

Han, Y., Vimolmangkang, S., Soria-Guerra, R.E., Rosales-Mendoza, S., Zheng, D., Lygin, A.V. and Korban, S.S. (2010) Ectopic expression of apple F3′H genes contributes to anthocyanin accumulation in the Arabidopsis tt7 mutant grown under nitrogen stress. *Plant Physiol.* **153**, 806–820.

Hashimoto, F., Tanaka, M., Maeda, H., Shimizu, K. and Sakata, Y. (2000) Characterization of cyanic flower color of Delphinium cultivars. *J. Jpn. Soc. Hortic. Sci.* **69**, 428–434.

Hernández, I., Alegre, L., Van Breusegem, F. and Munné-Bosch, S. (2009) How relevant are flavonoids as antioxidants in plants? *Trends Plant Sci.* **14**, 125–132.

Higo, K., Ugawa, Y., Iwamoto, M. and Korenaga, T. (1999) Plant cis-acting regulatory DNA elements (PLACE) database: 1999. *Nucleic Acids Res.* **27**, 297–300.

Holton, T.A. and Cornish, E.C. (1995) Genetics and biochemistry of anthocyanin biosynthesis. *Plant Cell*, **7**, 1071–1083.

Horsch, R.B., Fry, J., Hoffmann, N., Neidermeyer, J., Rogers, S.G. and Fraley, R.T. (1989) Leaf disc transformation. In *Plant Molecular Biology Manual* (Gelvin, S., Schilperoort, R. A., eds), pp. 63–71. The Netherlands: Kluwer Academic Publishers.

Ito, M., Iwase, M., Kodama, H., Lavisse, P., Komamine, A., Nishihama, R., Machida, Y. and Watanabe, A. (1998) A novel cis-acting element in promoters of plant B-Type cyclin genes activates M phase-specific transcription. *Plant Cell*, **10**, 331–341.

Jaakola, L. (2013) New insights into the regulation of anthocyanin biosynthesis in fruits. *Trends Plant Sci.* **18**, 477–483.

Jeong, S.T., Goto-Yamamoto, N., Kobayashi, S. and Esaka, M. (2004) Effects of plant hormones and shading on the accumulation of anthocyanins and the expression of anthocyanin biosynthetic genes in grape berry skins. *Plant Sci.* **167**, 247–252.

Jiang, R., Tian, J., Song, T., Zhang, J. and Yao, Y. (2014) The Malus crabapple transcription factor McMYB10 regulates anthocyanin biosynthesis during petal coloration. *Sci. Hortic.* **166**, 42–49.

Katsumoto, Y., Fukuchi-Mizutani, M., Fukui, Y., Brugliera, F., Holton, T.A., Karan, M., Nakamura, N., Yonekura-Sakakibara, K., Togami, J., Pigeaire, A., Tao, G.Q., Nehra, N.S., Lu, C.Y., Dyson, B.K., Tsuda, S., Ashikari, T., Kusumi, T., Mason, J.G. and Tanaka, Y. (2007) Engineering of the rose flavonoid biosynthetic pathway successfully generated blue-hued flowers accumulating delphinidin. *Plant Cell Physiol.* **48**, 1589–1600.

Koes, R., Verweij, W. and Quattrocchio, F. (2005) Flavonoids: a colorful model for the regulation and evolution of biochemical pathways. *Trends Plant Sci.* **10**, 236–242.

Kuromori, T., Hirayama, T., Kiyosue, Y., Takabe, H., Mizukado, S., Sakurai, T., Akiyama, K., Kamiya, A., Ito, T. and Shinozaki, K. (2004) A collection of 11 800 single-copy Ds transposon insertion lines in Arabidopsis. *Plant J.* **37**, 897–905.

Lebowitz, R.J. (1985) The genetics and breeding of coleus. *Plant Breed. Rev.* **3**, 343–360.

Li, Y.Y., Mao, K., Zhao, C., Zhao, X.Y., Zhang, H.L., Shu, H.R. and Hao, Y.J. (2012) MdCOP1 ubiquitin E3 ligases interact with MdMYB1 to regulate light-induced anthocyanin biosynthesis and red fruit coloration in apple. *Plant Physiol.* **160**, 1011–1022.

Lightbourn, G.J., Griesbach, R.J., Novotny, J.A., Clevidence, B.A., Rao, D.D. and Stommel, J.R. (2008) Effects of anthocyanin and carotenoid combinations on foliage and immature fruit color of Capsicum annuum L. *J. Hered.* **99**, 105–111.

Lin-Wang, K., Bolitho, K., Grafton, K., Kortstee, A., Karunairetnam, S., McGhie, T.K., Espley, R.V., Hellens, R.P. and Allan, A.C. (2010) An R2R3 MYB transcription factor associated with regulation of the anthocyanin biosynthetic pathway in Rosaceae. *BMC Plant Biol.* **10**, 50.

Lo Piero, A.R., Puglisi, I., Rapisarda, P. and Petrone, G. (2005) Anthocyanins accumulation and related gene expression in red orange fruit induced by low temperature storage. *J. Agric. Food Chem.* **53**, 9083–9088.

Mori, K., Sugaya, S. and Gemma, H. (2005) Decreased anthocyanin biosynthesis in grape berries grown under elevated night temperature condition. *Sci. Hortic.* **105**, 319–330.

Morohashi, K., Casas, M.I., Falcone Ferreyra, M.L., Mejia-Guerra, M.K., Pourcel, L., Yilmaz, A., Feller, A., Carvalho, B., Emiliani, J., Rodriguez, E., Pellegrinet, S., McMullen, M., Casati, P. and Grotewold, E. (2012) A genome-wide regulatory framework identifies maize pericarp color1 controlled genes. *Plant Cell*, **24**, 2745–2764.

Nguyen, P. and Cin, V.D. (2009) The role of light on foliage colour development in coleus (Solenostemon scutellarioides (L.) Codd). *Plant Physiol. Biochem.* **47**, 934–945.

Peters, D.J. and Constabel, C.P. (2002) Molecular analysis of herbivore-induced condensed tannin synthesis: cloning and expression of dihydroflavonol reductase from trembling aspen (Populus tremuloides). *Plant J.* **32**, 701–712.

Ravaglia, D., Espley, R.V., Henry-Kirk, R.A., Andreotti, C., Ziosi, V., Hellens, R.P., Costa, G. and Allan, A.C. (2013) Transcriptional regulation of flavonoid biosynthesis in nectarine (Prunus persica) by a set of R2R3 MYB transcription factors. *BMC Plant Biol.* **13**, 68.

Revilla, E. and Ryan, J.-M.A. (2000) Analysis of several phenolic compounds with potential antioxidant properties in grape extracts and wines by high-performance liquid chromatography–photodiode array detection without sample preparation. *J. Chromatogr. A* **881**, 461–469.

Rowan, D.D., Cao, M., Lin-Wang, K., Cooney, J.M., Jensen, D.J., Austin, P.T., Hunt, M.B., Norling, C., Hellens, R.P., Schaffer, R.J. and Allan, A.C. (2009) Environmental regulation of leaf colour in red 35S:PAP1 Arabidopsis thaliana. *New Phytol.* **182**, 102–115.

Shen, H., Zhang, J., Yao, Y., Tian, J., Song, T., Geng, J. and Gao, J. (2012) Isolation and expression of McF3H gene in the leaves of crabapple. *Acta Physiol. Plant*, **34**, 1353–1361.

Sun, C., Huang, H., Xu, C., Li, X. and Chen, K. (2013) Biological activities of extracts from Chinese bayberry (Myrica rubra Sieb. et Zucc.): a review. *Plant Foods Hum. Nutr.* **68**, 97–106.

Takos, A.M., Jaffe, F.W., Jacob, S.R., Bogs, J., Robinson, S.P. and Walker, A.R. (2006) Light-induced expression of a MYB gene regulates anthocyanin biosynthesis in red apples. *Plant Physiol.* **142**, 1216–1232.

Teng, S., Keurentjes, J., Bentsink, L., Koornneef, M. and Smeekens, S. (2005) Sucrose-specific induction of anthocyanin biosynthesis in Arabidopsis requires the MYB75/PAP1 gene. *Plant Physiol.* **139**, 1840–1852.

Tian, J., Shen, H., Zhang, J., Song, T. and Yao, Y. (2011) Characteristics of chalcone synthase promoters from different leaf-color malus crabapple cultivars. *Sci. Hortic.* **129**, 449–458.

Tian, J., Pei, H., Zhang, S., Chen, J., Chen, W., Yang, R., Meng, Y., You, J., Gao, J. and Ma, N. (2014) TRV-GFP: a modified Tobacco rattle virus vector for efficient and visualizable analysis of gene function. *J. Exp. Bot.* **65**, 311–322.

Tohge, T., Nishiyama, Y., Hirai, M.Y., Yano, M., Nakajima, J.-I., Awazuhara, M., Inoue, E., Takahashi, H., Goodenowe, D.B., Kitayama, M., Noji, M., Yamazaki, M. and Saito, K. (2005) Functional genomics by integrated analysis of metabolome and transcriptome of Arabidopsis plants over-expressing an MYB transcription factor. *Plant J.* **42**, 218–235.

Ubi, B.E., Honda, C., Bessho, H., Kondo, S., Wada, M., Kobayashi, S. and Moriguchi, T. (2006) Expression analysis of anthocyanin biosynthetic genes in apple skin: effect of UV-B and temperature. *Plant Sci.* **170**, 571–578.

Wang, L., Li, X., Zhao, Q., Jing, S., Chen, S. and Yuan, H. (2008) Identification of genes induced in response to low-temperature treatment in tea leaves. *Plant Mol. Biol. Rep.* **27**, 257–265.

Wang, Z., Meng, D., Wang, A., Li, T., Jiang, S. and Cong, P. (2013) The methylation of PcMYB10 promoter is associated with Green skinned sport in 'Max Red Bartlett' pear. *Plant Physiol.* **162**, 885–896.

Winkel-Shirley, B. (2001) Flavonoid biosynthesis. A colorful model for genetics, biochemistry, cell biology, and biotechnology. *Plant Physiol.* **126**, 485–493.

Xie, X.B., Li, S., Zhang, R.F., Zhao, J., Chen, Y.C., Zhao, Q., Yao, Y.X., You, C.X., Zhang, X.S. and Hao, Y.J. (2012) The bHLH transcription factor MdbHLH3 promotes anthocyanin accumulation and fruit colouration in response to low temperature in apples. *Plant, Cell Environ.* **35**, 1884–1897.

Xu, Q., Yin, X.R., Zeng, J.K., Ge, H., Song, M., Xu, C.J., Li, X., Ferguson, I.B. and Chen, K.S. (2014) Activator- and repressor-type MYB transcription factors are involved in chilling injury induced flesh lignification in loquat via their interactions with the phenylpropanoid pathway. *J. Exp. Bot.* **65**, 4349–4359.

9

Metabolic engineering of terpene biosynthesis in plants using a trichome-specific transcription factor *MsYABBY5* from spearmint (*Mentha spicata*)

Qian Wang[1], Vaishnavi Amarr Reddy[1,2], Deepa Panicker[1,†], Hui-Zhu Mao[1], Nadimuthu Kumar[1], Chakravarthy Rajan[1], Prasanna Nori Venkatesh[1], Nam-Hai Chua[3] and Rajani Sarojam[1,*]

[1]Temasek Life Sciences Laboratory, 1 Research Link, National University of Singapore, Singapore City, Singapore
[2]Department of Biological Sciences, National University of Singapore, Singapore City, Singapore
[3]Laboratory of Plant Molecular Biology, The Rockefeller University, New York, NY, USA

*Correspondence
email rajanis@tll.org.sg
†Present address: Singapore Centre on Environmental Life Sciences Engineering, Nanyang Technological University, Singapore 637551, Singapore.

Keywords: spearmint, secondary metabolism, terpene, sweet basil, transcription factor, YABBY.

Summary

In many aromatic plants including spearmint (*Mentha spicata*), the sites of secondary metabolite production are tiny specialized structures called peltate glandular trichomes (PGT). Having high commercial values, these secondary metabolites are exploited largely as flavours, fragrances and pharmaceuticals. But, knowledge about transcription factors (TFs) that regulate secondary metabolism in PGT remains elusive. Understanding the role of TFs in secondary metabolism pathway will aid in metabolic engineering for increased yield of secondary metabolites and also the development of new production techniques for valuable metabolites. Here, we isolated and functionally characterized a novel *MsYABBY5* gene that is preferentially expressed in PGT of spearmint. We generated transgenic plants in which *MsYABBY5* was either overexpressed or silenced using RNA interference (RNAi). Analysis of the transgenic lines showed that the reduced expression of *MsYABBY5* led to increased levels of terpenes and that overexpression decreased terpene levels. Additionally, ectopic expression of *MsYABBY5* in *Ocimum basilicum* and *Nicotiana sylvestris* decreased secondary metabolite production in them, suggesting that the encoded transcription factor is probably a repressor of secondary metabolism.

Introduction

The genus *Mentha*, a member of *Lamiaceae* family, includes species that are widely used as medicinal and aromatic herbs. The essential oils produced by these plants find wide usage in food, flavour, cosmetic and pharmaceutical industries (Champagne and Boutry, 2013; Sinha *et al.*, 2013). Plants produce these volatile essential oils as secondary metabolites which have important roles in plant defence, plant-to-plant communication and pollination (Gershenzon *et al.*, 2000). In mint, these essential oils are produced in specialized nonphotosynthetic glandular trichomes known as peltate glandular trichomes (PGT) which are found on the aerial surface of the plants. The PGT are dedicated to the production and storage of large amounts of volatile secretions (Champagne and Boutry, 2013; Croteau *et al.*, 2000; Lange and Turner, 2013). In the case of spearmint (*Mentha spicata*), the essential oil is dominated mainly by two monoterpenes, limonene and carvone. Monoterpenes are the C10 type of terpenoids and are generally colourless, lipophilic and volatile. They are responsible for the characteristic aromas and flavours of essential oils, floral scents and resin of aromatic plants (Loza-Tavera, 1999). Given their economic importance, strategies to metabolically engineer monoterpenes to increase yield is of considerable interest.

Varietal improvement in cultivated spearmint or peppermint varieties has been challenging because these varieties are sterile hybrids making classical breeding approach unfeasible. Hence, metabolic engineering provides an alternative method to improve essential oil yield and composition. Plants synthesize terpenes either by the mevalonate (MVA) pathway in the cytosol or by the 2-C-methyl-D-erythritol 4-phosphate (MEP) pathway in plastids. Both pathways provide the precursors for terpene biosynthesis (Vranová *et al.*, 2013) and have been well investigated. The MEP pathway in plastids is mainly responsible for producing monoterpenes and diterpenes, whereas the MVA pathway generates sesquiterpenes and triterpenes (Dubey *et al.*, 2003). Studies have shown that under certain conditions an exchange of precursor metabolites can occur between the cytosolic MVA and plastid MEP pathways. Analysis of terpenoid production in many different plants has shown that under specific ecological conditions synthesis of monoterpenes, diterpenes, sesquiterpenes and polyterpenes can occur from precursors produced by both pathways. Metabolic intermediates like isopentenyl diphosphate, geranyl diphosphate or geranylgeranyl diphosphate can be transported across plastidial membranes (Hemmerlin *et al.*, 2012; Vranová *et al.*, 2013).

Apart from the precursor pathways, the downstream monoterpene biosynthetic pathways in both spearmint and peppermint are also well characterized (Lange *et al.*, 2011). One of the strategies to increase yield in peppermint was to manipulate genes that code for enzymes involved in the monoterpene pathway, for example genes for limonene synthase and limonene hydroxylase (Diemer *et al.*, 2001; Mahmoud *et al.*, 2004), but overexpression of these genes did not enhance oil yields significantly. In their pioneering work, Mahmoud and Croteau

(2001) and Lange *et al.* (2011) evaluated the efficacy of overexpressing genes encoding enzymes involved in precursor pathways on oil yields in peppermint (*Mentha piperita*). Most encouraging results were obtained in plants where two genes were manipulated simultaneously, the gene encoding 1-deoxy-d-xylulose-5-phosphate reductoisomerase was overexpressed and the gene encoding menthofuran synthase was down-regulated. Oil yields in these transgenic plants increased up to 61% over wild-type controls while reducing the undesirable side-products (+)-menthofuran and its intermediate (+)-pulegone (Lange *et al.*, 2011). Recently, increase in monoterpene formation was achieved by introducing a noncanonical substrate neryl diphosphate (NPP), but the additional NPP had to be eliminated to avoid adverse impact on plant growth (Gutensohn *et al.*, 2014).

It is increasingly evident that transcription factors (TFs) which are regulators of structural genes can activate or repress multiple genes in a metabolic pathway (Grotewold, 2008; Iwase *et al.*, 2009). Manipulation of such TFs can be more effective for engineering pathways rather than changing the expression of genes for individual enzymes involved, because plant metabolic pathways are complex comprising of multiple genes encoding various enzymes (Broun and Somerville, 2001). The effectiveness of using TFs to modulate metabolic pathways has been validated in a few studies (Butelli *et al.*, 2008; Luo *et al.*, 2008; Schwinn *et al.*, 2006). Although the enzymatic pathway leading to the synthesis of spearmint monoterpenes is well defined (Croteau *et al.*, 1991; Lange *et al.*, 2011; Muñoz-Bertomeu *et al.*, 2008), the developmental regulation of this secondary metabolite pathway still remains elusive. Few TFs have been reported from other plants that are involved in regulating terpene biosynthesis. They are *Artemisia annua*, *AaWRKY1*, *AaERF1*, *AaERF2*, *AaORA1* and *AabZIP1* (Lu *et al.*, 2013; Ma *et al.*, 2009; Yu *et al.*, 2012; Zhang *et al.*, 2015), cotton *GaWRKY1* (Xu *et al.*, 2004), *TaWRKY1* from *Taxus chinensis* (Wang *et al.*, 2001), rubber *EREBP1* and *HbWRKY1* (Chen *et al.*, 2012; Zhou *et al.*, 2012) and *OsTGAP1* in rice (Miyamoto *et al.*, 2014).

To investigate the genes involved in PGT formation and secondary metabolism in spearmint, we had performed comparative RNA-Seq analysis of different tissues of spearmint, namely PGT, leaf devoid of PGT (leaf-PGT) and leaf in a previous study (Jin *et al.*, 2014). This led to the identification of many TF transcripts that were significantly more abundant in PGT when compared to leaf-PGT and a *YABBY* transcript was among the top candidates. We cloned the full-length cDNA of this transcript and sequence analysis showed that it is similar to YABBY5 subfamily of proteins. *YABBY* genes constitute a group of plant-specific TFs that are known to play important roles in various aspects of vegetative and floral development in plants (Bonaccorso *et al.*, 2012; Bowman, 2000). In this study, we report the engineering of spearmint plants for higher yields by suppressing this glandular trichome-enriched TF *MsYABBY5*. The resulting *MsYABBY5* RNAi lines showed an increase in monoterpene production which ranged from 20% to 77%. This is the first report of a transcription factor regulating monoterpene production in mint plants and assigns a new role for *YABBY* genes in plant secondary metabolism. Ectopic expression of *MsYABBY5* in sweet basil (*Ocimum basilicum*), an aromatic herb similar to mint, and in *Nicotiana sylvestris* resulted in decreased secondary metabolite production in them. Essential oil of sweet basil has compounds derived from both terpene and phenylpropanoid pathways, whereas *N. sylvestris* produces mainly diterpenes. As MsYABBY5 could affect metabolites derived from different metabolic pathways, it suggests that it regulates an upstream step in plant secondary metabolism. We further found that MsYABBY5 probably regulates terpene synthesis through a regulatory network that involves *MsWRKY75*.

Results

MsYABBY5 shows high expression in spearmint PGT

Mint leaves have PGT on both surfaces (Figure 1A). From the RNA-Seq data of leaves, we identified four *YABBY*-like transcripts that showed high expression levels. Of these, only *MsYABBY5* was preferentially expressed in PGT, whereas the others were more enriched in leaf tissues. The differential expression pattern of these transcripts as observed by RNA Seq was further validated by quantitative RT-PCR (qRT-PCR) (Figure 1B). Full-length open reading frames (ORFs) of all these four *YABBYs* including *MsYABBY5* were amplified from leaf cDNA using RACE. All the four cloned ORFs contained a conserved C_2C_2 zinc finger domain located at N-terminus and a helix-loop-helix motif (YABBY domain) at the C terminus which is similar to the HMG box motif. These two domains are highly conserved among all YABBY proteins (Figure 2A). As we were interested in TFs involved in regulating secondary metabolism in mint, we focussed on *MsYABBY5*. *In situ* hybridization also confirmed the PGT-specific expression of *MsYABBY5*, as no signal was observed in the leaf tissue (Figure 1C). The ORF of *MsYABBY5* encoded a polypeptide of 190 amino acids. BLAST analysis showed that *MsYABBY5* has highest sequence similarity to *Antirrhinum PROLONGATA YABBY*-like transcription factor. We generated a phylogenetic tree based on amino acids sequences of YABBY proteins from different plants. The results revealed that MsYABBY5 and MsYABBY6 belonged to the YABBY5 subfamily, whereas the other two, MsYABBY2 and MsYABBY4, are members of the YABBY2 subfamily (Figure 2B).

Subcellular localization of MsYABBY5 protein

To examine the subcellular localization patterns of YABBY proteins, cDNAs of all the four *MsYABBYs* were fused in-frame to cDNA encoding the yellow-fluorescent protein (YFP) and the fusion genes were transiently expressed in tobacco by agroinfiltration. All the MsYABBYs except MsYABBY5 showed nuclear localization. Interestingly, MsYABBY5 showed both nuclear and cytoplasmic localization (Figure 3A). Online software prediction programs indicated that MsYABBY5 contained a potential transmembrane domain (http://dgpred.cbr.su.se/index.php?p=fullscan) at the amino terminal and participated in the secretory pathway (http://www.cbs.dtu.dk/services/TargetP-1.1/output.php). To investigate this, Golgi markers were used for colocalization experiment which showed that MsYABBY5 localized to Golgi (Figure 3B). To further assess this localization pattern, tobacco leaves were treated with Brefeldin A (BFA). BFA treatment in tobacco results in the complete disappearance of Golgi apparatus and disrupts the secretory system (Robinson and Ritzenthaler, 2006). After treatment with 50 µg/mL BFA for 3 h, MsYABBY5 was found to exhibit nuclear localization only, while both nuclear and cytosolic distribution was still observed in the control plants (treated with 1 : 1000 dilution of DMSO in ddH$_2$O) (Figure 3C).

To directly observe the distribution pattern of MsYABBY5 protein, we performed immunogold labelling of native MsYABBY5 in PGT of spearmint. A 14-amino acid peptide that is presumably located on the surface of the predicted three-

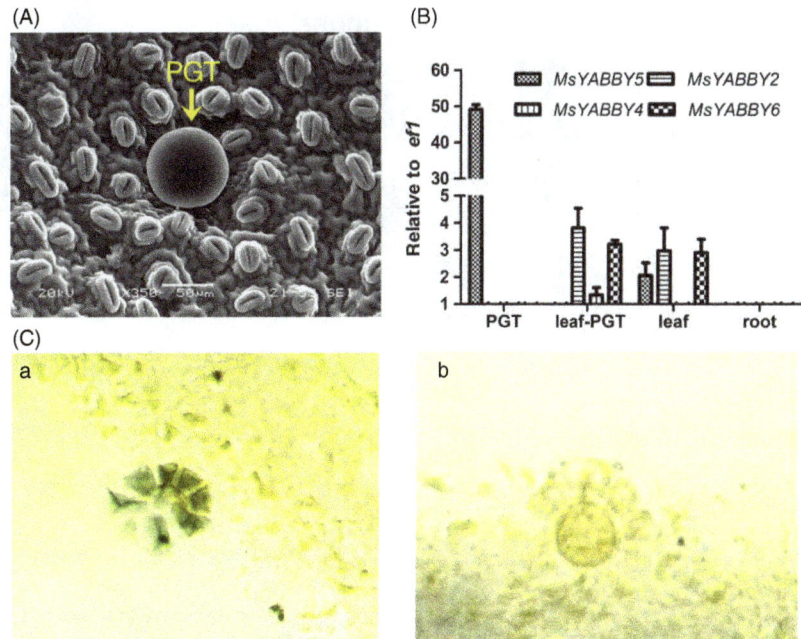

Figure 1 Validation of *MsYABBY* genes expression pattern in spearmint. (A) Spearmint leaf showing peltate glandular trichome (PGT) on upper leaf surface as visualized under scanning electron microscope. (B) qRT-PCR analysis of *MsYABBY* genes in different tissues. PGT, peltate glandular trichome; leaf-PGT, leaves where PGT were brushed away. The housekeeping gene *elongation factor 1 (ef1)* was used as control. (C) *In situ* hybridization: antisense (a) and sense (b) probe detection of *MsYABBY5*.

dimensional structure of MsYABBY5 was used as an antigen to raise specific antibody. Western blot analysis showed that the antibody exclusively bound to MsYABBY5 but not to MsYABBY2, MsYABBY4 or MsYABBY6 (Figure S1A,B). MsYABBY5 proteins conjugated with gold particles were observed to accumulate inside the nucleus and also present in the cytoplasm. As the cell organelles were not clearly distinguishable in the TEM sections, the localization to cell organelles could not be verified by immunostaining (Figure S1C).

Analysis of the promoter region of *MsYABBY5*

A 1116-bp genomic DNA fragment upstream of the putative start codon of *MsYABBY5* was cloned by genome walking. Bioinformatics analysis of this region revealed the presence of many *cis*-acting regulatory elements apart from the common TATA and CAAT box (http://bioinformatics.psb.ugent.be/webtools/plant-care/html/) (Figure 4A). Four light-responsive motifs, including two Box4, one TCT and one Sp1 motifs, were found in the promoter sequence. Regulation of terpene biosynthesis by light is known in many plants (Cordoba *et al.*, 2009). With respect to hormones, two *cis*-acting elements, CGTCA-motif and TGACG-motif, involved in the MeJA-responsiveness were found and one for gibberellin *cis* elements, TATC box, was found within the sequence (Zhou *et al.*, 2012; Zhu *et al.*, 2014). Further, tissue-specific expression pattern of the cloned promoter was analysed. The 1116-bp promoter fragment was fused with a β-glucuronidase (GUS) reporter gene and transformed into *Nicotiana benthamiana* plants. The transgenic plants showed trichome-specific expression pattern in leaves and stems of tobacco plants (Figure 4B,C). No staining was observed in flowers or roots. Hence, this promoter is potentially a glandular trichome-specific promoter. Additionally, this promoter was used to drive *MsYABBY5* cDNA fused to a cyan-fluorescent protein reporter gene in basil and tobacco. The fluorescence was observed exclusively in PGT of basil plants and head cells of the glandular trichomes of tobacco, but subcellular localization was difficult to decipher (Figure 5).

Silencing of *MsYABBY5* increases monoterpene production in spearmint

To examine the function of *MsYABBY5* in spearmint PGT, an RNAi construct targeting a specific region of *MsYABBY5* was generated and transformed into wild-type spearmint using *Agrobacterium tumefaciens*-mediated T-DNA transfer. Many transgenic lines were generated, of which four independent transgenic lines analysed by Southern blotting for transgene integration were selected for further characterization (Figure S2). All these RNAi plants showed a reduction in *MsYABBY5* transcripts (Figure 6A). No significant changes were observed in the expression of other leaf-specific *YABBY* genes (*MsYABBY2*, *MsYABBY4* and *MsYABBY6*), suggesting that the RNAi construct was specific to *MsYABBY5* and did not target other *YABBY* transcripts (Figure S3). The RNAi transgenic plants appeared phenotypically similar to WT plants. Scanning electron microscopy analysis revealed no phenotypical changes in either the number or the structure of PGT.

Gas chromatography–mass spectrometry (GC–MS) analysis was performed on these transgenic plants to evaluate the quality and quantity of the volatiles produced. Young WT spearmint leaves contain an abundance of both limonene and carvone monoterpenes. Limonene is the first committed step towards carvone production. Limonene is converted to carvone by a two-step reaction. In our greenhouse conditions, we observed that in WT spearmint, the productions of limonene and carvone were about 1.47 ± 0.11 and 2.10 ± 0.25 µg/mg fresh leaf, respectively. Upon GC–MS analysis, all the four transgenic lines showed a significant increase (20%–77%) in total monoterpene production (limonene and carvone) (Figure 6B,C). The RNAi lines were tested for the expression levels of enzymes involved in carvone production (limonene synthase, limonene 6-hydroxylase and carveol dehydrogenase) by qRT-PCR; however, no major changes were observed. Transcripts for all enzymes of the MEP precursor pathways including geranyl diphosphate synthase small and big subunits were also investigated, but no significant changes in

(A)

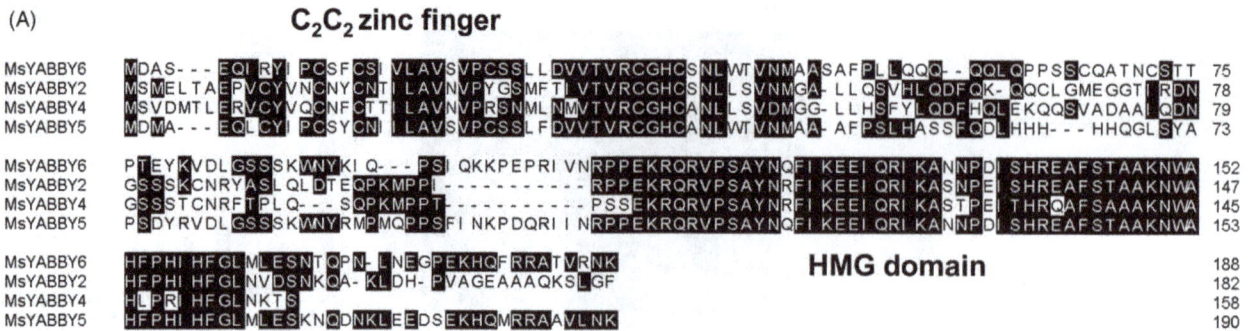

(B)

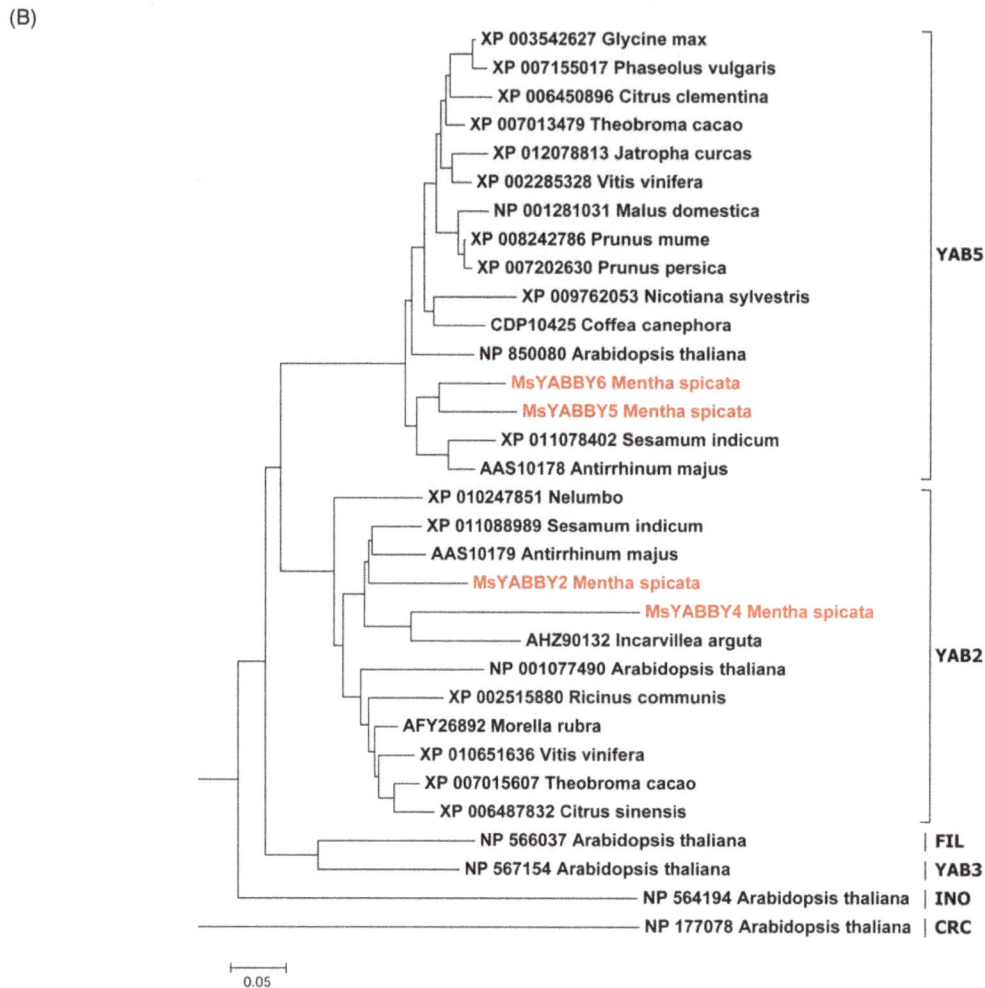

Figure 2 Amino acid sequence alignment (A) and phylogenetic tree analysis (B) of MsYABBYs.

their expression levels were found. These results suggest that increase in monoterpene production is probably not due to the transcriptional activation of biosynthetic genes. *MsYABBY5* might be acting upstream to regulate flux into the terpene pathway.

Overexpression of *MsYABBY5* results in decrease in monoterpene production

To gain further insight into the role of *MsYABBY5* in secondary metabolism, we overexpressed this gene in spearmint under the control of a CaMV *35S* promoter. Four independent lines confirmed by southern hybridization were selected for further

characterization. The results of qRT-PCR showed high *MsYABBY5* expression levels in all the transgenic plants (Figure 6D). GC–MS analysis of young leaves showed a reduction in total monoterpene production, which ranged from 23% to 52% (Figure 6E,F). The observed phenotype in RNAi and overexpression transgenic lines suggests that *MsYABBY5* might be a repressor of secondary metabolism in spearmint.

Possible downstream target of MsYABBY5 to regulate secondary metabolism

To better understand MsYABBY5's biological functions and signalling pathways, it is essential to know its downstream target

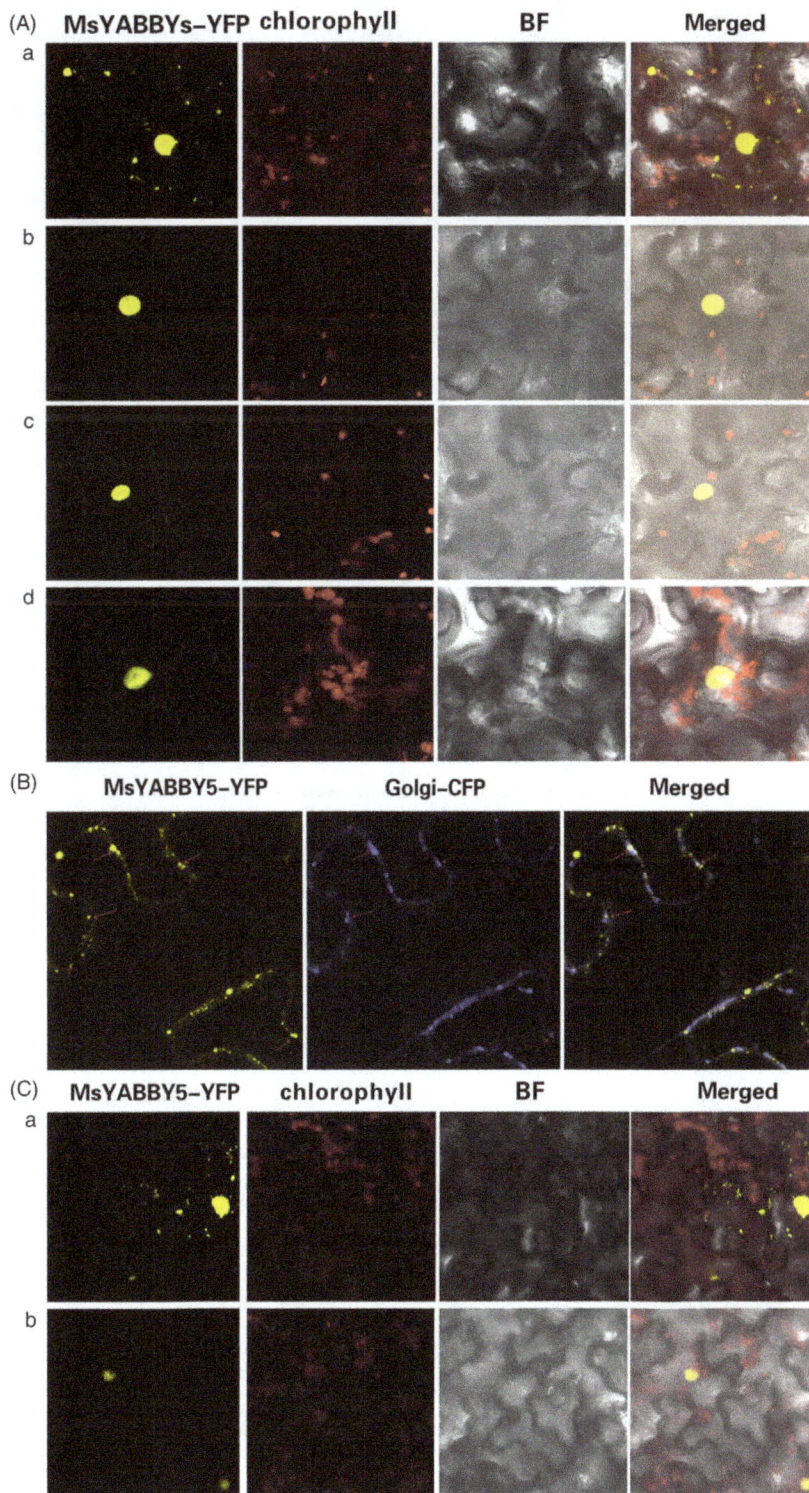

Figure 3 Subcellular localization of MsYABBYs in *Nicotiana benthamiana.* (A) MsYABBY5 showed both nuclear and cytoplasmic expression, while other MsYABBYs were found in nucleus only. (a) MsYABBY5 was localized to both nucleus and cytoplasm. (b) MsYABBY6. (c) MsYABBY2. (d) MsYABBY4. (B) MsYABBY5 protein colocalization with Golgi marker. (C) BFA treatment leads to nuclear localization of MsYABBY5 protein in *N. benthamiana.* (a) Mock group treated with DMSO. (b) Test group treated with 50 µg/mL BFA for 3 h.

genes. There is not much known about the molecular mechanism of regulation by *YABBYs* and their direct target genes. As genes encoding enzymes in the biosynthetic pathways leading to monoterpene production were not significantly changed, we decided to investigate the expression of genes for transporters. Transporters play a key role in plant cellular metabolism (Fischer, 2011; Flügge *et al.*, 2011). From our previous RNA-Seq data, we had identified several transcripts in PGT which were involved in transport of carbon and ATP (Jin *et al.*, 2014). One such transcript

MsNTT similar to the plastidic ATP/ADP transporter-like proteins showed differential expression in overexpression and RNAi plants. All the RNAi lines showed increased *MsNTT* transcript levels compared with WT plants or *35S::GFP* control plants and reduced expression was observed in the overexpression plants (Figure 7A, B). Plastidic ATP/ADP transporter facilitates the movement of ATP across the plastid inner membrane and can determine the rate of metabolic activity in plastids (Flügge *et al.*, 2011; Neuhaus *et al.*, 1997). MsNTT was seen localized to the plastid membrane

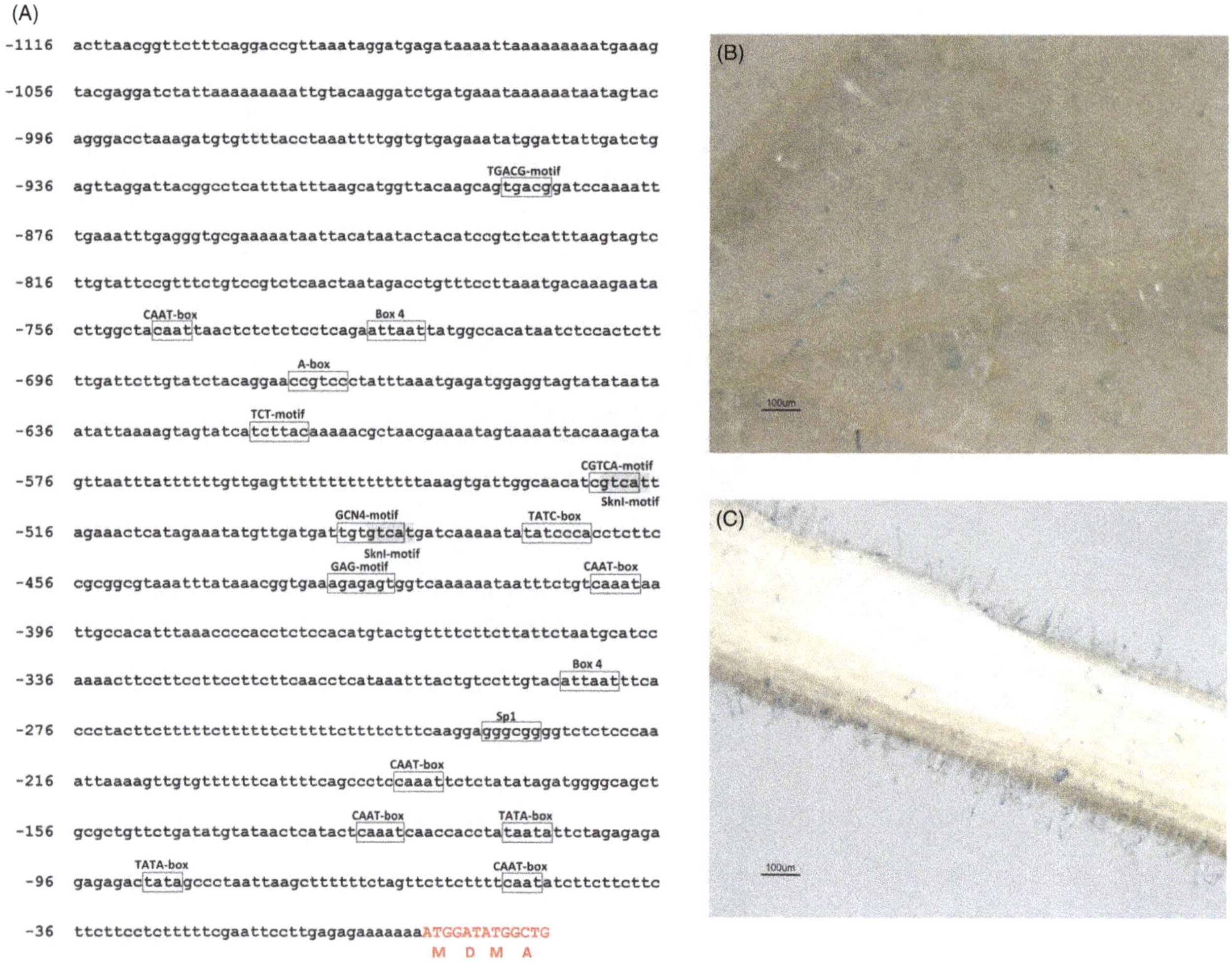

(A)

```
-1116  acttaacggttctttcaggaccgttaaataggatgagataaaattaaaaaaaaatgaaag

-1056  tacgaggatctattaaaaaaaaattgtacaaggatctgatgaaataaaaaataatagtac

-996   agggacctaaagatgtgttttacctaaattttggtgtgagaaatatggattattgatctg
                                                TGACG-motif
-936   agttaggattacggcctcatttatttaagcatggttacaagcagtgacggatccaaaatt

-876   tgaaatttgagggtgcgaaaaataattacataatactacatccgtctcatttaagtagtc

-816   ttgtattccgtttctgtccgtctcaactaatagacctgtttccttaaatgacaaagaata
              CAAT-box                     Box 4
-756   cttggctacaattaactctctctcctcagaattaattatggccacataatctccactctt
                            A-box
-696   ttgattcttgtatctacaggaaccgtccctatttaaatgagatggaggtagtatataata
              TCT-motif
-636   atattaaaagtagtatcatcttacaaaaacgctaacgaaaatagtaaaattacaaagata
                                                    CGTCA-motif
-576   gttaatttattttttgttgagtttttttttttttttaaagtgattggcaacatcgtcatt
                                                    SknI-motif
              GCN4-motif               TATC-box
-516   agaaactcatagaaatatgttgatgattgtgtcatgatcaaaaatatatcccacctcttc
              GAG-motif                            CAAT-box
-456   cgcggcgtaaatttataaacggtgaaagagagtggtcaaaaaataatttctgtcaaataa

-396   ttgccacatttaaaccccacctctccacatgtactgtttcttcttattctaatgcatcc
                                                Box 4
-336   aaaacttccttccttccttcttcaacctcataaatttactgtccttgtacattaatttca
                                        Sp1
-276   ccctacttcttttttcttttttcttttttcttttctttcaaggagggcgggtctctcccaa
                                CAAT-box
-216   attaaaagttgtgttttttcattttcagccctccaaattctctatatagatggggcagct
              CAAT-box                TATA-box
-156   gcgctgttctgatatgtataactcatactcaaatcaaccacctataatattctagagaga
        TATA-box                            CAAT-box
-96    gagagactatagcccctaattaagctttttttctagttcttcttttcaatatcttcttcttc

-36    ttcttcctcttttcgaattccttgagagaaaaaaaATGGATATGGCTG
                                          M  D  M  A
```

(B)

(C)

Figure 4 Ms*YABBY5* promoter analysis and expression pattern. (A) *cis*-acting regulatory elements in the 5′UTR (−1116 bp) region of *MsYABBY5*. (B and C) Trichome-specific GUS expression pattern observed in *Nicotiana benthamiana* leaves and stems *of* plants transformed with p*MsYABBY5::GUS*.

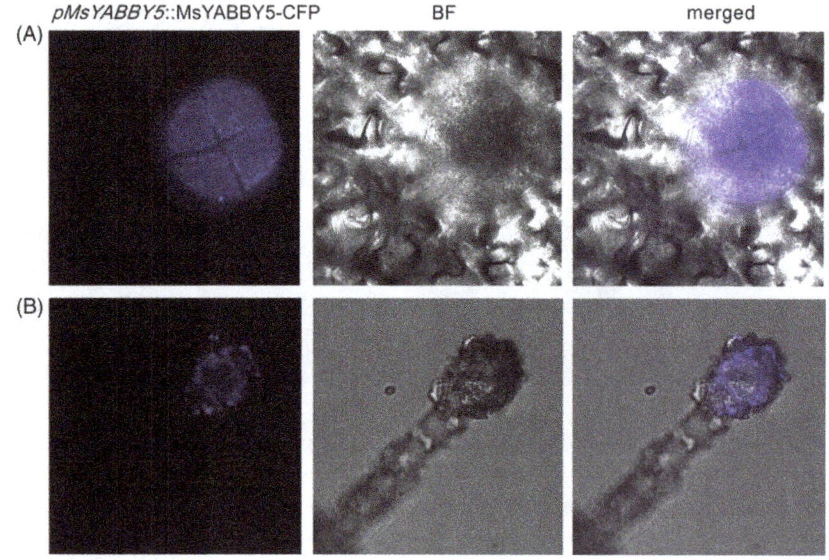

p*MsYABBY5*::MsYABBY5-CFP   BF   merged

(A)

(B)

Figure 5 Localization of MsYABBY5 under its native promoter in sweet basil (A) and tobacco (B). p*MsYABBY5*:: MsYABBY5-CFP showed exclusive expression in PGT of plants.

(Figure 7C). Enhanced energy import into the plastids can be one of the reasons the knock-down plants showed a higher metabolic activity.

In a recent study, ChIP-Seq and RNA-Seq methods were used to identify YABBY-regulated genes during various stages of soya bean seedling development. The major candidate genes regu-

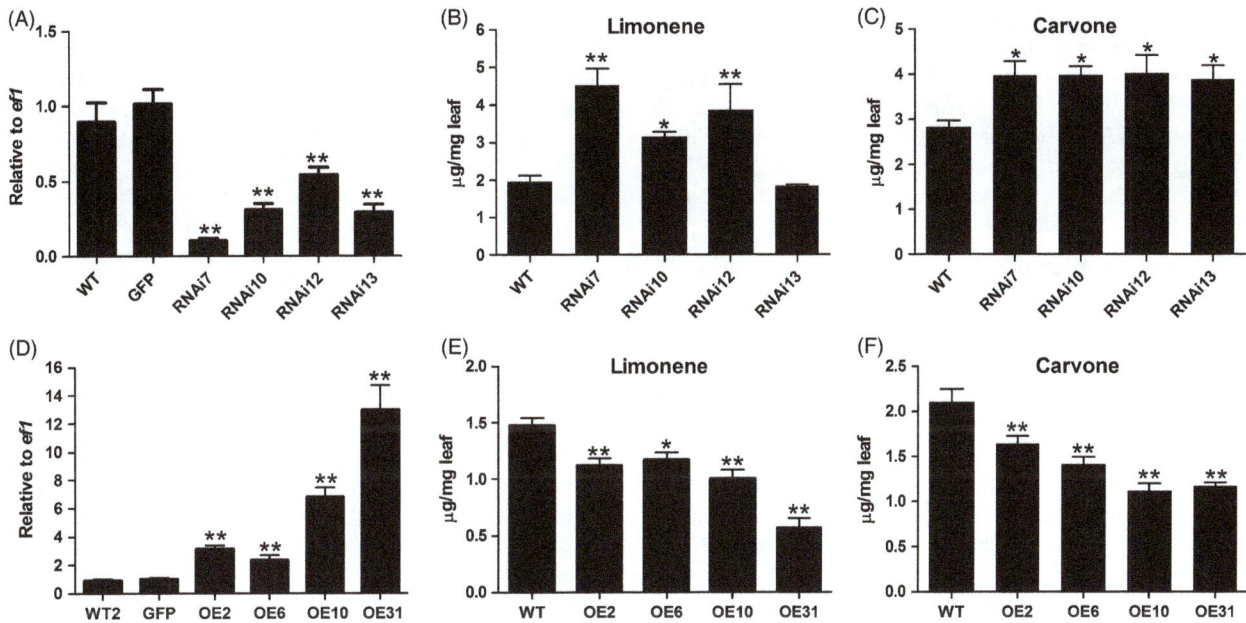

Figure 6 Transcript level of *MsYABBY5* and monoterpene production in *MsYABBY5* RNAi and overexpression plants. (A) *MsYABBY5* transcripts level in RNAi plants. (B) Limonene production in RNAi plants. (C) Carvone production in RNAi plants. (D) *MsYABBY5* transcripts level in overexpression plants. (E) Limonene production in overexpression plants. (F) Carvone production in overexpression plants. Gene expression is presented as relative to *ef1*. Leaves from the second node (2–3 cm) were harvested and used for analysis. Results of terpene production are presented as mean ± SD. *P < 0.05; **P < 0.01.

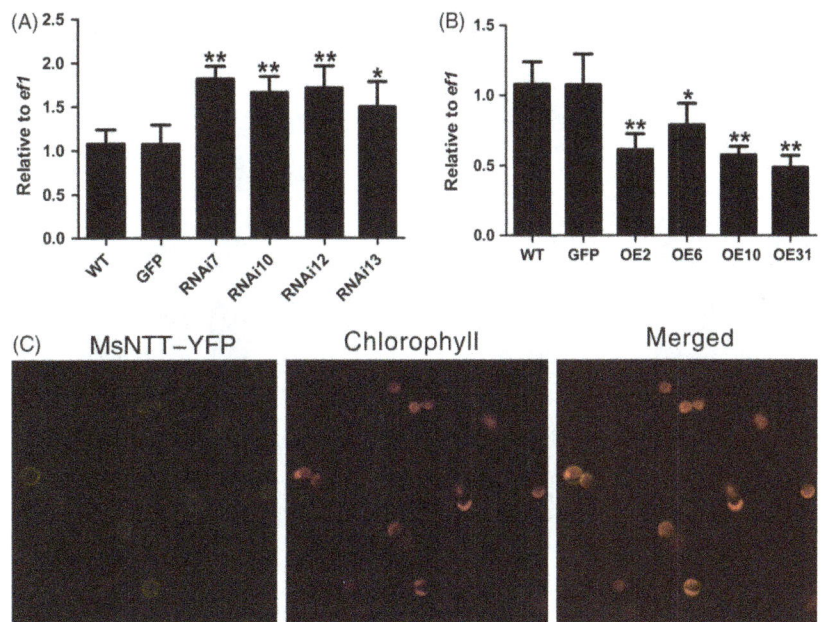

Figure 7 *MsNTT* expression and localization. Transcript levels of *MsNTT* in *MsYABBY5* RNAi (A) and overexpression plants (B) Leaves from the second node (2–3 cm) were harvested and used for qPCR analysis. Gene expression was normalized against the house keeping gene *ef1*. *P < 0.05; **P < 0.01. (C) MsNTT was localized to the chloroplast membrane in *Nicotiana benthamiana*.

lated by YABBY were found to be fatty acid desaturase, APETALA2 (AP2) and WRKY transcription factor (Shamimuzzaman and Vodkin, 2013). Recent research shows the emergence of WRKY TFs as key regulators of terpene production (Patra et al., 2013). From our RNA-Seq data, we identified a WRKY transcript *MsWRKY75*, which was enriched in PGTs (Figure 8A). The level of this transcript showed reduction in *MsYABBY5* RNAi lines, but no significant increase was observed in overexpression lines (Figure 8B). To check whether MsYABBY5 can bind to the promoter regions of *MsNTT* and *MsWRKY75*, we performed EMSA using the purified recombinant His tagged MsYABBY5

protein (Figure 8C). About ~1 kb promoter regions of both the genes were cloned and they were divided into four overlapping fragments and screened by EMSA. No binding was observed with *MsNTT* promoter, but a protein–DNA complex with reduced mobility was observed when recombinant MsYABB5 was incubated with *MsWRKY75* probe (Figure 8D). Interestingly, only the fragment of −909 to −555 bp region of *MsWRKY75* promoter was found to bind with MsYABBY5. DNA binding specificity was further confirmed by competition experiments using 10–100-fold excess unlabelled probe which led to the disappearance of DNA/ protein complex. To further determine whether MsYABBY5 protein can regulate the *MsWRKY75* promoter in plants, transient expression assays in *N. benthamiana* were performed. Leaves were coinfilterated with reporter *MsWRKY75 promoter::GUS* and effector *35S::YFP* or *35S::MsYABBY5*. Promoter activity of *MsWRKY75::GUS* was significantly enhanced in 35S::*MsYABBY5* expressing leaves when compared to *35S::YFP* (Figure 8E). This suggests that MsYABBY5 activates *MsWRKY75* which probably represses terpene production in spearmint.

Ectopic expression of *MsYABBY5* affects secondary metabolism in tobacco and sweet basil

To understand the functions of *MsYABBY5* in other plants, the gene was ectopically expressed in tobacco and sweet basil. High transcript levels of *MsYABBY5* were detected in transgenic tobacco and sweet basil plants (Figure S4). *Nicotiana sylvestris* glandular trichomes mainly produce diterpenes which are generally derived from the same MEP pathway as the monoterpenes. Ectopic expression of *MsYABBY5* was found to reduce cembranoids (CBT-diol) production in *N. sylvestris* by 29.5%–47.1% (Figure 9A). Sweet basil essential oil produced in PGT consists of both terpenes and phenylpropanoids. To explore whether *MsYABBY5* has an effect on secondary metabolites originating from different metabolic pathways in PGT, we ectopically expressed *MsYABBY5* in sweet basil. Three independent sweet basil transgenic lines confirmed by southern hybridization were selected for further characterization. The results of qRT-PCR showed high expression levels of *MsYABBY5* in all the transgenic plants (data not shown). GC–MS analysis on T-2 plants showed that the total production of both monoterpene (eucalyptol, β-

ocimene and linalool) and sesquiterpene (α-bergamotene, γ-muurolene and copaene) decreased (Figure 9B,C). Besides terpene production, phenylpropanoid production was also affected. Eugenol, which is the dominant compound, with a production of 1.93 ± 0.58 µg/mg fresh leaf, showed a significant reduction ($P < 0.05$) in transgenic plants (Figure 9D). This suggests that transcriptional regulators can govern fluxes in multiple metabolic pathways. Additionally, the sweet basil transgenic lines also showed curled leaf and delayed flowering (about 2–3 weeks delay) when compared to WT plants sown at the same time (Figure S5).

Discussion

Glandular trichomes are found on the aerial surface of approximately 30% of vascular plants. As they can synthesize and store a large amount of secondary metabolites, they are aptly termed as 'tiny chemical factories' of plants. But very few studies have focussed on TFs that regulate glandular trichome-specific metabolic pathways (Wang, 2014), which will greatly facilitate metabolic engineering efforts to increase yield or develop plant platforms to produce high value compounds. Studies in understanding the transcriptional control of secondary metabolite production show the expression of both activators and repressors is necessary to fine-tune the flux, timing and the level of structural gene expression in a pathway (Albert *et al.*, 2014; Cavallini *et al.*, 2015; Patra *et al.*, 2013). In this study, we isolated a *YABBY* gene that is preferentially expressed in spearmint PGT and appears to be a negative regulator of secondary metabolism. This is a novel

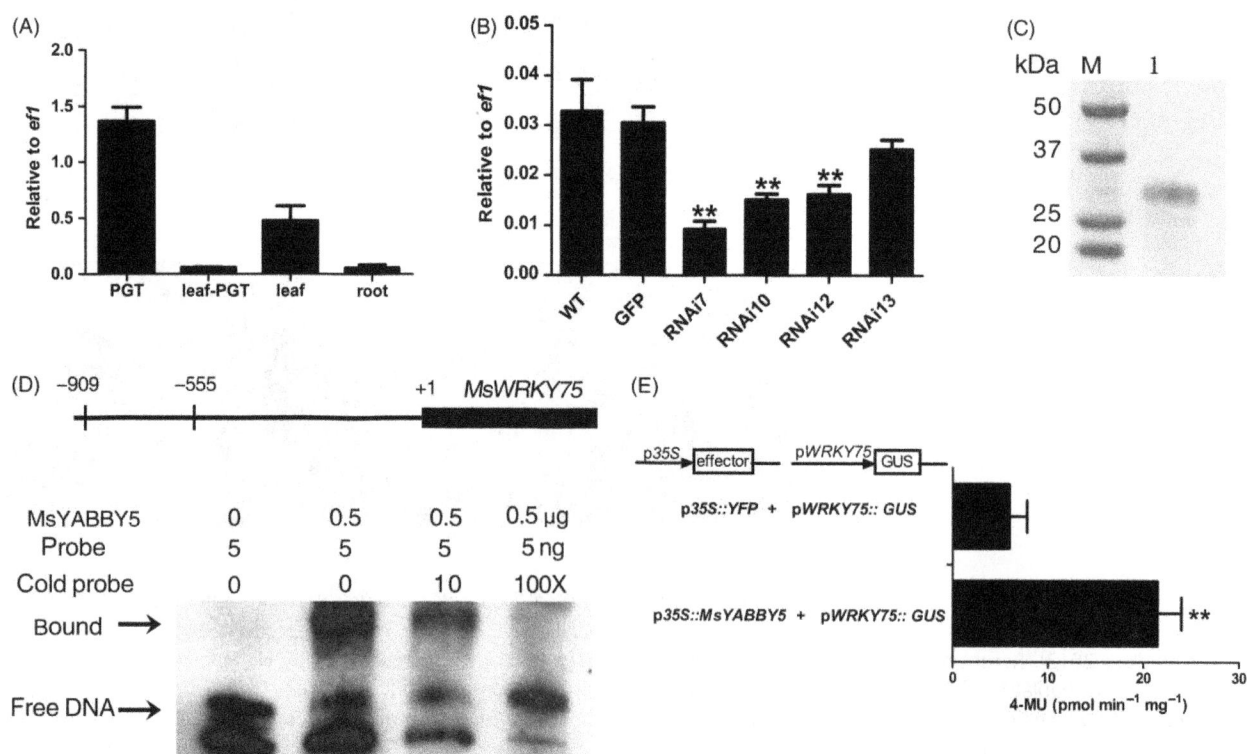

Figure 8 Transactivation of *MsWRKY75* by *MsYABBY5*. (A) Expression pattern of *MsWRKY75* in different tissues of wild-type spearmint. (B) Transcript levels of *MsWRKY75* in *MsYABBY5* RNAi plants. (C) Purification of recombinant His tagged MsYABBY5 expressed in *Escherichia coli*. (D) EMSA assay of MsYABBY5 binding ability to *MsWRKY75* promoter. Upper panel shows the fragment −909 to −555 of *MsWRKY75* promoter region, (E) transactivation of *MsWRKY75* by MsYABBY5 in *Nicotiana Benthamiana*. Results are presented as mean ± SD. **, $P < 0.01$.

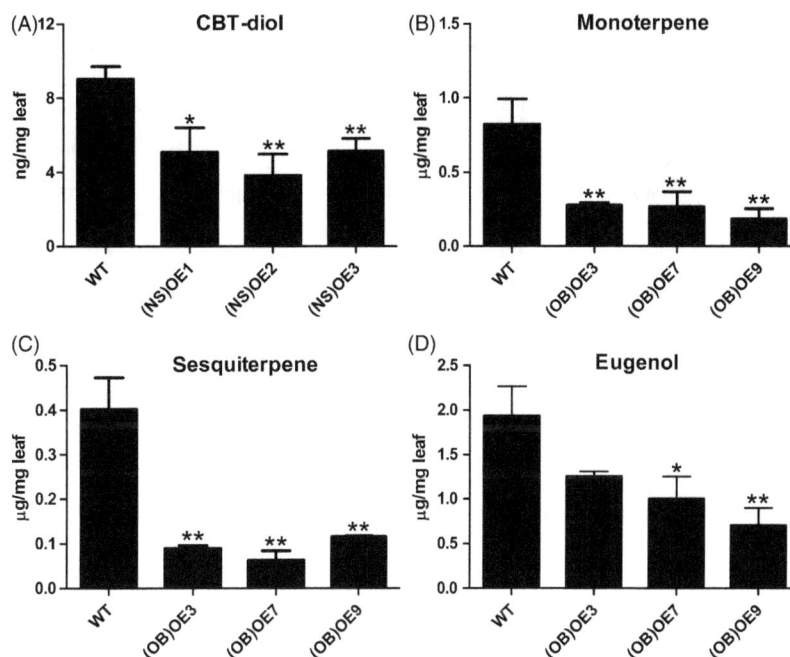

Figure 9 Ectopic expression of *MsYABBY5* decreased terpene production in *Nicotiana sylvestris* and sweet basil. (A) CBT-diol production in wild-type and transgenic *N. sylvestris* overexpressed with *MsYABBY5*. Volatiles production of total monoterpenes (B), sesquiterpenes (C) and eugenol (D) in wild-type and transgenic basil overexpressed with *MsYABBY5*. Leaves from the second node (2–4 cm) were harvested and used for GC–MS analysis. Results of terpenes and phenylpropanoid production are presented as mean ± SD. *$P < 0.05$; **$P < 0.01$.

function for the plant-specific *YABBY* family of TFs. In Arabidopsis, *YABBY* gene family promotes several aspects of leaf, shoot and flower development (Eshed et al., 1999; Goldshmidt et al., 2008; Golz et al., 2004; Stahle et al., 2009), but how they mediate these effects at the molecular level largely remains unknown. Single mutants of *yabby5* in Arabidopsis showed no morphological defects, but significantly enhanced the phenotype of *yab1yab3* double mutant (Sarojam et al., 2010). In rice, the functions of *YABBY* genes are divergent from their Arabidopsis homologs. Rice *YAB1* is required for gibberellin-mediated repression of *GA3ox2* gene which is involved in the synthesis of gibberellin (Dai et al., 2007).

Studies have revealed that YABBYs are bifunctional TFs acting as either repressors or activators (Bonaccorso et al., 2012; Stahle et al., 2009), but their direct downstream target genes are not well known. The increase in monoterpene production observed in *MsYABBY5* RNAi lines was not due to an increase in transcripts level of the structural genes involved in the pathway. Additionally, no significant changes were observed in transcript level of genes encoding enzymes in the precursor MEP pathway. The difference between metabolite and transcript levels can be attributed to either post-transcriptional modification, protein stability or enhanced flux into the metabolic pathway (Xie et al., 2008). In peppermint, it was shown that most of the biosynthetic enzymes leading to monoterpene production including limonene synthase were regulated at the level of gene expression (McConkey et al., 2000). Many primary metabolic pathways like glycolysis, the TCA cycle, pentose phosphate pathway and shikimate pathway provides carbon, ATP and precursors compounds to diverse secondary metabolic pathways. Transcription factors can affect the synthesis of a particular secondary metabolite by regulating metabolic enzymes in these primary pathways (Aharoni and Galili, 2011). Both mint and basil PGT are nonphotosynthetic organs and need to import ATP and carbon to sustain their high metabolic activities, which too can be possibly controlled by TFs. The fact that *MsYABBY5* expression was able to affect metabolite production in tobacco and sweet basil plants suggests that this gene might be probably functioning upstream regulating flux into

metabolic pathways. The MEP pathway and shikimate pathway leading to mono/diterpene and phenylpropanoid precursor production, respectively, are both localized in plastids making direct interactions between these pathways possible. *Production of anthocyanin pigment (PAP1)*, a MYB transcription factor from Arabidopsis, is an activator of the phenylpropanoid pathway (Li et al., 2010). Recently it was shown that ectopic expression of *PAP1* in rose plants, led to an increase in floral volatile compounds originating from both phenylpropanoid and terpenoid pathways. Transcriptional activation of only few biosynthetic genes was observed, whereas the rest of the increase was attributed to enhanced flux in both pathways (Ben Zvi et al., 2012). Interactions between phenylpropanoid and terpenoid pathways have also been shown in tomato mutants (Enfissi et al., 2010), as well as in Ipomoea flowers (Majetic et al., 2010), but the mechanism still remains to be elucidated.

In a recent study, ChIP-Seq and RNA-Seq methods were used to identify YABBY-regulated genes during various stages of soya bean seedling development. About 96 potential genes were found to be either up-regulated or down-regulated by YABBY. One of the major candidate genes regulated by YABBY was found to be WRKY transcription factor (Shamimuzzaman and Vodkin, 2013). WRKY TFs play important roles in regulation of plant stress response and secondary metabolism both as activators and repressors (Schluttenhofer and Yuan, 2015). Research has shown that WRKYs can activate structural genes involved in monoterpene (Spyropoulou et al., 2014), sesquiterpene (Ma et al., 2009) and diterpene (Qiu et al., 2008), as well as phenylpropanoid production (Wang et al., 2007). On the other hand, they can act as negative regulators too. In rice, *OsWRKY76* repressed terpene and phenylpropanoid synthesis but increased cold stress tolerance (Yokotani et al., 2013). In rubber, *HbWRKY1* negatively regulates a gene involved in natural rubber synthesis (Zhou et al., 2012).

In our study, we found that *MsYABBY5* negatively regulates the process of terpene biosynthesis and activates *MsWRKY75* in spearmint. MsWRKY75 may be a negative transcriptional regulator of genes involved in terpene synthesis. *MsYABBY5* expression overlaps with *MsWRKY75* in PGTs. Our RNA seq was performed on PGTs isolated from young leaves at a stage where terpene production is very active. The high expression of both of these TFs at this stage does indicate an important role for them. In *MsYABBY5* overexpression lines, no significant increase in *MsWRKY75* was observed; this can be due to a threshold of minimal expression required for *MsWRKY75* activation. *MsWRKY75* probably is one of the many downstream targets regulated by *MsYABBY5* to control secondary metabolism. Transcriptome analysis of *MsYABBY5* RNAi lines can provide us with more candidate genes that are potentially regulated by *MsYABBY5*. Our data reveal a regulatory network involving *YABBY5* and *WRKY* that controls terpene production in spearmint. As information regarding terpene pathway regulation is minimal, where *YABBY5* and *WRKY75* fall in the regulatory hierarchy to govern terpene biosynthesis and how they function with other TFs in the network remains to be explored. Considering the phenotypes observed in transgenic plants, it seems plausible that *MsYABBY5* controls an upstream event in metabolite production.

We found increased levels of transcript for the plastidic ATP/ADP transporter in transgenic lines making more secondary metabolites and decreased levels in transgenic lines producing less metabolite. In potato, overexpression and suppression of the plastidic ATP/ADP transporter led to an increase and decrease, respectively, in the amount of tuber starch produced (Tjaden *et al.*, 1998). Further overexpression of both NTT and glucose-6-phosphate/phosphate translocator which supplies carbon skeletons to the plastids significantly increased total starch content in potato (Zhang *et al.*, 2008). This suggested that import of both energy and carbons into amyloplasts is a rate-limiting step for starch formation. Most of the plastid proteins are encoded in the nucleus and transported to plastids. *MsYABBY5* does not seem to directly regulate the expression of *MsNTT* to control secondary metabolite production, but the effect can be indirect due to changes in secondary metabolism brought by *MsYABBY5*.

Immunostaining and fluorescence assays showed MsYABBY5 localized to both nucleus and cytoplasm. Cell fractionation and protein gel blot assays can be further performed to test the association of MsYABBY5 with intracellular membranes. There are several reports of TFs especially those involved in stress response to rapidly translocate from the cytoplasm to the nucleus in response to external signals. Some examples from the mammalian systems are STAT (signal transducer and activator of transcription), NF-κB (nuclear factor of immunoglobulin kappa B cells), NFAT (nuclear factor of activated T cells) and steroid receptors proteins (Beals *et al.*, 1997; McBride *et al.*, 2002). In the case of plants, ER membrane-associated bZIP and NAC089 TFs which are responsible for mediating ER-related plant immunity and abiotic stress responses, respectively, were detected in the cytoplasm (Che *et al.*, 2010; Moreno *et al.*, 2012; Yang *et al.*, 2014). Abscisic acid (ABA) is a vital plant hormone that plays important roles in stress response. In Arabidopsis, a WRKY protein, AtWRKY40 functions as negative regulator of ABA signalling by directly repressing expression of many ABA-responsive genes. WRKY40 was shown to localize to both nucleus and cytosol in wild-type plant cells that has ABA at physiological concentrations and ABA was essential for its cytosolic distribution (Shang *et al.*, 2010). Secondary metabolites produced by plants are also related to abiotic and biotic stress responses of plants. Further studies are required to assess the change in expression levels and localization patterns of

MsYABBY5 and WRKY75 in response to stress conditions and its significance.

Experimental procedures

Plant material and transformation

Commercial spearmint variety (*M. spicata*) and sweet basil (*O. basilicum*) were tested for their secondary metabolites by GC–MS and grown in greenhouse under natural light conditions. Spearmint plants were propagated using stem cuttings, whereas basil plants were propagated from seeds. *Agrobacterium*-mediated transformation of spearmint was performed according to the previously published protocol (Niu *et al.*, 1998, 2000). *Agrobacterium*-mediated transformation of sweet basil was performed by the following procedure. *O. basilicum* seeds were sterilized by washing in 40% Clorox for 3 min followed by several rinses with sterile water. The sterile seeds were imbibed overnight and kept at 4 °C. The following day the seeds were dissected under a dissection microscope to harvest the mature embryos. The dissected embryos were precultured in dark for 1 day in cocultivation medium (CC). *Agrobacterium* EHA105 strain was used for transformation. Enhanced green-fluorescent protein gene along with kanamycin was used as a selection marker. The precultured embryos were immersed in agrobacterium culture and sonicated for 15 s, four times. After sonication, the embryos were immersed in fresh *Agrobacterium* solution and vacuum-infiltrated for 3 min. After infection for 30 min, the embryos were placed in CC medium [MS salts + myo-inositol 100 mg/L sucrose (30 g/L) + BA (0.4 mg/L) + IBA (0.4 mg/L) + cefotaxime (150 mg/l)] for 3 days. After 3 days, the embryos were washed multiple times with sterile-distilled water containing cefotaxime (150 mg/L). The washed embryos were kept in CC media for 3–4 weeks in dark for shoot induction. After 3–4 weeks, GFP-positive shoots were selected and transferred to light. The well-grown shoots were transferred to elongation media [MS salts + sucrose (30 g/L) + BA (3 mg/L) + IAA (0.5 mg/L) + cefotaxime (150 mg/L)] and kept for 2–3 weeks. The shoots were hardened on basal media and allowed for root formation. Plantlets with well-developed roots were transferred to soil and grown under glasshouse conditions before further analysis. Tobacco transformation was done as previously described by Gallois and Marinho (1995).

RNA extraction and quantitative RT-PCR

Total RNA was extracted from different tissues (PGT, leaf-PGT, leaf and root) of spearmint using an RNeasy® Plus Mini kit from Qiagen. Reverse transcription reaction and quantitative RT-PCR (qRT-PCR) were carried out as described in Jin *et al.* (2014). Expression levels of target genes were represented as mean ± SD. Approximately 1 μg RNA was employed to synthesize first strand cDNA.

In situ hybridization

In situ hybridization assay was performed according to the method described by Javelle *et al.* (2011) with some minor modifications. Briefly, samples were fixed in 4% paraformaldehyde fixative and subjected to vacuum for 30 min on ice. After that, the vials were kept overnight at 4 °C. Next day, samples were dehydrated with ethanol series and embedded in Paraplast (McCormick Scientific, St Louis, MO) until use. The blocks were sectioned at 10 μm and mounted on Probe-on Plus slides (Fisher Scientific, singapore). For probe synthesis, the *MsYABBY5* cDNA was inserted into a pGEM®-T vector (Promega, Wisconsin). Sense and antisense probes were synthesized by T7 and SP6 RNA polymerase (Roche, Basel, Switzerland), respectively.

Cloning and vector construction

Promoter cloning of MsYABBY5

Genomic DNA was isolated from young leaves of spearmint using CTAB method. The flanking sequence of *MsYABBY5* gene was amplified using a GenomeWalker™ Universal kit. The −1116 bp flanking region of the gene was ligated with pGEM®-T vector. The resulting product was transformed into *Escherichia coli* XL1-Blue and sequenced. The promoter was amplified with Phusion® High-Fidelity DNA Polymerase (NEB) and subcloned into a gateway donor vector pENTR™/D-TOPO® (Invitrogen, California). Then, the recombinant plasmid was introduced into destination vectors pBGWFS7 by LR recombination. The destination plasmid was further transformed into *A. tumefaciens* EHA105 by heat shock and used to generate transgenic tobacco lines. Sequences of all primers used in this study are listed in Table S1.

Full-length cloning of all four YABBYs

Full length of all four YABBYs cDNAs was obtained by performing 3′ and 5′ RACE using the SMARTer™ RACE cDNA amplification kit from Clontech. For sequencing full-length ORFs, the purified fragments were ligated with pGEM®-T vector. The resulting product was transformed into *E. coli* XL1-Blue.

Overexpression and RNAi vector construction

To overexpress or silence *MsYABBY5*, sequences were amplified with Phusion® High-Fidelity DNA Polymerase (NEB). The purified fragments were inserted into a gateway donor vector pENTR™/D-TOPO® (Invitrogen). Then, the recombinant plasmids were introduced into destination vectors pK7WG2D for overexpression in spearmint and sweet basil by LR recombination. For *MsYABBY5* RNAi, four primers with restriction enzymes located at flanking region were used to amplify the fragment showing low similarity to other *YABBY* genes. The purified PCR product was cloned into the donor vector and subsequently introduced into pK7WG2D by LR recombination. The *MsYABBY5* gene was driven by *35S* promoter in both overexpression and RNAi plants. All destination plasmids harbouring the target genes were transformed into *A. tumefaciens* EHA105 by heat shock and used for spearmint and basil transformation.

Subcellular localization

YABBY and *MsNTT* ORFs were amplified and inserted into the pENTR™/D-TOPO®. The donor vectors harbouring ORFs were introduced into pBADC/YFP vector by LR recombination. For testing expression pattern of *MsYABBY5* promoter, the 5′UTR sequence was amplified and inserted into pENTR™/D-TOPO®. Subsequently, the plasmid was transformed into pBGWFS7 by LR recombination. All destination plasmids harbouring target genes were transformed into *A. tumefaciens* EHA105 by heat shock. The recombinant *A. tumefaciens* EHA strains were used for plant transformation. Subcellular localization pattern of YABBY proteins was performed as described in Jin *et al.* (2014). Briefly, the recombinant *A. tumefaciens* EHA strains were grown in LB medium overnight at 28 °C. After centrifugation at 4000 × *g*, 4 °C for 15 min, cell pellets were collected and resuspended in MMA solution (10 mM MES, 10 mM MgCl$_2$, 100 μM acetosyringone) to OD$_{600}$ = 1. The solution was then injected into

N. benthamiana leaves. After that, plants were kept at 28 °C for 2 days. Leaf samples were collected and viewed with an upright confocal microscope (Zeiss, Jena, Germany).

Southern blotting

A total of 15 μg genomic DNA was digested overnight with *Eco*RI at 37 °C. Next day, digestion product was electrophoresed on a 1.2% (w/v) agarose gel at 50 V for 4 h. After that, the gel was transferred to a nylon membrane and hybridized by the CaMV *35S* promoter probe using a DIG DNA labelling and detection kit (Roche). DNA probe against *35S* promoter was generated using PCR DIG probe synthesis kit from Roche (Hart and Basu, 2009).

Immunogold labelling

A 14-AA peptide of MsYABBY5 showing low similarity to other MsYABBYs was used as antigen for antibody synthesis (GenScript, Piscataway, NJ). Specificity test of the antibody was performed by Western blotting). Leaf samples from spearmint were fixed for 3 h in 4% paraformaldehyde/0.5% glutaraldehyde in 0.1 M phosphate buffer (pH 7.2) and rinsed in 0.1 M phosphate buffer (pH 7.2) for three times followed by dehydration in ethanol. After that, the samples were infiltrated with and embedded in LR White. Ultrathin sections around 90 nm were prepared with Leica Ultracut UCT microtome equipped with diamond knives and collected on uncoated, 300-mesh nickel grids. The procedure of labelling and washing was performed according to the protocol described by Skepper and Powell (2008) with some minor modifications. Briefly, the sections were incubated for 4 h on drops of antimint YAB5 antibody (produced in rabbit) with 1 : 100 dilutions in PBSG buffer [1% (w/v) gelatin in PBS buffer]. After that, sections were rinsed on drops of TBST (50 mM Tris, 150 mM NaCl, 0.05% Tween 20) for ten times, 2 min for each time. Then, sections were incubated for 1 h on drops of goat anti-rabbit antibody conjugated with 10 nm gold particles with 1 : 100 dilutions in PBSG buffer [1% (w/v) gelatin in PBS buffer]. The sections were further rinsed in TBST buffer for 10 times, 2 min each time, and in ddH$_2$O for 30 s. Subsequently, samples were counterstained by applying the grid on drops of uranyl acetate and lead citrate. Finally, sections were extensively rinsed in ddH$_2$O and viewed at 120 kV with a transmission electron microscope (JEOL JEM-1230, Japan).

Electrophoretic mobility shift assay (EMSA)

MsYABBY5 was expressed in *E. coli* BL21 (DE3) and induced by 1 mM isopropyl-β-thiogalactopyranoside (IPTG) for 6 h. Then, the recombinant protein was purified using 6× His tagged Ni-NTA agarose (Qiagen, Hilden, Germany) and used for EMSA. The 5′ ends of probes used for EMSA were labelled with biotin (Table S1). The assay was performed using a LightShift Chemiluminescent EMSA kit (Thermo, Waltham, MA) according to the manufacturer's instructions.

Transactivation activity assay

The ~1 kb promoter region of *MsWRKY75* was amplified and inserted into pENTR™/D-TOPO®. The resulting plasmid was transformed into pBGWFS7 by LR recombination and further introduced into *A. tumefaciens* EHA. Leaves of *N. benthamiana* were agroinfiltrated with effector and reporter at a ratio of 1 : 1. Two days after infiltration, leaves were harvested to isolate crude protein. GUS quantitative assay was performed as described by Li *et al.* (2014). Each assay was performed in triplicate.

Gas chromatography–mass spectrometry analysis

In case of mint, each transgenic plant was propagated clonally before GC–MS studies were conducted on them. For basil and tobacco plants, the analysis was performed on T-1 plants. Terpene and phenylpropanoid production in leaves of spearmint, sweet basil and tobacco were determined using a GC–MS method as described in Jin *et al.* (2014). Camphor was used as an internal standard.

Statistical analysis

Data are indicated as 'mean $\pm$ SD' of three biological replicates each performed in triplicates. Statistical significance between transgenic plants and WT was analysed using a two-tailed Student's *t*-test and indicated by asterisks. $*P < 0.05$; $**P < 0.01$.

Competing interests

The authors declare that they have no competing interests.

Acknowledgements

This research was funded by a grant from Singapore National Research Foundation (Competitive Research Programme Award No: NRF-CRP8-2011-02). We thank Dr Jin Jingjing for transcriptome data analysis, Mr. Xuezhi Ouyang for TEM sample preparation and Dr Jun-Lin Yin for support in GC–MS analysis.

References

Aharoni, A. and Galili, G. (2011) Metabolic engineering of the plant primary–secondary metabolism interface. *Curr. Opin. Biotechnol.* **22**, 239–244.

Albert, N.W., Davies, K.M., Lewis, D.H., Zhang, H., Montefiori, M., Brendolise, C., Boase, M.R. *et al.* (2014) A conserved network of transcriptional activators and repressors regulates anthocyanin pigmentation in eudicots. *Plant Cell*, **26**, 962–980.

Beals, C.R., Sheridan, C.M., Turck, C.W., Gardner, P. and Crabtree, G.R. (1997) Nuclear export of NF-ATc enhanced by glycogen synthase kinase-3. *Science*, **275**, 1930–1933.

Ben Zvi, M.M., Shklarman, E., Masci, T., Kalev, H., Debener, T., Shafir, S., Ovadis, M. *et al.* (2012) PAP1 transcription factor enhances production of phenylpropanoid and terpenoid scent compounds in rose flowers. *New Phytol.* **195**, 335–345.

Bonaccorso, O., Lee, J.E., Puah, L., Scutt, C.P. and Golz, J.F. (2012) *FILAMENTOUS FLOWER* controls lateral organ development by acting as both an activator and a repressor. *BMC Plant Biol.* **12**, 176.

Bowman, J.L. (2000) The YABBY gene family and abaxial cell fate. *Curr. Opin. Plant Biol.* **3**, 17–22.

Broun, P. and Somerville, C. (2001) Progress in plant metabolic engineering. *Proc. Natl Acad. Sci. USA*, **98**, 8925–8927.

Butelli, E., Titta, L., Giorgio, M., Mock, H.P., Matros, A., Peterek, S., Schijlen, E.G. *et al.* (2008) Enrichment of tomato fruit with health-promoting anthocyanins by expression of select transcription factors. *Nat. Biotechnol.* **26**, 1301–1308.

Cavallini, E., Matus, J.T., Finezzo, L., Zenoni, S., Loyola, R., Guzzo, F., Schlechter, R. *et al.* (2015) The phenylpropanoid pathway is controlled at different branches by a set of R2R3-MYB C2 repressors in grapevine. *Plant Physiol.* **167**, 1448–1470.

Champagne, A. and Boutry, M. (2013) Proteomic snapshot of spearmint (*Mentha spicata* L.) leaf trichomes: a genuine terpenoid factory. *Proteomics*, **13**, 3327–3332.

Che, P., Bussell, J.D., Zhou, W., Estavillo, G.M., Pogson, B.J. and Smith, S.M. (2010) Signaling from the endoplasmic reticulum activates brassinosteroid signaling and promotes acclimation to stress in *Arabidopsis*. *Sci. Signal.* **3**, ra69.

Chen, Y.Y., Wang, L.F., Dai, L.J., Yang, S.G. and Tian, W.M. (2012) Characterization of HbEREBP1, a wound-responsive transcription factor gene in laticifers of *Hevea brasiliensis* Muell. Arg. *Mol. Biol. Rep.* **39**, 3713–3719.

Cordoba, E., Salmi, M. and León, P. (2009) Unravelling the regulatory mechanisms that modulate the MEP pathway in higher plants. *J. Exp. Bot.* **60**, 2933–2943.

Croteau, R., Karp, F., Wagschal, K.C., Satterwhite, D.M., Hyatt, D.C. and Skotland, C.B. (1991) Biochemical characterization of a spearmint mutant that resembles peppermint in monoterpene content. *Plant Physiol.* **96**, 744–752.

Croteau, R., Kutchan, T.M. and Lewis, N.G. (2000) Natural products (secondary metabolites). In *Biochemistry and Molecular Biology of Plants*. (Buchanan, B., Gruissem, W. and Jones, R., eds), pp. 1250–1268. Rockville, MD: American Society of Plant Physiologists.

Dai, M., Zhao, Y., Ma, Q., Hu, Y., Hedden, P., Zhang, Q. and Zhou, D.X. (2007) The rice YABBY1 gene is involved in the feedback regulation of gibberellin metabolism. *Plant Physiol.* **144**, 121–133.

Diemer, F., Caissard, J.-C., Moja, S., Chalchat, J.-C. and Jullien, F. (2001) Altered monoterpene composition in transgenic mint following the introduction of 4S-limonene synthase. *Plant Physiol. Biochem.* **39**, 603–614.

Dubey, V.S., Bhalla, R. and Luthra, R. (2003) An overview of the non-mevalonate pathway for terpenoid biosynthesis in plants. *J. Biosci.* **28**, 637–646.

Enfissi, E.M., Barneche, F., Ahmed, I., Lichtlé, C., Gerrish, C., McQuinn, R.P., Giovannoni, J.J. *et al.* (2010) Integrative transcript and metabolite analysis of nutritionally enhanced *DE-ETIOLATED1* downregulated tomato fruit. *Plant Cell*, **22**, 1190–1215.

Eshed, Y., Baum, S.F. and Bowman, J.L. (1999) Distinct mechanisms promote polarity establishment in carpels of *Arabidopsis*. *Cell*, **99**, 199–209.

Fischer, K. (2011) The import and export business in plastids: transport processes across the inner envelope membrane. *Plant Physiol.* **155**, 1511–1519.

Flügge, U.I., Häusler, R.E., Ludewig, F. and Gierth, M. (2011) The role of transporters in supplying energy to plant plastids. *J. Exp. Bot.* **62**, 2381–2392.

Gallois, P. and Marinho, P. (1995) Leaf disk transformation using Agrobacterium tumefaciens-expression of heterologous genes in tobacco. In *Plant gene transfer and expression protocols*, Methods in Molecular Biology, Vol. 49 (Jones, H., ed.), pp. 39–48. Totowa, NJ: Humana Press.

Gershenzon, J., McConkey, M.E. and Croteau, R.B. (2000) Regulation of monoterpene accumulation in leaves of peppermint. *Plant Physiol.* **122**, 205–213.

Goldshmidt, A., Alvarez, J.P., Bowman, J.L. and Eshed, Y. (2008) Signals derived from *YABBY* gene activities in organ primordia regulate growth and partitioning of *Arabidopsis* shoot apical meristems. *Plant Cell*, **20**, 1217–1230.

Golz, J.F., Roccaro, M., Kuzoff, R. and Hudson, A. (2004) *GRAMINIFOLIA* promotes growth and polarity of *Antirrhinum* leaves. *Development*, **131**, 3661–3670.

Grotewold, E. (2008) Transcription factors for predictive plant metabolic engineering: are we there yet. *Curr. Opin. Biotechnol.* **19**, 138–144.

Gutensohn, M., Nguyen, T.T., McMahon, R.D., Kaplan, I., Pichersky, E. and Dudareva, N. (2014) Metabolic engineering of monoterpene biosynthesis in tomato fruits via introduction of the non-canonical substrate neryl diphosphate. *Metab. Eng.* **24**, 107–116.

Hart, S.M. and Basu, C. (2009) Optimization of a digoxigenin-based immunoassay system for gene detection in *Arabidopsis thaliana*. *J. Biomol. Tech.* **20**, 96–100.

Hemmerlin, A., Harwood, J.L. and Bach, T.J. (2012) A raison d'être for two distinct pathways in the early steps of plant isoprenoid biosynthesis? *Prog. Lipid Res.* **51**, 95–148.

Iwase, A., Matsui, K. and Ohme-Takag, M. (2009) Manipulation of plant metabolic pathways by transcription factors. *Plant Biotechnol.* **26**, 29–38.

Javelle, M., Marco, C.F. and Timmermans, M. (2011) *In situ* hybridization for the precise localization of transcripts in plants. *J. Vis. Exp.* **57**, 3328.

Jin, J., Panicker, D., Wang, Q., Kim, M.J., Liu, J., Yin, J.L., Wong, L. *et al.* (2014) Next generation sequencing unravels the biosynthetic ability of Spearmint (*Mentha spicata*) peltate glandular trichomes through comparative transcriptomics. *BMC Plant Biol.* **14**, 292.

Lange, B.M. and Turner, G.W. (2013) Terpenoid biosynthesis in trichomes – current status and future opportunities. *Plant Biotechnol. J.* **11**, 2–22.

Lange, B.M., Mahmoud, S.S., Wildung, M.R., Turner, G.W., Davis, E.M., Lange, I., Baker, R.C. *et al.* (2011) Improving peppermint essential oil yield and composition by metabolic engineering. *Proc. Natl Acad. Sci. USA*, **108**, 16944–16949.

Li, X., Gao, M.J., Pan, H.Y., Cui, D.J. and Gruber, M.Y. (2010) Purple canola: *Arabidopsis* PAP1 increases antioxidants and phenolics in *Brassica napus* leaves. *J. Agric. Food Chem.* **58**, 1639–1645.

Li, R., Weldegergis, B.T., Li, J., Jung, C., Qu, J., Sun, Y., Qian, H., *et al.* (2014) Virulence factors of Geminivirus interact with MYC2 to subvert plant resistance and promote vector performance. *Plant Cell*, **26**, 4991–5008.

Loza-Tavera, H. (1999) Monoterpenes in essential oils – biosynthesis and properties. *Adv. Exp. Med. Biol.* **464**, 49–62.

Lu, X., Jiang, W.M., Zhang, L., Zhang, F., Zhang, F.Y., Shen, Q., Wang, G.F. *et al.* (2013) AaERF1 positively regulates the resistance to Botrytis cinerea in *Artemisia annua*. *PLoS ONE*, **8**, e57657.

Luo, J., Butelli, E., Hill, L., Parr, A., Niggeweg, R., Bailey, P., Weisshaar, B. *et al.* (2008) AtMYB12 regulates caffeoyl quinic acid and flavonol synthesis in tomato: expression in fruit results in very high levels of both types of polyphenol. *Plant J.* **56**, 316–326.

Ma, D., Pu, G., Lei, C., Ma, L., Wang, H., Guo, Y., Chen, J. *et al.* (2009) Isolation and characterization of AaWRKY1, an *Artemisia annua* transcription factor that regulates the amorpha-4,11-diene synthase gene, a key gene of artemisinin biosynthesis. *Plant Cell Physiol.* **50**, 2146–2161.

Mahmoud, S.S. and Croteau, R.B. (2001) Metabolic engineering of essential oil yield and composition in mint by altering expression of deoxyxylulose phosphate reductoisomerase and menthofuran synthase. *Proc. Natl Acad. Sci. USA*, **98**, 8915–8920.

Mahmoud, S.S., Williams, M. and Croteau, R. (2004) Cosuppression of limonene-3-hydroxylase in peppermint promotes accumulation of limonene in the essential oil. *Phytochemistry*, **65**, 547–554.

Majetic, C.J., Rausher, M.D. and Raguso, R.A. (2010) The pigment-scent connection: do mutations in regulatory vs. structural anthocyanin genes differentially alter floral scent production in *Ipomoea purpurea*? *S. Afr. J. Bot.* **76**, 632–642.

McBride, K.M., Banninger, G., McDonald, C. and Reich, N.C. (2002) Regulated nuclear import of the STAT1 transcription factor by direct binding of importin-α. *EMBO J.* **21**, 1754–1763.

McConkey, M.E., Gershenzon, J. and Croteau, R.B. (2000) Developmental regulation of monoterpene biosynthesis in the glandular trichomes of peppermint. *Plant Physiol.* **122**, 215–222.

Miyamoto, K., Matsumoto, T., Okada, A., Komiyama, K., Chujo, T., Yoshikawa, H., Nojiri, H. *et al.* (2014) Identification of target genes of the bZIP transcription factor OsTGAP1, whose overexpression causes elicitor-induced hyperaccumulation of diterpenoid phytoalexins in rice cells. *PLoS ONE*, **9**, e105823.

Moreno, A.A., Mukhtar, M.S., Blanco, F., Boatwright, J.L., Moreno, I., Jordan, M.R., Chen, Y. *et al.* (2012) IRE1/bZIP60-mediated unfolded protein response plays distinct roles in plant immunity and abiotic stress responses. *PLoS ONE*, **7**, e31944.

Muñoz-Bertomeu, J., Ros, R., Arrillaga, I. and Segura, J. (2008) Expression of spearmint limonene synthase in transgenic spike lavender results in an altered monoterpene composition in developing leaves. *Metab. Eng.* **10**, 166–177.

Neuhaus, H.E., Thom, E., Möhlmann, T., Steup, M. and Kampfenkel, K. (1997) Characterization of a novel eukaryotic ATP/ADP translocator located in the plastid envelope of *Arabidopsis thaliana* L. *Plant J.* **11**, 73–82.

Niu, X., Lin, K., Hasegawa, P.M., Bressan, R.A. and Weller, S.C. (1998) Transgenic peppermint (*Mentha × piperita* L.) plants obtained by cocultivation with *Agrobacterium tumefaciens*. *Plant Cell Rep.* **17**, 165–171.

Niu, X., Li, X., Veronese, P., Bressan, R.A., Weller, S.C. and Hasegawa, P.M. (2000) Factors affecting *Agrobacterium tumefaciens*-mediated transformation of peppermint. *Plant Cell Rep.* **19**, 304–310.

Patra, B., Schluttenhofer, C., Wu, Y., Pattanaik, S. and Yuan, L. (2013) Transcriptional regulation of secondary metabolite biosynthesis in plants. *Biochim. Biophys. Acta*, **1829**, 1236–1247.

Qiu, D., Xiao, J., Xie, W., Liu, H., Li, X., Xiong, L. and Wang, S. (2008) Rice gene network inferred from expression profiling of plants overexpressing OsWRKY13, a positive regulator of disease resistance. *Mol. Plant*, **1**, 538–551.

Robinson, D.G. and Ritzenthaler, C. (2006) Imaging the early secretory pathway in BY-2 cells. In *Tobacco BY-2 Cells: From Cellular Dynamics to Omics*, Vol. **58** (Nagata, T., Matsuoka, K. and Inzé, D., eds), pp. 135–151. Biotechnology in Agriculture and Forestry. Heidelberg: Springer-Verlag.

Sarojam, R., Sappl, P.G., Goldshmidt, A., Efroni, I., Floyd, S.K., Eshed, Y. and Bowman, J.L. (2010) Differentiating *Arabidopsis* shoots from leaves by combined YABBY activities. *Plant Cell*, **22**, 2113–2130.

Schluttenhofer, C. and Yuan, L. (2015) Regulation of specialized metabolism by WRKY transcription factors. *Plant Physiol.* **167**, 295–306.

Schwinn, K., Venail, J., Shang, Y., Mackay, S., Alm, V., Butelli, E., Oyama, R. *et al.* (2006) A small family of *MYB*-regulatory genes controls floral pigmentation intensity and patterning in the genus *Antirrhinum*. *Plant Cell*, **18**, 831–851.

Shamimuzzaman, M. and Vodkin, L. (2013) Genome-wide identification of binding sites for NAC and YABBY transcription factors and co-regulated genes during soybean seedling development by ChIP-Seq and RNA-Seq. *BMC Genom.* **14**, 477.

Shang, Y., Yan, L., Liu, Z.Q., Cao, Z., Mei, C., Xin, Q., Wu, F.Q. *et al.* (2010) The Mg-chelatase H subunit of *Arabidopsis* antagonizes a group of WRKY transcription repressors to relieve ABA-responsive genes of inhibition. *Plant Cell*, **22**, 1909–1935.

Sinha, R., Bhattacharyya, D., Majumdar, A.B., Datta, R., Hazra, S. and Chattopadhyay, S. (2013) Leaf proteome profiling of transgenic mint infected with *Alternaria alternata*. *J. Proteomics*, **93**, 117–132.

Skepper, J.N. and Powell, J.M. (2008) Immunogold staining of epoxy resin sections for transmission electron microscopy (TEM). *Cold Spring Harbor Protocols*, **6**, pdb-prot5015.

Spyropoulou, E.A., Haring, M.A. and Schuurink, R.C. (2014) RNA sequencing on *Solanum lycopersicum* trichomes identifies transcription factors that activate terpene synthase promoters. *BMC Genom.* **15**, 402.

Stahle, M.I., Kuehlich, J., Staron, L., von Arnim, A.G. and Golz, J.F. (2009) YABBYs and the transcriptional corepressors LEUNIG and LEUNIG_HOMOLOG maintain leaf polarity and meristem activity in *Arabidopsis*. *Plant Cell*, **21**, 3105–3118.

Tjaden, J., Möhlmann, T. and Kampfenkel, K. (1998) Altered plastidic ATP/ADP-transporter activity influences potato (*Solanum tuberosum* L.) tuber morphology, yield and composition of tuber starch. *Plant J.* **16**, 531–540.

Vranová, E., Coman, D. and Gruissem, W. (2013) Network analysis of the MVA and MEP pathways for isoprenoid synthesis. *Annu. Rev. Plant Biol.* **64**, 665–700.

Wang, G. (2014) Recent progress in secondary metabolism of glandular trichome. *Plant Biotechnol.* **31**, 353–361.

Wang, C., Wu, J. and Mei, X. (2001) Enhancement of taxol production and excretion in *Taxus chinensis* cell culture by fungal elicitation and medium renewal. *Appl. Microbiol. Biot.* **55**, 404–410.

Wang, H., Hao, J., Chen, X., Hao, Z., Wang, X., Lou, Y., Peng, Y. *et al.* (2007) Overexpression of rice WRKY89 enhances ultraviolet B tolerance and disease resistance in rice plants. *Plant Mol. Biol.* **65**, 799–815.

Wang, Y., Guo, D., Li, H.L. and Peng, S.Q. (2013) Characterization of HbWRKY1, a WRKY transcription factor from *Hevea brasiliensis* that negatively regulates *HbSRPP*. *Plant Physiol. Biochem.* **71**, 283–289.

Xie, Z., Kapteyn, J. and Gang, D.R. (2008) A systems biology investigation of the MEP/terpenoid and shikimate/phenylpropanoid pathways points to multiple levels of metabolic control in sweet basil glandular trichomes. *Plant J.* **54**, 349–361.

Xu, Y.H., Wang, J.W., Wang, S., Wang, J.Y. and Chen, X.Y. (2004) Characterization of GaWRKY1, a cotton transcription factor that regulates the sesquiterpene synthase gene (+)-δ-cadinene synthase-A. *Plant Physiol.* **135**, 507–515.

Yang, Z.T., Wang, M.J., Sun, L., Lu, S.J., Bi, D.L., Sun, L., Song, Z.T. *et al.* (2014) The membrane-associated transcription factor NAC089 controls ER-stress-induced programmed cell death in plants. *PLoS Genet.* **10**, e1004243.

Yokotani, N., Sato, Y., Tanabe, S., Chujo, T., Shimizu, T., Okada, K., Yamane, H. *et al.* (2013) WRKY76 is a rice transcriptional repressor playing opposite roles in blast disease resistance and cold stress tolerance. *J. Exp. Bot.* **64**, 5085–5097.

Yu, Z.X., Li, J.X., Yang, C.Q., Hu, W.L., Wang, L.J. and Chen, X.Y. (2012) The jasmonate-responsive AP2/ERF transcription factors AaERF1 and AaERF2 positively regulate artemisinin biosynthesis in *Artemisia annua* L. *Mol. Plant*, **5**, 353–365.

Zhang, L., Häusler, R.E., Greiten, C., Hajirezaei, M.R., Haferkamp, I., Neuhaus, H.E., Flügge, U.I. *et al.* (2008) Overriding the co-limiting import of carbon and energy into tuber amyloplasts increases the starch content and yield of transgenic potato plants. *Plant Biotechnol. J.* **6**, 453–464.

Zhang, Q., Zhu, J., Ni, Y., Cai, Y. and Zhang, Z. (2012) Expression profiling of HbWRKY1, an ethephon-induced WRKY gene in latex from *Hevea brasiliensis* in responding to wounding and drought. *Trees*, **26**, 587–595.

Zhang, F.Y., Fu, X.Q., Lv, Z.Y., Lu, X., Shen, Q., Zhang, L., Zhu, M.M. *et al.* (2015) A basic leucine zipper transcription factor, Aabzip1, connects abscisic acid signaling with artemisinin biosynthesis in *Artemisia annua*. *Mol. Plant*, **8**, 163–175.

Zhou, M., Wu, L., Liang, J., Shen, C. and Lin, J. (2012) Expression analysis and functional characterization of a novel cold-responsive gene *CbCOR15a* from *Capsella bursa-pastoris*. *Mol. Biol. Rep.* **39**, 5169–5179.

Zhu, W., Zhang, L., Lv, H., Zhang, H., Zhang, D., Wang, X. and Chen, J. (2014) The dehydrin *wzy2* promoter from wheat defines its contribution to stress tolerance. *Funct. Integr. Genomics*, **14**, 111–125.

10

The enhancement of tolerance to salt and cold stresses by modifying the redox state and salicylic acid content via the *cytosolic malate dehydrogenase* gene in transgenic apple plants

Qing-Jie Wang, Hong Sun, Qing-Long Dong, Tian-Yu Sun, Zhong-Xin Jin, Yu-Jin Hao and Yu-Xin Yao*

State Key Laboratory of Crop Biology, College of Horticulture Science and Engineering, Shandong Agricultural University, Tai-An, Shandong, China

*Correspondence

email yaoyx@sdau.edu.cn

Keywords: apple (*Malus domestica* B.), *cytosolic malate dehydrogenase* gene, overexpression, salt and cold tolerance, redox state, salicylic acid.

Summary

In this study, we characterized the role of an apple *cytosolic malate dehydrogenase* gene (*MdcyMDH*) in the tolerance to salt and cold stresses and investigated its regulation mechanism in stress tolerance. The *MdcyMDH* transcript was induced by mild cold and salt treatments, and *MdcyMDH*-overexpressing apple plants possessed improved cold and salt tolerance compared to wild-type (WT) plants. A digital gene expression tag profiling analysis revealed that *MdcyMDH* overexpression largely altered some biological processes, including hormone signal transduction, photosynthesis, citrate cycle and oxidation–reduction. Further experiments verified that *MdcyMDH* overexpression modified the mitochondrial and chloroplast metabolisms and elevated the level of reducing power, primarily caused by increased ascorbate and glutathione, as well as the increased ratios of ascorbate/dehydroascorbate and glutathione/glutathione disulphide, under normal and especially stress conditions. Concurrently, the transgenic plants produced a high H_2O_2 content, but a low O_2^- production rate was observed compared to the WT plants. On the other hand, the transgenic plants accumulated more free and total salicylic acid (SA) than the WT plants under normal and stress conditions. Taken together, *MdcyMDH* conferred the transgenic apple plants a higher stress tolerance by producing more reductive redox states and increasing the SA level; *MdcyMDH* could serve as a target gene to genetically engineer salt- and cold-tolerant trees.

Introduction

Plant tissues possess multiple isoforms of malate dehydrogenase (MDH, EC 1.1.1.37), which belongs to the group of oxidoreductases and catalyses the interconversion of malate and oxaloacetate (OAA) coupled to the reduction or oxidation of the NAD(H) or NADP(H) pool. NAD-dependent MDHs are located in the cytosol and in organelles, including plastids, mitochondria, peroxisomes and microbodies; additionally, chloroplasts contain an NADP-dependent MDH (Gietl, 1992; Scheibe, 2004). In the *Arabidopsis* genome, eight putative NAD-MDH isoforms have been identified, where two are mitochondrial MDH (mMDH), two are peroxisomal MDH, one is a plastidial MDH, and three have no detectable target sequences and are thought to be cytosolic MDH (cyMDH) (Beeler *et al.*, 2014).

The redox state of the cells is especially critical during exposure to abiotic stresses, which are known to induce oxidative stress at the cellular level; under these conditions, regulation of multiple redox and reactive oxygen species (ROS) signals in plants requires a high degree of coordination and balance between the signalling and metabolic pathways in different cellular compartments (Suzuki *et al.*, 2012). In green tissues, an important aspect is the chloroplast/cytosol/mitochondrion cooperation under stress to modulate cell redox homeostasis (Raghavendra and Padmasree, 2003). MDH reaction is involved in central metabolism and redox homeostasis between

organelle compartments (Tomaz *et al.*, 2010). MDH isozymes in the chloroplast inner envelope membrane, stroma, mitochondrial membrane and cytosol are the components of malate/OAA and malate/aspartate (Asp) shuttles (malate valves) (Gietl, 1992; Taniguchi and Miyake, 2012). When plants are subjected to abiotic stresses, changes in the activities of the enzymes of the malate valves and expression levels of the MDH isoforms can be observed (Scheibe, 2004), and malate valves play major roles in reductant export under stress conditions (Taniguchi and Miyake, 2012). Chloroplast NADP-MDH makes the link between the redox states of the chloroplast and the cytosol and other cell compartments, such as peroxisomes (Heyno *et al.*, 2014). In *Arabidopsis*, malate/OAA transporter knockout impairs the malate valve function of chloroplasts, and the knockout plants show enhanced photoinhibition and ROS accumulation in response to a high-intensity light, due to a greater accumulation of reducing equivalents in the stroma (Lemaire *et al.*, 1996). Regarding mMDH, its repression modifies the ascorbate-mediated link between the energy-generating processes of respiration and photosynthesis (Nunes-Nesi *et al.*, 2005). Additionally, mMDH lowers leaf respiration and alters photorespiration and plant growth in *Arabidopsis* (Tomaz *et al.*, 2010). In contrast, cyMDH is thought to function as a key player in the transfer of reducing equivalents from the chloroplast or mitochondria to other destinations in plant cells (Gietl, 1992; Hara *et al.*, 2006; Scheibe, 2004).

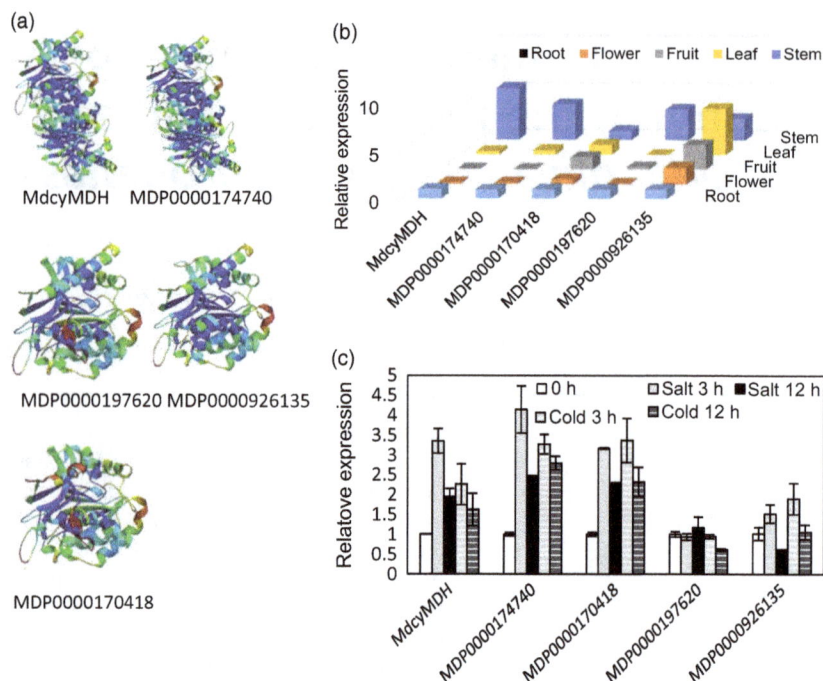

Figure 1 Predicted protein tertiary structures of the different cyMDH isoforms (a) and expression patterns of the *cyMDH* genes in different tissues (b) and in response to 8 °C temperature and 50-mM salt in the leaves of apple *in vitro* shoot cultures (c). Protein tertiary structures were predicted using the SWISS-MODEL (http://swissmodel.expasy.org/). Values represent the means ± SD of three replicates.

On the other hand, plant hormones play central roles in the ability of plants to adapt to abiotic stresses by mediating a wide range of adaptive responses (Peleg and Blumwald, 2011). For example, evidence of the role of salicylic acid (SA) in the oxidative damage generated by NaCl and osmotic stress was reported, and the NPR1-dependent SA signalling pathway was found to play pivotal roles in enhancing salt and oxidative stress tolerance in *Arabidopsis* (Borsani *et al.*, 2001; Jayakannan *et al.*, 2015); in contrast, ABA-dependent and ABA-independent signalling pathways in response to osmotic stress were widely investigated (Yoshida *et al.*, 2014). Additionally, interactions between hormone and redox signalling pathways can control plant growth and cross-tolerance to stress (Bartoli *et al.*, 2013). Up to now, evidence that cyMDH and other MDHs can enhance plant stress tolerance by regulating redox and hormone levels, as well as from their interactions, is still lacking.

Because most of the arable lands are used for growing grain crops, fruit trees are increasingly grown on marginal land. Because fruit trees cannot be rotated to avoid stresses, in contrast to annual crops, they often encounter multiple environmental stresses simultaneously. It has become an emerging challenge for scientists to clarify how fruit trees respond to and grow against adverse environments and to improve the stress tolerance of fruit trees by genetic modification. Presently, it has been reported that *cyMDH* confers superior manganese tolerance by mediating malate synthesis and secretion and inhibiting ROS generation in *Stylosanthes guianensis* (Chen *et al.*, 2015). Our previous study also showed that *MdcyMDH* conferred transgenic tomato plants high tolerance to cold and salt stresses alongside with decreased ROS levels (Yao *et al.*, 2011a). However, more research is needed to unravel the mechanism of *MdcyMDH* in regulating abiotic stresses. Therefore, this study is aimed to assess the potential application of *MdcyMDH* in engineering salt- and cold-tolerant apple trees and to better understand the mechanism of *MdcyMDH*-mediated abiotic stress tolerance by modifying redox homeostasis and hormone levels.

Results

The identification of *cyMDH* genes and their expression in different tissues and under cold and salt treatments in apple

In our previous study (Yao *et al.*, 2008), the *MdcyMDH* (DQ221207) ORF was shown to encode a 332-aa polypeptide which shared over 90% similarity with cyMDHs from other species, but we failed to locate this gene in the present apple genome database. In contrast, the other four *cyMDH* isoforms were identified from the entire apple genome, and their deduced proteins possessed the highly conserved domains of cytosolic malate dehydrogenase (Figure S1). Highly similar protein tertiary structures were observed between MdcyMDH and MDP0000174740, as well as between MDP0000197620 and MDP0000926135 (Figure 1a). Transcripts of the five *cyMDH* genes could be detected in all tissues, but they exhibited varying expression levels in different tissues; *MdcyMDH*, *MDP0000 174740* and *MDP0000197620* were expressed primarily in the stems and roots (Figure 1b,c). *MdcyMDH*, *MDP0000174740* and *MDP0000170418* expression were induced by 50 mM NaCl and temperature of 8 °C (Figure 1c). Therefore, MdcyMDH and MDP0000174740 most likely possessed similar functions in the light of their highly similar tertiary structures, spatial expressions and expression responses to salt and cold in apple.

Improved tolerance to salt and cold in transgenic apple plants via *MdcyMDH* overexpression

To further characterize the biological function of the *MdcyMDH* gene, the *MdcyMDH*-overexpressing lines were obtained by agrobacterium-mediated transformation and confirmed by PCR detection of the target gene (Figure 2a,b). Lines 5 and 7, with high *MdcyMDH* expression level compared to the wild-type (WT) plants, were selected to be further assayed (Figure 2b). When subjected to 50 mM salt and 8 °C stresses, no obvious phenotype differences were observed between the transgenic and WT apple

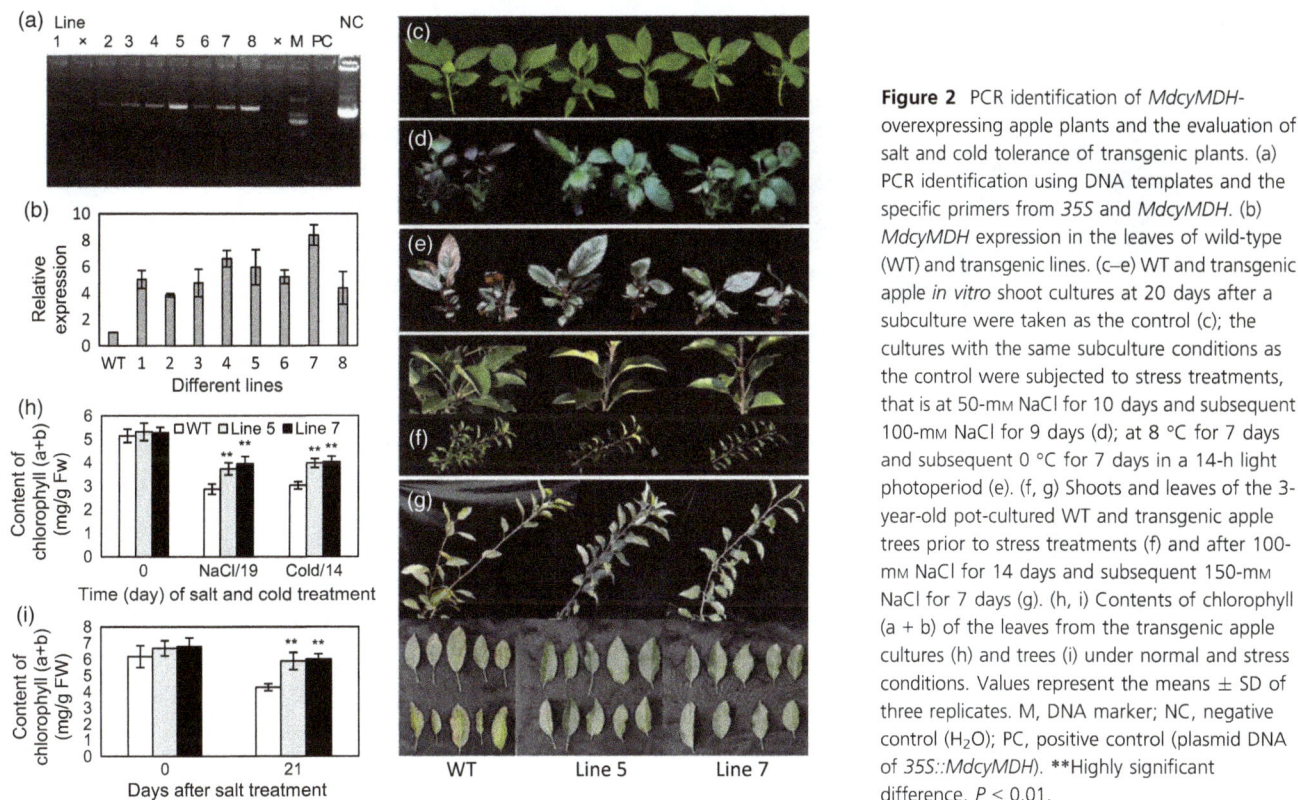

Figure 2 PCR identification of *MdcyMDH*-overexpressing apple plants and the evaluation of salt and cold tolerance of transgenic plants. (a) PCR identification using DNA templates and the specific primers from *35S* and *MdcyMDH*. (b) *MdcyMDH* expression in the leaves of wild-type (WT) and transgenic lines. (c–e) WT and transgenic apple *in vitro* shoot cultures at 20 days after a subculture were taken as the control (c); the cultures with the same subculture conditions as the control were subjected to stress treatments, that is at 50-mm NaCl for 10 days and subsequent 100-mm NaCl for 9 days (d); at 8 °C for 7 days and subsequent 0 °C for 7 days in a 14-h light photoperiod (e). (f, g) Shoots and leaves of the 3-year-old pot-cultured WT and transgenic apple trees prior to stress treatments (f) and after 100-mm NaCl for 14 days and subsequent 150-mm NaCl for 7 days (g). (h, i) Contents of chlorophyll (a + b) of the leaves from the transgenic apple cultures (h) and trees (i) under normal and stress conditions. Values represent the means ± SD of three replicates. M, DNA marker; NC, negative control (H₂O); PC, positive control (plasmid DNA of *35S::MdcyMDH*). **Highly significant difference, $P < 0.01$.

in vitro shoot cultures at 10 days after the treatments; after nine more days of 100 mm NaCl and seven more days of 0 °C treatment, both the transgenic and WT apple cultures showed the necrosis phenotypes, but the transgenic plants exhibited less severe phenotypes (Figure 2d,e). In addition, after 3 weeks of NaCl treatment, the leaves of the 3-year-old WT plants exhibited symptom of etiolation and marginal necrosis but the leaves of the transgenic plants exhibited near-to-normal colour (Figure 2g). Moreover, the leaves of the transgenic apple cultures and trees possessed a much higher chlorophyll content than the leaves of the WT plants after the stress treatments although the chlorophyll contents were reduced in the WT and transgenic plants when subjected to the stress treatments (Figure 2h,i). Therefore, *MdcyMDH* overexpression enhanced the tolerance of transgenic apple plants to salt and cold stresses.

Comparison of expression profiles between the WT and *MdcyMDH*-overexpressing plants

To explore the regulation mechanism of *MdcyMDH* in abiotic stresses, a digital gene expression (DGE) tag profiling analysis between the WT and mixed transgenic plants at 3 h after the cold treatment was conducted to quantify gene changes. It was found that 961 and 958 genes were up- and down-regulated by at least onefold in the transgenic lines, respectively, demonstrating that a massive transcriptional reprogramming took place in the *MdcyMDH*-overexpressing plants. According to the putative homology to sequences present in public databases, the differentially expressed genes were classified into 12 different cellular component categories; the cytosol contained the most differentially expressed genes (40.1%), followed by the membrane (33.3%), chloroplast (6.4%), nucleus (2.8%) and mitochondria (2.7%) (Figure 3a). All of the DGE genes are associated with 22 biological processes. The two processes of plant–pathogen interaction and plant hormone signal transduction contain the most DGE genes; additionally, photosynthesis, citrate cycle (TCA cycle) and the oxidation–reduction process were also the clearly changed biological processes (Figure 3b). Moreover, the expression levels of the 28 most differentially expressed annotated genes in the cold-treated shoot cultures were detected by qRT-PCR, which contained genes related to SA biosynthesis and signal transduction, chloroplast and mitochondrial metabolism, redox process and abiotic stresses; similar expression changes to DGE tag profiles were found, validating the reliability of the DGE genes (Figure 3c). Additionally, the 28 genes exhibited similar expression changes under salt treatment as that under cold treatment (Figure 3c). Besides, most of the above genes were induced by cold and/or salt in the WT and transgenic plants (Figure S2). Therefore, it is suggested that *MdcyMDH* improved the stress tolerance of the transgenic apple

Figure 3 Cell component (a) and biological process (b) of the digital gene expression (DGE) genes classified by Gene Ontology and the most differentially expressed genes related to SA biosynthesis and signalling, mitochondrial and chloroplast metabolism, and redox and abiotic stress tolerance (c). In the heat map of panel (c), the ID and annotation of each gene are provided. Fold changes from qRT-PCR at 3 h after cold and salt treatments were used to validate the DGE genes and to determine the expression changes of the DGE genes induced by *MdcyMDH* overexpression under stress treatments. Fold change from qRT-PCR is calculated by comparing the relative expression values of the selected genes in the transgenic and wild-type (WT) plants. Data are presented as the means of three replicates.

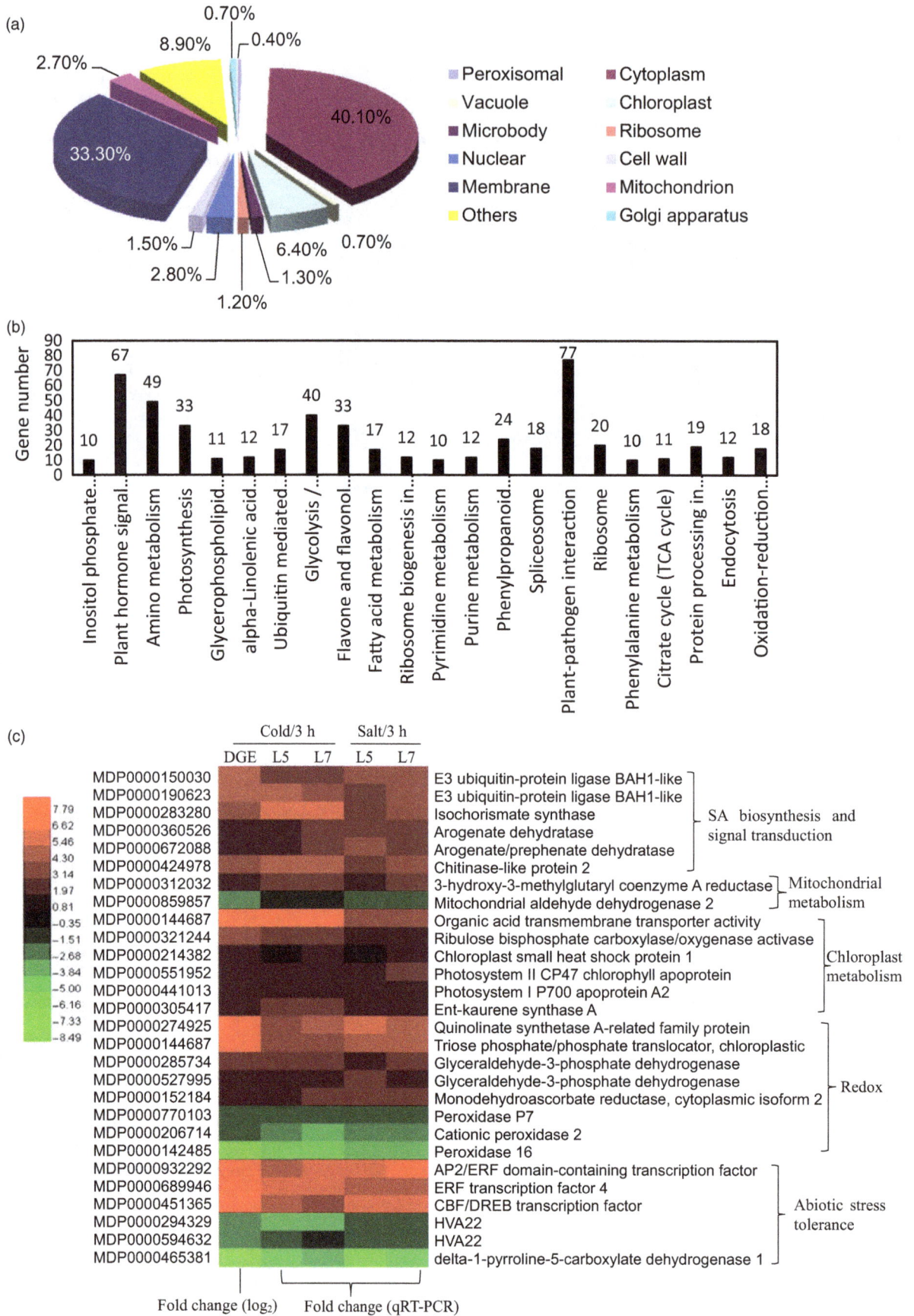

(a)

- Peroxisomal
- Vacuole
- Microbody
- Nuclear
- Membrane
- Others
- Cytoplasm
- Chloroplast
- Ribosome
- Cell wall
- Mitochondrion
- Golgi apparatus

0.70%
0.40%
8.90%
2.70%
40.10%
33.30%
1.50%
2.80%
1.20%
6.40%
1.30%
0.70%

(b)

(c)

plants at least partly via modifying hormone signal transduction, photosynthesis, TCA cycle and the oxidation–reduction process.

MdcyMDH overexpression modulated mitochondrial and chloroplast metabolisms

First, MDH activities in the cytosol, mitochondria and chloroplast were detected in the *MdcyMDH*-overexpressing apple *in vitro* shoot cultures (Figure 4a–c). The *MdcyMDH* overexpression increased the reductive activities of cyMDH and chMDH and the oxidative activity of mMDH compared to the WT plants; in contrast, the difference of activities was enlarged and reached a significant level under cold and salt stresses. In contrast, the oxidative activities of cyMDH and chMDH and the reductive activity of mMDH were generally not significantly changed in the transgenic plants compared to those in the WT plants. Moreover, the reductive activities of cyMDH and chMDH and the oxidative activity of mMDH were clearly induced by cold and salt treatment

in the WT and transgenic plants (Figure 4a–c). Therefore, it can be suggested that the increases in the cyMDH and chMDH reductive activities and the mMDH oxidative activity contributed favourably to the stress tolerance of apple plants in the transgenic plants. Moreover, the significantly increased malate content indicated that the reductive activity of MDH was strengthened against its oxidative activity in the transgenic plants under normal and stress conditions (Figure 5a).

Because the MDH activities in the cytosol, mitochondria and chloroplast were altered in the transgenic plants and MDH is one of the components of the malate/OAA shuttle between the cytosol and other organelles, it was speculated that the mitochondrial and chloroplast metabolisms would be changed. To verify this speculation, mitochondrial and chloroplast metabolisms were detected in the WT and transgenic plants under normal and stress conditions. On the one hand, malate-dependent oxygen consumption by isolated mitochondria was measured to assess the change in the mitochondrial activity. The results showed that the oxidation capacity of malate was significantly increased in the transgenic plants under cold and salt treatments (Figure 5b), which was consistent with the increased oxidative activity of mMDH (Figure 4b). On the other hand, parameters related to photosynthesis and chlorophyll fluorescence were determined to assess the changes in chloroplast metabolism in the transgenic plants. The results showed that the NaCl treatment induced a decrease in the photosynthetic rate in the WT and transgenic plants; nevertheless, the *MdcyMDH* overexpression produced an increase in the photosynthetic rate

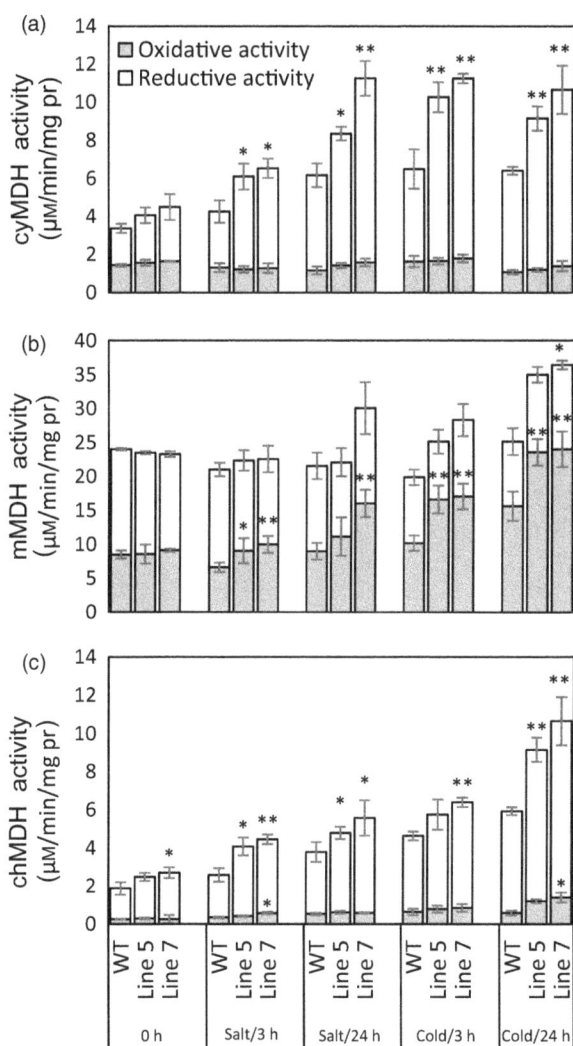

Figure 4 Modifications of reductive and oxidative activities of malate dehydrogenase (MDH) in the cytosol (a), mitochondria (b) and chloroplast (c) in the wild-type (WT) and transgenic plants in response 50 mм salt and 8 °C stresses. Data are presented as means ± SD (*n* = 3). The difference level was compared between the transgenic lines and WT at the same treatment time. *Significant difference, *P* < 0.05; **highly significant difference, *P* < 0.01.

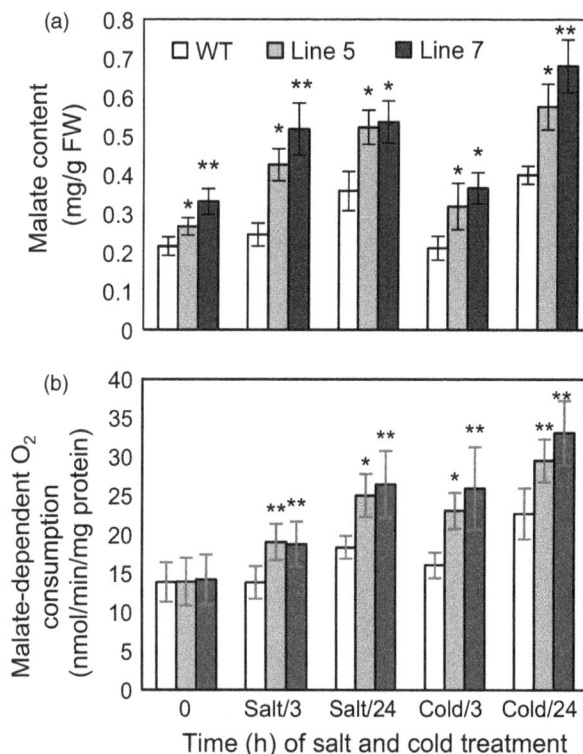

Figure 5 Malate content and malate-dependent O_2 consumption in the wild-type (WT) and transgenic plants under cold and salt treatments. Data are presented as means ± SD (*n* = 3). The difference level was compared between transgenic lines and WT at the same treatment time. *Significant difference, *P* < 0.05; **highly significant difference, *P* < 0.01.

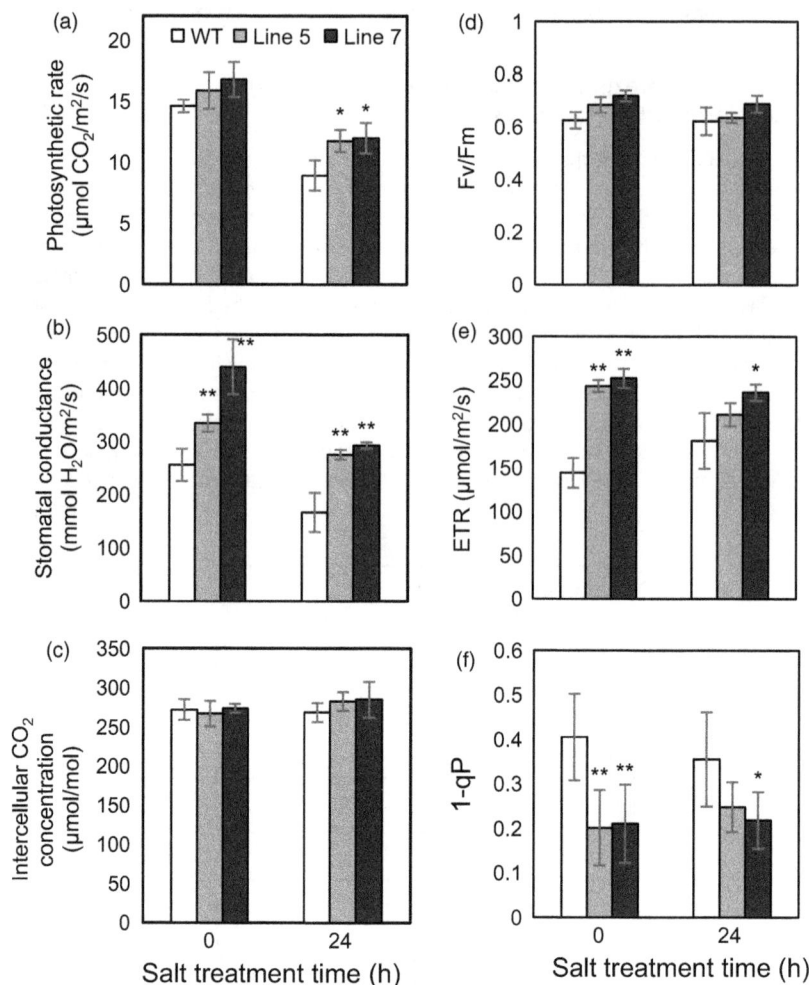

Figure 6 Changes of the photosynthetic rate (a), stomatal conductance (b), intercellular CO_2 concentration (c), maximum quantum yield of PSII (F_v/F_m) (d), photosynthetic electron transport rate (ETR) (e), and PSII excitation pressure (1−qP) (f) in the leaves of the wild-type (WT) and transgenic plants at 0 and 24 h after 50-mM NaCl treatment. Data are presented as means ± SD ($n = 3$). *Significant difference, $P < 0.05$; **highly significant difference, $P < 0.01$.

and a concomitant increase in the stomatal conductance (Figure 6a,b), suggesting the contribution of stomatal factors to the increased photosynthetic rate in the transgenic plants. However, similar intercellular CO_2 concentration in the leaves of the WT and transgenic plants also indicated altered activities of the mesophyll by the *MdcyMDH* overexpression (Figure 6c). Compared to the WT plants, *MdcyMDH* overexpression did not significantly influence the photoinhibition of PSII in the light of similar ratios of F_v/F_m under normal and salt treatment (Figure 6d; Björkman and Demmig, 1987); in contrast, *MdcyMDH* overexpression significantly increased the electron transport rate (ETR), which was used as a quantitative indicator of the electron transport beyond PSII (Maxwell and Johnson, 2000) (Figure 6e). Additionally, the transgenic plants significantly modified the redox level and excitation pressure of PSII, indicated by the altered values of 1−qP (Figure 6f; Maxwell and Johnson, 2000). Taken together, the mitochondrial and chloroplast metabolisms were modified in the *MdcyMDH*-overexpressing plants.

Overexpression of *MdcyMDH* increases the reducing power and modifies the ROS level under cold and salt stresses

The three most abundant antioxidants in plant cells, that is NAD(P)H, ascorbate and glutathione (Queval and Noctor, 2007), and their reductive and oxidized forms were determined in the WT and transgenic plants under normal and stress conditions (Figure 7). *MdcyMDH* overexpression increased the NADH levels

and the ratios of NADH/NAD+ although generally not in a statistically significant manner under normal and stress conditions (Figure 7a). In contrast, the NADPH content was slightly enhanced but the ratios of NADPH/NADP+ were almost unchanged (Figure 7b,e). Moreover, accumulation of both reduced ascorbate (AsA) and oxidized ascorbate (DHA) was induced by cold and salt stresses (Figure 7c). The contents of both AsA and DHA increased in the transgenic plants compared to the WT plants; additionally, the increments in AsA and AsA/DHA reached statistical significance at most of the treatment points (Figure 7c,e). Similarly, salt and cold induced the accumulation of oxidized glutathione (GSSG) and especially reduced glutathione (GSH); *MdcyMDH* overexpression significantly promoted the GSH content and GSH/GSSG under stress conditions (Figure 7d,e). Taken together, *MdcyMDH* overexpression increased the reducing power predominantly via enhancing the levels of AsA and GSH, as well as the ratios of AsA/DHA and GSH/GSSG.

On the other hand, the transgenic lines generated a lower O_2^{-} production rate but a higher H_2O_2 content than the WT plants under normal and stress conditions (Figure 7f).

The level of SA is elevated in the transgenic plants

To examine whether *MdcyMDH* overexpression impacted hormone levels, the contents of ABA, IAA, GA3 and SA were determined. It was found that IAA and GA3 were not significantly altered in the transgenic plants. In contrast, the ABA level was slightly decreased, while the SA level was largely increased in the

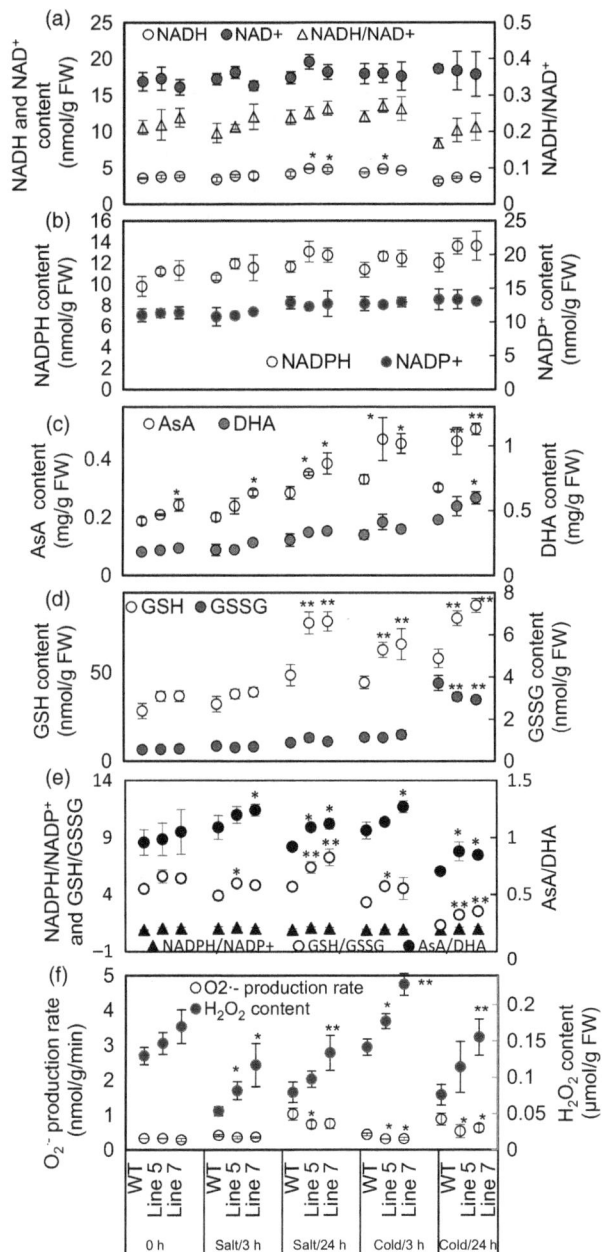

Figure 7 Assays of redox couples and reactive oxygen species (ROS) in the wild-type (WT) and transgenic plants under 50 mM NaCl and 8 °C treatments. (a) The contents of the reduced form of NADH and the oxidized form of NAD⁺, as well as the ratio of NADH/NAD⁺. (b–d) The contents of the reduced form of redox couples (NADPH, AsA and GSH) and the oxidized form of redox couples (NADP⁺, DHA and GSSG). (e) The ratios of the reduced and oxidative forms of redox couples. (f) The contents of O_2^- and H_2O_2. Data are presented as means ± SD ($n = 3$). *Significant difference, $P < 0.05$; **highly significant difference, $P < 0.01$. The analysis of the significant difference was performed between the transgenic lines and WT plants at the same treatment point.

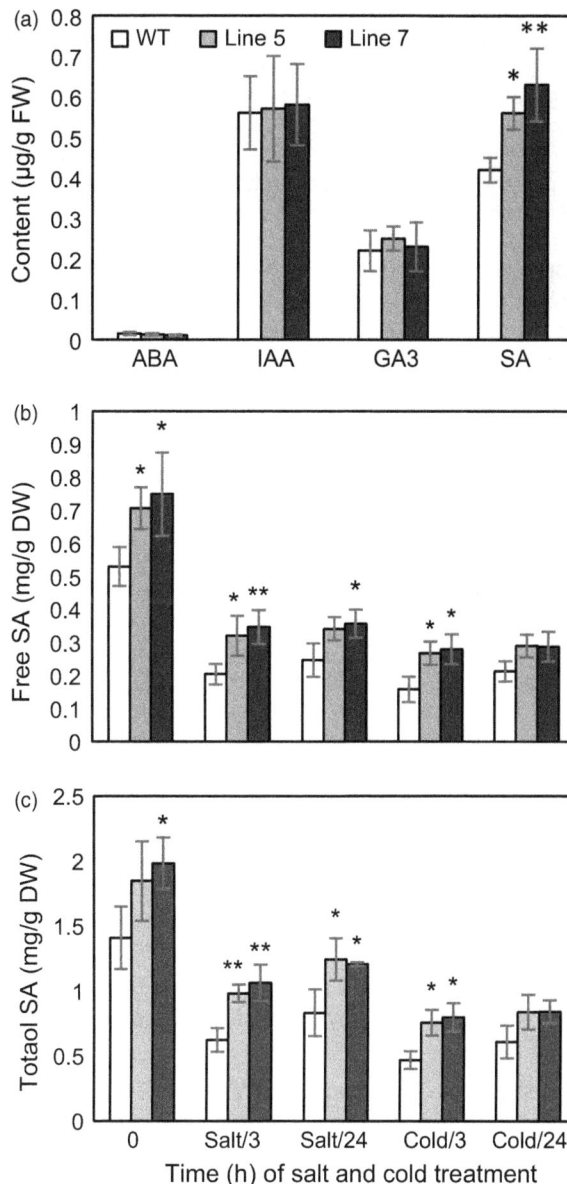

Figure 8 Hormone content in the leaves of the transgenic and wild-type (WT) plants. (a) Contents of ABA, IAA, GA₃ and SA in the leaves under normal growth conditions. (b) Total and free SA contents in the leaves under salt and cold treatments. Data are presented as means ± SD ($n = 3$). *Significant difference, $P < 0.05$; **highly significant difference, $P < 0.01$.

transgenic plants (Figure 8a). Free and total SA contents were further evaluated in the WT and transgenic plants under stress conditions. Compared to the WT controls, *MdcyMDH* overexpression elevated the free and total SA contents, and the increments reached a significant level at some time points under the stress treatments (Figure 8b,c). Free SA and total SA showed

a fast decline induced by salt and cold temperature, but the decline was not intensified by the extension of the salt and cold treatments in the WT and transgenic plants. Moreover, *MdcyMDH* overexpression did not impact the ratio of free SA in the total SA (Figure 8b,c).

Discussion

cyMDH reaction favours malate synthesis under normal conditions and especially under abiotic stresses

The kinetic properties of recombinant cyMDH proteins showed that the cyMDH reaction favoured malate production *in vitro* in apple (Yao *et al.*, 2011b), wheat (Ding and Ma, 2004) and pineapple (Cuevas and Podestá, 2000). Additionally, it was demonstrated that *MdcyMDH* promoted malate accumulation in *MdcyMDH*-overexpressing apple callus (Yao *et al.*, 2011b) and leaves (Figure 5a). Similarly, malate synthesis was promoted through the overexpression of *cyMDH* in tobacco and *Stylosanthes guianensis* (Chen *et al.*, 2015; Wang *et al.*, 2010). Therefore, regeneration of NAD$^+$ and malate inside the cytosol was the most primary role of cyMDH, although the enzyme kinetics of the MDH reaction *in vivo* were affected by substrate/product ratios and the NAD$^+$ redox state (Tomaz *et al.*, 2010). Moreover, cyMDH reductive activity increased in response to cold and salt stresses in the WT and transgenic plants (Figure 4a), and cyMDH conferred superior manganese tolerance by increasing malate synthesis in *Stylosanthes guianensis* (Chen *et al.*, 2015). Therefore, cyMDH primarily catalysed malate synthesis under stresses, and the reaction positively contributed to the tolerance of abiotic stresses.

Because cyMDH primarily catalysed the reductive reaction, more NADH was supposed to be converted into NAD$^+$. However, a high ratio of NADH/NAD$^+$ was found in the transgenic plants (Figure 7a), which can be explained, at least partly, by the following facts. First, NADH and NAD$^+$ were used to generate other antioxidants and provide electron cofactors for other oxidoreductases and hence were tied to every other redox pair and large amounts of oxidoreductases (Hashida *et al.*, 2009); second, large redox differences in the NAD(H) pools existed between cell compartments (Igamberdiev and Gardeström, 2003); third, *MdcyMDH* overexpression affected the expression of genes involved in NAD(H) metabolism, such as *quinolinate synthetase A* and *dihydrolipoamide dehydrogenase* (Figure 3c; Ceciliani *et al.*, 2000). Therefore, the sole reaction catalysed by MdcyMDH could not determine the NAD(H) level of the cell, and an enhanced NADH/NAD$^+$ might result from the wide metabolism modification mediated by *MdcyMDH* overexpression.

MdcyMDH enhances the tolerance of the transgenic plants to cold and salt by improving the cell reducing power

Abiotic stresses, such as drought, salt and low temperature, can damage cells and lead to oxidative stress by producing certain deleterious chemical entities called ROS, which include H_2O_2, O_2^-, hydroxyl radical (OH$^-$), etc. Enzymatic and nonenzymatic antioxidants can be employed to directly scavenge ROS and free radicals. In our previous study, *MdcyMDH* overexpression increased SOD and CAT activities in the transgenic apple callus and tomato plants (Yao *et al.*, 2011b). In this study, the significantly increased nonenzymatic antioxidants (AsA and GSH) in the transgenic plants favoured the direct scavenging of ROS (Figure 7c,d; Noctor and Foyer, 1998). In contrast, AsA/DHA, GSH/GSSH, NADH/NAD$^+$ and NADPH/NADP$^+$ couples were placed at the heart of the cell's redox metabolism (Potters *et al.*, 2010). Presently, the regulatory roles of cellular redox couples have predominantly come from studies using rather severe stress treatments that induced large amounts of ROS or programmed cell death (Potters *et al.*, 2010). For example, severe salt stress caused a 10-fold decrease in the GSH/GSSH ratio in *Arabidopsis* (Borsani *et al.*, 2001). In contrast, the redox state and responses to mild stress remain a conundrum (Potters *et al.*, 2010). Here, mild cold and salt stresses generated increased ratios of AsA/DHA, GSH/GSSG and NAD(P)H/NAD(P)$^+$ (Figure 7a,e). Similarly, CO_2-induced heat stress alleviation can be associated with the reduction of oxidative stress through increasing AsA/DHA and GSH/GSSG ratios in tomato plants (Li *et al.*, 2015). Therefore, it is

suggested that the increased ratio of AsA/DHA, GSH/GSSG and NAD(P)H/NAD(P)$^+$ contributed favourably to the mild oxidative stress tolerances.

Redox compounds are primarily derived from the mitochondrial and chloroplast metabolisms (Chew *et al.*, 2003; Green and Fry, 2005). The modified chloroplast and mitochondrial metabolisms (Figures 5 and 6) might be one of the factors affecting the level of redox compounds in the transgenic plants. Similarly, it was reported that *mMDH* can alter photosynthetic, photorespiration and leaf respiration and affect ascorbate content in tomato and *Arabidopsis* (Nunes-Nesi *et al.*, 2005; Tomaz *et al.*, 2010); plastid *NAD-MDH* can alter glutathione levels and the respiration rate in *Arabidopsis* (Beeler *et al.*, 2014). Additionally, modifications in the mMDH and chMDH activities were found in the *MdcyMDH* transgenic plants (Figure 4b,c). Therefore, it is speculated that MDHs from different cell compartments might cooperate to modify the chloroplast and mitochondrial metabolisms via malate valves. Moreover, another shuttle system (triose-phosphate/3-phosphoglycerate shuttle) between the chloroplast and the cytosol is suggested to be activated by *MdcyMDH* overexpression, based on the large expression modification of the key genes in this shuttle system, that is triose-phosphate/phosphate translocator and glyceraldehyde-3-phosphate dehydrogenase (Figure 3c; Guo *et al.*, 2012; Taniguchi and Miyake, 2012). Therefore, MdcyMDH might modify chloroplast and mitochondrial metabolisms and hence the redox state via different shuttle systems.

MdcyMDH increases cold and salt tolerance by modifying the redox signal

The redox pairs within the cellular stress response network are more complex and do not simply act as a scavenger; redox metabolism and its associated signalling are important machinery during abiotic stress (Munné-Bosch *et al.*, 2013). It is suggested that ascorbate and glutathione pools have precise and distinct roles in redox signalling (Munné-Bosch *et al.*, 2013) and participate in the signal transduction pathways (Ramel *et al.*, 2012). Changes in the glutathione content and in the ratio of GSH/GSSG clearly modify redox- and H_2O_2-dependent gene regulation (Han *et al.*, 2013). In this study, the increased ratios of AsA/DHA and GSH/GSSG (Figure 7e) may affect the redox signalling in the transgenic plants. On the other hand, ROS, such as H_2O_2 and O_2^-, are proposed to function in signalling between different organelles and the nucleus; the cytosol should be considered as a hub where cross-communication between divergent ROS signals occurs (Choudhury *et al.*, 2013). However, ROS accumulation at high levels causes oxidative stress leading to cell death (Choudhury *et al.*, 2013). The dual function of ROS implies that its cellular concentration has to be tightly controlled (Apel and Hirt, 2004). In this study, the H_2O_2 level was increased, but O_2^- was decreased by *MdcyMDH* overexpression (Figure 7f); this modification possibly altered the ROS signal, which was also suggested by the two largely up-regulated *glyceraldehyde 3-phosphate dehydrogenase* genes, related to H_2O_2 signal transduction (Figure 3c; Guo *et al.*, 2012); the altered redox signalling might positively contribute to the increased cold and salt tolerance in the transgenic plants.

It is proposed that malate valves play an essential role in the regulation of catalase activity and the accumulation of a H_2O_2 signal by transmitting the redox state of the chloroplast to other cell compartments (Heyno *et al.*, 2014). Therefore, the increase in H_2O_2 content in the transgenic plants may be influenced by the altered MdcyMDH-mediated regulation of malate valves. Additionally, the expression of three peroxidases was largely down-regulated (Figure 3c), suggesting a possible decrease in H_2O_2 scavenging, which led to the increase of H_2O_2 (Passardi et al., 2004). Moreover, O_2^{-} can be rapidly converted to H_2O_2 by SOD1 or SOD2 (Murphy, 2009); the decreased O_2^{-} (Figure 7f) might contribute to the increase of H_2O_2 in the transgenic plants.

The promoted interaction of redox and SA favours the tolerance of the transgenic plants to cold and salt stresses

It is known that the SA signalling pathway is pivotal for enhanced salt and oxidative stress tolerance in *Arabidopsis* (Jayakannan et al., 2015). *MdcyMDH* overexpression might affect the stress tolerance of the transgenic plants via the SA signalling pathway based on the increased free and total SA (Figure 8b,c). SA is synthesized by two distinct pathways: The phenylalanine ammonia-lyase pathway in the cytoplasm and the isochorismate (ICS1) pathway in the chloroplast. The ICS1 pathway is now thought to be the predominant pathway responsible for SA synthesis (Jayakannan et al., 2015). In the transgenic plants, an *ICS1* gene, marker gene in the ICS1 pathway and two *arogenate dehydratase* genes involved in the biosynthesis of phenylalanine were largely transcriptionally up-regulated (Figure 3c), implying that *MdcyMDH* increased SA biosynthesis by promoting the above pathways. The defectiveness in the expression of the *ICS1* gene is hypersensitive to salt stress in *Arabidopsis* (Asensi-Fabado and Munné-Bosch, 2011), and salt- and cold-induced *ICS1* expression was found in apple plants (Figure S2), implying that this pathway was essential for salt and cold tolerance in plants. Noticeably, two E3 ubiquitin-protein ligase BAH1-like genes were largely enhanced by *MdcyMDH* overexpression (Figure 3c), and they were largely induced by cold and salt (Figure S2); this type of gene is involved in the negative regulation of SA accumulation independent of *ICS1* (Kim et al., 2010) and might be one the factors decreasing the SA level when the transgenic and WT plants were subjected to stresses (Figure 8b, c). Taken together, *MdcyMDH* increased the SA level by several SA metabolism pathways including synthesis and degradation, which are likely to maintain the most suitable SA level under stress.

The modification of the redox state might be an important factor affecting the SA level. It has been found that an elevated NAD^+ level did influence SA turnover and biosynthesis via downstream *ICS1* genes (Pétricaq et al., 2012). Other antioxidants, such as ascorbate and glutathione, are also central redox regulators of the SA signalling pathway (Bartoli et al., 2013). Moreover, ROS can influence the SA metabolism, and the H_2O_2-dependent activation of SA-dependent pathways is well characterized (Munné-Bosch et al., 2013). On the other hand, the tight correlation between SA, H_2O_2 and GSH contents was observed in plants, implying an essential role of SA in the acclimation processes and in regulating the redox homeostasis of the cell (Mateo et al., 2006). Taken together, the complex network of cross-communication between oxidants and antioxidants in the redox signalling hub and the different hormone signalling pathways regulates plant survival upon exposure to stress (Bartoli et al., 2013), which might be the key mechanism of *MdcyMDH*-mediated abiotic stress tolerance.

Thus to conclude, *MdcyMDH* overexpression improved the tolerance of the transgenic apple plants to salt and cold conditions by the following possible mechanisms. First, the reducing power was enhanced, and the level of O_2^{-} was decreased; second, changes in the redox couples and the level of H_2O_2 contributed favourably to stress tolerance as redox signals; third, the stress-resistant pathway mediated by SA was strengthened; finally, the widely modified gene expression in the transgenic plants implicated that redox compounds and SA participated in the chloroplast-to-nucleus and mitochondria-to-nucleus retrograde communication, which served to coordinate gene expression in the nuclear and cytoplasmic genomes and played an important role in the stress acclimation of plants (Cela et al., 2011; Ramel et al., 2012).

Experimental procedures

Plant materials and growth conditions

'Gala' apple *in vitro* shoot cultures were used for the gene expression assays, genetic transformation and other analyses. The 'Gala' cultures were grown on MS subculture media containing 0.6 mg/L of 6-BA and 0.2 mg/L of IAA at 25 °C under a 16-h photoperiod with a light intensity of 600 μmol/m²/s. Three-year-old pot-cultured WT and transgenic apple trees were used for the salt treatment, photosynthesis and chlorophyll fluorescence assays.

RNA extraction and gene expression analysis with real-time quantitative PCR

Total RNAs were extracted from apple *in vitro* shoot cultures using TRIzol Reagent (Invitrogen, Carlsbad, CA), according to the manufacturer's instructions. Two micrograms of total RNAs was used to synthesize first-strand cDNA. For real-time quantitative PCR, the specific primer pairs are given in Table S1. The reactions were performed using SYBR Green MasterMix (SYBR Premix EX Taq TM, Dalian, China), as described by the manufacturer. Real-time quantitative PCRs were performed using a BIO-RAD iQ5 (Hercules, CA) instrument. The expression values obtained were normalized against *18s* rRNA by the cycle threshold (C_t) $2^{-\Delta\Delta C_t}$ method (Software IQ5 2.0) (Bio-Rad, Hercules, CA).

Agrobacterium-mediated genetic transformation of *MdcyMDH* into apple *in vitro* shoot cultures

The *MdcyMDH* ORF was obtained by PCR using the forward primer 5'-GG<u>TCTAGA</u>ATGGCGAAAGAACCAGTTC-3' and the reverse primer 5'-CG<u>TCGACA</u>GTCGAAGTGTCCGAATAGAAT-3'. The forward primer contained a *Xba* I digestion site, and the reverse primer contained a *Sal*I site (both underlined). Subsequently, the PCR product was cloned into the pMD18-T vector (TaKaRa, Dalian, China). *MdcyMDH* was double-digested with *Xba*I and *Sal*I, and then ligated into the pBI121 vector, downstream of the *CaMV 35S* promoter. The resultant constructs were introduced into the *Agrobacterium* strain LBA4404 and transformed into apple leaves from *in vitro* shoot cultures, as described by our previous study (Feng et al., 2012). The specific primers from *35s* (5'-GACGCACAATCCCACTATCC-3') and *MdcyMDH* (5'-GGATCCAGAGAGGCAAGAGTA-3') were used in PCR to determine whether *MdcyMDH* was integrated into the apple plants. A plasmid DNA containing the *pBI121::MdcyMDH* construct and H_2O was used as positive and negative controls.

Enzyme extraction and isolation of mitochondria and chloroplast

Cytosolic enzyme extractions were performed according to our previous study (Yao et al., 2011b). Mitochondria and chloroplast were isolated mechanically from 15 and 10 g of leaves of apple *in vitro* shoot cultures, respectively, and then purified by Percoll gradient centrifugation as described by Osmond and Makino (1991). Mitochondrial integrity was detected by measuring the cytochrome c oxidase activity with and without 0.05% Triton X-100. In the tested samples, the membrane integrity exceeded 86%. The intactness of the purified chloroplast fraction was over 84% as judged by the ferricyanide test. The purities of the cytosolic, mitochondria and chloroplast fractions were verified by measuring the activities of maker enzymes located in other organelles (Table S2). Protein amounts were quantified using dye-binding assay with bovine serum albumin (Bradford, 1976).

MDH activity assays and malate-dependent respiratory rates

MDH reductive activity was measured in 1 mL of a reaction mixture containing 50 mM Tris–HCl (pH 7.8), 2 mM $MgCl_2$, 0.5 mM EDTA, 0.2 mM NADH, 2 mM OAA and 50 μL of extract at 30 °C. The reaction was initiated by adding OAA. MDH oxidative activity was measured in 1 mL of a reaction mixture containing 50 mM Tris–HCl (pH 8.9), 2 mM $MgCl_2$, 0.5 mM EDTA, 0.2 mM NAD^+ and 25 mM malate. The reaction was initiated by adding malate. For each reaction, 5 min of spectrophotometric change at 340 nm was monitored automatically at 40-s intervals, and the activities were calculated according to the slopes of the obtained lines. The assay conditions were not optimized with respect to the concentration of each component of the reaction mixture. However, linear relationships between the activity and time and amount of the extract were found.

Malate-dependent respiratory rates on isolated intact mitochondria were performed according to the method of Tomaz et al. (2010).

Determination of malate content

Malate content was determined using a capillary electrophoresis system (Beckman P/ACE, Palo Alto, CA) as described in our previous study (Yao et al., 2007).

Measurement of photosynthesis and chlorophyll content

Photosynthesis was measured using a portable photosynthesis system (CIRAS-2; PPS Co. Ltd., Hitchin, UK). Photosynthesis was measured three times for each selected leaf, so that nine measurements were made for each plant. Measurements were made between 8:30 and 11:00 h on a sunny day. During each measurement, CO_2 concentration was maintained at 396 ± 21 μL/L by the CIRAS system at an air temperature of approximately 25 °C and a relative humidity of 85 ± 0.9%. Chlorophyll content was determined according to the methods described by Zhu et al. (1990).

Measurement of chlorophyll fluorescence

Chlorophyll fluorescence was measured with a FMS-2 pulse-modulated fluorometer (Hansatech, Norfolk, UK). The light-fluorescence measurement protocol was as follows: The light-adapted leaves were continuously illuminated by actinic light at 100 μmol/m^2/s from the FMS-2 light source; steady-state fluorescence (F_s) was recorded after 2 min of illumination; and 0.8 s of saturating light of 8000 μmol/m^2/s was imposed to obtain a maximum fluorescence in the light-adapted state (F_m). The actinic light was then turned off, and the minimum fluorescence in the light-adapted state (F_0) was determined by 3 s of illumination with far-red light. The following parameters were then calculated as follows (Maxwell and Johnson, 2000):

$$\text{Quantum yield of PSII}, \Phi PSII = (F_m - F_s)/F_m$$

$$\text{Electron transport rate}, ETR = \Phi PSII \times PFD \times 0.5 \times 0.84,$$
$$PFD = 1000$$

$$\text{Maximum quantum yield of PSII}, F_v/F_m = 1 - (F_0/F_m).$$

Extractions and assays of redox couples and ROS

Extractions and determinations of the four redox couples, that is $NADH/NAD^+$, $NADPH/NADP^+$, reduced ascorbate (AsA)/oxidized ascorbate (DHA) and reduced glutathione (GSH)/oxidized glutathione (GSSG), were performed as described by Queval and Noctor (2007). Hydrogen peroxide (H_2O_2) and superoxide radical (O_2^-) were extracted and measured according to the methods described by Zhu et al. (1990).

Hormone extraction and determination

Lyophilized leave (0.5 g) was ground into a powder. The powdered samples were extracted three times in 5 mL of cold 80% methanol (v/v) mixed with 30 μg/mL of sodium diethyldithiocarbamate. The extracts were centrifuged at 8000 $\boldsymbol{g}$ and 4 °C for 10 min. The supernatant was concentrated to dryness under vacuum. The residue was dissolved in 4 mL of 0.4 M phosphate buffer (pH 8.0). The solution was mixed with 4 mL of trichloromethane to remove the pigment. After centrifuging at 8000 $\boldsymbol{g}$ and 4 °C for 10 min, the aqueous phase was collected and supplemented with polyvinylpyrrolidone to remove the phenolics, and then centrifuged at 8000 $\boldsymbol{g}$ and 4 °C for 10 min. The supernatant was extracted with 4 mL of ethyl acetate (pH 3.0) twice. The upper phase was collected and concentrated to dryness under vacuum. The residue was dissolved in 1 mL of 0.5% (v/v) acetic acid/methanol (55 : 45, v/v) and finally filtered through a 0.45 μm filter. The obtained solution was used to determine ABA, GA3, IAA and SA. Extractions of free and conjugated SA were performed as described by Fragnière et al. (2011).

The extracted hormones were quantified using a LC-ESI-MS (Thermo, San Jose, CA) instrument. Ten microlitres of the sample was injected into a Thermo Scientific Hypersil Gold column (50 × 2.1 mm, 1.9 μm) in the Thermo Scientific Ultimate 3000 HPLC system (Thermo, San Jose, CA). The HPLC solvents were as follows: A, 0.04% acetate acid in water and B, methanol (0.4 mL/min). The two mobile phases were used in the gradient mode under the following time/concentration (min/%) of B: 0.0/20, 0.5/20, 2.5/90, 3.5/90, 3.6/20 and 5.0/20. Detection and quantification were performed using the TSQ Quantum Access MAX system (Thermo). SA was detected in the ESI negative mode and selected reaction monitoring with the following parameters: parent mass by charge (m/z) of 263.1, daughter mass by charge (m/z) of 153.0 and collision energy of 14 eV. The parameters for the ion source were set as follows: ion spray voltage of 3000 V, temperature of 350 °C, collision gas pressure of 1.5 mTorr, sheath gas of 25 arbitrary units and auxiliary gas of 15 arbitrary units. The amount of free and conjugated SA was calculated in mg/g DW with reference to the amount of internal standard (ortho-anisic acid).

Sequence-based DGE

Sample preparation and sequencing were performed with the Illumina Gene Expression Sample Prep kit and Solexa Sequencing Chip (flowcell) according to the manufacturer's introductions on the Illumina Cluster Station and Illumina HiSeq™ 2000 system (Illumina Inc., San Diego, CA). The sequencing lengths of raw reads were 49 bp in each line of the flowcell tunnel. The raw sequences were transformed into 17 bp clean tags, and tag counting was carried out using the Illumina Pipeline.

All tags were annotated using the apple genome database (http://genomics.research.iasma.it/). Briefly, a preprocessed database containing all possible CATG + 17-base tag sequences was created. All clean tags were mapped to the reference sequence, and only 1 bp mismatch was considered. Clean tags mapped to the reference sequences from multiple genes were filtered. The remainder clean tags were designed as unambiguous clean tags. The number of unambiguous clean tags for each gene was calculated, and then normalized to TPM (number of transcripts per million clean tags).

Finally, a rigorous algorithm developed by the Beijing Genomics Institute (BGI) referring to 'the significance of digital gene expression profiles' (false discovery rate < 0.001) (Audic and Claverie, 1997) was used to identify differentially expressed genes between two samples, and absolute value of log_2ratio ≥ 1 (minimum of twofold difference) was used as the threshold to judge the significance of gene expression differences.

Acknowledgements

This work was supported by the Special Financial Grant from the China Postdoctoral Science Foundation (2012T50623), the General Financial Grant from the China Postdoctoral Science Foundation (2011M501157), China's Agricultural Research System (CARS-30-ZP-06), and the Changjiang Scholars and Innovative Research Team in University (IRT1155).

References

Apel, K. and Hirt, H. (2004) Reactive oxygen species: metabolism, oxidative stress, and signal transduction. *Annu. Rev. Plant Biol.* **55**, 373–399.

Asensi-Fabado, M. and Munné-Bosch, S. (2011) The aba3-1 mutant of *Arabidopsis thaliana* withstands moderate doses of salt stress by modulating leaf growth and salicylic acid levels. *J. Plant Growth Regul.* **30**, 456–466.

Audic, S. and Claverie, J.M. (1997) The significance of digital gene expression profiles. *Genome Res.* **7**, 986–995.

Bartoli, C.G., Casalongué, C.A., Simontacchi, M., Marquez-Garcia, B. and Foyer, C.H. (2013) Interactions between hormone and redox signalling pathways in the control of growth and cross tolerance to stress. *Environ. Exp. Bot.* **94**, 73–88.

Beeler, S., Liu, H.C., Stadler, M., Schreier, T., Eicke, S., Lue, W.L., Truernit, E. et al. (2014) Plastidial NAD-Dependent malate dehydrogenase is critical for embryo development and heterotrophic metabolism in *Arabidopsis*. *Plant Physiol.* **164**, 1175–1190.

Björkman, O. and Demmig, B. (1987) Photon yield of O_2 evolution and chlorophyll fluorescence characteristics at 77K among vascular plants of diverse origins. *Planta*, **170**, 489–504.

Borsani, O., Valpuesta, V. and Botella, M.A. (2001) Evidence for a role of salicylic acid in the oxidative damage generated by NaCl and osmotic stress in *Arabidopsis* seedlings. *Plant Physiol.* **126**, 1024–1030.

Bradford, M.M. (1976) A rapid and sensitive method for the quantitation of microgram quantities of protein utilizing the principle of protein-dye binding. *Anal. Biochem.* **72**, 248–254.

Ceciliani, F., Caramori, T., Ronchi, S., Tedeschi, G., Mortarino, M. and Galizzi, A. (2000) Cloning, overexpression, and purification of *Escherichia coli* quinolinate synthetase. *Protein Expr. Purif.* **18**, 64–70.

Cela, J., Chang, C. and Munné-Bosch, S. (2011) Accumulation of g- rather than a-tocopherol alters ethylene signaling gene expression in the vte4 mutant of *Arabidopsis thaliana*. *Plant Cell Physiol.* **52**, 1389–1400.

Chen, Z.J., Sun, L.L., Liu, P.D., Liu, G.D., Tian, J. and Liao, H. (2015) Malate synthesis and secretion mediated by a manganese-enhanced malate dehydrogenase confers superior manganese tolerance in *Stylosanthes guianensis*. *Plant Physiol.* **167**, 176–188.

Chew, O., Whelan, J. and Millar, A.H. (2003) Molecular definition of the ascorbate/glutathione cycle in *Arabidopsis* mitochondria reveals dual targeting of antioxidant defenses in plants. *J. Biol. Chem.* **278**, 46869–46877.

Choudhury, S., Panda, P., Sahoo, L. and Panda, S.K. (2013) Reactive oxygen species signaling in plants under abiotic stress. *Plant Signal. Behav.* **8**, 23681.

Cuevas, I.C. and Podestá, F.E. (2000) Purification and physical and kinetic characterization of an NAD^+-dependent malate dehydrogenase from leaves of pineapple (*Ananas comosus*). *Physiol. Plant.* **108**, 240–248.

Ding, Y. and Ma, Q.H. (2004) Characterization of a cytosolic malate dehydrogenase cDNA which encodes an isozyme toward oxaloacetate reduction in wheat. *Biochimie*, **86**, 509–518.

Feng, X.M., Zhao, Q., Zhao, L.L., Qiao, Y., Xie, X.B., Li, H.F., Yao, Y.X. et al. (2012) The cold-induced basic helix-loop-helix transcription factor gene *MdCIbHLH1* encodes an ICE-like protein in apple. *BMC Plant Biol.* **12**, 22.

Fragnière, C., Serrano, M., Abou-Mansour, E., Métraux, J. and L'Haridon, F. (2011) Salicylic acid and its location in response to biotic and abiotic stress. *FEBS Lett.* **585**, 1847–1852.

Gietl, C. (1992) Malate dehydrogenase isoenzymes: cellular locations and role in the flow of metabolites between the cytoplasm and cell organelles. *Biochim. Biophys. Acta*, **1100**, 217–234.

Green, M.A. and Fry, S.C. (2005) Vitamin C degradation in plant cells via enzymatic hydrolysis of 4-O-oxalyl-L-threonate. *Nature*, **433**, 83–87.

Guo, L., Devaiah, S.P., Narasimhan, R., Pan, X.Q., Zhang, Y.Y., Zhang, W.H. and Wang, X.M. (2012) Cytosolic glyceraldehyde-3-phosphate dehydrogenases interact with phospholipase Dδ to transduce hydrogen peroxide signals in the *Arabidopsis* response to stress. *Plant Cell*, **24**, 2200–2212.

Han, Y., Chaouch, S., Mhamdi, A., Queval, G., Zechmann, B. and Noctor, G. (2013) Functional analysis of *Arabidopsis* mutants points to novel roles for glutathione in coupling H_2O_2 to activation of salicylic acid accumulation and signaling. *Antioxid. Redox Signal.* **18**, 2106–2121.

Hara, S., Motohashi, K., Arisaka, F., Romano, P.G.N., Hosoya-Matsuda, N., Kikuchi, N., Fusada, N. et al. (2006) Thioredoxin-h1 reduces and reactivates the oxidized cytosolic malate dehydrogenase dimer in higher plants. *J. Biol. Chem.* **281**, 32065–32071.

Hashida, S.N., Takahashi, H. and Uchimiya, H. (2009) The role of NAD biosynthesis in plant development and stress responses. *Ann. Bot.* **103**, 819–824.

Heyno, E., Innocenti, G., Lemaire, S.D., Issakidis-Bourguet, E. and Krieger-Liszkay, A. (2014) Putative role of the malate valve enzyme NADP–malate dehydrogenase in H_2O_2 signalling in *Arabidopsis*. *Philos. Trans. R. Soc. Lond. B Biol. Sci.* **369**, 20130228.

Igamberdiev, A.U. and Gardeström, P. (2003) Regulation of NAD-and NADP-dependent isocitrate dehydrogenases by reduction levels of pyridine nucleotides in mitochondria and cytosol of pea leaves. *Biochim. Biophys. Acta*, **1606**, 117–125.

Jayakannan, M., Bose, J., Babourina, O., Shabala, S., Massart, A., Poschenrieder, C. and Rengel, Z. (2015) The NPR1-dependent salicylic acid signalling pathway is pivotal for enhanced salt and oxidative stress tolerance in *Arabidopsis*. *J. Exp. Bot.* **66**, 1865–1875.

Kim, Y.H., Bae, J.M. and Huh, G.H. (2010) Transcriptional regulation of the cinnamyl alcohol dehydrogenase gene from sweetpotato in response to plant developmental stage and environmental stress. *Plant Cell Rep.* **29**, 779–791.

Lemaire, M., Issakidis, E., Ruelland, E., Decottignies, P. and Miginiac-Maslow, M. (1996) An active-site cysteine of sorghum leaf NADP-malate dehydrogenase studied by site-directed mutagenesis. *FEBS Lett.* **382**, 137–140.

Li, X., Ahammed, G.J., Zhang, Y.Q., Zhang, G.Q., Sun, Z.H., Zhou, J., Zhou, Y.H. et al. (2015) Carbon dioxide enrichment alleviates heat stress by improving cellular redox homeostasis through an ABA-independent process in tomato plants. *Plant Biol.* **17**, 81–89.

Mateo, A., Funck, D., Mühlenbock, P., Kular, B., Mullineaux, P.M. and Karpinski, S. (2006) Controlled levels of salicylic acid are required for optimal photosynthesis and redox homeostasis. *J. Exp. Bot.* **57**, 1795–1807.

Maxwell, K. and Johnson, G.N. (2000) Chlorophyll fluorescence-a practical guide. *J. Exp. Bot.* **51**, 659–668.

Munné-Bosch, S., Queval, G. and Foyer, C.H. (2013) The impact of global change factors on redox signaling underpinning stress tolerance. *Plant Physiol.* **161**, 5–19.

Murphy, M. (2009) How mitochondria produce reactive oxygen species. *Biochem. J.* **417**, 1–13.

Noctor, G. and Foyer, C.H. (1998) Ascorbate and glutathione: keeping active oxygen under control. *Annu. Rev. Plant Biol.* **49**, 249–279.

Nunes-Nesi, A., Carrari, F., Lytovchenko, A., Smith, A.M.O., Loureiro, M.E., Ratcliffe, R.G., Sweetlove, L.J. *et al.* (2005) Enhanced photosynthetic performance and growth as a consequence of decreasing mitochondrial malate dehydrogenase activity in transgenic tomato plants. *Plant Physiol.* **137**, 611–622.

Osmond, B. and Makino, A. (1991) Effects of nitrogen nutrition on nitrogen partitioning between chloroplasts and mitochondria in pea and wheat. *Plant Physiol.* **96**, 355–356.

Passardi, F., Longet, D., Penel, C. and Dunand, C. (2004) The class III peroxidase multigenic family in rice and its evolution in land plants. *Phytochemistry*, **65**, 1879–1893.

Peleg, Z. and Blumwald, E. (2011) Hormone balance and abiotic stress tolerance in crop plants. *Curr. Opin. Plant Biol.* **14**, 290–295.

Pétriacq, P., de Bont, L., Hager, J., Didierlaurent, L., Mauve, C., Guérard, F., Noctor, G. *et al.* (2012) Inducible NAD overproduction in *Arabidopsis* alters metabolic pools and gene expression correlated with increased salicylate content and resistance to *Pst-AvrRpm 1*. *Plant J.* **70**, 650–665.

Potters, G., Horemans, N. and Jansen, M.A.K. (2010) The cellular redox state in plant stress biology-A charging concept. *Plant Physiol. Biochem.* **48**, 1–9.

Queval, G. and Noctor, G. (2007) A plate reader method for the measurement of NAD, NADP, glutathione, and ascorbate in tissue extracts: application to redox profiling during *Arabidopsis* rosette development. *Anal. Biochem.* **363**, 58–69.

Raghavendra, A.S. and Padmasree, K. (2003) Beneficial interactions of mitochondrial metabolism with photosynthetic carbon assimilation. *Trends Plant Sci.* **8**, 546–553.

Ramel, F., Birtic, S., Ginies, C., Soubigou-Taconnat, L., Triantaphylidès, C. and Havaux, M. (2012) Carotenoid oxidation products are stress signals that mediate gene responses to singlet oxygen in plants. *Proc. Natl Acad. Sci. USA*, **109**, 5535–5540.

Scheibe, R. (2004) Malate valves to balance cellular energy supply. *Physiol. Plant.* **120**, 2126.

Suzuki, N., Koussevitzky, S., Mittler, R. and Miller, G. (2012) ROS and redox signalling in the response of plants to abiotic stress. *Plant Cell Environ.* **35**, 259–270.

Taniguchi, M. and Miyake, H. (2012) Redox-shuttling between chloroplast and cytosol: integration of intra-chloroplast and extra-chloroplast metabolism. *Curr. Opin. Plant Biol.* **15**, 252–260.

Tomaz, T., Bagard, M., Pracharoenwattana, I., Lindén, P., Lee, C.P., Carroll, A.J., Stroher, E. *et al.* (2010) Mitochondrial malate dehydrogenase lowers leaf respiration and alters photorespiration and plant growth in *Arabidopsis*. *Plant Physiol.* **154**, 1143–1157.

Wang, Q.F., Zhao, Y., Yi, Q., Li, K.Z., Yu, Y.X. and Chen, L.M. (2010) Overexpression of malate dehydrogenase in transgenic tobacco leaves: enhanced malate synthesis and augmented Al-resistance. *Acta Physiol. Plant*, **32**, 1209–1220.

Yao, Y.X., Li, M., Liu, Z., Hao, Y.J. and Zhai, H. (2007) A novel gene, screened by cDNA-AFLP approach, contributes to lowering the acidity of fruit in apple. *Plant Physiol. Biochem.* **45**, 139–145.

Yao, Y.X., Hao, Y.J., Li, M., Pang, M.L., Liu, Z. and Zhai, H. (2008) Gene cloning expression and enzyme activity assay of a cytosolic malate dehydrogenase from apple fruits. *Acta Horticult. Sin.* **35**, 181–188 (in Chinese, with English abstract).

Yao, Y.X., Dong, Q.L., Zhai, H., You, C.X. and Hao, Y.J. (2011a) The functions of an apple *cytosolic malate dehydrogenase* gene in growth and tolerance to cold and salt stresses. *Plant Physiol. Biochem.* **49**, 257–264.

Yao, Y.X., Li, M., Zhai, H., You, C.X. and Hao, Y.J. (2011b) Isolation and characterization of an apple *cytosolic malate dehydrogenase* gene reveal its function in malate synthesis. *J. Plant Physiol.* **168**, 474–480.

Yoshida, T., Mogami, J. and Yamaguchi-Shinozaki, K. (2014) ABA-dependent and ABA-independent signaling in response to osmotic stress in plants. *Curr. Opin. Plant Biol.* **21**, 133–139.

Zhu, G.L., Zhang, H.W. and Zhang, A.Q. (1990) *Plant Physiological Experiments*, 1st edn. Beijing: Peking University Press.

11

Antisense suppression of *LOX3* gene expression in rice endosperm enhances seed longevity

Huibin Xu[1,2,3,4,5], Yidong Wei[1,2,3,4,5], Yongsheng Zhu[1,2,3,4,5], Ling Lian[1,2,3,4,5], Hongguang Xie[1,2,3,4,5], Qiuhua Cai[1,2,3,4,5], Qiushi Chen[1,2,3,4,5,6], Zhongping Lin[7], Zonghua Wang[6], Huaan Xie[1,2,3,4,5]* and Jianfu Zhang[1,2,3,4,5]*

[1]*Rice Research Institute, Fujian Academy of Agricultural Sciences, Fuzhou, China*
[2]*Incubator of National Key Laboratory of Fujian Crop Germplasm Innovation and Molecular Breeding between Fujian and Ministry of Sciences and Technology, Fuzhou, China*
[3]*Key Laboratory of Germplasm Innovation and Molecular Breeding of Hybrid Rice for South China, Ministry of Agriculture, Fuzhou, China*
[4]*South-China Base of National Key Laboratory of Hybrid Rice of China, Fuzhou, China*
[5]*National Engineering Laboratory of Rice, Fuzhou, Fujian, China*
[6]*Fujian Agriculture and Forestry University, Fuzhou, China*
[7]*School Of Life Sciences, Peking University, Beijing, China*

*Correspondence

email
jianfzhang@163.com

email huaanxie@163.com

Keywords: seed longevity, rice, LOX3, artificial ageing.

Summary

Lipid peroxidation plays a major role in seed longevity and viability. In rice grains, lipid peroxidation is catalyzed by the enzyme lipoxygenase 3 (LOX3). Previous reports showed that grain from the rice variety DawDam in which the *LOX3* gene was deleted had less stale flavour after grain storage than normal rice. The molecular mechanism by which *LOX3* expression is regulated during endosperm development remains unclear. In this study, we expressed a *LOX3* antisense construct in transgenic rice (*Oryza sativa* L.) plants to down-regulate *LOX3* expression in rice endosperm. The transgenic plants exhibited a marked decrease in *LOX* mRNA levels, normal phenotypes and a normal life cycle. We showed that LOX3 activity and its ability to produce 9-hydroperoxyoctadecadienoic acid (9-HPOD) from linoleic acid were significantly lower in transgenic seeds than in wild-type seeds by measuring the ultraviolet absorption of 9-HPOD at 234 nm and by high-performance liquid chromatography. The suppression of *LOX3* expression in rice endosperm increased grain storability. The germination rate of TS-91 (antisense *LOX3* transgenic line) was much higher than the WT (29% higher after artificial ageing for 21 days, and 40% higher after natural ageing for 12 months). To our knowledge, this is the first report to demonstrate that decreased *LOX3* expression can preserve rice grain quality during storage with no impact on grain yield, suggesting potential applications in agricultural production.

Introduction

Rice (*Oryza sativa* L.) is one of the most important cereal crops, providing food for more than half of the world population. Although the production of rice is increasing owing to advances in rice breeding, losses caused by rice grain deterioration during storage remain substantial. In recent years, the average annual rice grain yield in China was approximately 500 billion kg, but nearly 3%, or 15 billion kg of the yield, was lost due to seed ageing during storage. Seed ageing is the consequence of two opposing processes: an inevitable tendency towards destabilization and deterioration over time and a counteracting *in vivo* detoxification system that delays, prevents and repairs the damage caused by the former (McDonald, 1999). High-quality seeds are undamaged and produce vigorous and viable seedlings under different environmental conditions (Dickson, 1980). Because seed deterioration presents a serious problem for agriculture, multiple approaches are needed to analyse the seed ageing process to improve seed quality.

Many characteristics of ageing are shared by all organisms, but seeds are distinctive in that they are in a state of very low metabolic activity and are relatively protected from the external environment (Smith and Berjak, 1995). To achieve homeostasis in the face of environmental stress, seeds make use of a highly efficient repair system and a variety of mechanisms to directly delay or prevent potential damage (Veselovsky and Veselova, 2012).

Lipid peroxidation is a major factor influencing seed longevity and viability. Some end products of lipid peroxidation such as malondialdehyde (MDA) and acetaldehyde are chemically reactive molecules which, when they accumulate, cause cell damage by reacting with macromolecules. In addition to these well-known end products, intermediate products also cause varying degrees of damage in plant cells and tissues. In particular, 4-hydroxy-2-nonenal, an aldehyde by-product of the peroxidation of fatty acids, has toxic effects on neurons in culture (Lovell *et al.*, 1997). Linolenic acid (LNA) and linoleic acid (LA) are the most important polyunsaturated fatty acids (White *et al.*, 1961), and both are vulnerable to lipid peroxidation in enzymatic and nonenzymatic reactions resulting in the decomposition of cell structure and formation of cytotoxic products. Peroxidation of LNA and LA also gives rise to volatile decomposition products that are the primary contributors to stale flavour and reduce rice grain quality during storage.

Lipoxygenases (LOX: EC1.13.11.12) are a family of nonhaeme iron-containing fatty acid dioxygenases that catalyze the addition of oxygen to polyunsaturated fatty acids containing a (Z), (Z)-1,4-pentadiene group such as LNA and LA to generate conjugated unsaturated fatty acid hydroperoxides (Axelrod et al., 1981). LNA and LA are the most common substrates for LOXs in plants. LOXs exhibit high positional specificities and add oxygen at either end of the pentadiene group in LNA and LA to form 9- or 13-hydroperoxy fatty acids. Hydroperoxy fatty acids, which appear to be nonbioactive, subsequently form bioactive products via different metabolic pathways that function as phytoalexins and as signalling molecules (Feussner and Wasternack, 2002).

Previous studies on LOXs have focused mainly on their functions in development and in defence responses to biotic stress (Marla and Singh, 2012), but their roles in seed longevity and viability remain elusive. LOXs affect seed storage stability because they delay decomposition of polyunsaturated fatty acids in seeds, thereby preserving intact cell membranes and seed structure and reducing the formation of cytotoxic derivatives. Three LOX isozymes, LOX1, LOX2 and LOX3, have been identified in rice embryos with activities localized in the bran fraction (Shastry and Raghavendra Rao, 1975) and the embryo (Yamamoto et al., 1980). LOX3 is the predominant isozyme (Ida et al., 1983). The rice variety DawDam, which lacks the LOX3 gene, exhibited a lower level of unsaturated fatty acid peroxidation during seed storage than a normal variety and a noticeably decreased level of the stale flavour caused by lipid peroxidation, which reduces stored grain quality (Suzuki et al., 1999). However, progress on understanding seed longevity has been limited due to the complexity of the lipid metabolic pathway. To clarify the role of LOX3 in rice seed viability and its potential application for improving seed storage stability, we performed physiological and molecular analysis of rice seeds from transgenic rice plants expressing antisense LOX3 under the control of a rice endosperm-specific promoter.

Seed ageing is a complex process during which a series of physiological and biochemical reactions occur, including signal transduction, reactivation of metabolism and redox homeostasis regulation (Parkhey et al., 2012). Because all of these processes are catalyzed or mediated by different proteins, determining the protein profile in seeds during storage would be of great value. Integrating the proteomic data into a network of metabolic or regulatory pathways would enable us to assess the roles of individual proteins at a systemic level (He et al., 2011). Previous rice seed proteomics studies were limited mainly to two-dimensional polyacrylamide gel electrophoresis (2D-PAGE) analysis (Görg et al., 2004) and multidimensional protein identification technology (MudPIT) (Link et al., 1999). Because these approaches use complementary and independent separation methods for the resolution of proteomic components, the integration of data sets obtained with these technologies should provide improved proteomic coverage (Schulze and Usadel, 2010). Alternative methods such as in-solution peptide labelling combined with liquid chromatography coupled with tandem mass spectrometry (LC-MS/MS) are also now available (Zieske, 2006). In this study, we used the isobaric tags for relative and absolute quantification (iTRAQ) method (Wiese et al., 2007) to assess proteomic changes in antisense LOX3 transgenic rice and wild-type (WT) seeds before and after artificial ageing for 30 days. Our results provide comprehensive information at the protein level and improve the understanding of the mechanism of rice seed ageing and the roles played by relevant enzymes.

Results

Molecular analysis of transgenes in plants

A LOX3 antisense gene construct was introduced into rice using Agrobacterium tumefaciens-mediated transformation. T_0 transgenic and WT plants were grown in a greenhouse (Figure 1a). DNA blot analysis of T_0 seedlings using a gene-specific probe

Figure 1 Agrobacterium-mediated transformation of rice and complementation tests of transgenic plants. (a) Phenotypes of antisense LOX3 transgenic (TS-91) and wild-type (WT; Hang 1) rice plants in the greenhouse. (b) Southern hybridization analysis of transgenic plants. (c) RT-PCR analysis showing bar gene expression in transgenic T_0 plants. The actin1 gene was used as a control for equal loading.

showed that the phosphinothricin acetyltransferase bialaphos resistance (*bar*) marker gene was introduced into the rice genome with different copy numbers in different transgenic lines (Figure 1b). Reverse transcription–polymerase chain reaction (RT-PCR) analysis indicated that *bar* transcripts were present in five transformants (TS-58, TS-61, TS-63, TS-91 and TS-105) (Figure 1c). The transgene segregated in a 3 : 1 ratio in the T_1 lines in agreement with results of the DNA blot analysis (Table S1). The transgenic plants exhibited normal phenotypes, a normal life cycle, grew to maturity, flowered and set seed. T_4-generation transgenic plants were used for experiments in this study.

Verification of LOX3 suppression at the protein and RNA levels

Western blot analysis revealed a marked decrease in LOX3 expression in the endosperm of the transgenic seeds demonstrating that LOX3 expression was reduced successfully using the antisense technique. In particular, LOX3 protein levels were down-regulated in the TS-91 transgenic line compared with WT plants. In comparison, no LOX3 protein was detected in the *LOX* deletion variety DawDam (Figure 2a). Analysis using quantitative RT-PCR confirmed that *LOX3* transcript levels decreased in different degree in T_4 antisense *LOX3* transgenic lines TS-58, TS-61, TS-91 and TS-105, comparing with the WT (Figure 2b). Total RNA was isolated from fresh seeds of WT plants and the different T_4 transgenic lines at 28 days after heading. *LOX3* mRNA levels were significantly lower in TS-58, TS-61 and TS-105 than in the WT (Figure 2b). Three LOX isozymes, LOX1, LOX2 and LOX3, have been identified in rice embryos. Quantitative RT-PCR analysis revealed that when *LOX3* was down-regulated, the expression of *LOX1* and *LOX2* was also down-regulated, especially *LOX1* (Figure 2c). These results indicated that the expression of the *LOX3* antisense gene not only down-regulated *LOX3* expression, but also *LOX1* and *LOX2*, relative to the WT. It may delay the peroxidation rate of rice seed lipid to a greater extent.

Detection of LOX3 activities in transgenic rice seeds

We determined the LOX3 activities in seeds from the TS-58, TS-61, TS-91 and TS-105 transgenic lines and in WT and DawDam seeds using KI–I_2 staining of starch (Wan *et al.*, 2000). Stained seeds of the transgenic lines were much lighter in colour than WT seeds, and the seeds of the *LOX3* deletion variety DawDam were colourless after staining (Figure 3a). This result indicated that LOX3 activities had decreased in the transgenic lines. LOX activity is typically determined by measuring the ultraviolet (UV) absorption at 234 nm of the conjugated double bonds of the 9- and 13-hydroperoxyoctadecadienoic acid (HPOD) isomers produced from LA or LNA (Nuñez *et al.*, 2001). Among the three LOXs (LOX1, LOX2 and LOX3) in rice seeds, LOX3 is a 9-LOX that incorporates molecular oxygen at the 9 position of LA at pH 7.2 to form 9-hydroperoxyoctadecadienoic acid (9-HPOD). Therefore, we could specifically determine differences in LOX3 activity between WT and transgenic rice seeds by measuring the UV absorption of 9-HPOD at 234 nm. LOX3 activity was significantly lower in transgenic seeds than in WT seeds ($P < 0.01$) (Figure 3b). We also determined the LA content in seeds of transgenic lines and WT plants using high-performance LC (HPLC) (Figure 3c). Higher levels of LA were detected in the antisense *LOX3* transformants than in WT (Figure 3c), consistent with the results of the KI–I_2 starch staining method. These results showed that the expression

of the antisense *LOX3* gene in rice endosperm reduced LOX3 activity.

Heat-shock treatment and excision efficiency of the marker *bar* gene

We constructed the *Agrobacterium* vector pHG_1PL (Figure S1a) containing a Cre/loxP recombinase system under the control of a heat-shock-responsive promoter. The heat-shock-responsive promoter and the sequence encoding the selectable *bar* marker gene were flanked by two *loxP* sites (Figure S1b). Heat-shock treatment induced strong expression of the Cre recombinase from the heat-shock-responsive promoter causing excision of all sequences between the two *loxP* sites (Figure S1b). We subjected T_4 antisense *LOX3* transgenic lines to heat-shock treatment at different growth stages (Figure S2a,b) and demonstrated that the *bar* gene was excised using PCR analysis. The *bar* gene excision frequency after heat shock was significantly higher in transgenic plants at the four-leaf

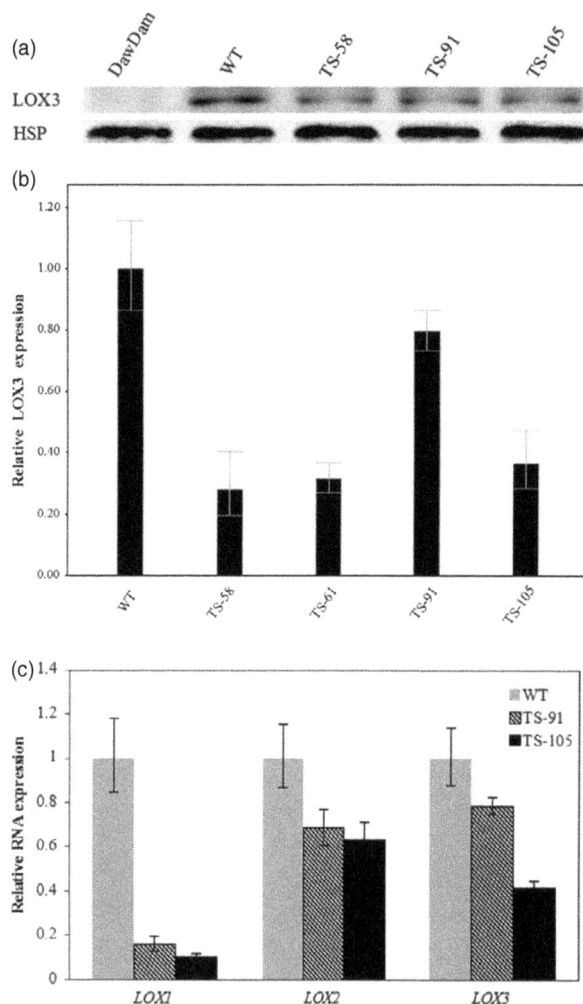

Figure 2 Verification of LOX3 suppression at the protein and RNA levels. (a) Western blot analysis of LOX3 expression levels in rice seeds collected from control and T_4 transgenic lines. (b) qRT-PCR analysis of *LOX3* expression in endosperm of WT and T_4 transgenic lines after heading for 28 days. (c) Comparison of *LOX1*, *LOX2* and *LOX3* expression level between WT and T_4 transgenic lines in 28-day-old rice seeds. *Actin1* expression was used as a control.

stage (83.3%) than in seeds after germination (24.67%). RT-PCR analysis detected no expression of the *bar* or *Cre* genes in TS91-1 to TS91-7 transgenic plants following heat-shock treatment (Figure S2c), indicating that the *bar* and *Cre* genes were excised. In addition, no hybridization signal corresponding to the *bar* gene was detected by Southern blot analysis in TS91-1 to TS91-5 transgenic plants or in the WT. However, a DNA fragment of the expected size was present in uninduced TS91 transgenic plants, indicating that the *bar* gene had been

excised from the genomes of the transgenic plants after heat shock (Figure S2d).

Longevity of transgenic and WT seeds

Seed longevity was measured as the percentage of germination after artificial or natural ageing treatments. TS-91 and TS-105 transgenic lines and WT plants were grown in the greenhouse. After harvest, seeds were dried to levels appropriate for storage

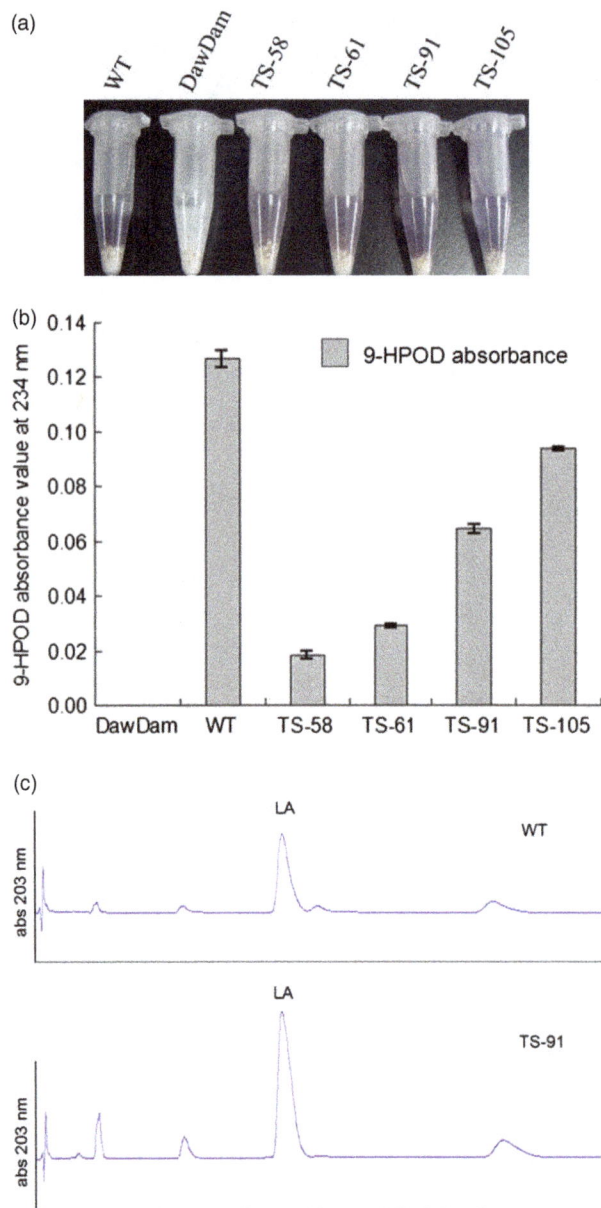

Figure 3 Detection of LOX3 activity in transgenic rice seeds. (a) LOX3 activities detected using the KI–I$_2$ starch staining method. (b) Production of lipid hydroperoxides in rice embryo. (c) HPLC chromatograms of linoleic acid (LA) in seeds of WT and transgenic plants. DawDam, *LOX3* deletion variety; Hang1, WT plants; HG$_1$PL, positive control plasmid; TS-58, TS-61, TS-63, TS-91 and TS-105, antisense *LOX3* transgenic lines. Error bars indicate the standard error of the mean from at least three independent experiments.

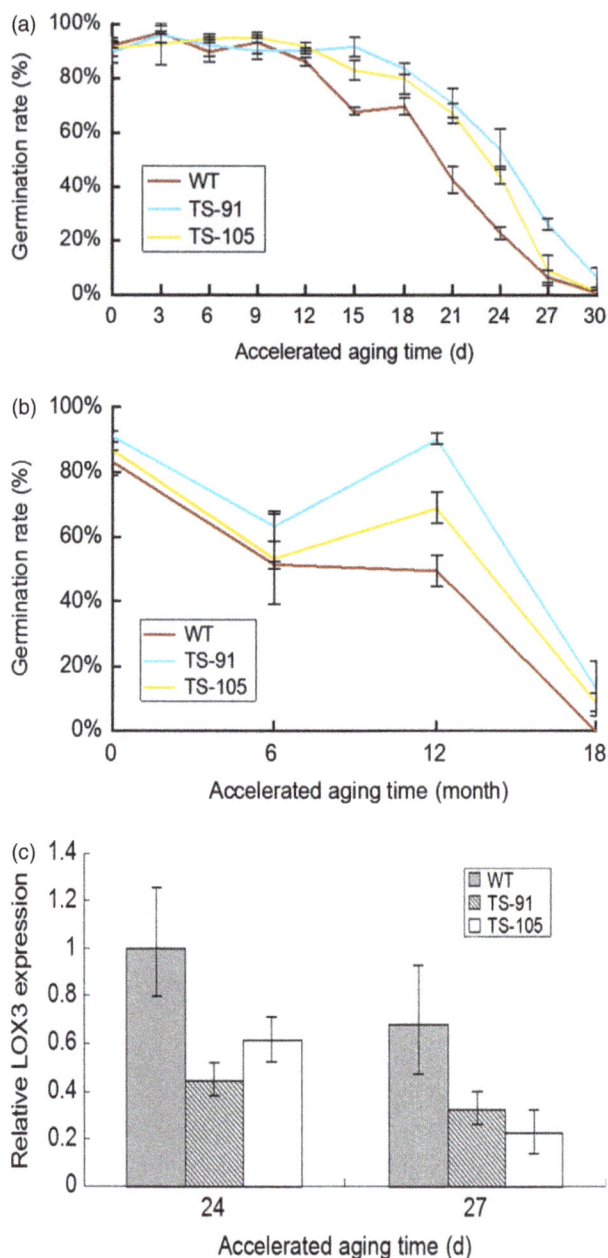

Figure 4 Variation in germinability and *LOX3* expression during artificial and natural ageing of rice seeds. (a) Variation in germinability after artificial ageing for 0–30 days. Each value is the mean of three replicates ± standard error of the mean. (b) Variation in germinability after natural ageing for 6, 12 and 18 months. Each value is the mean of three replicates ± standard error of the mean. (c) Determination of *LOX3* gene expression in transgenic lines by real-time PCR after artificial ageing for 24 and 27 days. The relative expression in a WT plant was used as a control.

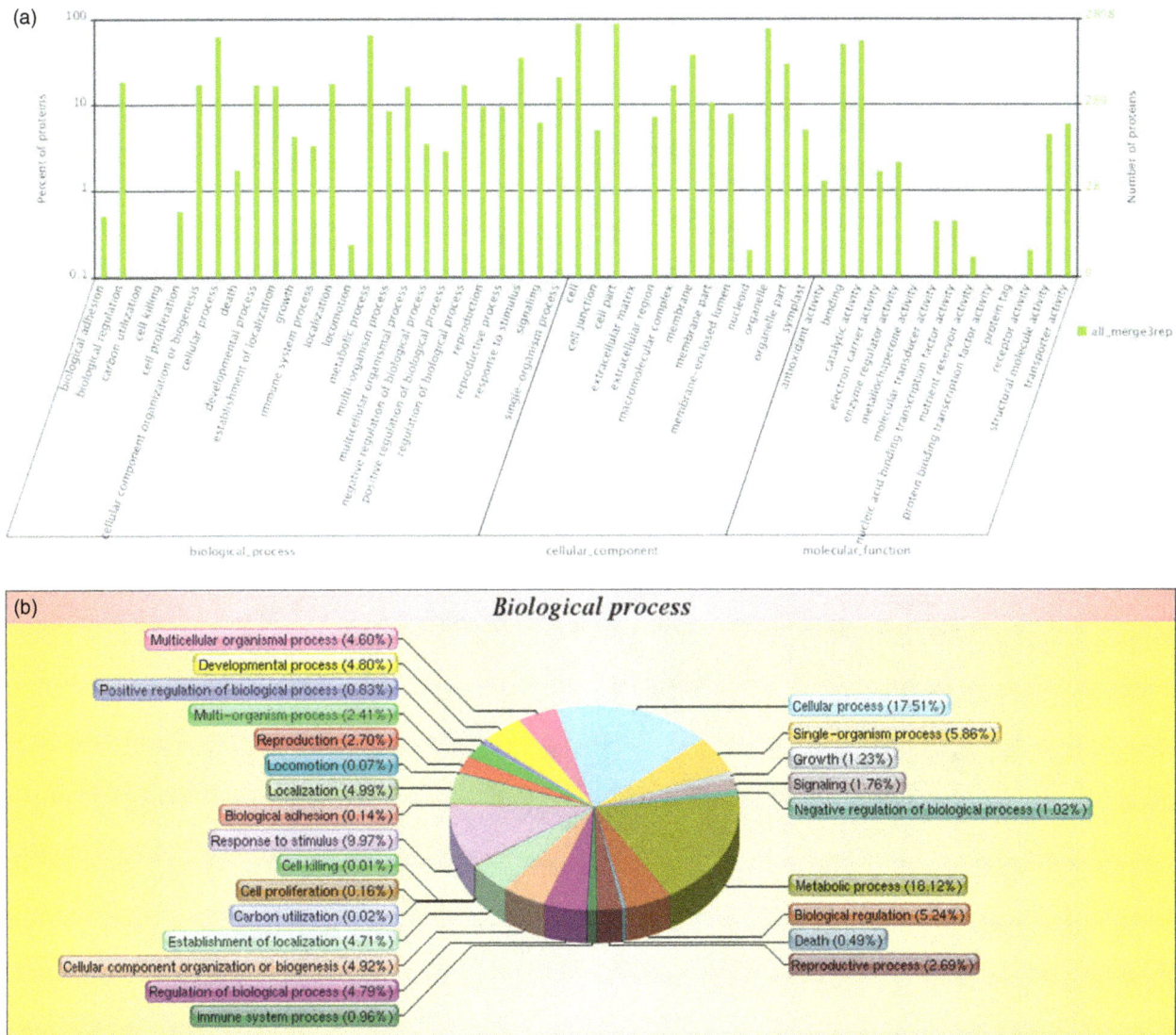

Figure 5 Proteomics profiling of rice seeds before and after artificial ageing. (a) Gene ontology (GO) annotation of the various proteins. Flash bar chart of various proteins represented in the three ontologies. GO term definitions are indicated on the x-axis. The numbers and percentages of various proteins mapped by each GO term are indicated on the y-axes and represent the abundance of each GO term. (b) Pie charts of overrepresented terms in the biological process ontology. (c) Pie charts of overrepresented terms in the cellular component ontology. (d) Pie charts of overrepresented terms in the molecular function ontology. Different colours represent different GO terms.

(Roberts, 1961). Before artificial ageing, the germinability of each sample was measured as described previously (Wu et al., 2010). The germinability of the TS-91 and TS-105 transgenic and WT seeds was not significantly different (92.33%, 89.33% and 91.33%, respectively, $P > 0.05$; Figure 4a). After artificial ageing for 12 days, the germination rates of the transgenic and WT seeds decreased somewhat, but the TS-105 germination rate was much higher than that of the WT ($P < 0.05$; Figure 4a). The WT germination rate decreased rapidly with extended ageing time, while the germination rates of the transgenic lines remained significantly higher than that of the WT. Overall, the transgenic lines and WT exhibited similar trends with little difference under both natural and artificial ageing conditions (Figure 4b). After storage for 12 months under natural conditions, the germination rate of the TS-91 and TS-105 transgenic lines increased (TS-91 reached 90%). However, after storage for

18 months, the germination rate decreased markedly with significant differences between the transgenic lines and the WT ($P < 0.01$, Figure 4b). Extension of the artificial ageing period resulted in a gradual decrease in the germination rate (Figure 4a). Although the artificial and natural ageing treatments exhibited some differences, the germination rates of the transgenic lines were always higher than those of the WT for both methods (Figure 4a,b). Analysis using quantitative RT-PCR confirmed that LOX3 transcript levels decreased significantly in antisense LOX3 transgenic lines compared with the WT (Figure 4c). Total RNA was isolated from the transgenic lines and WT plants with artificial ageing for 24 and 27 days. LOX3 mRNA levels were much higher in WT plants than the TS-91 and TS-105 transgenic lines (Figure 4c). These results demonstrated that suppression of LOX3 expression in rice endosperm enhanced seed longevity.

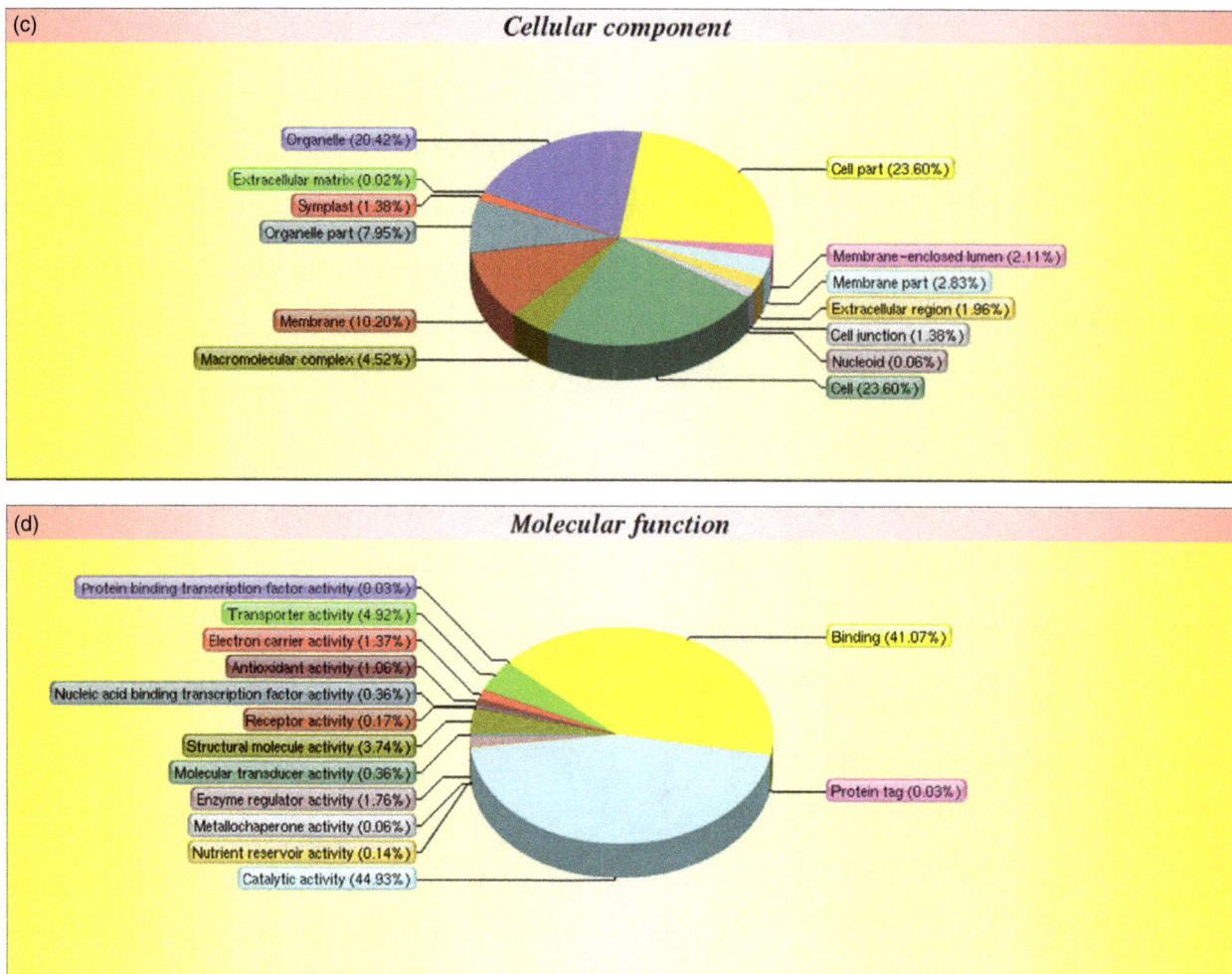

Figure 5 Continued

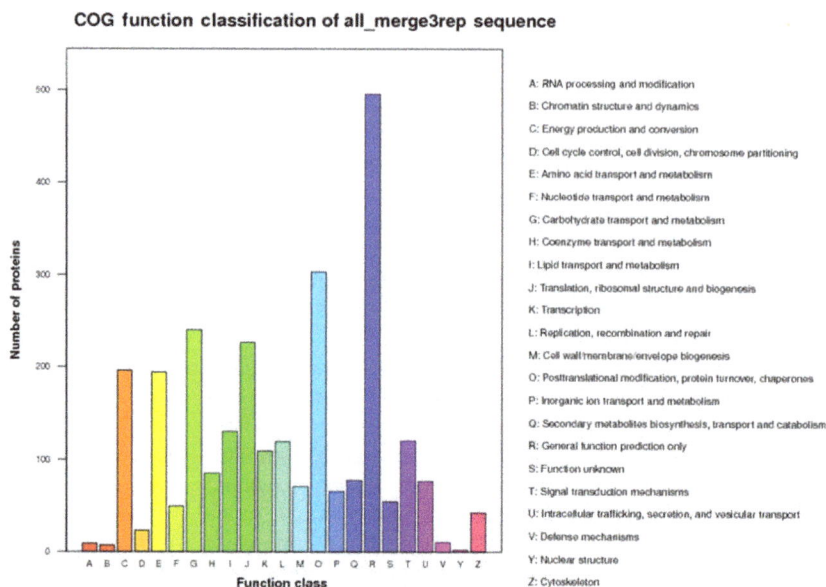

Figure 6 COG analysis of different expression proteins from transgenic rice seeds and WT after artificial ageing for 30 days. A–Z, different function classes.

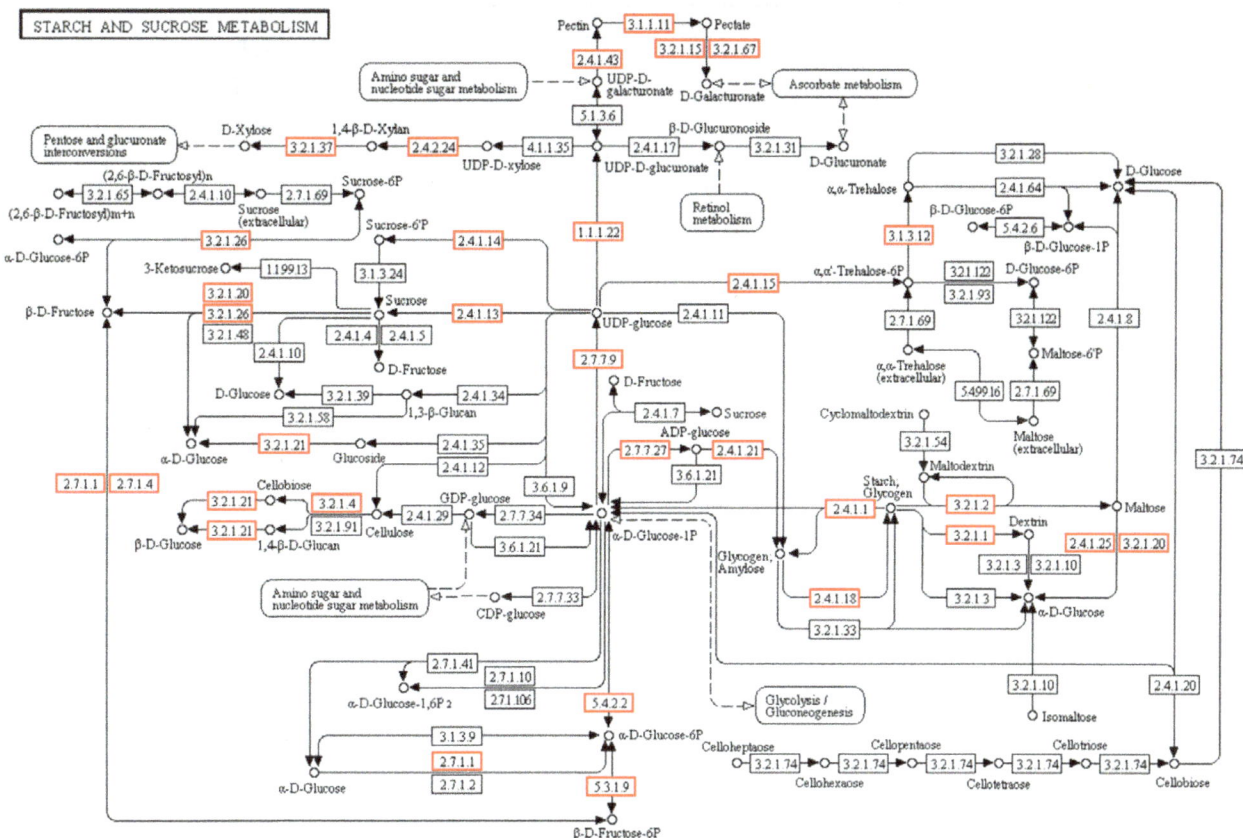

Figure 7 Starch and sucrose metabolic pathways affected by artificial ageing in transgenic rice seeds. Proteomic data were used to construct starch and sucrose metabolism. The numbers represent the protein IDs, and up-regulated proteins are indicated in red. Broken arrows indicate multiple steps between two compounds.

Construction of metabolic networks based on proteome-wide analysis of rice seeds

To combine our large-scale rice seed proteome analysis with quantitative information on proteins modulated during rice seed ageing, we used the iTRAQ method (Owiti et al., 2011). Protein extracts from rice seeds were analysed using the two complementary approaches: one-dimensional electrophoresis (1-DE) and LC-MS/MS. For 1-DE, approximately 100 µg of protein was loaded (Figure S3a), and then, all visualized slices were subjected to trypsin digestion. The various proteins were identified by LC-MS/MS analysis. Proteins that exhibited a fold change >1.2 with a P-value <0.05 were considered to be different proteins (Yang et al., 2013). In this study, 1200 different proteins were identified, of which 612 (51%) were up-regulated and 588 (49%) were down-regulated (Figure S3b). This represents the most comprehensive protein profile completed for rice seed ageing to date.

Based on cellular component ontology, the proteins were classified into the following categories: cell (23.60%), cell part (23.60%), organelle (20.42%), membrane (10.20%), organelle part (7.95%), macromolecular complex (4.52%), membrane part (2.83%), membrane-enclosed lumen (2.11%), extracellular region (1.96%), cell junction (1.38%), symplast (1.38%), nucleoid (0.06%) and extracellular matrix (0.02%) (Figure 5a, c). Based on molecular function ontology, the proteins were classified into functional categories including catalytic activity (44.93%), binding (41.07%), transporter activity (4.92%),

structural molecule activity (3.74%), enzyme regulator activity (1.76%) and electron carrier activity (1.37%) (Figure 5a,d). Based on biological process ontology, the proteins were classified into functional categories including metabolic processes (18.12%), cellular processes (17.51%), response to stimulus (9.97%), single-organism processes (5.86%), biological regulation (5.24%), localization (4.99%), cellular component organization or biogenesis (4.92%), developmental processes (4.80%), regulation of biological processes (4.79%), establishment of localization (4.71%), multicellular organismal processes (4.60%), reproduction (2.70%), reproductive processes (2.69%) and multi-organism processes (2.41%) (Figure 5a,b).

Gene ontology (GO) annotation showed that the proteins were located primarily in cells and organelles, played roles in metabolic process and had catalytic activity (Figure 5a). In addition, we classified the differentially expressed proteins on the basis of the ontology concept using the database of Clusters of Orthologous Groups (COGs) of proteins. The differentially expressed proteins were classified mainly into the following functional categories: energy production and conversion, amino acid transport and metabolism, carbohydrate transport and metabolism, lipid transport and metabolism, translation, ribosomal structure and biogenesis, post-translational modification, protein turnover, chaperones and general function prediction (Figure 6). These results reflected the gradual increase in metabolic activity that occurs in rice seeds during storage.

The most abundant proteins were those belonging to the class of proteins involved in metabolic pathways (33.24%, Tables S2

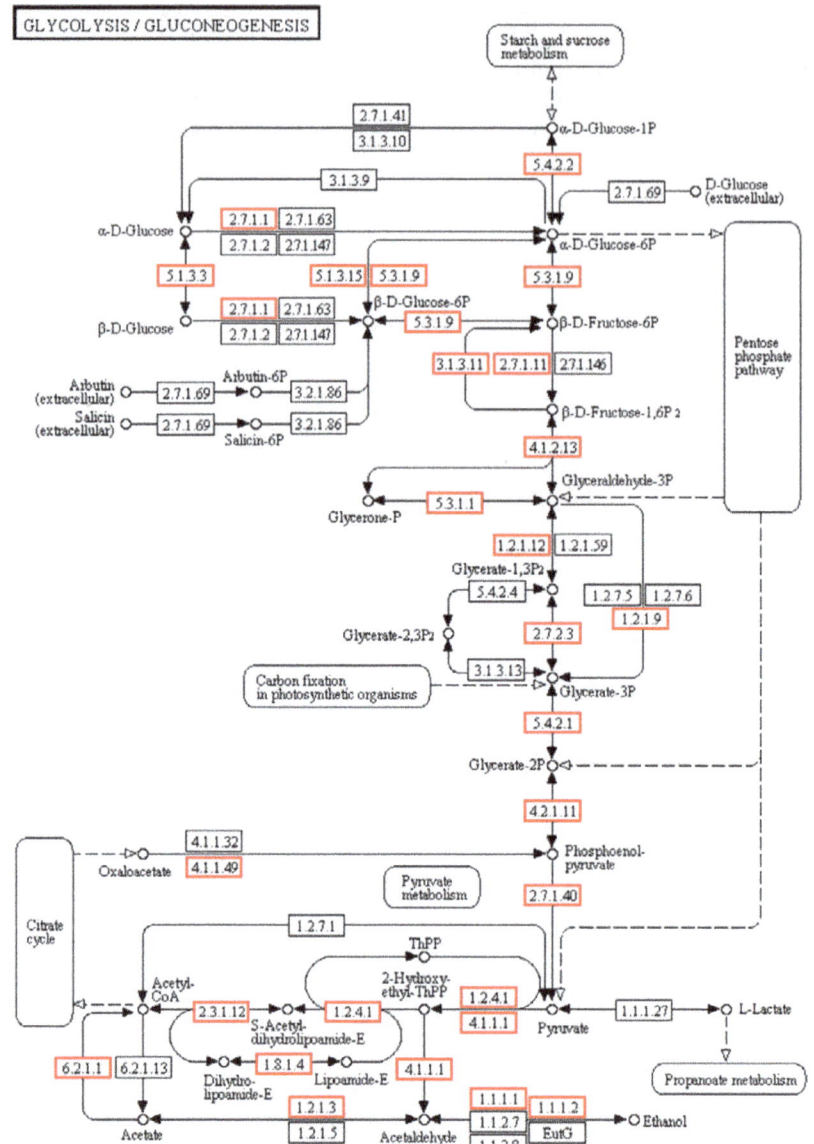

Figure 8 Glycolysis pathway affected by artificial ageing in transgenic rice seeds. Proteomic data were used to construct glycolysis metabolism. The numbers represent the protein IDs, and up-regulated proteins are indicated in red. Broken arrows indicate multiple steps between two compounds.

and S3), and the second most abundant category comprised proteins involved in the biosynthesis of secondary metabolites (19.91%). Some of the proteins detected were involved in glycolysis, starch and sucrose metabolism, oxidative phosphorylation and fatty acid metabolism. Starch constitutes the major source of reserves in mature rice seeds. All of the enzymes involved in starch degradation were identified in the proteomic analysis, 31 of which were up-regulated during seed storage (Figure 7 and Tables S2 and S3), including pectinesterase, α-1-4-galacturonosyl transferase and polygalacturonase. Glycolysis takes place after the degradation of starch or the phosphorylation of glucose. In total, 23 enzymes that catalyze all of the steps in the glycolysis pathway were identified as up-regulated in this experiment (Figure 8 and Tables S2 and S3), including phospho-glucomutase, hexokinase, 6-phosphofructokinase and pyruvate kinase. The final product of glycolysis is pyruvate, which can be transferred to mitochondria and used as substrate for the tricarboxylic acid (TCA) cycle. Lipase and lipolytic enzymes first catalyze the hydrolysis of triacylglycerol into glycerol and fatty acid, and then, fatty acids are degraded through two major

pathways: β-oxidation and the glyoxylate cycle. LNA and LA, the most important polyunsaturated fatty acids, are substrates for lipoxygenases. In the pathway of α-linolenic acid metabolism, 13 enzymes were up-regulated (Figure 9). When suppress LOX3 expression in rice endosperm with antisense RNA interference, the metabolism of LNA and LA were inhibited and some enzymes downstream were up-regulated (Figure 3b,c and Figure 9). Most of the enzymes required for oxidative phosphorylation and fatty acid metabolism were detected in this study (Figures S4 and S5 and Table S3). Of these, 10 enzymes involved in fatty acid metabolism and seven enzymes involved in oxidative phosphorylation were up-regulated. The enzymes that all took part will help to determine how the stocking reserves are mobilized during rice seed storing. From our results, the down-regulation of *LOX3* expression to reduce the rate of lipid oxidation can enhance rice seed longevity. Based on the results of analysis using the Kyoto Encyclopedia of Genes and Genomes (KEGG), we selected key enzymes involved in glycolysis, starch and sucrose metabolism, oxidative phosphorylation and fatty acid metabolism for future studies of their roles in rice seed longevity.

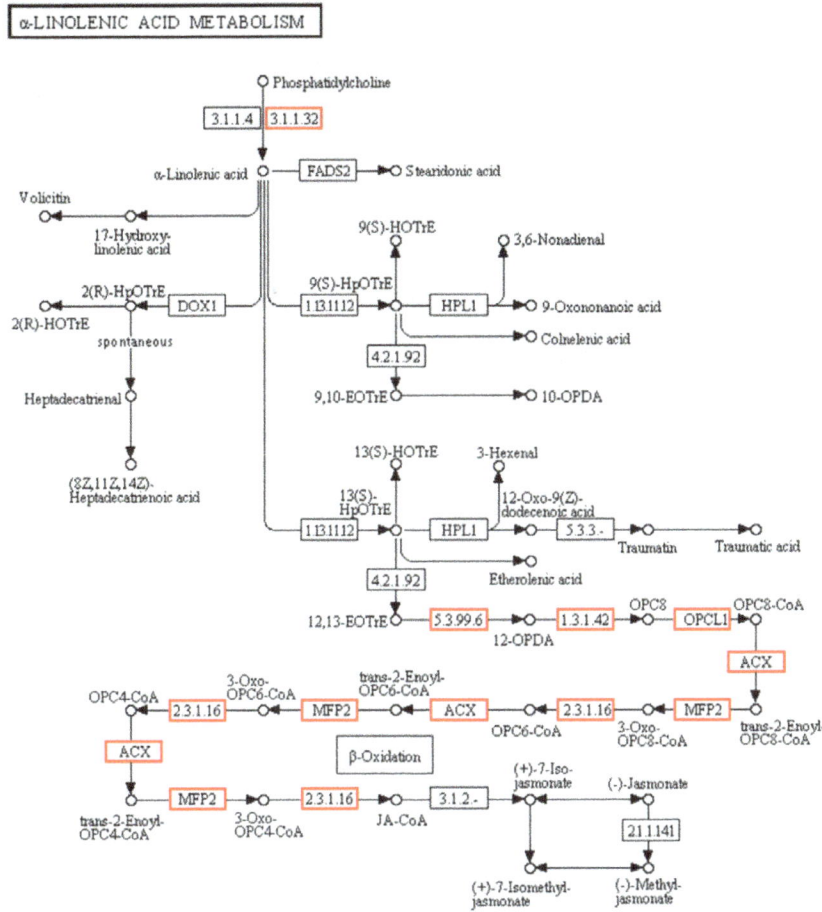

Figure 9 α-Linolenic acid metabolic affected by artificial ageing in transgenic rice seeds. Proteomic data were used to construct α-linolenic acid metabolism. The numbers represent the protein IDs, and up-regulated proteins are indicated in red. Broken arrows indicate multiple steps between two compounds.

Down-regulation of *LOX3* affects grain starch granule structure

Starch and storage proteins accumulate to high levels in rice endosperm and represent the primary form of stored resources. These resources are consumed gradually during rice seed storage. To investigate whether the introduction of the antisense *LOX3* gene into rice affected starch granule morphology, we performed scanning electron microscopy (SEM) analysis on transverse sections of endosperm from the transgenic and

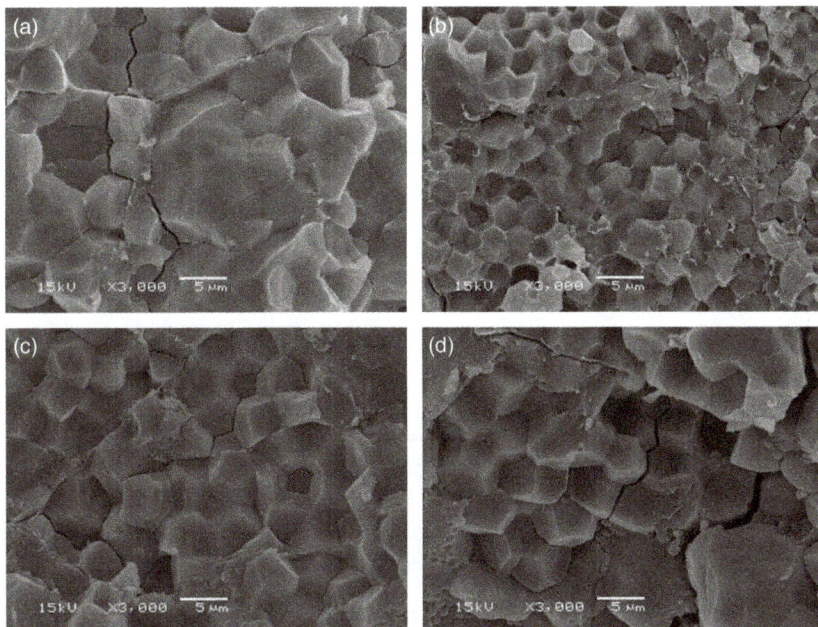

Figure 10 Scanning electron microscope images of starch granules in antisense *LOX3* transgenic plants and wild-type (WT) plants. (a) WT seeds without ageing. (b) WT seeds aged for 30 days. (c) Antisense *LOX3* transgenic seeds without ageing. (d) Antisense *LOX3* transgenic seeds aged for 30 days. Bar = 5 μm.

Table 1 Agronomic traits of antisense *LOX3* transgenic lines and wild-type plants in the field

Line	Plant height (cm)	Panicles per plant	Seed set rate (%)	1000-grain weight (g)
TS-58	134.6 ± 2.8*	8.6 ± 2.1	74.2 ± 0.1	27.1 ± 1.2
TS-61	130.0 ± 4.0	8.4 ± 2.1	53.4 ± 0.1**	26.1 ± 1.2
TS-63	129.6 ± 3.9	7.0 ± 0.0 *	78.7 ± 0.1	27.0 ± 2.0
TS-91	132.8 ± 2.3	10.3 ± 1.0	49.6 ± 0.1**	26.9 ± 1.6
TS-105	128.2 ± 4.7	8.2 ± 1.9	58.3 ± 0.1**	26.0 ± 1.2
Hang1 (WT)	130.2 ± 4.4	9.4 ± 1.8	81.2 ± 0.1	27.5 ± 2.5

All values are shown as the means ± standard error of the mean of three replicates. Values in a row with like symbols are significantly different. *$P < 0.05$; **$P < 0.01$.

WT seeds before and after artificial ageing. The sectioned endosperm appeared to be tightly packed before artificial ageing in both the antisense *LOX3* transgenic and WT seeds (Figure 10a,c). The tight structure was reflected in the structure of isolated starch granules, which were polygonal and of similar size with sharp edges. The granules were much more regular in the transformants (Figure 10a,c). After artificial ageing for 30 days, the starch granules isolated from the WT endosperm had rounded edges that differed in size from those of the transgenic endosperm, and the endosperm structure appeared to be loosely packed and fragile (Figure 10b). In contrast, the starch granule structure of the antisense *LOX3* transformant TS-91 exhibited little change with only a few starch granules becoming rounded (Figure 10d). Therefore, the morphological changes in the starch granules result from the down-regulation of the *LOX3* gene in the antisense transformants.

Agronomic traits and yield performance of antisense *LOX3* transformants in the field

In plants, the 9- and 13-hydroperoxide products of the LOX pathway can be metabolized further to compounds such as jasmonic acid (JA), methyl jasmonate, conjugated dienoic acids, traumatin and volatile aldehydes (Creelman and Mullet, 1997; Nemchenko et al., 2006). These compounds have physiological functions in growth and development and are involved in the defence responses to wounding and insect and pathogen attack (Siedow, 1991). To determine the phenotypic effects of expressing the antisense *LOX3* gene in rice endosperm, we conducted a detailed examination of the phenotypes of the transformants. The agronomic traits and yield performance of the transgenic lines were observed in the same plots. The transformants exhibited no significant differences from the WT at the vegetative stage (Figure 1a). Plant height, panicle number and 1000-grain weight in most of the transformants were very similar to those of the WT, except for the rate of seed set (Table 1). Although the grain filling ratio in most transgenic lines was less than in the WT (Table 1), testcross groups of homozygous lines showed no significant differences (Table S4). The timing of flowering, panicle number per plant and spikelet number per panicle were not significantly different between the transformants and the WT. These observations indicated that the introduction of the antisense transgene did not affect yield components and that this transgenic approach has great potential for application to agriculture.

Discussion

Seed deterioration during storage is a major problem in agricultural production. Seed longevity has been a major emphasis of research worldwide, but the mechanisms that regulate seed longevity remain unclear, and very few studies have addressed them specifically. The influences of environmental factors such as temperature and moisture content have been the main focus of previous studies (Ellis and Roberts, 1980). The possible involvement of oxidative processes in seed ageing was first considered in 1995 (Smith and Berjak, 1995). Since then, researchers have begun to study the molecular mechanism(s) of seed ageing. In this study, we focused on LOX3, which plays an important role in seed longevity and viability. We suppressed *LOX3* expression in rice endosperm using an antisense technique to determine the relationship between LOX3 and rice seed longevity. Our results indicated that decreased *LOX3* expression extended seed longevity significantly (Figure 4) and that LOX3 might be a key enzyme controlling seed ageing. These results are consistent with the results of previous reports (Suzuki et al., 1999). LOXs also play roles in plant development and defence responses to biotic stress (Marla and Singh, 2012). We showed that the introduction of an antisense *LOX3* gene expressed in rice endosperm did not affect agronomic traits significantly (Table 1).

While artificial ageing is widely used in seed longevity studies because it is less time-consuming and easier to control than natural ageing, it subjects seeds to an environment that is very different from natural conditions (Salama and Pearce, 1993). How well artificial ageing reflects natural ageing is a key problem in the study of rice seed longevity (Priestley and Leopold, 1983). Our comparison of antisense *LOX3* transgenic and WT seeds under artificial and natural ageing conditions (Figure 4a,b) showed that seed longevity during artificial ageing correlated well with that of during natural ageing, indicating that artificial ageing might be useful for estimating seed longevity under natural ageing conditions. However, we observed some differences between artificial and natural ageing. During the first 12 days of artificial ageing, the germination percentage of both the WT and the TS-91 and TS-105 transgenic lines showed little change, even exhibiting a rising trend in the first 3 days, suggesting that high temperature and high humidity can break rice seed dormancy and the deterioration of rice seeds with the superposition effect. Once the end products of LOX3 enzymatic activity accumulated to a certain level, their toxic effects became evident and the germination rate decreased rapidly (Figure 4a). Because the LOX3 activity was much lower in the transgenic lines, the germination rate was much higher than that of the WT (Figure 4a). After storage for 12 months under artificial ageing conditions, the germination rate of the TS-91 and TS-105 transgenic lines increased up to 90% for TS-91, and when storage was extended to 18 months, the germination rate decreased again (Figure 4b). With increased artificial ageing time, the germination percentage gradually decreased (Figure 4a). These results suggest that the mechanisms of artificial and natural ageing might differ and that further research is needed to clarify the mechanism(s) for better application in research on rice seed longevity.

The genetic basis of seed longevity remains unclear. However, with the availability of genome sequence information, investigators can now study living systems on an unprecedented scale. Network discovery is defined as the identification of natural relationships between molecules and associated properties that exist in a biological system (Baginsky et al., 2010). In a previous study, researchers showed that LOX3 was involved in the decomposition of polyunsaturated fatty acids and the production of volatile compounds in stored rice (Suzuki et al., 1999). To identify the metabolic pathways in which LOX3 was involved, we constructed metabolic networks using seeds from the TS-91 transgenic line collected before and after artificial ageing. Our goal was to provide a complete view of the biochemical reactions that occur in stored rice seeds and to enhance rice seed longevity. Most relevant studies in plants have focused on specific metabolic pathways such as amino acid and fatty acid metabolism. Systematic analysis based on protein profiling is essential to gain an accurate understanding of the metabolic and regulatory networks in rice seed ageing. In this study, we constructed metabolic and regulatory pathways active in stored rice seeds using a protein profiling strategy. Our results suggest that during rice seed storage, reserves including starch, storage proteins and lipids are mobilized. When LOX3 expression was suppressed in transgenic rice, some enzymes that take part in the main pathways, such as long-chain acyl-CoA synthetase and acyl-CoA oxidase (Table S3), were affected, resulting in enhanced seed longevity. Thus, LOX3 may act with other enzymes during lipid metabolism in rice. The set of metabolic pathways related to rice storage constructed as part of this study is currently the most comprehensive. The regulatory networks help to explain how cells regulate gene expression and how LOX3 functions in combination with other enzymes.

Criticism of transgenic crops based on possible hazards to the environment and to food safety limits their use primarily to research and greatly restricts their practical application for food agriculture (Yoder and Goldsbrough, 1994). Almost all transgenic plants contain at least two independent gene cassettes—a gene of interest and a selectable marker gene—both of which may have impacts on the environment and food safety. Several strategies have been developed to generate marker-free transgenic plants including co-transformation of two T-DNA molecules (McKnight et al., 1987), transposition-mediated removal of marker genes (Goldsbrough et al., 1993) and site-specific recombination (Corneille et al., 2001; Dale and Ow, 1990). Inducible site-specific recombination has been developed into a highly efficient system during recent years. The Cre/loxP site-specific recombination system consists of a gene cassette flanked by two loxP sequences in the same orientation that cause excision of the cassette when the Cre recombination enzyme is expressed (Dale and Ow, 1990). In the inducible site-specific recombination system, the Cre gene is under the control of an inducible promoter, which allows excision of the selectable marker after selection of transgenic plants. In this study, we used a strategy that employed a heat-inducible Cre/loxP recombination system to generate marker-free transgenic plants containing the LOX3 gene and its associated glutelin (GluB-1) promoter, both of which originate from rice (Figure S1). This approach may minimize potential impacts of transgenic crops, contribute to their safety and allow the production of transgenic crops for commercial use.

Heat-shock treatment of the transgenic plants at various vegetative growth stages showed that the efficiency of bar marker gene excision was highest (83.3%) at the four-leaf stage (Figure S2b) as a result of different levels of thermal tolerance at different stages of plant development. However, because marker gene removal in this heat-shock-dependent Cre-lox site-specific recombination system is temperature dependent, removal of the marker gene cannot be ensured in farm-grown plants. Other

methods for triggering marker gene excision under agricultural conditions await further study.

Experimental procedures

Transformation vector construction

A plasmid (pHG₁PL) containing an antisense LOX3 gene consisting of the rice glutelin (GluB-1) promoter, a rice LOX3 cDNA sequence and a nopaline synthase (NOS) 3′ terminator sequence was constructed (Figure S1a). A 1479-bp fragment of the rice GluB-1 promoter excised from the GluB-1 gene using HindIII and BamHI restriction enzymes was cloned into the HindIII and BamHI sites of the pCAMBIA1301 plasmid (supported by Fujian Normal University) into which the NOS 3′ terminator had been inserted previously between SacI and EcoRI sites. A 990-bp rice LOX3 small subunit cDNA fragment containing 304-bp 5′ and 341-bp 3′ noncoding sequences was amplified by PCR using appropriate primers, and the PCR product was cloned into the pMDᴛᴍ18-T vector (TaKaRa, Dalian, China). After verification of the nucleotide sequence and orientation, the LOX3 cDNA fragment was excised from the BamHI and SacI sites and inserted in the antisense orientation with respect to the GluB1 promoter to create the plasmid pHG₁PL. The bacterial phosphinothricin acetyltransferase bialaphos resistance gene (bar) contained in the pHG₁PL plasmid conferred resistance to the phosphinothricin-based herbicide bialaphos and was used as the selectable marker for rice transformation. Expression of the bar gene was controlled by the ubiquitin (Ubi) promoter, and bar was fused at its 3′ end to the NOS terminator.

Generation of transgenic rice

The pHG₁PL plasmid was introduced into mature rice (O. sativa L. cv. Hang1) embryos using Agrobacterium-mediated transformation. The transformed cells were selected initially on NB medium (Sivamani et al., 1996) containing 20 mg/L bialaphos (PhytoTech, Overland Park, KS). After 6–8 weeks, surviving cell clusters were regenerated. The plants that were generated were selected further on NB medium containing 20 mg/L bialaphos with a 16-h photoperiod at 26 °C for 3–5 weeks and then transferred to a greenhouse and grown with WT plants under natural sunlight.

The presence of the introduced gene in the regenerated plants was verified by PCR using the primers 5-GCCGCAGGAACCGCAGGAGTGGAC-3 and 5-TCAGATCTCGGTGACGGGCAGGACC-3. Genomic DNA was prepared from the bialaphos-resistant plants by the CTAB method (Cuellar et al., 2006) and used as a template for PCR. Total DNA was digested with the SalI restriction enzyme. After fractionation by 0.8% (w/v) agarose gel electrophoresis and immobilization on a nylon membrane, the DNA fragments were hybridized with a DNA probe containing a partial sequence of the Cre gene.

RNA isolation and quantitative RT-PCR analysis

Total RNA was isolated from seeds of transgenic plants (TS-58, TS-61, TS-91, TS-105) and WT using TRIzol reagent (Simms et al., 1993) after heading for 28 days. First-strand cDNA was synthesized from 2 µg of total RNA using a high-capacity cDNA reverse transcription kit (ABI, Foster City, CA) following the manufacturer's instructions. Real-time quantitative PCR was performed using the ABI PRISM 7500 sequence detection system (ABI) and a SYBR Green real-time PCR mix (Roche, Shanghai, China). Relative quantification of gene expression was performed using actin gene expression as a reference. Data were analysed using ABI

PRISM 7500 SDS software (ABI). The primer sequences used for RT-PCR amplification were 5'-TCGCATTGGCTCGCACCCT-3' and 5'-TCGTCTTCTTCAGCCGCACGAT-3' for *LOX3*, 5'-CCAAGGCT TATGTTGCTGTTA-3' and 5'-CCGCCGTTGATGAGTGT-3' for *LOX1*, 5'-CAACAAAGACTGACCCAAAT-3' and 5'-GGAGAACA CCCTCAACAATAG-3' for *LOX2*, 5'-AGACTACATACAACTCCAT CAT-3' and 5'-CACCACTGAGAACGATGT-3' for *actin*.

Determination of LOX3 activity

LOX3 activity was measured in transgenic (TS-58, TS-61, TS-91 and TS-105), WT and DawDam (*LOX3* deletion variety) seeds using the KI–I_2 starch staining method (Wan *et al.*, 2000). Under acidic conditions, I^- is oxidized to I_2 as the product of the reaction between LOX3 and LA to produce 9-HPOD, and the I_2 then reacts with starch. The level of LOX3 activity was determined by comparing the colours of different reaction solutions. The completely colourless solution indicated no LOX3 activity, slight yellow colour from green to red indicated different LOX3 activity, while darker colour indicated more LOX3 activity. Twelve seeds of each plant type were placed in 1.5-mL Eppendorf tubes and ground into powder to facilitate enzyme release. A 0.5-mL aliquot of a 0.2 mol/L (pH 8.2) borate/boric acid buffer solution was added and vortexed until all embryos were in full contact with the solution. The embryos were then incubated at room temperature for 1 h, after which 0.5 mL of LA was added. After an additional 10 min, 100 μL of KI solution and 100 μL of 1% starch were added and the embryos were placed in darkness for 6–14 h at 20–35 °C. The final colours of the solutions were observed.

LOX3 enzyme activities were also assayed in crude protein extracts from antisense *LOX3* transgenic and WT rice seeds. Homogenates were prepared by grinding flash-frozen tissue in a mortar in 100 mM potassium phosphate at pH 7.2, 10 mM KCl, 1 mM EDTA, 1 mM EGTA, 1 mM $MgCl_2$, 0.2 mM dodecylmaltoside and 400 mM sucrose on ice. Crude homogenates were vortexed for 1 min, left on ice for 40 min, clarified by centrifugation at 650 × *g* at 4 °C for 3 min and used directly in enzyme assays. Protein concentration was determined from a bovine serum albumin standard curve (Bradford Protein Assay; Bio-Rad, Hercules, CA). Inhibitors (or solvent-only control) were pre-incubated with enzyme extracts in a 1 mL final volume for 10 min prior to substrate addition. LOX3 activity was determined by measuring the UV absorption at 234 nm of the conjugated double bonds of 9-HPOD produced by LOX3 from LA.

LA extraction and assay

Rice seeds (2.00 g) were ground into powder, screened through a 0.28-mm mesh and placed in a distillation flask to which 50 mL of ethanol, 10 mL of 50% KOH solution and 5 mL of 10% ascorbic acid solution were added, refluxed and saponified for 1.5 h to generate LA salt. The salt was extracted twice with petroleum ether and adjusted to pH 4 to convert LA salt to LA and extracted with petroleum ether three more times. The petroleum ether was collected, and anhydrous Na_2SO_4 was added. After several hours, the Na_2SO_4 was filtered out, evaporated and a CH_3CN:THF (4 : 1) mixture was added to bring the volume to 10 mL. The resulting materials were analysed on a Gilson liquid chromatograph with a ZORBAXODS reversed-phase C-18 column (6 mm × 250 mm; Agilent Technologies, Santa Clara, CA). LA was eluted using the following gradient profile: 62% methyl cyanide, 3% tetramethylene oxide and 35% HAC with a flow rate of 1.2 mL/min. A Gilson 116 UV detector was used to monitor absorbance of the total LA hydroperoxides at 203 nm. Each determination was repeated three times.

Heat-shock induction

Rice seeds were germinated at 28 °C under a 12-h light/12-h dark cycle. Heat induction experiments were carried out using six consecutive heat treatments. Seeds after germination and plants at four-leave stage were used; they were both incubated at 45 °C for 6 h, allowed to recover for 20 h at 28 °C and then incubated a second time at 45 °C for 6 h. This treatment was repeated six times. Plants were analysed 2 weeks after the sixth heat treatment. The excision frequencies were compared, and the protocol that yielded the highest excision frequency of marker gene excision was used for further experiments.

Measurement of seed longevity

Seed longevity was determined by the rate of germination after several months of storage under natural conditions (27.5–28.5 °C, 65–70% humidity). We used seeds from TS-91 and TS-105 transgenic plants and WT plants. The germination rate after dormancy release reached 90% for all lines. Germination assays after ageing for 6, 12 and 18 months were performed according to Wu *et al.* (2010) over a 2-week period. The samples were sown on two layers of white germination paper in three replicates of 100 seeds per treatment. The seeds were germinated in an incubator at 30 °C, 100% relative humidity and 8 h of light per day. The numbers of germinated seeds were counted. Seeds without ageing treatment were used as a control.

The artificial ageing treatment used in this study was based on the method of Zeng *et al.* (Zeng *et al.*, 2002) with some modification. To accelerate ageing, seeds of TS-91 and TS-105 transgenic lines and WT plants were treated at 42 °C and 88% relative humidity for 0–30 days in a closed desiccator (Binder, Tuttlingen, Germany) with a thermostatic moisture regulator. One hundred seeds from each sample were treated with three replications of each treatment. After artificial ageing, germination assays were performed as described above.

Protein extraction from rice seeds and iTRAQ analysis

Rice seeds from WT plants and the TS-91 transgenic line were placed in an incubator at 42 °C and 88% relative humidity, and aged artificially for 30 days. Three biological replicates were performed, each with 10 g of seeds per genotype per treatment.

Total protein was extracted from control and artificially aged WT and TS-91 transgenic seeds (Saravanan and Rose, 2004). The tissue was ground into a fine powder under liquid nitrogen, extracted with 500 μL of lysis buffer (7 M urea, 2 M thiourea, 4% CHAPS and 40 mM Tris–HCl at pH 8.5) with 1 mM PMSF and 2 mM EDTA followed by addition of 10 mM DTT 5 min later. Samples were treated ultrasonically for 15 min, centrifuged for 20 min at 25 000 × *g* and the supernatants collected. An additional 10 mM DTT was added to the supernatant and incubated at 56 °C for 1 h followed by the addition of 55 mM IAM and incubation in darkness for 45 min. Two volumes of acetone was added, and the reaction was incubated at –20 °C for 2 h. Purified protein was collected by centrifugation at 25 000 × *g* for 20 min and dissolved in 200 μL of 0.5 M tetraethylammonium bromide (TEAB).

iTRAQ analysis was performed at the Beijing Genomics Institute (BGI, Shenzhen, China). Protein samples were digested using sequencing-grade trypsin (Promega, Madison, WI) at a ratio of 1:20 (w : w) for 4 h at 37 °C. Additional trypsin at the same ratio was added to the samples, and the protein was digested for 8 h at 37 °C. Peptides from each sample were dissolved in 0.5 M TEAB and labelled with iTRAQ Reagents-8plex Chemistry (Applied Biosystems, Beverly, MA) at room temperature according to the manufacturer's instructions. Samples of the TS-91 transgenic line collected before and after artificial ageing were labelled with reagents 113 and 117, while WT control samples were labelled with reagents 114 and 118. All four samples were combined and acidified to pH 3 using phosphoric acid.

After labelling the tryptic digests from various samples, the peptides were diluted with 4 mL buffer A (25 mM sodium dihydrogen phosphate buffer in 25% acetonitrile, pH 2.7) and fractionated by strong cation exchange (SCX) chromatography using a Shimadzu LC-20AB HPLC system (https://www.ssi.shimadzu.com). Four SCX fractions were collected in the presence of buffer A for 10 min at a flow rate of 1 mL/min, and the peptides were eluted with 5–35% buffer B (25 mM sodium dihydrogen phosphate buffer, 1 M KCl in 25% acetonitrile, pH 2.7) for 11 min. Absorbance at 214 nm was used to monitor the eluent. Samples were desalted using Strata-X cartridges and freeze-dried prior to LC-matrix-assisted laser desorption/ionization (LC-MALDI) analysis.

Samples were dissolved in buffer C (5% acetonitrile and 0.1% folic acid) to a concentration of 0.5 µg/µL. Peptide separation was performed on an LC-20AD chromatography system (Shimadzu) equipped with a Probot MALDI spotting device (ThermoFisher Scientific, San Jose, CA). Samples (8 µL) were injected and loaded directly onto a reverse-phase column at a flow rate of 8 nL/min for 4 min. Peptides were eluted at a flow rate of 300 nL/min for 50 min using the following gradient: 0–40 min, 2–35% buffer D (95% acetonitrile, 0.1% folic acid); 40–45 min, 35–80% buffer D; 45–49 min, 80% solvent D; and 49–50 min, buffer C.

Matrix-assisted laser desorption/ionization plates were analysed using a Q-Exactive instrument (ThermoFisher Scientific). Spectral peaks meeting threshold criteria were included in the acquisition list for the MS/MS spectra. The following threshold criteria and settings were used: mass range of 350–2000 Da and peptide collision-induced dissociation performed at a collision energy of 1.6 kV. (http://www.absciex.com/Documents/Downloads/Literature/mass-spectrometry-4370075B.pdf).

Differentially expressed proteins were mapped to Gene Ontology Terms using a reference database downloaded from the Web site resource (GO-Annotation@EBI) (Ashburner et al., 2000). The database of COGs is an attempt to phylogenetically classify proteins encoded in 21 complete genomes of archaea, bacteria and eukaryotes (http://www.ncbi.nlm.nih.gov/COG). The database was designed to classify proteins on the basis of the orthology concept—any group of at least three proteins that are more similar to each other than they are to any other proteins from the same genomes are most likely to belong to an orthologous family (Tatusov et al., 2000). The COG system can be employed for functional annotation of genes from new genomes and for research on genome evolution. KEGG is an updating system that computerizes the current knowledge of biochemical pathways and other types of molecular interactions and can be used as reference for systematic interpretation of sequence data (Kanehisa et al., 2012). To update the biological and functional properties of differentially expressed proteins, we submitted them to the COG (http://www.ncbi.nlm.nih.gov/COG/) and KEGG databases (http://www.genome.jp/kegg/pathway.html).

SEM analysis

Scanning electron microscopy observation was performed as described previously (Imai et al., 2006). Rice seed samples were cut longitudinally using a knife and placed on the stage of an S-3500N scanning electron microscope (Hitachi, Tokyo, Japan) to collect images. All procedures were performed according to the manufacturer's protocol.

Yield characteristics of transgenic rice

Five transgenic lines and the untransformed WT cultivar Hang1 were grown with three replications. Normal agronomic and fertilization practices were followed for optimal plant growth. Plant height (cm), panicle number, seed set rate and 1000-grain weight (g) data were collected and analysed. Treatment means were separated based on the least significant difference ($P < 0.05$).

Acknowledgements

We thank Prof. Jiayang Li at Chinese Academy of Agricultural Sciences for reading the manuscript; we also thank Dr. Kent D. Chapman and other laboratory members at Center for Plant Lipid Research, Department of Biological Sciences, University of North Texas (UNT) for technical support. The research is supported by grants from The Hi-Tech Research and Development (863) Program of China (2014AA10A603, 2014AA10A604) and The National Major Projects of Cultivated Transgenic New Crop Varieties Foundation of China (2008ZX001-006, 2011ZX001-006, 2011ZX001-004).

References

Ashburner, M., Ball, C.A., Blake, J.A., Botstein, D., Butler, H., Cherry, J.M., Davis, A.P., Dolinski, K., Dwight, S.S., Eppig, J.T., Harris, M.A., Hill, D.P., Issel-Tarver, L., Kasarskis, A., Lewis, S., Matese, J.C., Richardson, J.E., Ringwald, M., Rubin, G.M. and Sherlock, G. (2000) Gene Ontology: tool for the unification of biology. Nat. Genet. 25, 25–29.

Axelrod, B., Cheesbrough, T.M. and Laakso, S. (1981) Lipoxygenase from soybeans: EC 1.13. 11.12 Linoleate: oxygen oxidoreductase. Methods Enzymol. 71, 441–451.

Baginsky, S., Hennig, L., Zimmermann, P. and Gruissem, W. (2010) Gene expression analysis, proteomics, and network discovery. Plant Physiol. 152, 402–410.

Corneille, S., Lutz, K., Svab, Z. and Maliga, P. (2001) Efficient elimination of selectable marker genes from the plastid genome by the CRE-lox site-specific recombination system. Plant J. 27, 171–178.

Creelman, R.A. and Mullet, J.E. (1997) Biosynthesis and action of jasmonates in plants. Annu. Rev. Plant Biol. 48, 355–381.

Cuellar, W., Gaudin, A., Solórzano, D., Casas, A., Nopo, L., Chudalayandi, P., Medrano, G., Kreuze, J. and Ghislain, M. (2006) Self-excision of the antibiotic resistance gene nptII using a heat inducible Cre-loxP system from transgenic potato. Plant Mol. Biol. 62, 71–82.

Dale, E.C. and Ow, D.W. (1990) Intra-and intramolecular site-specific recombination in plant cells mediated by bacteriophage p1 recombinase. Gene, 91, 79–85.

Dickson, M. (1980) Genetic aspects of seed quality. HortScience, 15, 771–774.

Ellis, R. and Roberts, E. (1980) Improved equations for the prediction of seed longevity. Ann. Bot. 45, 13–30.

Feussner, I. and Wasternack, C. (2002) The lipoxygenase pathway. Annu. Rev. Plant Biol. 53, 275–297.

Goldsbrough, A.P., Lastrella, C.N. and Yoder, J.I. (1993) Transposition mediated re-positioning and subsequent elimination of marker genes from transgenic tomato. Nat. Biotechnol. 11, 1286–1292.

Görg, A., Weiss, W. and Dunn, M.J. (2004) Current two-dimensional electrophoresis technology for proteomics. *Proteomics*, **4**, 3665–3685.

He, D., Han, C., Yao, J., Shen, S. and Yang, P. (2011) Constructing the metabolic and regulatory pathways in germinating rice seeds through proteomic approach. *Proteomics*, **11**, 2693–2713.

Ida, S., Masaki, Y. and Morita, Y. (1983) The isolation of multiple forms and product specificity of rice lipoxygenase. *Agric. Biol. Chem.* **47**, 637–641.

Imai, K.K., Ohashi, Y., Tsuge, T., Yoshizumi, T., Matsui, M., Oka, A. and Aoyama, T. (2006) The A-type cyclin CYCA2; 3 is a key regulator of ploidy levels in *Arabidopsis* endoreduplication. *Plant Cell*, **18**, 382–396.

Kanehisa, M., Goto, S., Sato, Y., Furumichi, M. and Tanabe, M. (2012) KEGG for integration and interpretation of large-scale molecular data sets. *Nucleic Acids Res.* **40**, D109–D114.

Link, A.J., Eng, J., Schieltz, D.M., Carmack, E., Mize, G.J., Morris, D.R., Garvik, B.M. and Yates, J.R. 3rd. (1999) Direct analysis of protein complexes using mass spectrometry. *Nat. Biotechnol.* **17**, 676–682.

Lovell, M., Ehmann, W., Mattson, M. and Markesbery, W. (1997) Elevated 4-hydroxynonenal in ventricular fluid in Alzheimer's disease. *Neurobiol. Aging*, **18**, 457–461.

Marla, S.S. and Singh, V. (2012) LOX genes in blast fungus (*Magnaporthe grisea*) resistance in rice. *Funct. Integr. Genomics*, **12**, 265–275.

McDonald, M. (1999) Seed deterioration: physiology, repair and assessment. *Seed Sci. Technol.* **27**, 177–237.

McKnight, T.D., Lillis, M.T. and Simpson, R.B. (1987) Segregation of genes transferred to one plant cell from two separate Agrobacterium strains. *Plant Mol. Biol.* **8**, 439–445.

Nemchenko, A., Kunze, S., Feussner, I. and Kolomiets, M. (2006) Duplicate maize 13-lipoxygenase genes are differentially regulated by circadian rhythm, cold stress, wounding, pathogen infection, and hormonal treatments. *J. Exp. Bot.* **57**, 3767–3779.

Nuñez, A., Foglia, T.A. and Piazza, G.J. (2001) Characterization of lipoxygenase oxidation products by high-performance liquid chromatography with electron impact-mass spectrometric detection. *Lipids*, **36**, 851–856.

Owiti, J., Grossmann, J., Gehrig, P., Dessimoz, C., Laloi, C., Hansen, M.B., Gruissem, W. and Vanderschuren, H. (2011) iTRAQ-based analysis of changes in the cassava root proteome reveals pathways associated with post-harvest physiological deterioration. *Plant J.* **67**, 145–156.

Parkhey, S., Naithani, S. and Keshavkant, S. (2012) ROS production and lipid catabolism in desiccating *Shorea robusta* seeds during aging. *Plant Physiol. Biochem.* **57**, 261–267.

Priestley, D.A. and Leopold, A.C. (1983) Lipid changes during natural aging of soybean seeds. *Physiol. Plant.* **59**, 467–470.

Roberts, E. (1961) The viability of rice seed in relation to temperature, moisture content, and gaseous environment. *Ann. Bot.* **25**, 381–390.

Salama, A.M. and Pearce, R.S. (1993) Ageing of cucumber and onion seeds: phospholipase D, lipoxygenase activity and changes in phospholipid content. *J. Exp. Bot.* **44**, 1253–1265.

Saravanan, R.S. and Rose, J.K. (2004) A critical evaluation of sample extraction techniques for enhanced proteomic analysis of recalcitrant plant tissues. *Proteomics*, **4**, 2522–2532.

Schulze, W.X. and Usadel, B. (2010) Quantitation in mass-spectrometry-based proteomics. *Annu. Rev. Plant Biol.* **61**, 491–516.

Shastry, B. and Raghavendra Rao, M. (1975) Studies on lipoxygenase from rice bran. *Cereal Chem.* **52**, 597–603.

Siedow, J.N. (1991) Plant lipoxygenase: structure and function. *Annu. Rev. Plant Biol.* **42**, 145–188.

Simms, D., Cizdziel, P.E. and Chomczynski, P. (1993) TRIzol: a new reagent for optimal single-step isolation of RNA. *Focus*, **15**, 532–535.

Sivamani, E., Shen, P., Opalka, N., Beachy, R.N. and Fauquet, C.M. (1996) Selection of large quantities of embryogenic calli from indica rice seeds for production of fertile transgenic plants using the biolistic method. *Plant Cell Rep.* **15**, 322–327.

Smith, M. T. and Berjak, P. (1995) Deteriorative changes associated with the loss of viability of stored desiccation-tolerant and desiccation-sensitive seeds. Pp. 701–746. In: Kigel J. and Galili G. (eds), *Seed development and germination*. Marcel Dekker Inc., New York.

Suzuki, Y., Ise, K., Li, C., Honda, I., Iwai, Y. and Matsukura, U. (1999) Volatile components in stored rice [*Oryza sativa* (L.)] of varieties with and without lipoxygenase-3 in seeds. *J. Agric. Food Chem.* **47**, 1119–1124.

Tatusov, R.L., Galperin, M.Y., Natale, D.A. and Koonin, E.V. (2000) The COG database: a tool for genome-scale analysis of protein functions and evolution. *Nucleic Acids Res.* **28**, 33–36.

Veselovsky, V. and Veselova, T. (2012) Lipid peroxidation, carbohydrate hydrolysis, and Amadori-Maillard reaction at early stages of dry seed aging. *Russ. J. Plant Physiol.* **59**, 811–817.

Wan, J.M., Jiang, L., Wang, C.M. and Zhu, S.S.. (2000) *Methods of screening storable rice varieties*. CN00132248.6.

White, H., Quackenbush, F. and Probst, A. (1961) Occurrence and inheritance of linolenic and linoleic acids in soybean seeds. *J. Am. Oil Chem. Soc.* **38**, 113–117.

Wiese, S., Reidegeld, K.A., Meyer, H.E. and Warscheid, B. (2007) Protein labeling by iTRAQ: a new tool for quantitative mass spectrometry in proteome research. *Proteomics*, **7**, 340–350.

Wu, F.X., Zhu, Y.S., Xie, H.G., Zhang, J.F. and Xie, H.A. (2010) Preliminary study on storability of Chinese micro-core collections of rice [J]. *J. Chinese Cereal Oil. Assoc.* **25**, 124–128.

Yamamoto, A., Fujii, Y., Yasumoto, K. and Mitsuda, H. (1980) Product specificity of rice germ lipoxygenase. *Lipids*, **15**, 1–5.

Yang, L.T., Qi, Y.P., Lu, Y.B., Guo, P., Sang, W., Feng, H., Zhang, H.X. and Chen, L.S. (2013) iTRAQ protein profile analysis of *Citrus sinensis* roots in response to long-term boron-deficiency. *J. Proteomics*, **93**, 179–206.

Yoder, J.I. and Goldsbrough, A.P. (1994) Transformation systems for generating marker-free transgenic plants. *Biotechnology*, **12**, 263–267.

Zeng, D.L., Qian, Q. and Yasukumi, K. (2002) Study on storability and morphological index in rice (*Oryza sativa L.*) under artificial ageing. *Acta Agronomica Sinica*, **28**, 551–554.

Zieske, L.R. (2006) A perspective on the use of iTRAQ™ reagent technology for protein complex and profiling studies. *J. Exp. Bot.* **57**, 1501–1508.

12

Development of a novel-type transgenic cotton plant for control of cotton bollworm

Zhen Yue[†], Xiaoguang Liu[†], Zijing Zhou[†], Guangming Hou, Jinping Hua* and Zhangwu Zhao*

College of Agriculture and Biotechnology, China Agricultural University, Beijing, China

*Correspondence

email zhaozw@cau.edu.cn
[†]These authors contributed equally to this work.

Keywords: transgenic plant, neuropeptide F, cotton budworm, *Helicoverpa armigera*.

Summary

The transgenic Bt cotton plant has been widely planted throughout the world for the control of cotton budworm *Helicoverpa armigera* (Hubner). However, a shift towards insect tolerance of Bt cotton is now apparent. In this study, the gene encoding neuropeptide F (NPF) was cloned from cotton budworm *H. armigera*, an important agricultural pest. The *npf* gene produces two splicing mRNA variants—*npf1* and *npf2* (with a 120-bp segment inserted into the *npf1* sequence). These are predicted to form the mature NPF1 and NPF2 peptides, and they were found to regulate feeding behaviour. Knock down of larval *npf* with dsNPF *in vitro* resulted in decreases of food consumption and body weight, and dsNPF also caused a decrease of glycogen and an increase of trehalose. Moreover, we produced transgenic tobacco plants transiently expressing dsNPF and transgenic cotton plants with stably expressed dsNPF. Results showed that *H. armigera* larvae fed on these transgenic plants or leaves had lower food consumption, body size and body weight compared to controls. These results indicate that NPF is important in the control of feeding of *H. armigera* and valuable for production of potential transgenic cotton.

Introduction

The cotton bollworm (*Helicoverpa armigera*) is an important agricultural pest and is responsible for great losses of cotton production. More than one and a half billion US dollars was lost in China in 1992 due to its outbreak. The transgenic *Bacillus thuringiensis* (Bt) cotton expressing the Cry1Ac toxin protein is being used for insect control. The report from the International Service for the Acquisition of Agricultural-biotechnology applications showed that the planting area of the transgenic Bt cotton continuously increased in the last 15 years and attained 160 million hm² in 2011, about half of the total cotton planting area in the world (James, 2012). A shift towards insect tolerance of Bt cotton is now apparent, indicating that potential resistance from the target pest—the cotton bollworm—has become a major threat for sustainable planting of Bt cotton (Kong-Ming, 2007). For example, a report from the Chinese Agriculture Department showed that the yield of Bt cotton decreased around 10% in China in 2009 due to the development of resistance by cotton bollworm. Recent monitoring records indicated that the increased tolerance of cotton bollworm was apparent year by year in the intensive planting area of Bt cotton. So, it is imperative to develop new types of transgenic cotton (Zhang et al., 2010).

Insect brain neuropeptides are important regulators of physiology and behaviour. Neuropeptide F (NPF), a family member of neuropeptide Y (NPY) of vertebrates (Huang et al., 2011; Nuss et al., 2010; Roller et al., 2008) because of their similar function and similar signalling path via G protein-coupled receptors (Garczynski et al., 2005), has pleiotropic functions. However, because of fast evolution, their peptide sequences and structures differ greatly among animal species.

The NPFs have not been widely reported in insects. NPF was first isolated from *Helicoverpa zea* (Huang et al., 1998), followed by *Drosophila melanogaster* (Brown et al., 1999), *Schistocerca grearia* (De Loof et al., 2001), *Locusta migratoria* (Clynen et al., 2006), *Aedes aegypti* (Stanek et al., 2002), *Anopheles gambiae* (Garczynski et al., 2005), *Bombyx mori* (Roller et al., 2008), *Reticulitermes flavipes* (Nuss et al., 2010) and *Helicoverpa assulta* (Liu et al., 2013). Extensive studies on NPF have focused on the fruit fly (*D. melanogaster*) as a model. In brief, NPF exerts diverse regulatory roles in feeding (Lingo et al., 2007; Wu et al., 2005), ethanol sensitivity (Wen et al., 2005), learning and memory (Krashes et al., 2009), aggression (Dierick and Greenspan, 2007), locomotor circadian rhythms and sleep (He et al., 2013a,b). However, there are no reports on NPF about the regulation of feeding behaviour in important agricultural and economical pests. Does NPF exist in *H. armigera*? Does it regulate feeding behaviour? Might it be used for controlling pests in the field through transgenic biotechnology? To answer these questions, we identified and cloned the NPF gene from the cotton bollworm (*H. armigera*), analysed its feeding function, and found a new way to efficiently control this pest using transgenic crops expressing dsNPF RNAi.

Results

Identification and cloning of *Harmnpfs* and formation of the mature peptides

The constructed non-normalized cDNA libraries from the brain tissues of *H. armigera* were isolated and collected with CHROMA SPIN-400 columns (Clontech, Laboratories, Inc., Mountain View, CA). The insert fragments of recombinant plasmids from bacterial colonies ranging from 400 to 1200 bp were separated on a 1% agarose gel (Biowest, Gene Company LTD, Chai Wan, Hong Kong), and the percentage of recombinants selected from 150 independent clones (133 positive clones) was 95%. Subsequently, the two *npf*s in *H. armigera* were identified, isolated and cloned (Figure 1a), and the sequences were deposited in the GenBank (*Harmnpf1* accession number HQ613404; *Harmnpf2* accession number HQ416718). *Harmnpf1* and *Harmnpf2* contain an ORF of 246 and 366 bp, respectively (Figure 1b), in which a 120-bp insertion segment between 135th and 136th nucleotides of *Harmnpf1* forms *Harmnpf2*. The predicted mature neuropep-

tides *HarmNPF1* and *HarmNPF2* are composed of 30 and 10 amino acids, respectively, formed by a series of proteolytic processes and post-translational modifications such as TrimKR and Amidation (Figure 1c,d) through sequential action of three enzymes (prohormone convertases for TrimKR; peptidyl-hydroxylating monooxygenase and peptidyl-hydroxyglycine a-amidating lyase for amidation; McVeigh *et al.*, 2005).

Temporal and spatial expression of *Harmnpf*

Spatial expressions of *Harmnpf1* and *Harmnpf2* mRNA in selected tissues were investigated by RT-PCR. Results showed that the

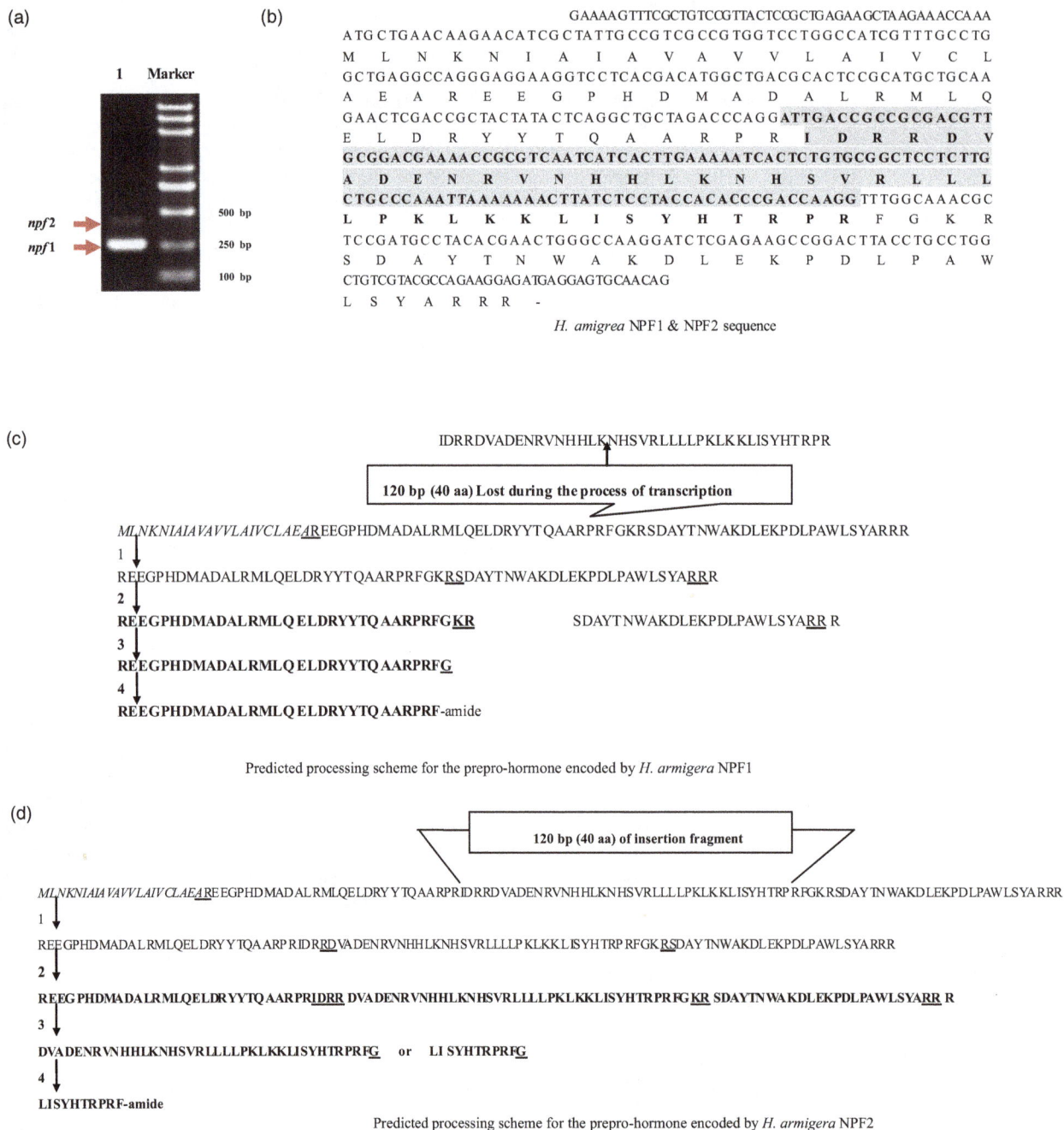

Figure 1 Structure of *Harmnpf1*/*Harmnpf2*. (a) The *Harmnpf1* and *Harmnpf2* mRNAs by PCR, in which Marker indicates standard molecular weight and 1 indicates PCR bands representing *Harmnpf1* and *Harmnpf2*. (b) Nucleotide sequences for the ORF encoding *Harmnpf1* and *Harmnpf2*. *Harmnpf2* includes a 120-bp sequence (red colour fragment) not found in *Harmnpf1*. (c, d) Separately indicates the predicted processing scheme for the preprohormone that is encoded by *Harmnpf1* and *Harmnpf2*.

transcript levels of *Harmnpf1* were detected in almost all tissues, and transcript levels of *Harmnpf2* were mainly expressed in midgut (MG), suboesophageal ganglia (SG) and thoracic ganglia (T) (Figure S1). Moreover, the expression levels of *Harmnpf* (*Harmnpf2* was selected for the qPCR throughout the manuscript) at different larval stages were also measured, and results showed that expression levels of *Harmnpf* during the first instar (just after emerging as larvae from egg shells) and last instar (the period for gluttony) were much higher than the other stages (Figure 2a). When larvae were starved for 24 h at beginning of the last instar, their feeding was stimulated and promoted with significant increases of the *Harmnpf* levels (Figure 2b,c). These results together suggest that *npf* is involved in modulating feeding behaviour.

Larval feeding inhibition by injection of dsNPF

The dsNPF was designed with consensus primers of *npf1* and *npf2*. Injection of dsNPF RNAi to the fifth instars significantly reduced larval *npf* mRNA levels (Figure 3a). Also, the relative consumption rate of food (Figure 3b; Table 1) and the relative growth rate (Figure 3c; Table 1) in the dsNPF-treated larvae were significantly lower than in control larvae ($P < 0.001$ and $P < 0.001$, respectively, compared to dsGFP RNAi controls). The amounts of food consumption in dsNPF RNAi-treated larvae from 24 to 72 h after injection were significantly lower than in the controls, demonstrating a 33% ($P < 0.001$), 26% ($P < 0.001$) and 33% ($P < 0.001$) decrease at 24, 48 and 72 h, respectively (Figure 3d; Table 2). And the larval net weight was also significantly reduced by 35% ($P < 0.001$), 22.89% ($P < 0.001$) and

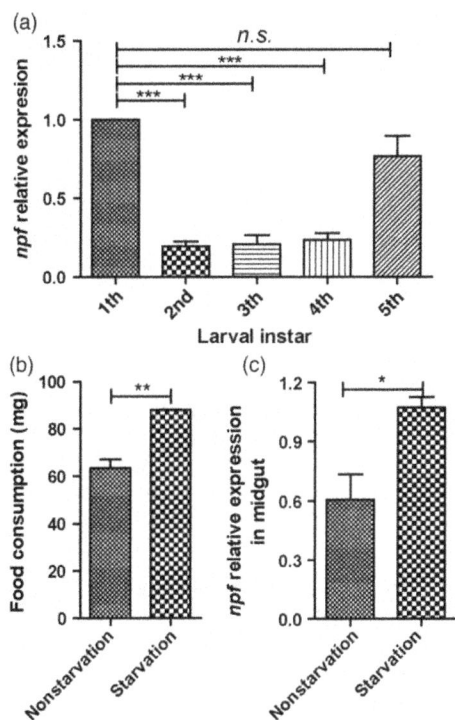

Figure 2 Relative expressions of *Harmnpf* at larval instar and effects of starvation treatment. (a) The relative expression level of *Harmnpf* at different larval instar stages ($n = 30$). n.s., no significant difference. (b) One-hour food consumption after 24-h starvation or no starvation. (c) The NPF relative expression levels after 24-h starvation treatment or no starvation ($n = 10$). All data are expressed as the mean ± SD of three replicates. Differences at "*"$P < 0.05$, "**"$P < 0.01$ and "***"$P < 0.001$.

21.45% ($P < 0.001$) at 24, 48 and 72 h, respectively (Figure 3e; Table 1). These results together indicate that NPF significantly regulates feeding.

Larval feeding inhibition on dsNPF transgenic tobaccos

The dsNPF was concatenated into the pGreen-HY104 vector (Figure S2a), identified by PCR and transfected into tobacco protoplasts by osmosis of tobacco leaf for transient expression of dsNPF. After 1 week, the protoplasts were separately identified by PCR for transfected dsNPF (Figure S2b), and the successful transgenic tobaccos, which were identified and determined by northern blot analysis (Figure S3a), were assayed for larval feeding after 2 weeks. When each larva was separately added to a transgenic tobacco leaf expressing dsNPF in a *Petri* dish, results showed that larval feeding was significantly reduced compared to dsGFP controls (Figure 4a). The areas of leaf eaten by larvae at 24 h ($P < 0.05$), 48 h ($P < 0.001$) and 72 h ($P < 0.01$) were significantly less than those of controls (Figure 4b; Table 2). When each larva was directly applied to a transgenic tobacco plant, the results were similar to those above; the larvae fed more on transgenic dsGFP controls than on transgenic tobacco plants (Figure 4c). These results together indicate that NPF in *H. armigera* regulates larval feeding behaviour.

Larval feeding inhibition on transgenic cotton plants expressing dsNPF

The dsNPF cloned in the DH5a plasmid was transformed into the cotton 2047B through hypodermic injection for stable expression of dsNPF. Their seeds were grown until the two-leaf stage used for the identification of the positive cottons by PCR. The identified NPF-positive and the dsGFP transgenic cotton as control were further grown. Larvae were added to the mature transgenic cotton leaves, further identified by northern and Southern blots (Figure S3a,b), and incubated in *Petri* dishes (one larva/one leaf/petri dish) for feeding assays. Results showed that the leaf area of the dsNPF transgenic cotton eaten by larva was much less (Figure 5a,e; Table 3) than that of the dsGFP transgenic cotton as a control ($P < 0.001$; Figure 5b,e; Table 3). The larval body size and weight after feeding transgenic leaves were also significantly smaller than for the controls ($P < 0.01$; Figure 5c,f,g; Table 3). Moreover, the *npf* level in the larvae after feeding dsNPF cotton leaves was significantly lower than that in control larvae ($P < 0.001$; Figure 5d). All these results indicate that NPF regulates feeding behaviour, and the dsNPF transgenic cotton is a potential and efficient biotechnology for field control of *H. armigera*.

Effects of dsNPF on regulating energy metabolism

To understand how *npf* regulates feeding, we applied dsNPF to fifth-instar larvae and measured their body trehalose, glycogen and total lipid. Results showed that dsNPF caused significant increases of trehalose and decreases of glycogen (Figure 6a,b). However, it had no effects on total lipid (Figure S4). These results suggest that *npf* regulates feeding behaviour to reduce the metabolism of glycogen to produce trehalose, potentially because feeding provides nutrients that reduce the need for metabolism of stored glycogen.

Discussion

In this study, we found that *Harmnpf1* and *Harmnpf2* were encoded by the same gene, which has been recognized as an

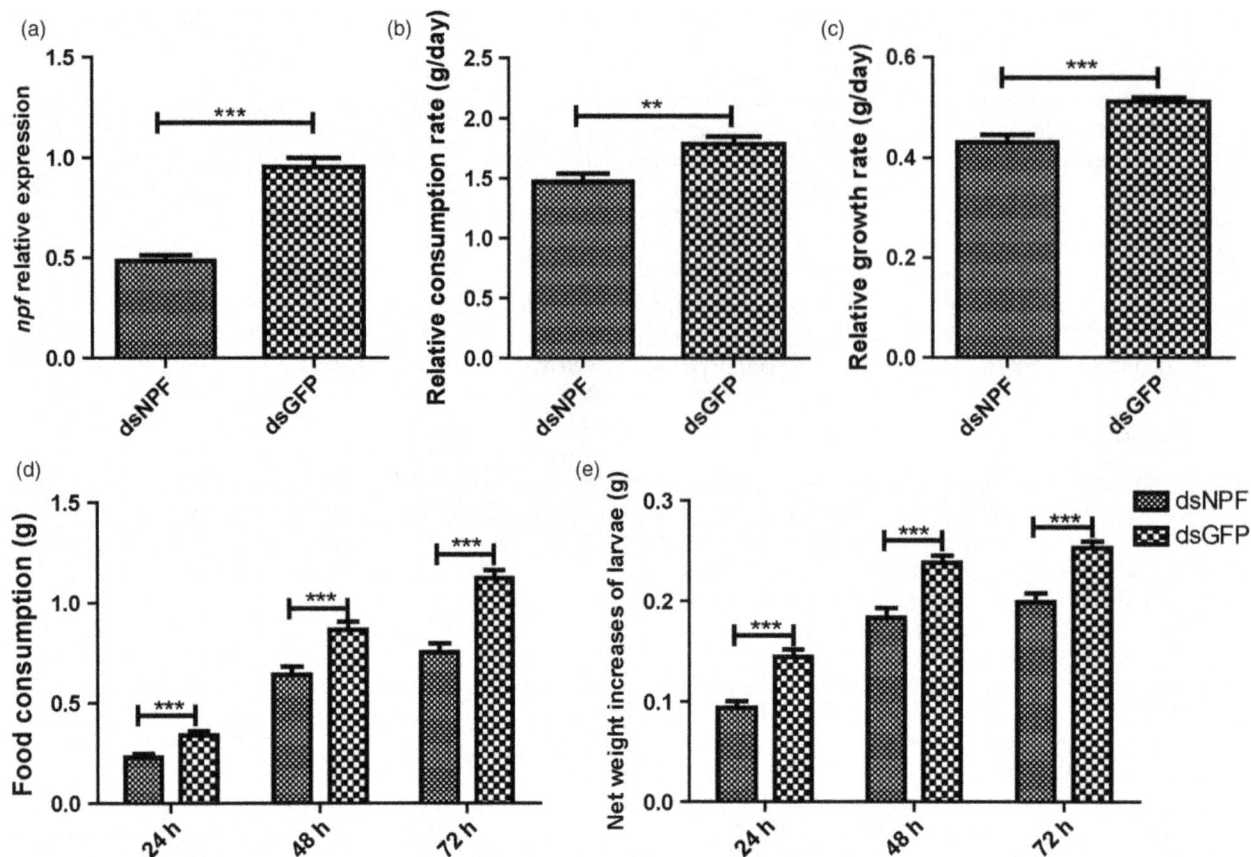

Figure 3 Impact of feeding on larvae treated with dsNPF. The dsNPF or dsGFP was injected to fifth-instar larvae at the first day. (a) The expressions of NPF transcripts after injection. Green Fluorescent Protein (GFP) is a negative a control. (b) Larval relative consumption rates. (c) Larval relative growth rates. (d) Larval food consumption. (e) Larval net weight increase. All data are expressed as the mean ± SD of three replicates. Differences at "*"$P < 0.05$, "**"$P < 0.01$ and "***"$P < 0.001$.

interesting aspect in gene regulation (Leff and Rosenfeld, 1986). The biological significance for alternative splicing is to produce a diversity of proteins (Black, 2003). The insect NPF was first purified in 1998 and completely sequenced in 2011 (Huang et al., 2011). The precursors of NPFs include a signal peptide, a propeptide and a C-terminal peptide (Maule et al., 1995)—a result consistent with our findings.

We found that the expression levels of H. armigera npf1 and npf2 are different, because npf2 levels were lower than those of npf1 (Figure 1a). This case is similar to adipokinetic hormone (AKH) in locust (Vroemen et al., 1997). npf is highly expressed in midgut, and is also expressed in brain, suboesophageal ganglion, foregut and hindgut, as has been previously reported for R. flavipes, H. zea and D. melanogaster (Huang et al., 2011; Nuss et al., 2010; Veenstra and Sellami, 2008). In addition, npf is highly expressed at the larval 1st instar (just after emerging as larvae from egg shells) and the 5th instar (the larval gluttony stage), which suggests that the first-instar larvae also are an important feeding period for their growth and development, in addition to the fifth-instar gluttony stage. From previous reports, NPF has important roles in resisting low temperature and noxious food in D. melanogaster (Lingo et al., 2007; Wu et al., 2005).

The feeding regulation of NPF has been assessed in D. melanogaster, a Dipteran insect. However, the possible roles of NPFs in regulating feeding in Lepidoptera have not yet been reported, even though many of them are important economic and

agricultural pests. In this study, we demonstrated that the npf of H. armigera is associated with increased feeding when npf is increased after food deprivation. On the contrary, knocking npf down by both applications of dsNPF (the transgenic tobaccos transiently expressing dsNPF and the transgenic cottons stably expressing dsNPF) had impacts on feeding inhibition, with reduced relative food consumption rate and growth rate, and smaller body size and weight. The damage to the transgenic cotton plants by cotton bollworms seems to be effectively reduced by 74%. Importantly, the NPF RNAi results in high mortality of this insect, in which all individuals did not normally pupate or emerge because of undeveloped bodies (Table 3). Therefore, npf is very critical for feeding regulation of H. armigera. This new type of biotechnological product could bring about a potential control of the cotton bollworm, H. armigera.

In D. melanogaster, a previous report showed that the NPF system is regulated by insulin, through the InR (insulin receptor)/ PI3K/S6K pathway (Wu et al., 2005). However, how NPF regulates feeding is still unclear. In this study, we found that changed larval feeding behaviour by the dsNPF may cause changed metabolism, with a decrease of glycogen and an increase of trehalose. In other words, the decrease in feeding which results from NPF knockdown causes changes in glycogen and trehalose levels. Presumably with less feeding, more stored glycogen has to be used to break down to trehalose for energy supply of H. armigera.

Table 1 Effect of feeding dsNFP to larvae

	Injected dsNPF	Injected dsGFP	P value	Decrease %
NPF relative expression level (%)	0.48 ± 0.03 n = 3	0.95 ± 0.07 n = 3	0.013	49.25
Relative consumption rate (g/day)	1.47 ± 0.07 n = 38	1.79 ± 0.06 n = 37	0.001	17.61
Relative growth rate (g/day)	0.43 ± 0.02 n = 38	0.51 ± 0.01 n = 36	<0.0001	15.77
Food consumption (g)				
24 h	0.23 ± 0.02 n = 39	0.34 ± 0.02 n = 35	<0.0001	32.58
48 h	0.64 ± 0.04 n = 39	0.87 ± 0.04 n = 35	<0.0001	25.96
72 h	0.75 ± 0.05 n = 39	1.12 ± 0.04 n = 35	<0.0001	32.94
Net weight increases of larvae (g)				
24 h	0.10 ± 0.01 n = 36	0.14 ± 0.01 n = 34	<0.0001	35.00
48 h	0.18 ± 0.01 n = 36	0.24 ± 0.01 n = 34	<0.0001	22.89
72 h	0.20 ± 0.01 n = 36	0.25 ± 0.01 n = 34	<0.0001	21.45

Table 2 Larval feeding inhibition on dsNPF transgenic tobaccos

Area of leaf eaten (cm^2)	dsGFP n = 9	dsNPF n = 9	P value	Decrease %
24 h	21.50 ± 1.92	16.36 ± 0.96	0.044	23.80
48 h	37.13 ± 2.56	18.41 ± 1.94	<0.0001	50.41
72 h	23.11 ± 1.87	16.36 ± 1.04	0.013	29.23

Experimental procedures

Insects rearing

The *H. armigera* colony used in this study was established from insects collected in Zhengzhou, Henan Province, and has been maintained in the laboratory since 2007. They were reared at 25 °C and 65% relative humidity (RH) under a photoperiod of 16L : 8D with an artificial diet described by Wu and Gong (1997).

Construction of cDNA library

Brains from fourth-instar larvae were dissected in PBS treated with diethyl pyrocarbonate (DEPC), then frozen immediately in liquid nitrogen and stored at −80 °C until use. Total RNA was extracted with TRIzon reagent (TIANGEN, Beijing, China). For the specific methodology, see Liu *et al.* (2013).

Full-length cDNA cloning of *npf*

The full-length *npf* cDNAs were obtained by the rapid amplification of cDNA ends (RACE) technique using the DNA from the cDNA libraries as PCR templates. For 5′ RACE, the degenerate reverse primers (5′ RACE1F) were designed according to the conserved region (NPF: QAARPRFGKR). Plasmid libraries (0.1 μL)

were used as templates with *LA* Taq® (Takara, Dalian, China) using forward primer 5′ SMART; reverse primer 5′ RACE1F. For the specific methodology, see Liu *et al.* (2013).

Identifying open reading frames (ORF) of *HarmNPF*

The full-length genes were amplified using the specific primers (FSP1 and NPF-R), which were designed according to the 5′ and 3′ RACE sequences (Table S1). Conventional PCR was applied using 1 μL of cDNA (first Strand cDNA synthesized above). Reaction mixtures contained 1.25 U (5 U/L) *LA* Taq®, 2.5 μL 10× *LA* buffer solution and 4 μL dNTP (2.5 mM each) in a final volume of 25 μL. For the touchdown PCR conditions, please see Liu *et al.* (2013). The longest band for each gene was eluted from the gel using the E.Z.N.A® Gel Extraction kit (OMEGA, Bio-Tek, Inc., Norcross, GA, USA), and then directly cloned into the pGM-T vector (TIANGEN, Beijing, China). The clone was sequenced twice in both directions using M13 forward and M13 reverse primers.

Bioinformatic analysis

The *H. armigera npf* signal peptides were predicted using the online program SignalP 3.0: http://www.cbs.dtu.dk/services/SignalP/ (Bendtsen *et al.*, 2004). Prohormone cleavage sites were predicted based on previously established protocol (Southey *et al.*, 2008) using the website: http://neuroproteomics.scs.illinois.edu/cgi-bin/neuropred.py (Southey *et al.*, 2006). For the specific methodology, see Liu *et al.* (2013).

Spatiotemporal expressions of *Harmnpf*

To monitor transcriptional levels of *Harmnpf* from different larval tissues, we explored the semi-quantitative reverse transcription–PCR (RT-PCR). Total RNA was extracted from the brains (Br), suboesophageal ganglia (SG), thoracic ganglia (T), abdominal ganglia (A), foregut (FG), midgut (MG) and anterior hindgut (HG) as described. Tissue distribution of *Harmnpf* was investigated by RT-PCR according to Chen *et al.* (2007) and Yang *et al.* (2010). For the specific methodology, see Liu *et al.* (2013).

To determine the expression of *Harmnpf* at different developmental stages of larvae, total RNA from larval whole bodies was prepared using TRIzol (TIANGEN), and cDNA was further synthesized using the RealMasterMix System for qRT-PCR. The qRT-PCR was used to quantify the levels of *npf* (*Harmnpf2* was selected for the qPCR throughout the manuscript) and actin as a control. The experiments were performed in triplicates. Data analysis was performed using ABI Stepone software (Applied Biosystem, Foster, CA).

Relationship between food deprivation and *npf*

To further determine that *npf* impacts larval feeding, we designed a food deprivation experiment. After 24 h of normal feeding, the fifth-instar larvae were reared with agarose (food deprivation) and normal food separately. The larvae midguts were collected for quantitative analysis of *npf* expression after food deprivation for 24 h. The qRT-PCR method and analysis were the same as above.

Synthesis of double-strand RNA (dsRNA)

The dsNPF was produced by *in vitro* transcription using the T7 RiboMAX™ expression system (Promega, Madison, Wisconsin, USA). To produce DNA templates for the synthesis of both dsNPF1 and dsNPF2, a T7 RNA polymerase promoter was added to the 5′-end of the DNA sequence using PCR with their specific

(a)

dsNPF   dsGFP

(b)

(c)

dsNPF   dsGFP

Figure 4 Larvae feeding inhibition by transgenic tobacco expressing dsNPF. (a) The larvae feeding on transgenic tobacco leaves and controls. (b) The leaf areas produced by larvae feeding on transgenic and control leaves separately at 24, 48 and 72 h ($n = 9$). (c) Damages of leaves by feeding whole tobacco plants at 72 h ($n = 9$). All data were expressed as the mean $\pm$ SD of three replicates. Differences at "*"$P < 0.05$, "**"$P < 0.01$ and "***"$P < 0.001$.

primers for dsNPF-F & dsNPF-R in Table S1. PCR products were purified by TIANgel Midi Purification Kit (Cat. #DP209-02). The PCR products were quantified and used as templates to prepare dsRNA using RiboMAX™ kit. To anneal, RNA reactions were incubated at 70 °C for 10 min, and then slowly cooled down to room temperature (~20 min). The annealed dsRNA was treated with A RQ1 RNase-Free DNase and precipitated by adding 0.1 volume of 3M Sodium Acetate (pH 5.2) and 1 volume of isopropanol. The dsRNA sample was resuspended in nuclease-free water and stored at −20 or −70 °C.

Effects of *NPF* by applications of dsNPF RNAi

Ten micrograms of dsNPF RNAi or dsGFP RNAi was respectively injected into the lateral intersegmental membrane between the third and fourth abdominal segment of the selected fifth-instar larvae, and the incision was sealed immediately with wax at the injection point. Each group was performed in triplicates with 15 individuals at each repeat ($n = 45$ larvae/group). Both controls and treatments were reared with a certain amount of artificial diets renewed every day. After 24/48/72 h, the treated and control larvae were observed separately, with measurements including larval weight, remainder of the artificial diet and faeces. As food is fresh, it is also necessary to set a blank experiment as a control, measuring the weight change of diet caused by the change of water content. Larval food consumption is calculated by the following formulae from Scriber and Slansky (1981).

$$I = w - \left(L + \frac{aW + bL}{2}\right)$$

I is the food ingested (food consumption).
W is the initial weight of the food in experimental group.
L is the final weight of the food in experimental group.

Table 3 Larval feeding inhibition on transgenic cotton plants expressing dsNPF

	dsGFP $n = 9$	dsNPF $n = 9$	P value	Decrease %
Area of leaf eaten (cm²)	18.280 ± 1.25	4.73 ± 0.99	<0.0001	74.13
Body length of larvae (cm)	2.27 ± 0.05	1.77 ± 0.05	<0.0001	21.87
Body weight of larvae (g)	0.34 ± 0.01	0.24 ± 0.02	0.003	29.76
Mortality rate after closion (%)	0	100	<0.0001	—

Relative consumption rate (RCR) $= \frac{I}{\bar{B} \times T}$.

Relative growth rate (RGR) $= \frac{B}{\bar{B} \times T}$.

$\bar{B}$ = mean weight during the time period.

B is the assimilated food used for growth (biomass gained).

T is the duration of feeding period (days).

$a = \dfrac{\text{the initial weight of the food in control group} - \text{the final weight of the food in control group}}{\text{the initial weight of the food in control group}}$.

$b = \dfrac{\text{the initial weight of the food in control group} - \text{the final weight of the food in control group}}{\text{the final weight of the food in control group}}$.

Construction of plasmids

Plasmids were constructed using standard cloning techniques. dsRNAi constructs were prepared by adding appropriate restriction sites to the ends of the primers used to perform PCR amplification with DNA polymerase (TIANGEN) and primers (p-NPF-F and p-NPF-R) in Table S1. The PCRs began with 94 °C denaturation for 3 min, then 35 cycles of denaturation at 94 °C for 30 s, 56 °C annealing for 30 s, and 72 °C extension for 1 min. The PCR products and pGreen-HY104 vector were digested separately with restriction enzymes HindIII and EcoRI. They then were further purified, ligated and transformed into DH5a. The newly constructed plasmid was named dsNPF-pGreen-HY104 plasmid. The control plasmid dsGFP-pGreen-HY104 was constructed with the same method above, using primers for dsGFP.

Preparation of transgenic tobacco

Tobacco seeds were planted in sterilized culture medium and were transplanted to the aseptic nutritive bowl after 7 days. Then, they were cultured at 25 °C with 16-h light and 20 °C with 8-h dark. Four to five-leaf-stage tobacco plants were chosen for infiltration with agrobacterium (*Agrobacterium tumefaciens*)

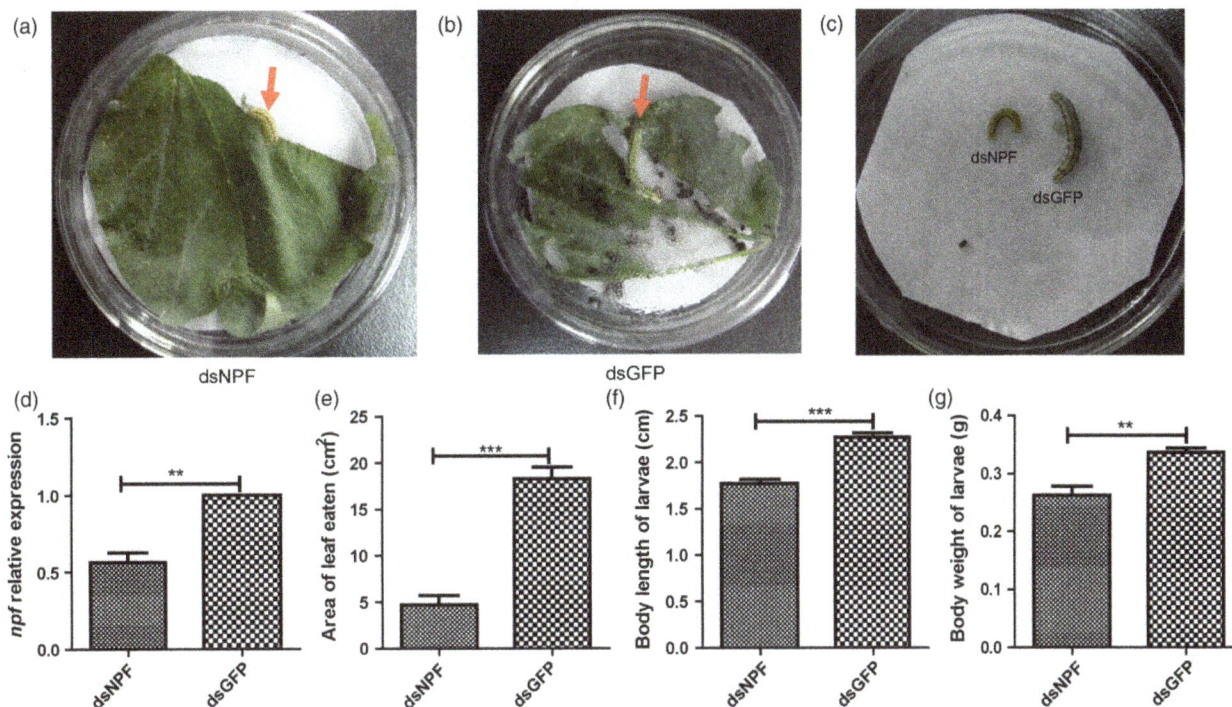

Figure 5 Larvae feeding inhibition by transgenic cotton expressing dsNPF. Ten larvae were released on top of mature leaves to feed on the transgenic plant for 3 days (n = 9). (a) Larvae feeding on the dsNPF transgenic cotton leaf. (b) Larvae feeding on the dsGFP (Green Fluorescent Protein) transgenic cotton leaf (a negative control). (c) Comparison of larvae growth with feeding on transgenic cottons. (d) the NPF expression levels of larvae fed on transgenic cottons. (e) A leaf area comparison of larval feeding on dsNPF and dsGFP RNAi transgenic cottons. (f) A comparison of larvae body length after feeding on the transgenic cottons. (g) A comparison of larvae body weight after feeding on the transgenic cottons. Differences at "*"P < 0.05, "**"P < 0.01 and "***"P < 0.001.

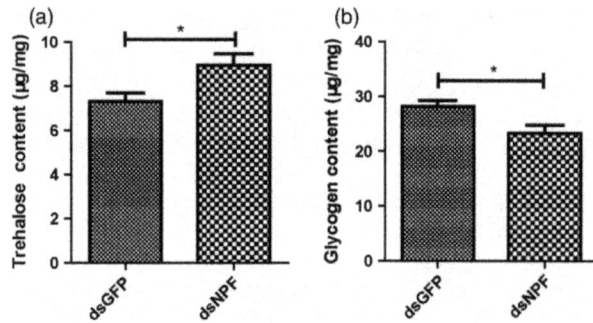

Figure 6 Larvae energy metabolism affected by applications of dsNPF. Glycogen and trehalose were tested at 72 h after injection of dsNPF RNAi ($n = 30$). (a) Trehalose content in dsNPF or dsGFP RNAi-treated larvae ($P < 0.05$). (b) Glycogen content in dsNPF RNAi-treated larvae compared with control counterparts ($P < 0.05$). Differences at "*" $P < 0.05$.

containing the constructed plasmids (dsNPF-pGreen-HY104 and dsGFP-pGreen-HY104). For the detailed methodology of transgenic tobacco, see Yang et al. (2000).

Cotton Preparation of transgenic cotton and larval feeding

The cotton strain 2047B was grown in the field until flowering. After self-pollination for 24 h, the dsNPF-pGreen-HY104 plasmid was injected into the ovary by hypodermic syringe. The injected bolls continuously grew to produce mature seeds. These seeds were grown until two leaves had appeared. Genomic DNA from one was analysed to identity the positive strains by PCR using the primers (NPF-F and NPF-R) in Table S1.

For larvae feeding on cotton plants in the field, 10 larvae were randomly released on the top of mature leaves to feed on each plant for 3 days. The treatments were performed in triplicates with three individual plants for each repeat ($n = 9$ plants/treatment). The cotton plants were mesh-enclosed to prevent insects climbing to other plants. The treatment was performed in triplicates with three individual plants for each repeat ($n = 9$ plants/treatment).

For larvae feeding on cotton leaves in the laboratory, the leaf area was scanned with a scanner (HP Deskjet 1050), and the leaf was then placed in a Petri dish with moist filter paper and a fifth-instar larva was allowed to feed on the leaf. After every 24 h, the eaten leaf area was calculated by ImageJ (leaf was changed every 24 h). The feeding leaf area was calculated as the fresh leaf area (S_0) minus the eaten leaf area (S_1). The treatment was performed in triplicates with three larval individuals for each repeat ($n = 9$ larvae/treatment).

Northern blot analysis

For the northern blot hybridization, total RNA was extracted from transgenic plants (cotton and tobacco) leaves with TRIzol® reagent. Forty micrograms of total RNA from each sample was heated at 65 °C for 15 min, cooled on ice and loaded on 1.3% w/v agarose gels, electrophoresed in denaturing buffer containing formaldehyde at 50 V for 2 h and visualized using UV. RNAs were blotted on to nylon membranes with 10 × SSC and cross-linked to membranes by UV cross-linking. A α^{32}P-UTP labelled full-length NPF riboprobe was generated by in vitro

transcription (Maxscript kit; Ambion, Austin, TX, USA). Probe was added and hybridized overnight at 65 °C. Membranes were washed in wash buffer (0.1 × SSC, 0.1% SDS) once, then washed 3 times with 100 mL of prewarmed wash buffer for 20 min, each in a hybridization oven at 68 °C, after which hybridization signals were detected by X-ray film (Kodak, Rochester, NY, USA).

Southern blot analysis

Total genomic DNA was isolated from 1-week transgenic cotton leaves. Fifty micrograms of the genomic DNA was digested with Hind III, and the DNA fragments were separated by electrophoresis in a 1.0% w/v agarose gel in 1 × TAE buffer for 12 h at 25 V. The gel was sequentially subjected to denaturation buffer (1.5 M NaCl and 0.5 M NaOH for 30 min) and neutralization buffer (1.5 M NaCl and 1 M Tris base for 30 min). The DNA was transferred to a nylon membrane using 10 × SSC buffer and UV cross-linked. npf coding regions labelled with α^{32}P-CTP were generated using a Rediprime Labelling Kite (Prime-a-Gene Labelling System; Promega). The probe was added to hybridization buffer (Rapid-hyb buffer) and incubated overnight at 65 °C. The hybridization buffer was eluted and the membrane rinsed with wash buffer (0.5 × SSC, 0.2% SDS) once. Then, the membrane was washed 3 times with 100 mL of prewarmed wash buffer for 20 min each in a hybridization oven at 68 °C. The radioactivity signal on the membrane was detected by X-ray film (Kodak).

Determination of total lipid, glycogen, trehalose

The microseparation of glycogen, trehalose and total lipid used the method described by Van Handel (1965) with a slight modification by Zhou et al. (2004). Whole-body homogenates of each individual were used to extract glycogen, trehalose and total lipid, respectively. Glycogen and trehalose were measured using the anthrone method with glycogen and trehalose as standards, respectively (Sigma Chemical, St.Louis, MO, USA). Total lipid was quantified by the vanillin assay. Each independent experiment was performed with triplicates with 10 individuals in total for each replicate.

Statistical analysis

All data were statistically analysed by one-way ANOVA using the Statistical Package for the Social Sciences (SPSS), version 11.5 for Windows. More than two group data were analysed with one-way ANOVA followed by the Tukey–Kramer HSD Test as the post hoc test.

Acknowledgements

We thank Prof. Jeffrey Price (University of Missouri—Kansas City) and Prof. Reddy Palli (University of Kentucky) for manuscript revision. This work was supported by the National Basic Research Program from Ministry of Science and Technology of the People's Republic of China ('973' Programme Grant number 2012CB114100) and the National Natural Science Foundation of China (Grant number 30870339) to Z. Zhao.

References

Bendtsen, J.D., Jensen, L.J., Blom, N., Von Heijne, G. and Brunak, S. (2004) Feature-based prediction of non-classical and leaderless protein secretion. *Protein Eng. Des. Sel.* **17**, 349–356.

Black, D.L. (2003) Mechanisms of alternative pre-messenger RNA splicing. *Annu. Rev. Biochem.* **72**, 291–336.

Brown, M.R., Crim, J.W., Arata, R.C., Cai, H.N., Chun, C. and Shen, P. (1999) Identification of a *Drosophila* brain-gut peptide related to the neuropeptide Y family. *Peptides*, **20**, 1035–1042.

Chen, X., Yang, X., Kumar, N.S., Tang, B., Sun, X., Qiu, X., Hu, J. *et al.* (2007) The class A chitin synthase gene of Spodoptera exigua: molecular cloning and expression patterns. *Insect Biochem. Mol. Biol.* **37**, 409–417.

Clynen, E., Huybrechts, J., Verleyen, P., De Loof, A. and Schoofs, L. (2006) Annotation of novel neuropeptide precursors in the migratory locust based on transcript screening of a public EST database and mass spectrometry. *BMC Genom.* **7**, 201.

De Loof, A., Baggerman, G., Breuer, M., Claeys, I., Cerstiaens, A., Clynen, E., Janssen, T. *et al.* (2001) Gonadotropins in insects: an overview. *Arch. Insect Biochem. Physiol.* **47**, 129–138.

Dierick, H.A. and Greenspan, R.J. (2007) Serotonin and neuropeptide F have opposite modulatory effects on fly aggression. *Nat. Genet.* **39**, 678–682.

Garczynski, S.F., Crim, J.W. and Brown, M.R. (2005) Characterization of neuropeptide F and its receptor from the African malaria mosquito, *Anopheles gambiae*. *Peptides*, **26**, 99–107.

He, C., Cong, X., Zhang, R., Wu, D., An, C. and Zhao, Z. (2013a) Regulation of circadian locomotor rhythm by neuropeptide Y-like system in *Drosophila melanogaster*. *Insect Mol. Biol.* **22**, 376–388.

He, C., Yang, Y., Zhang, M., Price, J.L. and Zhao, Z. (2013b) Regulation of sleep by neuropeptide Y-like system in *Drosophila melanogaster*. *PLoS One*, **8**, e74237.

Huang, Y., Brown, M.R., Lee, T.D. and Crim, J.W. (1998) RF-amide peptides isolated from the midgut of the corn earworm, *Helicoverpa zea*, resemble pancreatic polypeptide. *Insect Biochem. Mol. Biol.* **28**, 345–356.

Huang, Y., Crim, J.W., Nuss, A.B. and Brown, M.R. (2011) Neuropeptide F and the corn earworm, *Helicoverpa zea*: a midgut peptide revisited. *Peptides*, **32**, 483–492.

James, C. (2012) The development trend of the 2011 global biotechnology/commercialization of genetically modified crops. *China Biotechnol.* **32**, 1–14.

Kong-Ming, W. (2007) Environmental impact and risk management strategies of Bt cotton commercialization in China. *Chin. J. Agric. Biotechnol.* **4**, 93–97.

Krashes, M.J., DasGupta, S., Vreede, A., White, B., Armstrong, J.D. and Waddell, S. (2009) A neural circuit mechanism integrating motivational state with memory expression in *Drosophila*. *Cell*, **139**, 416–427.

Leff, S.E. and Rosenfeld, M.G. (1986) Complex transcriptional units: diversity in gene expression by alternative RNA processing. *Annu. Rev. Biochem.* **55**, 1091–1117.

Lingo, P.R., Zhao, Z. and Shen, P. (2007) Co-regulation of cold-resistant food acquisition by insulin-and neuropeptide Y-like systems in *Drosophila melanogaster*. *Neuroscience*, **148**, 371–374.

Liu, X., Zhang, Y., Zhou, Z. and Zhao, Z. (2013) Cloning and sequence analysis of neuropeptide F from the oriental tobacco budworm *Helicoverpa assulta* (Guenée). *Arch. Insect Biochem. Physiol.* **84**, 115–129.

Maule, A.G., Halton, D.W. and Shaw, C. (1995) Neuropeptide F: a ubiquitous invertebrate neuromediator? *Hydrobiologia*, **305**, 297–303.

McVeigh, P., Kimber, M., Novozhilova, E. and Day, T. (2005) Neuropeptide signalling systems in flatworms. *Parasitology*, **131**, S41–S55.

Nuss, A.B., Forschler, B.T., Crim, J.W., TeBrugge, V., Pohl, J. and Brown, M.R. (2010) Molecular characterization of neuropeptide F from the eastern subterranean termite *Reticulitermes flavipes* (Kollar) (Isoptera: Rhinotermitidae). *Peptides*, **31**, 419–428.

Roller, L., Yamanaka, N., Watanabe, K., Daubnerová, I., Žitňan, D., Kataoka, H. and Tanaka, Y. (2008) The unique evolution of neuropeptide genes in the silkworm *Bombyx mori*. *Insect Biochem. Mol. Biol.* **38**, 1147–1157.

Scriber, J. and Slansky, F. Jr. (1981) The nutritional ecology of immature insects. *Annu. Rev. Entomol.* **26**, 183–211.

Southey, B.R., Amare, A., Zimmerman, T.A., Rodriguez-Zas, S.L. and Sweedler, J.V. (2006) NeuroPred: a tool to predict cleavage sites in neuropeptide precursors and provide the masses of the resulting peptides. *Nucleic Acids Res.* **34**, W267–W272.

Southey, B.R., Sweedler, J.V. and Rodriguez-Zas, S.L. (2008) Prediction of neuropeptide cleavage sites in insects. *Bioinformatics*, **24**, 815–825.

Stanek, D.M., Pohl, J., Crim, J.W. and Brown, M.R. (2002) Neuropeptide F and its expression in the yellow fever mosquito, *Aedes aegypti*. *Peptides*, **23**, 1367–1378.

Van Handel, E. (1965) Microseparation of glycogen, sugars, and lipids. *Anal. Biochem.* **11**, 266–271.

Veenstra, J.A. and Sellami, A. (2008) Regulatory peptides in fruit fly midgut. *Cell Tissue Res.* **334**, 499–516.

Vroemen, S.F., Van Marrewijk, W.J., De Meijer, J., Van den Broek, A.T.M. and Van der Horst, D.J. (1997) Differential induction of inositol phosphate metabolism by three adipokinetic hormones. *Mol. Cell. Endocrinol.* **130**, 131–139.

Wen, T., Parrish, C.A., Xu, D., Wu, Q. and Shen, P. (2005) Drosophila neuropeptide F and its receptor, NPFR1, define a signaling pathway that acutely modulates alcohol sensitivity. *Proc. Natl Acad. Sci. USA*, **102**, 2141–2146.

Wu, K. and Gong, P. (1997) A new and practical artificial diet for the cotton boll-worm*. *Insect Sci.* **4**, 277–282.

Wu, Q., Zhao, Z. and Shen, P. (2005) Regulation of aversion to noxious food by Drosophila neuropeptide Y-and insulin-like systems. *Nat. Neurosci.* **8**, 1350–1355.

Yang, Y., Li, R. and Qi, M. (2000) In vivo analysis of plant promoters and transcription factors by agroinfiltration of tobacco leaves. *Plant J.* **22**, 543–551.

Yang, J., Zhu, J. and Xu, W.-H. (2010) Differential expression, phosphorylation of COX subunit 1 and COX activity during diapause phase in the cotton bollworm, *Helicoverpa armigera*. *J. Insect Physiol.* **56**, 1992–1998.

Zhang, Q., Guo, F., Liang, G.-M. and Guo, Y.-Y. (2010) Research progress of the development of resistance of target insects and resistance management strategy. *J. Environ. Entomol.* **2**, 020.

Zhou, G., Flowers, M., Friedrich, K., Horton, J., Pennington, J. and Wells, M.A. (2004) Metabolic fate of [^{14}C]-labeled meal protein amino acids in *Aedes aegypti* mosquitoes. *J. Insect Physiol.* **50**, 337–349.

13

Rootstock-to-scion transfer of transgene-derived small interfering RNAs and their effect on virus resistance in nontransgenic sweet cherry

Dongyan Zhao and Guo-qing Song*

Plant Biotechnology Resource and Outreach Center, Department of Horticulture, Michigan State University, East Lansing, MI, USA

*Correspondence

email songg@msu.edu

Keywords: grafting, hairpin RNA, rootstock-to-scion transfer, small interfering RNAs, transgenic tree, virus resistance.

Summary

Small interfering RNAs (siRNAs) are silencing signals in plants. Virus-resistant transgenic rootstocks developed through siRNA-mediated gene silencing may enhance virus resistance of nontransgenic scions *via* siRNAs transported from the transgenic rootstocks. However, convincing evidence of rootstock-to-scion movement of siRNAs of exogenous genes in woody plants is still lacking. To determine whether exogenous siRNAs can be transferred, nontransgenic sweet cherry (scions) was grafted on transgenic cherry rootstocks (TRs), which was transformed with an RNA interference (RNAi) vector expressing short hairpin RNAs of the genomic RNA3 of Prunus necrotic ringspot virus (PNRSV-hpRNA). Small RNA sequencing was conducted using bud tissues of TRs and those of grafted (rootstock/scion) trees, locating at about 1.2 m above the graft unions. Comparison of the siRNA profiles revealed that the PNRSV-hpRNA was efficient in producing siRNAs and eliminating PNRSV in the TRs. Furthermore, our study confirmed, for the first time, the long-distance (1.2 m) transfer of PNRSV-hpRNA-derived siRNAs from the transgenic rootstock to the nontransgenic scion in woody plants. Inoculation of nontransgenic scions with PNRSV revealed that the transferred siRNAs enhanced PNRSV resistance of the scions grafted on the TRs. Collectively, these findings provide the foundation for 'using transgenic rootstocks to produce products of nontransgenic scions in fruit trees'.

Introduction

With a history of over three thousand years, grafting has been used to produce woody stock/scion plants for multiple purposes, such as asexual propagation, disease/pest resistance, altered tree architecture and vigour, increased hardiness to abiotic stresses, precocity and higher yields (Aloni *et al.*, 2010; Gonçalves *et al.*, 2006; Koepke and Dhingra, 2013; Martinez-Ballesta *et al.*, 2010). The potential use of transgenic rootstocks to improve the performance of conventional nontransgenic scion varieties is a desirable strategy to address the concerns about transgene flow and exogenous protein production in most transgenic organisms.

Small interfering RNA (siRNA)-induced gene silencing provides a potential avenue to prevent viruses from their normal replicating and inducing pathogenesis in plants (Baulcombe, 2004; Brodersen and Voinnet, 2006; Fusaro *et al.*, 2006; Kalantidis, 2004; Kusaba, 2004; Matzke and Birchler, 2005; Mlotshwa *et al.*, 2002). It has been reported that siRNAs were silencing signals that are mobile in plants (Ali *et al.*, 2013; Bai *et al.*, 2011; Chitwood and Timmermans, 2010; Dunoyer *et al.*, 2010; Flachowsky *et al.*, 2012; Kasai *et al.*, 2011, 2013; Liang *et al.*, 2012; Molnar *et al.*, 2010; Palauqui *et al.*, 1997; Silva *et al.*, 2011). However, until recently, long-distance transfer of siRNAs was supported mostly by phenotype observation, Northern blot or quantitative reverse-transcription PCR (qRT-PCR) analysis in annual model plants. Using high-throughput sequencing, profiles of transported siRNAs (i.e. scion-to-root and root-to-scion) have been documented in *Arabidopsis thaliana* (Dunoyer *et al.*, 2010; Molnar *et al.*, 2010). In woody plants, such a profile has not been

reported. A recent study showed that small RNA (sRNA) sequencing was an effective approach in profiling small RNAs derived from a vector expressing a hairpin RNA in woody plants (Zhao and Song, 2014). This provides us a powerful tool to study sRNA transfer.

Prunus necrotic ringspot virus (PNRSV) is a major pollen-disseminated llarvirus that adversely affects many Prunus species (Barbara *et al.*, 1978). In a previous study, we found that the siRNAs derived from a hairpin sequence of partial PNRSV coat protein (PNRSV-hpRNA) were extremely efficient in protecting transgenic cherry rootstocks (TRs) from PNRSV damage (Song *et al.*, 2013). To investigate whether these siRNAs can be transferred and exert systemic silencing effect in nontransgenic scions, we conducted grafting experiments (cherry rootstock/scion), sequenced and profiled PNRSV-specific siRNAs in the samples of 1) PNRSV-free and PNRSV-inoculated transgenic rootstocks (TRs); 2) PNRSV-free and PNRSV-inoculated sweet cherry trees grafted on nontransgenic rootstocks (NRs) (herein, NR/scion); and 3) PNRSV-free and PNRSV-inoculated sweet cherry trees grafted on TRs (herein, TR/scion). Our results reveal, for the first time, the transfer of transgene-derived siRNAs from the TRs to the nontransgenic scions in grafted trees. Moreover, the transferred siRNAs from some transgenic events could enhance virus resistance of the TR/scion trees.

Materials and methods

PNRSV-hpRNA construct

The PNRSV genome consists of three RNA sequences (RNA1, RNA2, and RNA3) (Cui *et al.*, 2013). The pART27-PNRSV-hpRNA vector contains inverted repeat sequences of 414-nt of the coat protein region of the Michigan PNRSV isolate CB7 (GenBank: EF495168.1) (Figure 1a). The hpRNA in the construct is part of the RNA3 sequence. The pART27-PNRSV-hpRNA was transformed to cherry rootstock 'Gisela 6' using *Agrobacterium tumefaciens*-mediated transformation (Song and Sink, 2006; Song *et al.*, 2013).

Plant materials

Cherry rootstocks, that is, nontransgenic 'Gisela 6' and transgenic 'Gisela 6' containing PNRSV-hpRNA from 11 independent transgenic events (TR1–TR11), were used. *In vitro* shoots, 10 per transgenic event, were rooted and grown in greenhouse in 2009. One-year-old bud woods of PNRSV-free sweet cherry 'Emperor Francis' were bud-grafted on NRs and TRs in May of 2010 (Figure 1b).

Grafting for virus inoculation

A PNRSV strain Fulton G and a combined PNRSV/PDV (CH39) isolate were stored separately in mature trees of 'Bing' sweet cherry on *Prunus mahaleb* L. rootstocks. The infected 1-year-old bud wood was obtained from the Clean Plant Program of the Northwest, Irrigated Agriculture Research and Extension Center at Washington State University (Prosser, Washington). Virus inoculation for rootstock trees was initially conducted through bud grafting of individual virus-containing bud onto each tree at about 25 cm above the soil surface. For rootstock/scion trees, PNRSV inoculation was conducted on the bark of scions. As no grafted bud survived, chips of bark were peeled off the virus-containing donor bud wood and one chip was grafted onto the scion of each receiver tree. Noninfected control trees were self-grafted using their buds and bark. All inoculation was conducted in the spring of 2011. Successful inoculation was confirmed by triple-antibody sandwich enzyme-linked immunosorbent assay (TAS-ELISA) using a commercial kit for the PNRSV detection (Agdia Inc., Elkhart, Indiana). The grafted trees were grown in a secured greenhouse and evaluated periodically for visual symptoms of virus infection after successful bark grafting was confirmed (Figure 1c). PNRSV symptoms were investigated in

the fall of 2013 and the spring of 2014. ELISA was conducted on all the trees prior to RNA isolation for sequencing, and all PNRSV-positive trees were referred to as PNRSV-infected trees.

RNA sample preparation and sequencing

Bud tissues were collected from branches at a bark location about 1.5 m above the soil for rootstock trees or 1.2 m above the graft unions for rootstock/scion trees. Total RNA was isolated from bud tissues, 0.2 g for each sample, using a cetyltrimethylammonium bromide (CTAB) methodZamboni *et al.*, 2008). The samples were purified using miRNeasy Mini Kit (Qiagen, Valencia, CA). Integrity of the RNA samples was assessed using the Agilent RNA 6000 Pico Kit (Agilent Technologies, Inc. Waldbronn, Germany). Small RNA libraries (*n* = 23) were constructed using the Illumina TruSeq® Small RNA Sample Prep Kit (Illumina, Inc., Hayward, CA), which were pooled and sequenced (50-bp single end reads) in two lanes using the Illumina HiSeq2500 platform at the Research Technology Support Facility of Michigan State University (East Lansing, MI).

Analysis of the sRNA pools

The quality of sequencing reads was assessed using the FastQC program (http://www.bioinformatics.babraham.ac.uk/projects/fastqc/), which showed that the per base quality scores range from 26 to 40 across the 50 nucleotide (nt) read length. Adapter sequences were removed using the cutadapt program (http://code.google.com/p/cutadapt/), and only the reads longer than 19 nt were retained. The filtered reads were converted from fastq format to fasta for BLASTN searches. As the hpRNA in the pART27-PNRSV-hpRNA vector is part of the RNA3 sequence, we used a derived sequence of the PNRSV genomic RNA3 of CB7, as a reference to map the sRNA reads. The full-length PNRSV RNA3 sequence was deduced by aligning PNRSV-CB7 (GenBank: EF495168.1) with other full-length RNA3 sequences (GenBank: NC_004364.1 and JN416776.1), and the resulting sequence was referred to as PNRSV-mCB7 (1959 nt). BLASTN searches (word size = 7, e-value = 100, max_target_seqs = 100 000 000) were conducted using the PNRSV-mCB7 sequence against the filtered RNA sequencing reads. Only the

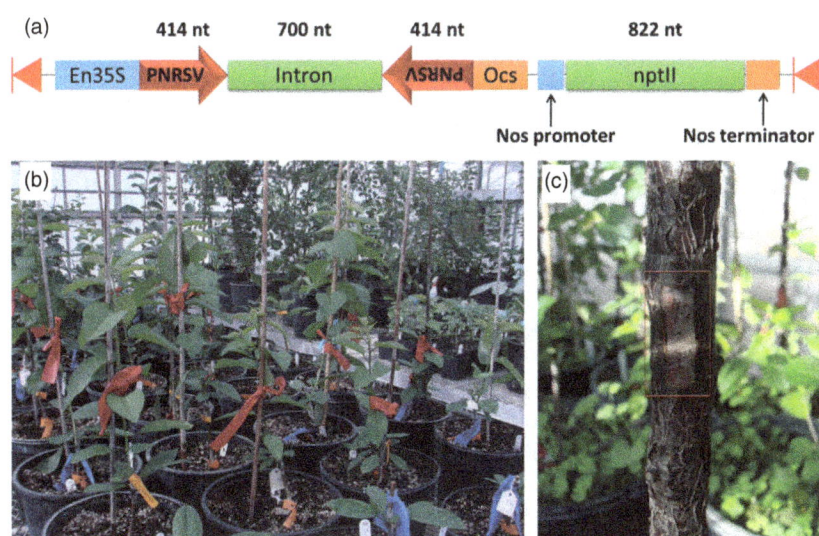

(a)

414 nt   700 nt   414 nt   822 nt

En35S PNRSV Intron PNRSV Ocs nptII

Nos promoter   Nos terminator

Figure 1 (a) Schematic representation of the T-DNA region of the pART27-PNRSV-hpRNA. En35S = the cauliflower mosaic virus (CaMV) 35S promoter plus an enhancer; Ocs = Ocs terminator; *npt*II = neomycin phosphotransferase gene; PNRSV, a 414-nt sequence derived from the coat protein of Prunus necrotic ringspot virus (GenBank: EF495168.1). Bud grafting sweet cherry 'Emperor Francis' on transgenic rootstocks from different transgenic events in 2010 (b) and PNRSV inoculation on a scion through bark grafting in 2011 (c).

reads with 100% identity to the PNRSV-mCB7 sequence were retained for further analysis. Matched reads were mapped to PNRSV-mCB7 sense strand. All these analyses were conducted using the resources at the High Performance Computing Center of Michigan State University.

Results

Hairpin RNA-derived and virus-induced siRNA profiles are distinct

To compare the profiles of hpRNA-derived and virus-induced siRNAs, the sequencing reads from two TRs (TR1 and PNRSV-inoculated TR2) of different transgenic events and scions of two grafted trees (NR/scion and PNRSV-infected NR/scion) were quality-filtered and mapped to a reference genomic RNA3 of PNRSV, PNRSV-mCB7, from which the hpRNA was derived (Table 1). The analyses revealed that hpRNA is efficient in producing siRNAs. For the TR1, 1933 reads/million sequencing reads (reads/MR) of PNRSV-mCB7-specific siRNAs (20–24 nt) were detected, whereby 99.5% (1924 reads/MR) were mapped to the PNRSV-hpRNA region (Table 1). Similarly, in the PNRSV-inoculated TR2, 2858 reads/MR of PNRSV-mCB7-specific siRNAs were detected, of which 99.9% (2856 reads/MR) were from the PNRSV-hpRNA region (Table 1). In TR1, the most prevalent hpRNA-specific siRNAs were 24-nt siRNAs (54%) followed by 21-nt ones (33%) (Table 1, Figure 2). The production of neither type of siRNAs across the PNRSV-hpRNA region was random, which was reflected by different abundance of siRNAs at different positions (Figures 3a and 4a). For example, the most abundant 24-nt siRNAs appeared at 1468–1491 nt of the PNRSV-mCB7 sequence and accounted for 42.8% (444 reads/MR) of the total PNRSV-hpRNA-derived 24-nt siRNAs (Figure 3a). In contrast, the most abundant 21-nt siRNAs was at 1266–1286 nt, constituting only 8% (51 reads/MR) of the total mapped 21-nt siRNAs (Figure 4a). A similar pattern of the PNRSV-mCB7-specific siRNAs was observed in the TR2 (Figures 3b and 4b, Table 1). The siRNA profiles of TR1 and TR2 suggested that hpRNA was very effective in producing 24-nt siRNAs in TRs.

In contrast, the profiles of naturally occurring PNRSV-specific siRNAs (20–24 nt), which was derived from a 3-year-old, PNRSV-infected NR/scion (PNRSV positive in ELISA assay), were different from those of PNRSV-hpRNA-derived siRNAs from TR1 and TR2 (Table 1, Figures 2, 3d, and 4d). We detected 761 reads/MR siRNAs across the PNRSV-mCB7 sequence in the PNRSV-infected NR/scion (28 months after inoculation), of which only 24.7% (188 reads/MR) were mapped to the PNRSV-hpRNA sequence, which was in contrast to over 99% of those detected in the TR1 (1924 reads/MR) and PNRSV-inoculated TR2 (2856 reads/MR) (Table 1). Even at the hpRNA region, the fractions of siRNAs with different sizes were different between the PNRSV-infected NR/scion and the TRs. In the PNRSV-infected NR/scion, 21-nt siRNAs were the most prevalent, constituting over half (61.6%) of the total PNRSV-mCB7 siRNAs, and few (0.47%) 24-nt siRNAs were detected (Figure 2). This suggested that RNase III Dicer-like 4 (DCL4), which is involved in 21-nt siRNA production, plays a major role (Chitwood and Timmermans, 2010). In addition, the fractions of 22-nt siRNAs for PNRSV-infected NR/scion (32.5%) and PNRSV-infected TR4/scion (37.6%) were much higher than those in both TR1 (7.6%) and TR2 (6.5%), suggesting that DCL2-mediated 22-nt siRNA formation was more active when PNRSV was present (Figure 2). Collectively, in the entire PNRSV-mCB7 region, 21 and 22-nt siRNAs were the major siRNA classes in the PNRSV-infected NR/scion (Table 1).

Small interfering RNAs were efficient in eliminating PNRSV

No visible virus symptom was observed in the TR2 28 months after PNRSV inoculation, and ELISA test was no longer positive at the time of our sampling for sRNA sequencing, suggesting the absence/low presence of viruses. The virus resistance of TR2 was also supported by the siRNA profile of PNRSV-inoculated TR2, where 99.9% of the siRNAs were mapped to the PNRSV-hpRNA region and almost none in the other regions. If PNRSV was present, we would see some siRNAs in the non-hpRNA regions of PNRSV-mCB7 as was observed in the virus-infected NR/scion trees (Table 1, Figures 3b, d and 4b, d).

Small interfering RNAs were transferred from TR to nontransgenic scion

To determine whether the PNRSV-hpRNA-derived siRNAs could be transferred from transgenic rootstocks to nontransgenic scions, we sequenced and analysed sRNA pools derived from scions (1.2 m above the graft union) of a PNRSV-free TR3/scion and a PNRSV-infected TR4/scion. Indeed, the siRNA profiles of both trees exhibited characteristics similar to those of the TR1 and TR2 and revealed that hpRNA-derived siRNAs in TR3 and TR4 were transported to the nontransgenic sweet cherry scions, regardless of the sequencing errors (10.1 reads/MR, 20–24 nt) as determined in a PNRSV-free NR/scion (a negative control) (Table 1). In the TR3/scion, 4.56 reads/MR of the PNRSV-hpRNA-specific 24-nt siRNAs were detected; in contrast, only 0.60 reads/MR were detected in the negative control tree (NR/scion) and 0.85 reads/MR in the PNRSV-infected NR/scion (Table 1, Figure 3c–e). Furthermore, a total of 2.1 reads/MR (45.4%) of the PNRSV-hpRNA-specific 24-nt siRNAs were detected at 1468–1491 nt of the PNRSV-mCB7 sequence in the PNRSV-free scion of the TR3/scion; in contrast, only 0.2 reads/MR (27.3%) were detected in the negative control (Figure 3c, e). Similarly, 1.9 reads/MR (26.3%) hpRNA-specific 24-nt siRNAs were detected at the same position in the PNRSV-infected TR4/scion, while only 0.1 reads/MR (3.0%) were detected in the PNRSV-infected NR/scion (Figure 3d, f). The predominance of the 24-nt siRNAs at 1468–1491 nt in the TR1, PNRSV-inoculated TR2, TR3/scion, and PNRSV-infected TR4/scion provided compelling evidence for the transfer of hpRNA-derived siRNAs from the TRs to the nontransgenic scions (Figure 3a, b, e, f). Similarly, difference in the amount and distribution patterns of 21-nt siRNAs observed in the TR3/scion and the NR/scion also suggested the transfer of hpRNA-derived siRNAs (Figure 4).

In addition, the ratios of 24 and 21-nt siRNAs in different siRNA pools also substantiated the transfer of hpRNA-derived siRNAs (Figures 2–4, Table 1). For example, in the PNRSV-free trees, the ratios of 24-nt/21-nt PNRSV-specific siRNAs in the scions of the TR3/scion and the NR/scion trees were 1.14 and 0.41, respectively; in the PNRSV-infected TR4/scion and PNRSV-infected NR/scion trees, the ratios were 0.0465 and 0.0076, respectively. In both cases, the TR/scion trees showed higher ratios of 24-nt/21-nt siRNAs than the NR/scion trees. Apparently, rootstock-to-scion transfer of siRNAs from TR3 and TR4 could contribute to the higher ratios in the two TR/scion trees, as high ratios of 24-nt/21-nt PNRSV-hpRNA-specific siRNAs were observed in the TR1 (1.65) and the TR2 (1.67).

Table 1 Summary of small interfering RNAs (siRNAs) mapped to PNRSV genomic RNA3 mCB7* and PNRSV-hpRNA[†]

	TR1[‡]		PNRSV-inoculated TR2		NR[§]/scion[¶]**		PNRSV-infected NR/scion		TR3/scion[††]		PNRSV-infected TR4/scion	
Total reads	21 220 329		18 471 919		15 107 728		18 865 267		23 262 729		16 603 565	
Size of siRNAs	hpRNA	non-hpRNA region in mCB7	hpRNA	non-hpRNA region in mCB7	hpRNA	non-hpRNA region in mCB7	hpRNA	non-hpRNA region in mCB7	hpRNA	non-hpRNA region in mCB7	hpRNA	non-hpRNA region in mCB7
20 nt (reads/MR[‡‡])	50.66	0.75	79.15	0.32	0.33	0.73	11.45	49.14	0.34	0.47	45.65	64.75
21 nt (reads/MR)	627.65	5.04	924.27	1.46	1.46	4.50	111.26	357.54	4.00	3.10	67.27	299.69
22 nt (reads/MR)	145.80	2.21	185.42	0.65	0.40	1.59	61.06	155.31	1.25	1.29	75.04	271.21
23 nt (reads/MR)	63.24	0.05	96.96	0.16	0.07	0.33	3.29	8.32	0.30	0.26	8.31	12.53
24 nt (reads/MR)	1037.12	0.09	1570.00	0.05	0.60	0.13	0.85	2.70	4.56	0.09	3.13	3.98
Total reads/MR	1924.48	8.15	2855.79	2.65	2.85	7.28	187.91	573.01	10.45	5.20	199.42	652.15
%21-nt siRNA[§§]	32.61%	61.85%	32.36%	55.10%	51.16%	61.82%	59.21%	62.40%	38.27%	59.50%	33.74%	45.95%
%24-nt siRNA	53.89%	1.16%	54.98%	2.04%	20.93%	1.82%	0.45%	0.47%	43.62%	1.65%	1.57%	0.61%

PNRSV, Prunus necrotic ringspot virus.

*A deduced sequence of the PNRSV genomic RNA3.

[†]The 414-nt PNRSV coat protein sequence integrated into the RNAi construct. It was part of PNRSV-mCB7.

[‡]TR1, TR2, TR3 and TR4: independent transgenic events of 'Gisela 6' rootstock transformed with pART27-PNRSV-hpRNA construct.

[§]NR: nontransgenic 'Gisela 6' rootstock.

[¶]Scion: nontransgenic sweet cherry 'Emperor Francis'.

**NR/scion: nontransgenic sweet cherry 'Emperor Francis' grafted on NR.

[††]TR3/scion: nontransgenic sweet cherry 'Emperor Francis' grafted on TR3.

[‡‡]Numbers in the same row indicate the total mapped reads per million sequencing reads (20–24 nt).

[§§]Percentage of 21-nt mapped reads.

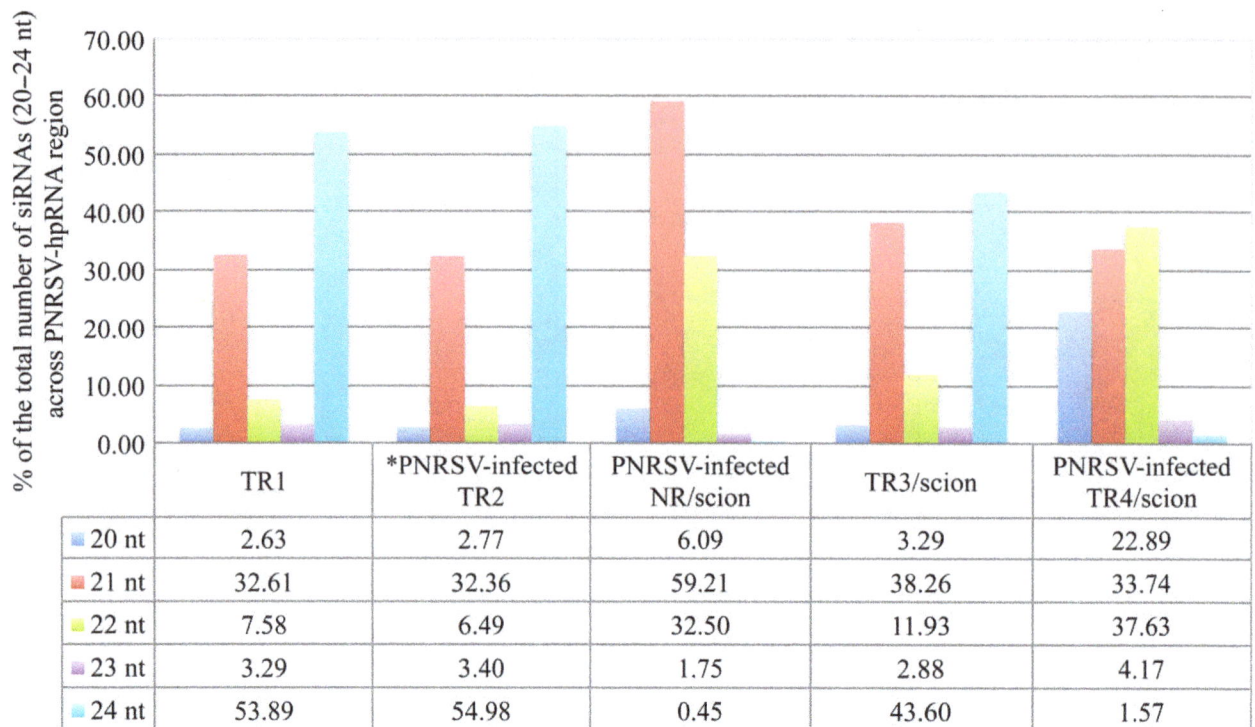

	TR1	*PNRSV-infected TR2	PNRSV-infected NR/scion	TR3/scion	PNRSV-infected TR4/scion
20 nt	2.63	2.77	6.09	3.29	22.89
21 nt	32.61	32.36	59.21	38.26	33.74
22 nt	7.58	6.49	32.50	11.93	37.63
23 nt	3.29	3.40	1.75	2.88	4.17
24 nt	53.89	54.98	0.45	43.60	1.57

Figure 2 Fractions of 20–24 nt siRNAs mapped to Prunus necrotic ringspot virus (PNRSV)-hpRNA region in different cherry trees. TR1, TR2, TR3 and TR4 were from independent transgenic events. NR: nontransgenic rootstock 'Gisela 6'. Scion: nontransgenic sweet cherry 'Emperor Francis'. PNRSV-inoculated TR2: the PNRSV-inoculated TR2 was PNRSV positive 1 year after inoculation. PNRSV-infected NR/scion: nontransgenic sweet cherry grafted on NR was infected by PNRSV isolate CH39 on the scion. TR3/scion: nontransgenic sweet cherry grafted on TR3. PNRSV-infected TR4/scion: nontransgenic sweet cherry grafted on transgenic rootstock was infected by CH39 on the scion. All PNRSV-inoculated trees, except *PNRSV-inoculated TR2, were PNRSV positive in ELISA test at the time of sampling for sequencing.

The transferred siRNAs enhanced virus resistance of the nontransgenic scions

To determine whether the transferred siRNAs from TRs were able to induce virus resistance of the nontransgenic scions, we conducted PNRSV inoculation experiments on the scions (Figure 5). Regardless of rootstocks (NRs or TRs), the grafted cherry trees grew normally when they were not inoculated (Figure 5b). Two years after PNRSV inoculation, two of three PNRSV-inoculated NR/scion trees died; only one was alive at the time (28 months after inoculation) of our sampling for sRNA sequencing. All NR/scion trees died within 3 years after inoculation with either CH39 or Fulton G, of which CH39 (a more virulent PNRSV isolate) caused tree death 1 year earlier than Fulton G. In contrast, for the TR/scion trees grafted onto transgenic rootstocks from seven independent transgenic events, all three trees were alive 3 years after the inoculation with the Fulton G and one (TR4/scion) of four TR/scion trees was alive after the inoculation with CH39 (Figure 5, Table 2). The expression levels of the PNRSV-hpRNA in different TR events could be responsible for the variations in PNRSV resistance among the TR/scion trees when challenged with CH39 (Song et al., 2013).

The survival of all Fulton G-infected trees and one CH39-infected tree grafted on different TRs demonstrated that the transferred siRNAs from the TRs (as confirmed by comparison of siRNA profiles among TRs, TR/scion and NR/scion) likely contributed to the enhanced virus resistance. In the TR4/scion that

survived after CH39 infection, 852 reads/MR 20–24 nt siRNAs were detected across the PNRSV-mCB7 sequence in the scion (Table 1). Compared with those in the PNRSV-infected NR/scion (0.1 reads/MR), the higher abundance of the PNRSV-mCB7-specific 24-nt siRNAs at 1468–1491 nt (1.9 reads/MR) suggested the transfer of the hpRNA-derived siRNAs from the TR4 to the nontransgenic scion (Figure 3d, f). On the other hand, the PNRSV-infected TR4/scion also contained high abundance of 21-nt siRNAs in the non-hpRNA regions, indicating the presence of PNRSV (Figure 4d, f). Furthermore, in the ELISA test, the PNRSV-infected NR/scion and the PNRSV-infected TR4/scion were PNRSV positive at the time of our sampling for sequencing (data not shown). These results indicated a coexistence stage of the transferred hpRNA-specific siRNAs and some naturally occurring PNRSV-mCB7-specific siRNAs generated upon virus infection.

Compared with the PNRSV-infected NR/scion, we found a higher percentage of 20-nt (22.9% versus 6.1%) and a lower percentage of 21-nt (33.7% versus 59.2%) siRNAs in the PNRSV-infected scion of the TR4/scion (Figure 2). This difference suggested the effectiveness of the transferred PNRSV-hpRNA-derived siRNAs in inhibiting the proliferation of PNRSV. Three years after PNRSV inoculation, the TR4/scion remained alive but showed symptoms of PNRSV damage and the PNRSV-infected NR/scion died. Collectively, these results demonstrated that transferrable hpRNA-derived siRNAs originated from TRs could enhance virus resistance in the TR/scion trees tested, which may be applicable in transgenic rootstock/scion trees.

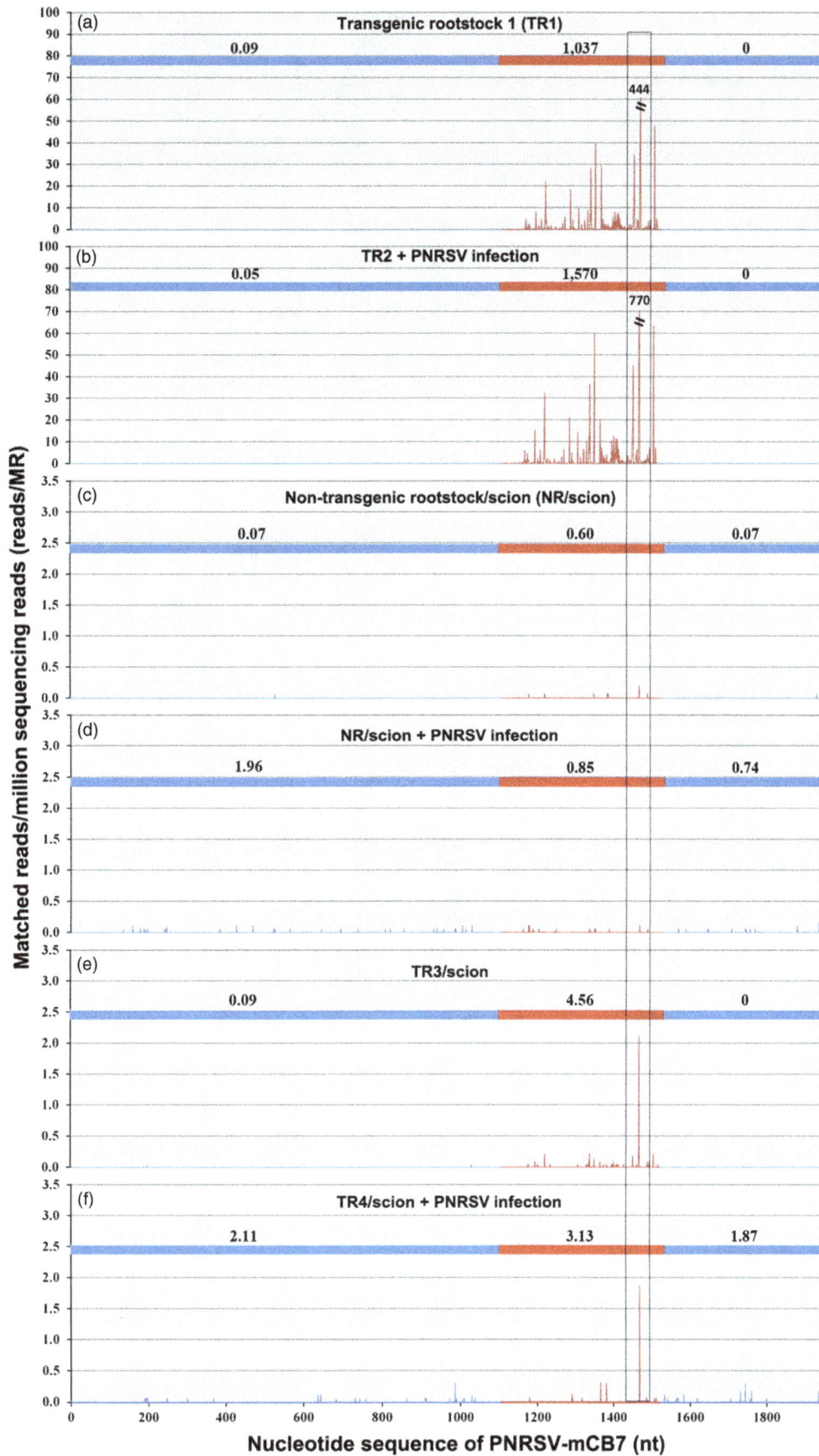

Discussion

To date, little has been carried out to investigate production and transportation of hpRNA-derived siRNAs for virus resistance in woody plants because of the challenges in genetic transformation and siRNA detection. In this study, PNRSV-specific siRNAs were profiled and compared among virus-free and virus-inoculated TRs, NR/scion and TR/scion cherry trees. We demonstrated, for the first time, the long-distance transfer of PNRSV-hpRNA-derived siRNAs from the transgenic rootstocks to the nontransgenic scions in grafted trees using sRNA sequencing technology.

Effectiveness of intron-spliced hpRNA is attributed to the high abundance of hpRNA-derived siRNAs

Previous studies demonstrated that hpRNAs containing a functional intron could reach 96%–100% efficiency in virus immunity or reduced expression of target genes, which is better than that containing either a random spacer fragment (e.g. GUS sequence) or an intron in the reverse orientation (Mroczka et al., 2010; Smith et al., 2000; Stoutjesdijk et al., 2002). In this study, the inverted PNRSV sequence (PNRSV-hpRNA) is flanking a functional intron from the pyruvate orthophosphate dikinase gene. Analysis of the siRNA profiles confirmed at the sequence level that the effectiveness of intron-spliced hpRNA is likely attributed to the large number of siRNAs generated by the hpRNA. In this case, the silencing effect may depend more on the 24-nt siRNA as they were overrepresented in the TRs (54%) and the PNRSV-free TR/scion (44%). In addition, the effectiveness of the hpRNA strategy in producing siRNAs is further supported by the few siRNAs of the nptII, which locates at non-hpRNA region next to the PNRSV-hpRNA in the same construct.

Characteristics (e.g. frequency, type and distribution) of naturally occurring virus-derived siRNAs in plants are often dependent on viruses and hosts. This has been well documented in several recent reports (Li et al., 2012; Mitter et al., 2013; Silva et al., 2011). For example, 22 and 23-nt were the major classes of cotton leafroll dwarf virus (CLRDV)-derived siRNAs in cotton plants (Silva et al., 2011). In four suspected virus-infected tomato samples, 21 and 22-nt were the major virus-specific siRNAs (Li et al., 2012). For tomato spotted wilt virus (TSWV), 24 and 21-nt siRNAs were major classes in tomato and tobacco (Mitter et al., 2013). In this study, naturally occurring PNRSV-specific siRNAs generated upon virus infection were mainly 21 and 22-nt classes; in contrast, for the PNRSV-free transgenic rootstocks, 24 and 21-nt siRNAs were the main PNRSV-hpRNA-derived siRNA species (Figure 2). This difference suggests that the presence of PNRSV promotes the production of 22-nt siRNAs. Similar to previous reports (Li et al., 2012; Mitter et al., 2013; Silva et al., 2011), neither the naturally occurring nor the hpRNA-derived siRNAs seem to be produced randomly (Figures 3 and 4), which is likely due to the preference of different DCLs in either initial formation or subsequent degradation of these siRNAs. More studies are

needed to find out the role of different siRNA species, for example 21 and 24-nt siRNAs.

In this study, the PNRSV-inoculated TR2 was PNRSV positive in our ELISA test 1 year after PNRSV inoculation. Twenty-eight months after inoculation, the effectiveness of the hpRNA-derived siRNAs in inhibiting/eliminating PNRSV in TR2 was readily seen by the presence of siRNAs mapped only to the hpRNA sequence but not to the non-hpRNA regions of the PNRSV RNA3 sequence (Figures 3b and 4b). In terms of the different numbers of hpRNA-specific siRNAs in TR1 and PNRSV-inoculated TR2, it is likely due to different insertion position(s) or copy numbers of the integrated hpRNA(s) from different transgenic events. Indeed, the PNRSV inoculation study revealed that transgenic 'Gisela 6' trees from different transgenic events exhibited different PNRSV resistance (G.-Q. Song, unpublished data). Regardless of the variations, continuous production of the PNRSV-specific siRNAs could contribute to the increased PNRSV resistance in the TRs.

Transfer of hpRNA-derived siRNAs from rootstock to scion

There is no question that high-throughput sequencing is an efficient way to verify presence/absence, abundance/distribution, and sequence features of exogenous siRNAs (i.e. hpRNA-derived siRNAs in TRs and naturally produced siRNAs in virus-infected trees). However, for the detection of low abundant siRNAs, sample contamination and sequencing errors could be a problem. In this study, cross-contamination among different samples is low because of the low number of siRNAs detected in the negative control. These matched siRNAs are likely due to sequencing errors, especially in the reads for barcodes, which would result in the assignment of reads to the wrong sample(s). Indeed, it is not uncommon to have mis-assignment of reads when multiple samples were combined and sequenced in a single lane (Kircher et al., 2011; Silva et al., 2011). Silva et al. (2011) found that out of a total of 20 047 294 reads, 1967 matched the genome sequence of CLRDV in the noninfected cotton plant. That is 98 reads/MR, which is much higher than what we got in the negative control plant (10 reads/MR). One may argue that the less reads detected in our study may be because less sRNA species we analysed (20–24 nt versus 18–26 nt in Silva et al., 2011). This possibility is low because the total number of 17–30-nt sRNAs in our negative control was ~13 reads/MR, which is still much lower than that from the noninfected cotton plant. However, the large number of total CLRDV sRNAs (640 325 reads) out of the total sequencing reads may explain the higher number of mapped reads in the noninfected cotton.

When we based merely on the total number of siRNAs, sequencing errors and the efficiency of long-distance transfer (1.2 m) of siRNAs may make it difficult to verify the transfer of siRNAs. Although limited replicates were performed for sRNA sequencing analysis, the distinct siRNA patterns between TRs and virus-infected NR/scions and similar siRNA patterns between TRs

Figure 3 The 24-nt siRNAs mapped to the Prunus necrotic ringspot virus (PNRSV) genomic RNA3 (PNRSV-mCB7) in cherry trees. x-axis is a schematic representation of the PNRSV-mCB7, the whole genomic RNA3 of PNRSV. The red traces/bars represent the partial coat protein sequence used in the hpRNA construct (414-nt, Figure 1a), and blue traces/bars are the rest of the PNRSV-mCB7 sequence. The blank box encompasses siRNAs at position 1468–1491 nt. PNRSV-mCB7-specific 24-nt siRNAs detected in TR1 (a), PNRSV-inoculated TR2 (b), NR/scion (c), PNRSV-infected NR/scion (d), TR3/scion (e) and PNRSV-infected TR4/scion (f). TR: transgenic rootstock 'Gisela 6'. TR1, TR2, TR3 and TR4 were from independent transgenic events. NR: nontransgenic rootstock 'Gisela 6'. Scion: nontransgenic sweet cherry 'Emperor Francis'. All PNRSV-inoculated trees, except TR2, were PNRSV positive in ELISA test at the time of sampling for sequencing.

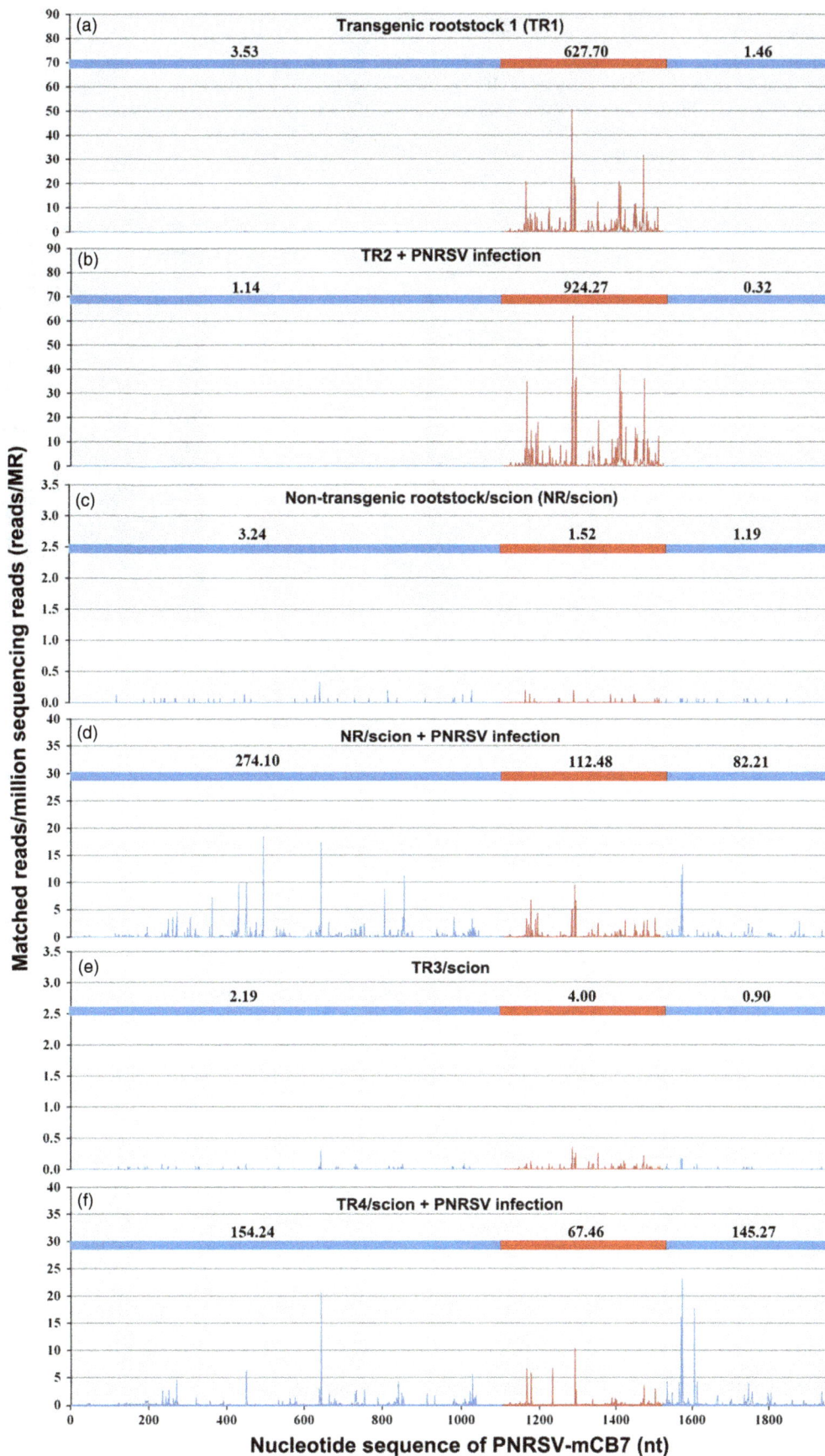

Figure 4 The 21-nt siRNAs mapped to the Prunus necrotic ringspot virus (PNRSV) genomic RNA3 (PNRSV-mCB7) in cherry trees. *x*-axis is a schematic representation of the PNRSV-mCB7, the whole genomic RNA3 of PNRSV. The red traces/bars represent the partial coat protein sequence used in the hpRNA construct (414-nt, Figure 1a), and blue traces/bars are the rest of the PNRSV-mCB7 sequence. *y*-axis represents the number of matched reads/million sequencing reads. PNRSV-mCB7-specific 21-nt siRNAs detected in TR1 (a), PNRSV-inoculated TR2 (b), NR/scion (c), PNRSV-infected NR/scion (d), TR3/scion (e) and PNRSV-infected TR4/scion (f). TR: transgenic rootstock 'Gisela 6'. TR1, TR2, TR3 and TR4 were from independent transgenic events. NR: nontransgenic rootstock 'Gisela 6'. Scion: nontransgenic sweet cherry 'Emperor Francis'. All PNRSV-inoculated trees, except TR2, were PNRSV positive in ELISA test at the time of sampling for sequencing.

Figure 5 A schematic showing the transfer of siRNAs and tree response to virus infection. (a) Schematic representation of the transfer of Prunus necrotic ringspot virus (PNRSV)-hpRNA-derived siRNAs from transgenic rootstock (TR) to nontransgenic scion. (b) Responses of rootstock/scion trees to infection of PNRSV isolate Fulton G. NR: nontransgenic rootstock 'Gisela 6'. Scion: nontransgenic sweet cherry 'Emperor Francis'. '+ PNRSV': inoculation with Fulton G 28-month before collecting the materials.

Table 2 Effect of PNRSV inoculation on sweet cherry scions grafted on transgenic and nontransgenic rootstocks 'Gisela 6'*

Tree ID	PNRSV isolate[†]	Tree response
TR6[†]/scion[‡]	No infection	Alive
TR7/scion[§]	No infection	Alive
NR/scion[¶]	No infection	Alive
TR6/scion	Fulton G	Alive
TR8/scion	Fulton G	Alive
TR5/scion	Fulton G	Alive
NR/scion	Fulton G	Dead
TR9/scion	CH39	Dead
TR4/scion	CH39	Alive
TR10/scion	CH39	Dead
TR11/scion	CH39	Dead
NR/scion	CH39	Dead

*PNRSV inoculation on the trunk of scion started in 2011 and tree responses were observed in 2013.

[†]TR: transgenic rootstock. TR6: transgenic event number 6, two TR6/scion trees were used in this study.

[‡]Scion: a nontransgenic sweet cherry 'Emperor Francis'.

[§]TR11/scion: nontransgenic sweet cherry 'Emperor Francis' grafted on TR11.

[¶]NR/scion: a nontransgenic sweet cherry 'Emperor Francis' grafted on nontransgenic rootstock (NR).

and TR/scions allowed us to confirm the transfer of siRNAs from TR to scion. Firstly, comparison of the siRNA profiles between TRs and PNRSV-infected NR/scion revealed that hpRNA produced predominantly 24-nt siRNAs whereas PNRSV-induced siRNAs were more abundant in 21-nt siRNAs (Figure 2). Secondly, there appeared a position (1468–1491 nt of the PNRSV-mCB7) where the hpRNA(s) produced the most abundant 24-nt siRNAs (the peak in the box of Figure 3). Similar to the TRs, both total 24-nt

siRNAs and those at 1468–1491 nt were also the most abundant in the scion of the PNRSV-free TR3/scion. Due to the presence of virus in the PNRSV-infected TR4/scion, 21-nt siRNAs were the most abundant. However, the predominance of the 24-nt siRNAs at 1468–1491 nt was still observed despite the small fraction of the total 24-nt siRNAs in the TR4/scion (Figure 3f). Taken together, these distinct features of the siRNA profiles among transgenic and nontransgenic trees enabled us to confirm the long-distance transfer of PNRSV-hpRNA-derived siRNAs.

Hairpin RNA is a rich source of siRNAs in plants. Mobility of the siRNAs across graft unions verified by sRNA sequencing, that is scion-to-root and root-to-scion in *Arabidopsis* and transgenic rootstock-to-scion in cherry, lays a foundation for using transgenic rootstocks or transgenic scions to produce products of nontransgenic parts in rootstock–scion plants (Dunoyer *et al.*, 2010; Molnar *et al.*, 2010). Continuous production of transportable PNRSV-specific siRNAs in TRs provides us an option to alleviate/cure PNRSV-infected trees (e.g. mature sweet cherry trees) by either stub grafting (with the TRs as rootstocks) or grafting the TRs (as scions) to the infected host. Here, we demonstrated the transfer of the siRNAs from rootstock to scion, and the transferred siRNAs seemed to enhance virus resistance of the nontransgenic scion. The effective/required amount of transferred siRNAs to exert a systemic protection of the nontransgenic scions from PNRSV damage remains to be determined. Of four grafted trees, TR4/scion was the only one survived after CH39 (more virulent than Fulton G) inoculation. Further investigation of all survived TR/scion trees after PNRSV infection is ongoing, which may shed light on the virus resistance induced by the transferred siRNAs.

Conclusion

In this study, nontransgenic cherry scions were grafted on both nontransgenic and transgenic rootstocks. Using sRNA sequencing, we developed profiles of transferred siRNAs (rootstock-to-scion) derived from an RNAi vector expressing a short hairpin RNA in the transgenic rootstocks. The analysis of viral siRNA profiles revealed, for the first time, that exogenous siRNAs produced in the transgenic rootstocks can be transferred to the nontransgenic scions over a long distance (about 1.2 m above graft unions). In addition, virus inoculation experiments indicated that the transferred PNRSV-hpRNA-specific siRNAs could enhance PNRSV resistance in nontransgenic scions. These findings demonstrate the potential of using transgenic rootstocks to produce improved products of nontransgenic scions of grafted trees.

Acknowledgements

We thank Dr Gregory A. Lang for kindly providing sweet cherry buds for grafting, Mr Aaron Walworth for the assistance of grafting for PNRSV inoculation, and Dr Jeff Landgraf and Mr Kevin M. Carr for small RNA sequencing. We also thank Dr Martin J. Bukovac and Dr James D. Kelly for their comments. This paper is dedicated to the memory of Dr Kenneth C. Sink (1937–2013) who initiated this project in 2006.

References

Ali, E.M., Kobayashi, K., Yamaoka, N., Ishikawa, M. and Nishiguchi, M. (2013) Graft transmission of RNA silencing to non-transgenic scions for conferring virus resistance in tobacco. *PLoS ONE*, **8**, e63257.

Aloni, B., Cohen, R., Karni, L., Aktas, H. and Edelstein, M. (2010) Hormonal signaling in rootstock-scion interactions. *Sci. Hortic.* **127**, 119–126.

Bai, S.L., Kasai, A., Yamada, K., Li, T.Z. and Harada, T. (2011) A mobile signal transported over a long distance induces systemic transcriptional gene silencing in a grafted partner. *J. Exp. Bot.* **62**, 4561–4570.

Barbara, D.J., Thresh, J.M. and Casper, R. (1978) Rapid detection and serotyping of prunus necrotic ringspot virus in perennial crops by enzyme-linked immunosorbent assay. *Ann. Appl. Biol.* **90**, 395–399.

Baulcombe, D. (2004) RNA silencing in plants. *Nature*, **431**, 356–363.

Brodersen, P. and Voinnet, O. (2006) The diversity of RNA silencing pathways in plants. *Trends Genet.* **22**, 268–280.

Chitwood, D.H. and Timmermans, M.C. (2010) Small RNAs are on the move. *Nature*, **467**, 415–419.

Cui, H.-G., Hong, N., Wang, G.-P. and Wang, A.-M. (2013) Genomic segments RNA1 and RNA2 of *Prunus necrotic ringspot virus* codetermine viral pathogenicity to adapt to alternating natural *Prunus* hosts. *Mol. Plant Microbe Interact.* **26**, 515–527.

Dunoyer, P., Brosnan, C.A., Schott, G., Wang, Y., Jay, F., Alioua, A., Himber, C. and Voinnet, O. (2010) An endogenous, systemic RNAi pathway in plants. *EMBO J.* **29**, 1699–1712.

Flachowsky, H., Tränkner, C., Szankowski, I., Waidmann, S., Hanke, M., Treutter, D. and Fischer, C.T. (2012) RNA-mediated gene silencing signals are not graft transmissible from the rootstock to the scion in greenhouse-grown apple plants *Malus* sp. *Int. J. Mol. Sci.* **13**, 9992–10009.

Fusaro, A.F., Matthew, L., Smith, N.A., Curtin, S.J., Dedic-Hagan, J., Ellacott, G.A., Watson, J.M., Wang, M.B., Brosnan, C., Carroll, B.J. and Waterhouse, P.M. (2006) RNA interference-inducing hairpin RNAs in plants act through the viral defence pathway. *EMBO Rep.* **7**, 1168–1175.

Gonçalves, B., Moutinho-Pereira, J., Santos, A., Silva, A.P., Bacelar, E., Correia, C. and Rosa, E. (2006) Scion-rootstock interaction affects the physiology and fruit quality of sweet cherry. *Tree Physiol.* **26**, 93–104.

Kalantidis, K. (2004) Grafting the way to the systemic silencing signal in plants. *PLoS Biol.* **2**, e224.

Kasai, A., Bai, S., Li, T. and Harada, T. (2011) Graft-transmitted siRNA signal from the root induces visual manifestation of endogenous post-transcriptional gene silencing in the scion. *PLoS ONE*, **6**, e16895.

Kasai, A., Sano, T. and Harada, T. (2013) Scion on a stock producing siRNAs of potato spindle tuber viroid (PSTVd) attenuates accumulation of the viroid. *PLoS ONE*, **8**, e57736.

Kircher, M., Sawyer, S. and Meyer, M. (2011) Double indexing overcomes inaccuracies in multiplex sequencing on the Illumina platform. *Nucleic Acids Res.* **40**, e3.

Koepke, T. and Dhingra, A. (2013) Rootstock scion somatogenetic interactions in perennial composite plants. *Plant Cell Rep.* **32**, 1321–1337.

Kusaba, M. (2004) RNA interference in crop plants. *Curr. Opin. Biotechnol.* **15**, 139–143.

Li, R., Gao, S., Hernandez, A.G., Wechter, W.P., Fei, Z. and Ling, K.-S. (2012) Deep sequencing of small RNAs in tomato for virus and viroid identification and strain differentiation. *PLoS ONE*, **7**, e3712.

Liang, D.C., White, R.G. and Waterhouse, P.M. (2012) Gene silencing in Arabidopsis spreads from the root to the shoot, through a gating barrier, by template-dependent, nonvascular, cell-to-cell movement. *Plant Physiol.* **159**, 984–1000.

Martinez-Ballesta, M.C., Alcaraz-Lopez, C., Muries, B., Mota-Cadenas, C. and Carvajal, M. (2010) Physiological aspects of rootstock-scion interactions. *Sci. Hortic.* **127**, 112–118.

Matzke, M.A. and Birchler, J.A. (2005) RNAi-mediated pathways in the nucleus. *Nat. Rev. Genet.* **6**, 24–35.

Mitter, N., Koundal, V., Williams, S. and Pappu, H. (2013) Differential expression of tomato spotted Wilt Virus-derived viral small RNAs in infected commercial and experimental host plants. *PLoS ONE*, **8**, e76276.

Mlotshwa, S., Voinnet, O., Mette, M.F., Matzke, M., Vaucheret, H., Ding, S.W., Pruss, G. and Vance, V.B. (2002) RNA silencing and the mobile silencing signal. *Plant Cell*, **14**, S289–S301.

Molnar, A., Melnyk, C.W., Bassett, A., Hardcastle, T.J., Dunn, R. and Baulcombe, D.C. (2010) Small silencing RNAs in plants are mobile and direct epigenetic modification in recipient cells. *Science*, **328**, 872–875.

Mroczka, A., Roberts, P.D., Fillatti, J.J., Wiggins, B.E., Ulmasov, T. and Voelker, T. (2010) An intron sense suppression construct targeting soybean *FAD2-1* requires a double-stranded RNA-producing inverted repeat T-DNA insert. *Plant Physiol.* **153**, 882–891.

Palauqui, J.C., Elmayan, T., Pollien, J.M. and Vaucheret, J. (1997) Systemic acquired silencing: transgene-specific post-translational silencing is transmitted by grafting from silenced stocks to non-silenced scions. *EMBO J.* **16**, 4738–4745.

Silva, T.F., Romanel, E.A.C., Andrade, R.R.S., Farinelli, L., Osteras, M., Deluen, C., Correa, R.L., Schrago, C.E.G. and Vaslin, M.F.S. (2011) Profile of small interfering RNAs from cotton plants infected with the polerovirus Cotton leafroll dwarf virus. *BMC Mol. Biol.* **12**, 40.

Smith, N.A., Singh, S.P., Wang, M.B., Stoutjesdijk, P.A., Green, A.G. and Waterhouse, P.M. (2000) Total silencing by intron-spliced hairpin RNAs. *Nature*, **407**, 319–320.

Song, G.-Q. and Sink, K.C. (2006) Transformation of Montmorency sour cherry (*Prunus cerasus* L.) and Gisela 6 (*P. cerasus* × *P. canescens*) cherry rootstock mediated by *Agrobacterium tumefaciens*. *Plant Cell Rep.* **25**, 117–123.

Song, G.-Q., Sink, K.C., Walworth, A.E., Cook, M.A., Allison, R.F. and Lang, G.A. (2013) Engineering cherry rootstocks with resistance to Prunus necrotic ring spot virus through RNAi-mediated silencing. *Plant Biotechnol. J.* **11**, 702–708.

Stoutjesdijk, P.A., Singh, S.P., Liu, Q., Hurlstone, C.J., Waterhouse, P.A. and Green, A.G. (2002) hpRNA-mediated targeting of the *Arabidopsis FAD2* gene gives highly efficient and stable silencing. *Plant Physiol.* **129**, 1723–1731.

Zamboni, A., Pierantoni, L. and de Franceschi, P. (2008) Total RNA extraction from strawberry tree (*Arbutus unedo*) and several other woody-plants. *IForest-Biogeosci. For.* **1**, 122–125.

Zhao, D. and Song, G.-Q. (2014) High-throughput sequencing as an effective approach in profiling small RNAs derived from a hairpin RNA expression vector in woody plants. *Plant Sci.* DOI: 10.1016/j.plantsci.2014.02.013 (in press).

14

Overexpression of *MdbHLH104* gene enhances the tolerance to iron deficiency in apple

Qiang Zhao, Yi-Ran Ren, Qing-Jie Wang, Yu-Xin Yao, Chun-Xiang You and Yu-Jin Hao*

National Key Laboratory of Crop Biology, National Research Center for Apple Engineering and Technology, College of Horticulture Science and Engineering, Shandong Agricultural University, Tai-An, Shandong, China

*Correspondence

email haoyujin@sdau.edu.cn

Keywords: apple, IVc subgroup bHLH transcription factor, plasma membrane H⁺-ATPase, iron deficiency, iron uptake.

Summary

Fe deficiency is a widespread nutritional disorder in plants. The basic helix-loop-helix (bHLH) transcription factors (TFs), especially Ib subgroup bHLH TFs which are involved in iron uptake, have been identified. In this study, an IVc subgroup bHLH TF MdbHLH104 was identified and characterized as a key component in the response to Fe deficiency in apple. The overexpression of the *MdbHLH104* gene noticeably increased the H^+-ATPase activity under iron limitation conditions and the tolerance to Fe deficiency in transgenic apple plants and calli. Further investigation showed that MdbHLH104 proteins bonded directly to the promoter of the *MdAHA8* gene, thereby positively regulating its expression, the plasma membrane (PM) H^+-ATPase activity and Fe uptake. Similarly, MdbHLH104 directly modulated the expression of three Fe-responsive bHLH genes, *MdbHLH38*, *MdbHLH39* and *MdPYE*. In addition, MdbHLH104 interacted with 5 other IVc subgroup bHLH proteins to coregulate the expression of the *MdAHA8* gene, the activity of PM H^+-ATPase and the content of Fe in apple calli. Therefore, MdbHLH104 acts together with other apple bHLH TFs to regulate Fe uptake by modulating the expression of the *MdAHA8* gene and the activity of PM H^+-ATPase in apple.

Introduction

As an essential mineral element for plants, iron is required for DNA synthesis, photosynthesis, nitrogen fixation, hormone synthesis and electron transport in the respiratory chain (Briat and Lobréaux, 1997). Although the total Fe content in earth generally satisfies plants' requirement, its availability is very low due to the unsuitable pH environments and low solubility in calcareous soils and anaerobic conditions, and approximately 30% of the world's soils are considered Fe limiting for plant growth (Korcak, 1988). Therefore, Fe deficiency is one of the major factors limiting plant growth and development (Guerinot and Yi, 1994). Thus, to maintain appropriate amounts of iron, plants have developed a number of sophisticated mechanisms to acclimate themselves to the surrounding conditions, including Fe limitation.

The need for the efficient acquisition of iron from soil has resulted in the evolution of two phylogenetically distinct uptake strategies, that is strategy I in dicotyledonous plants and strategy II in graminaceous monocots. Strategy I plants uptake Fe in a three-step process: the solubilization of Fe^{3+} complexes through rhizosphere acidification, the reduction of ferric (Fe^{3+}) into ferrous (Fe^{2+}) and lastly, the uptake of the ferrous into root cells (Marschner, 1995; Römheld and Marschner, 1986; Santi and Schmidt, 2009). In contrast, strategy II plants synthesize mugineic acids in root and chelate iron to form Fe^{3+}–MA complexes, which are transported into the cells of roots by yellow stripe transporters (Marschner and Römheld, 1986; Römheld and Marschner, 1986).

In strategy I plants, the transformation of Fe^{3+} to Fe^{2+} depends on ferric oxidoreductase ferric reductase oxidase 2 (FRO2), whereas the uptake of Fe^{2+} into root cells depends on iron transporter iron-regulated transporter 1 (IRT1). Both *FRO2* and *IRT1* genes are induced by Fe deficiency. It has been found that bHLH transcriptional factors (TFs) are involved in the regulation of

Fe acquisition and homeostasis. In *Arabidopsis*, genes of bHLH subgroups Ib and IVc are induced by iron-deficient conditions (Li *et al.*, 2006). The Ib subgroup bHLH TFs characterized with a function in Fe deficiency are FER in tomato and its homolog FIT (FER-like iron deficiency-induced transcription factor) in *Arabidopsis* (Ling *et al.*, 2002; Yuan *et al.*, 2005). FIT directly binds to the promoters of *FRO2* and *IRT1* and up-regulates their expression under Fe deficiency (Colangelo and Guerinot, 2004).

Prior to the reduction of Fe^{3+} to Fe^{2+}, Fe^{3+} complexes must be solubilized through rhizosphere acidification (Marschner, 1995). In dicotyledonous plants, the plasma membrane (PM) H^+-ATPase (EC 3.6.1.35) is responsible for the proton extrusion out of cells and the formation of rhizosphere acidification, which has a huge effect on the soluble of Fe in the vicinity of the roots (Dell'Orto *et al.*, 2000; Schmidt, 2003). In addition, the action of PM H^+-ATPase generates an electrochemical gradient, which constitutes a driving force for the transport of mineral nutrients, toxic ions, solutes and metabolites across the PM. Therefore, PM H^+-ATPase plays a crucial role in plants' responses to various environmental factors such as saline stress, low solution pH, nutrient supply and Fe deficiency (Niu *et al.*, 1993; Schubert and Yan, 1997; Yan *et al.*, 1998; Dell'Orto *et al.*, 2000; Palmgren, 2001).

In *Arabidopsis*, PM H^+-ATPases are encoded by 11 *AHA* genes, which are induced by various environmental stimuli. Among them, the expressions of *AHA2*, *AHA3*, *AHA4* and *AHA7* are up-regulated by Fe deficiency. In response to the absence of iron, *AHA2* is responsible for the major acidification activity, whereas *AHA7* may regulate root hair formation (Santi and Schmidt, 2009). Compared with the direct regulation of *FRO2* and *IRT1* by Ib bHLH TF FIT, it is largely unknown whether and how the PM H^+-ATPase gene is regulated by bHLH TF. Although *AHA2* expression is up-regulated in *FIT* overexpression plants than *fit-3* mutant in response to iron deficiency (Ivanov *et al.*, 2012; Long *et al.*, 2010), several evidences show that it appears not to be directly controlled by FIT, suggesting a different induction pathway for *AHA2* gene, compared with that for *IRT1* and *FRO2* genes (Ivanov *et al.*, 2012; Santi and Schmidt, 2009).

In addition to those of the Ib subgroup, IVc subgroup bHLHs influence the Fe chelate reductase activity and the acidification of rhizospheres to regulate plant growth and development under Fe-deficient conditions. Among them, PYE and IAA-Leu Resistant3 (ILR3, also named as bHLH105) target metal (or iron) homeostasis genes, which are involved in intracellular and long-distance metal (or iron) transport (Rampey *et al.*, 2006; Selote *et al.*, 2015). Meanwhile, PYE regulates the acidification of rhizospheres under Fe-deficient conditions (Long *et al.*, 2010; Selote *et al.*, 2015). Most recently, it is found in *Arabidopsis* that the mutations to IVc subgroup bHLH genes *AtbHLH104* and *AtbHLH105* greatly reduce the tolerance to Fe deficiency, whereas their overexpressions improve the tolerance and lead to an accumulation of excess Fe under soil-grown conditions. *AtbHLH104* also regulates the acidification of rhizospheres under Fe-deficient conditions (Zhang *et al.*, 2015). In chrysanthemum, *CmbHLH1*, which is highly similar to *AtbHLH105*, regulates Fe uptake via mediating the acidification of the rhizosphere by enhancing the transcription of the H^+-ATPase-encoding gene *CmHA* under iron-shortage conditions (Zhao *et al.*, 2014).

In this study, a Fe-responsive bHLH TF gene *MdbHLH104* was isolated from the apple. It was identified to encode an IVc bHLH subgroup member and was induced by Fe deficiency. After being genetically transformed into apple plant and calli, MdbHLH104 was characterized by a crucial function in Fe acquisition and the tolerance to Fe deficiency by directly binding to the promoter regions of the *MdAHA8*, *MdbHLH38*, *MdbHLH39* and *MdPYE* genes, thereby modulating PM H^+-ATPase activity. Finally, the potential utilization of *MdbHLH104* in the genetic improvement of fruit tree tolerance to iron deficiency is discussed.

Results

bHLH transcription factor MdbHLH104 is involved in responding to iron deficiency and promotes iron accumulation

BlastX search and phylogenetic tree analysis showed that there are 6 IVc subgroup bHLH TFs in apple (Figure S1a,b). Among them, the expression of the *MdbHLH104* gene was noticeably induced by Fe deficiency (Figure S1c). It is also highly expressed in root (Figure S1d). To characterize its function, an expression vector *35S::MdbHLH104-GFP* was constructed and transformed into apple with an *Agrobacterium*-mediated method. As a result, five independent transgenic apple lines were obtained (Figure S2). Three lines, L1, L2 and L3, were chosen for further investigation, whereas the wild-type (WT) apple was used as a control. Expression analysis and an immunoblot assay with an anti-GFP antibody showed that all three transgenic apple lines generated many more *MdbHLH104* transcripts and produced MdbHLH104-GFP fusion proteins (Figures 1a,b and S2d), indicating that *MdbHLH104* was overly expressed in apple.

To examine whether MdbHLH104 protein plays a role in response to iron starvation, three transgenic apple lines and the WT control were allowed to grow for 20 days under Fe-sufficient conditions and then shifted to Fe-deficient conditions for another 30 days. The results showed that the both transgenic and WT apple plantlets grew normally under Fe-sufficient conditions.

After being treated with Fe starvation, the WT control exhibited much more severe chlorosis in appearance (Figure 1c), which was also indicated as low chlorophyll contents in the unfolding young leaves than in three transgenic lines (Figure 1c). In contrast, the three transgenic lines showed much less chlorosis than the control (Figure 1c). Furthermore, the iron content was measured in the unfolding young leaves. The result indicated that three *35S::MdbHLH104-GFP* transgenic lines accumulated much higher iron than the WT control under iron-deficient conditions (Figure 1c). These results indicated that the overexpression of *MdbHLH104* confers remarkably increased tolerance to Fe deficiency in transgenic apple plantlets.

Plant root responds to iron deprivation by pumping out protons into the apoplast, which lowers the rhizosphere pH and solubilizes iron, thus increasing iron availability (Yi *et al.*, 1994). To test whether MdbHLH104 influences the rhizosphere pH in response to iron deprivation, the transgenic and WT apple plantlets grown under normal conditions were treated for 7 days on a Fe-deficient medium. Subsequently, they were shifted to a medium containing the pH indicator bromocresol purple for staining. The result showed that transgenic lines exhibited more obvious rhizosphere acidification, as indicated by the yellow colour of the medium around the roots, than the WT control under Fe-deficient conditions, and no phenotypic differences between the transgenic lines and WT control were revealed under Fe-sufficient conditions. Furthermore, the transport activity of PM H^+-ATPase was determined. The result indicated that the three transgenic lines exhibited a notably increased ATPase activity relative to the WT control under Fe-deficient conditions and no changes under Fe-sufficient condition (Figure 1e). These findings demonstrated that MdbHLH104 overexpression leads to an increased acidification of the rhizosphere in response to iron deficiency.

MdbHLH104 binds to the promoter of *MdAHA8* and activates its transcription

Among 11 *Arabidopsis* AHAs, AHA1 and AHA2 are two major PM H^+-ATPases responsible for rhizosphere acidification (Santi and Schmidt, 2009). Correspondingly, there are 18 *MdAHA* genes in the apple genome (http://genomics.research.iasma.it/). The phylogenetic tree demonstrated that 7 MdAHAs are close to AHA1 and AHA2 (Figure S3a). RT-PCR analysis showed that among them, only *MdAHA8* was noticeably up-regulated in the transgenic lines, compared with the WT control under Fe-deficient conditions (Figure 2a).

The bHLH transcription factors have been reported to be associated with the E-box (5'-CANNTG-3') or G-box (5'-CACGTG-3') cis element in the promoters of their target genes (Fisher and Goding, 1992). To elucidate how *MdAHA8* is regulated by MdbHLH104, its promoter region was searched for putative *cis* elements that are recognized by MdbHLH104. As a result, 6 E-box elements (CANNTG), that is P1 to P6, were found (Figures 2b and S3b). To verify whether MdbHLH104 binds to those elements, a chromatin immunoprecipitation PCR (ChIP-PCR) assay was conducted with an anti-GFP antibody and six pairs of primers specific to 6 E-box elements using *35S::GFP* and *35S::MdbHLH104-GFP* transgenic apple calli, which overexpressed *GFP* and *MdbHLH104-GFP*, respectively. A fragment of the actin promoter containing an E-box motif was used as a negative control. The ChIP-PCR assay demonstrated that only the P3-containing promoter regions of *MdAHA8*, but not the other 5 regions, were enriched by ChIP in the *35S::MdbHLH104-GFP* transgenic calli compared to the *35S::GFP* control (Figures S3c and 2b). In

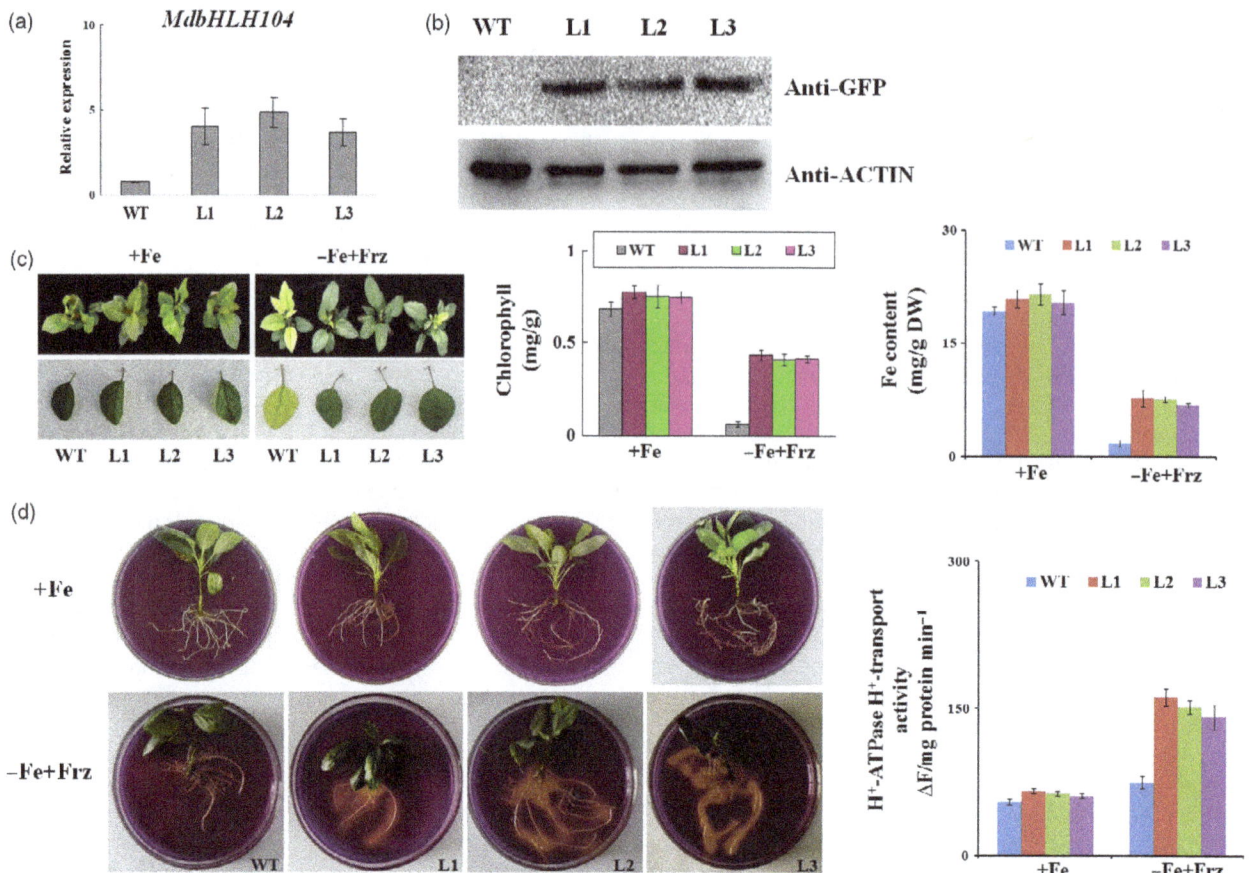

Figure 1 The phenotype of *MdbHLH104* transgenic apple plantlets under Fe-sufficient and deficient conditions. (a) Expression level of the *MdbHLH104* gene in the transgenic apple lines and the wild-type (empty vector WT control). (b) The level of the MdbHLH104-GFP fusion protein in *35S::MdbHLH104-GFP* transgenic apple lines, as determined by immunoblot analysis using an anti-GFP antibody. The anti-actin antibody was used as loading control. (c) The appearance, total chlorophyll contents and iron contents of *35S::MdbHLH104-GFP* transgenic apple lines and the WT control grown for 1 month on Fe-sufficient (+Fe) or Fe-deficient (−Fe + Frz) media. The data represent the means ± SD of three independent experiments. DW: dry weight. (d) The rhizosphere acidification and PM H$^+$-ATPase activity of wild-type and *35S::MdbHLH104-GFP* transgenic apple lines grown for 7 days on Fe-sufficient (+Fe) or Fe-deficient (−Fe + Frz) media. The yellow colour around the roots stained with bromocresol purple indicates rhizosphere acidification, and the plasma membrane vesicles were isolated for PM H$^+$-ATPase activity analysis. The data represent the means ± SD of three independent experiments.

addition, there are E-box *cis* elements in the other 6 MdAHAs, that is *MdAHA1*, *MdAHA3*, *MdAHA7*, *MdAHA9*, *MdAHA11* and *MdAHA12* (Figure S3b). However, ChIP-PCR assays demonstrated that none of them recruited MdbHLH104-GFP proteins (Figure S3c). These results provide *in vivo* evidence for the binding of MdbHLH104 to the *MdAHA8* promoter region around the P3 element.

To verify the direct binding of MdbHLH104 to the P3-containing recognition site in the *MdAHA8* promoter, an electrophoretic mobility shift assay (EMSA) was performed with an oligo-probe containing a P3 *cis* element using purified recombinant His-MdbHLH104 fusion protein. As a result, specific DNA–MdbHLH104 protein complexes were detected when the P3 (sequence)-containing sequence was used as a labelled oligo-probe. The formation of these complexes was reduced when increasing amounts of the unlabelled P3 competitor probe with the same sequence were added. This competition was not observed when the mutated version was used (Figure 2c). This specificity of competition verifies the physical interaction between the *MdAHA8* promoter region and MdbHLH104 that requires the P3 *cis* element.

To examine whether MdbHLH104 directly activates the *MdAHA8* promoter, a biochemical staining assay was performed using *GUS* as the reporter gene. The construct P$_{MdAHA8}$::*GUS* was genetically transformed into the WT apple calli, and then, *35S::MdbHLH104-GFP* was introduced into the transgenic calli containing P$_{MdAHA8}$::*GUS* (Figure S3d). The biochemical staining assay showed that the P$_{MdAHA8}$::*GUS*+*35S::MdbHLH104-GFP* double-transformed calli have higher GUS activity than the P$_{MdAHA8}$::*GUS* one (Figure 2d,e), indicating that MdbHLH104 is a positive regulator for the *MdAHA8* promoter.

Taken together, it may be concluded that MdbHLH104 activates the transcription of the *MdAHA8* gene by directly binding to the P3 *cis* element in its promoter. In addition, MdbHLH104 also binds to the promoters of *MdbHLH38*, *MdbHLH39* and *MdPYE* (Figure S4a–c).

MdbHLH104 modulates H$^+$-ATPase activity and Fe acquisition by regulating *MdAHA8* under Fe deficiency

Because it is difficult to obtain transgenic apple plants, particularly for those containing two or more exogenous genes, apple calli were thereafter used for genetic transformation and

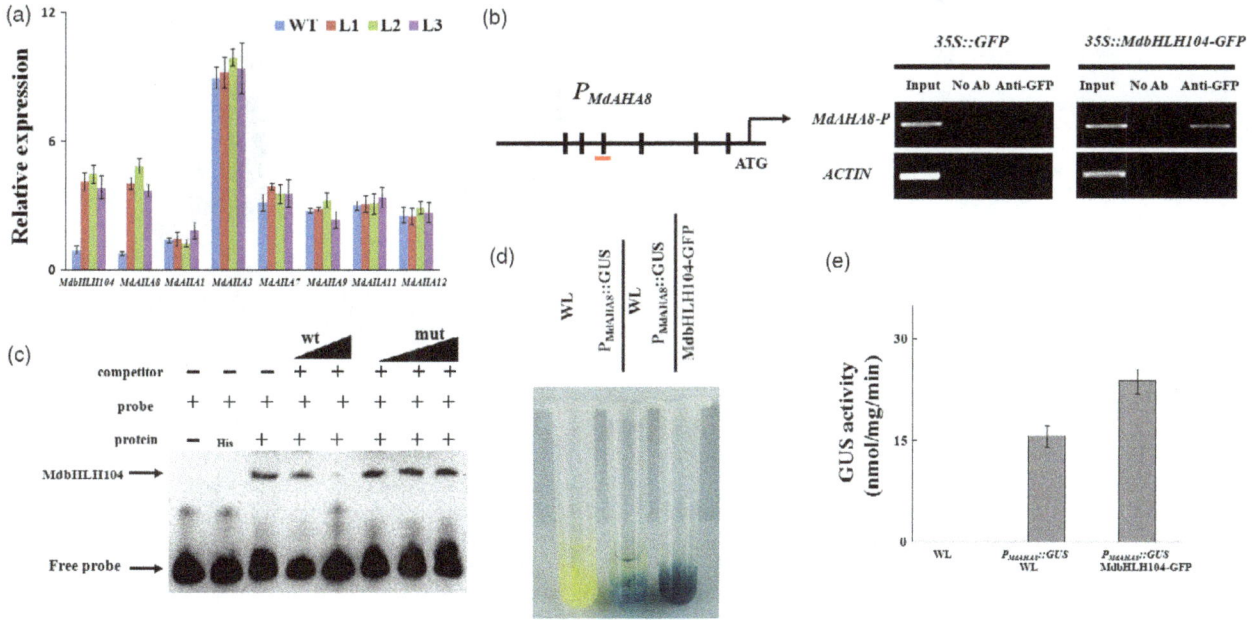

Figure 2 MdbHLH104 directly activates the expression of *MdAHA8* gene. (a) qRT-PCR assays for *MdbHLH104* and *MdAHA* genes in transgenic apple lines. (b) Illustration of the *MdAHA8* promoter region indicating the presence of E-box DNA motifs. Transverse lines show the positions of primers used in the ChIP-PCR experiment. ChIP assays were performed using the *35S::GFP* and *35S::MdbHLH104-GFP* apple calli. A region containing E-box in the actin promoter is negative control. (c) MdbHLH104 binds to the E-box motifs present in the *MdAHA8* promoter *in vitro*, as indicated by an EMSA method. The *MdAHA8* promoter fragment containing the E-box motifs was incubated with His-MdbHLH104 protein. Competition for MdbHLH104 binding was performed with 50× and 100× unlabelled probes (wt) or G-box-mutated probes (mut). His was used as the control. Mut indicates mutated probes. '+' indicates presence, and '−' indicates absence. (d) and (e) GUS staining assay and activity analysis of *MdAHA8* expression promoter using P_{MdAHA8}::*GUS* and *35S::MdbHLH104-GFP*+P_{MdAHA8}::*GUS* transgenic apple calli. GUS activity was measured using a 4-methylumbelliferyl-D-glucuronide assay. The data represent the means ± SD of three independent experiments.

further investigation. To examine whether apple calli can be used as a model system, apple transgenic calli containing *35S::GFP* and *35S::MdbHLH104-GFP*, respectively, were used to characterize the function of MdbHLH104 in modulating PM H^+-ATPase activity and Fe acquisition (Figure S5a). The result showed that MdbHLH104 increased the acidification of the apple calli and positively regulated the activity of PM H^+-ATPase (Figure S5b). In addition, the Fe^{2+} was analysed with a FerroZine method in the apple calli. The result showed that *35S::MdbHLH104-GFP* transgenic apple calli exhibited a deeper colour than the *35S::GFP* control. Correspondingly, the former did accumulate more Fe^{2+} than the latter under Fe-deficient conditions (Figure S5c). Therefore, MdbHLH104 regulated PM H^+-ATPase activity and Fe acquisition in the apple calli just as it did in the apple plant.

Using the apple calli, MdAHA8 was characterized with a function in PM H^+-ATPase activity and Fe acquisition. The full-length sense ORFs and antisense cDNA fragments of *MdAHA8* were used to construct expression vectors. Two *35S*-driven vectors, that is pIR-*MdAHA8* for overexpression and pIR-*MdAHA8-Anti* for suppression, were used for genetic transformation into apple calli. RT-PCR showed that transgenic calli pIR-*MdAHA8* and pIR-*MdAHA8-Anti* were obtained and used for the determination of H^+-ATPase activity and Fe acquisition (Figure 3a). The result showed that the pIR-*MdAHA8* transgenic calli exhibited higher and the pIR-*MdAHA8-Anti* calli exhibited lower PM H^+-ATPase activity than the WT control. As a result, as indicated by bromocresol purple staining into yellow colour, the pIR-*MdAHA8* transgenic calli pumped out more and the pIR-

MdAHA8-Anti calli pumped out less H^+ into the medium than the WT control under iron-sufficient and iron-shortage conditions (Figure 3b). Furthermore, pIR-*MdAHA8* transgenic calli accumulated more and the pIR-*MdAHA8-Anti* calli accumulated less PM H^+-ATPase activity than the WT control under iron-sufficient and iron-shortage conditions (Figure 3c). These findings indicated that MdAHA8 is involved in the regulation of PM H^+-ATPase activity.

Subsequently, the vector pIR-*MdAHA8-Anti* was genetically transformed into *35S::MdbHLH104-GFP* transgenic calli. As a result, a double transgenic calli that contained *35S::MdbHLH104-GFP* and pIR-*MdAHA8-Anti* were obtained and used for acidification assay and PM H^+-ATPase activity analysis. The result showed that the *35S::MdbHLH104-GFP*/pIR-*MdAHA8-Anti* calli exhibited acidification and PM H^+-ATPase activity that were noticeably reduced compared with the *35S::MdbHLH104-GFP* calli, but similar to the WT control under starvation conditions (Figure 4a,b). Furthermore, the *35S::MdbHLH104-GFP*+pIR-*MdAHA8-Anti* calli accumulated less Fe than *35S::MdbHLH104-GFP* calli, but were similar to the WT control under starvation conditions (Figure 4c). Therefore, *MdAHA8* is required for the MdbHLH104-mediated regulation of H^+-ATPase activity and Fe acquisition.

MdbHLH104 interacts with other apple IVc subgroup bHLH proteins to regulate MdAHA8 expression, PM H^+-ATPase activity and iron accumulation

In addition to MdbHLH104, apple contains 5 other IVc subgroup bHLH TFs, that is MdbHLH105, MdbHLH115, MdPYE, MdbHLH11

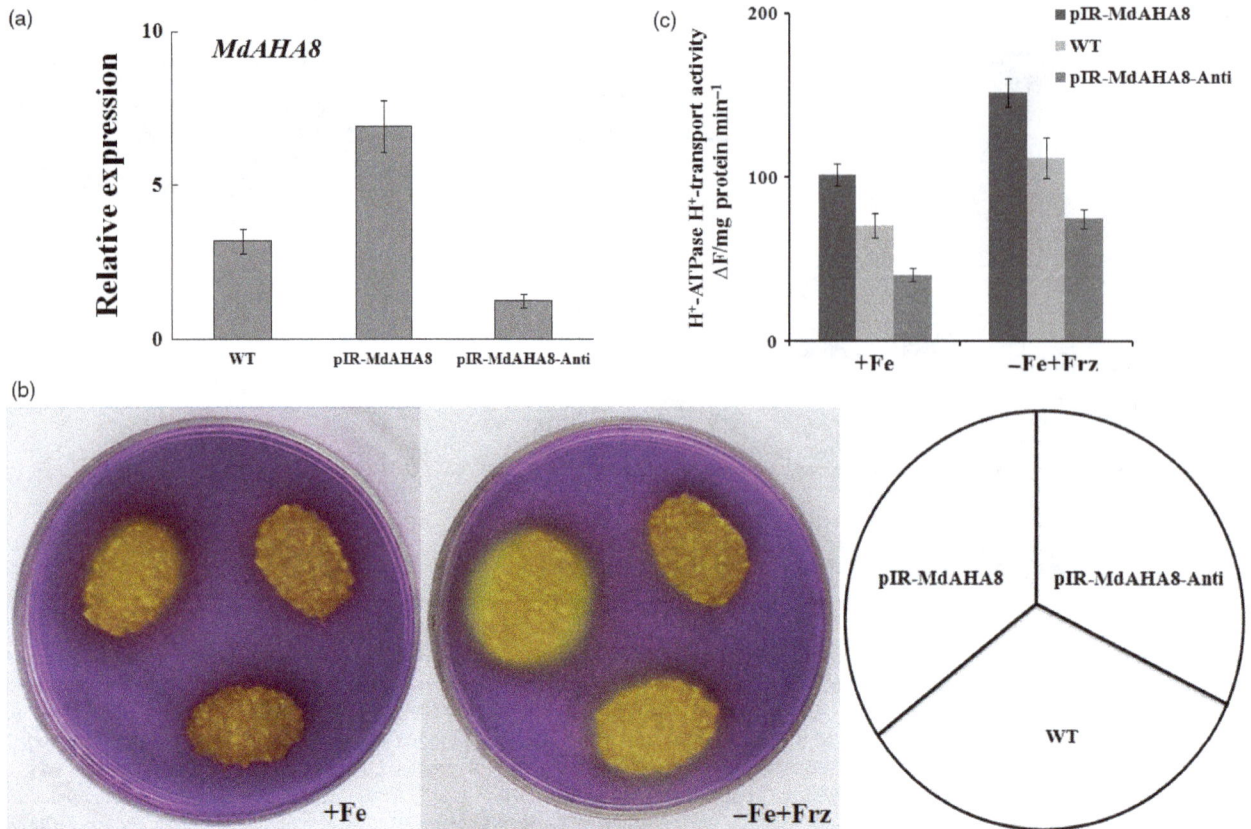

Figure 3 *MdAHA8* positively regulates PM H⁺-ATPase activity. (a) The relative expression of *MdAHA8* in wild-type, pIR-*MdAHA8* and pIR-*MdAHA8-Anti* transgenic apple calli. The data represent the means ± SD of three independent experiments. (b) Acidification analysis of *MdAHA8* transgenic apple calli treated on medium containing the pH indicator dye bromocresol purple. Acidification is indicated by yellow colour around the apple calli. The same is true below unless otherwise indicated. (c) PM H⁺-ATPase activity of wild-type, pIR-*MdAHA8* and pIR-*MdAHA8-Anti* transgenic apple calli grown for 7 days on Fe-sufficient (+Fe) or Fe-deficient (−Fe + Frz) media.

and MdbHLH121. To detect whether MdbHLH104 interacts with each of them, yeast two-hybrid assays and pull-down analysis were conducted. The full-length cDNA of MdbHLH104 was integrated into vector pGBT9 (BD-MdbHLH104) as bait, whereas that of each MdbHLH105, MdbHLH115, MdPYE, MdbHLH11 and MdbHLH121 into pGAD424 (AD-MdbHLHs) was integrated as preys. Positive X-α-gal activity was observed in yeasts that contained either pGBT9-MdbHLH104 plus each pGAD424-MdbHLHs grown on the -Trp/-Leu/-His/-Ade screening medium, but not in those containing pGBT9-MdbHLH104 plus the empty pGAD424 vector. The result indicated that MdbHLH104 interacted with the other IVc subgroup bHLH TFs MdbHLH105, MdbHLH115, MdPYE, MdbHLH11 and MdbHLH121 (Figure 5a). Furthermore, the interactions between MdbHLH104 and each apple IVc bHLH TFs were verified with pull-down analysis (Figure 5b).

Then, a yeast assay system was used to examine whether the interaction affects the function of MdbHLH104, thereby altering the activity of the *MdAHA8* gene promoter. The *MdAHA8* promoter region of 2510 bp upstream the start codon was cloned and fused with the reporter gene *GUS*, resulting in an expression cassette P*MdAHA8*::*GUS*. The cassette was inserted into the pBD-GAL4 vector, producing a plasmid pBD-P*MdAHA8*::*GUS*. Then, the coding sequence of MdbHLH104 was inserted into pBD-P*MdAHA8*::*GUS*, resulting in a plasmid pBD-*MdbHLH104*-P*MdAHA8*::*GUS*. Meanwhile, the coding sequences of 5 other IVc

subgroup bHLH TF genes were cloned into pAD-GAL4, respectively, to generate 5 plasmids, that is pAD-*MdbHLH115*, pAD-*MdbHLH11*, pAD-*MdbHLH121*, pAD-*MdPYE* and pAD-*MdbHLH105*. Subsequently, the pBD plasmids were genetically transformed alone or together with each pDA one into yeast cells. The histochemical assay showed the GUS activity was much higher in the transformants that contained pBD-*MdbHLH104*-P*MdAHA8*::*GUS* plus each of pAD-MdbHLHs plasmids than in those that contained it alone, indicating that the cotransformation of *MdPYE*, *MdbHLH105*, *MdbHLH115*, *MdbHLH11* and *MdbHLH121* promoted the function of MdbHLH104 to alter the activity of the *MdAHA8* promoter (Figure 6a).

To examine the biological function of the interaction between *MdbHLH104* and other IVc subgroup bHLH TFs in planta, expression vectors *35S::MdPYE*, *35S::MdbHLH105*, *35S::MdbHLH115*, *35S::MdbHLH11* and *35S::MdbHLH121* were constructed and cotransformed into *35S::MdbHLH104-GFP* transgenic calli, respectively. The resultant transgenic calli were used for ChIP-qPCR assays with anti-GFP antibody and primers specific to the promoter fragment of the *MdAHA8* gene. The results indicated a remarkable promotion of the recruitment of MdbHLH104 to the promoter fragment of *MdAHA8* (Figure 6b) when MdbHLH104 was cotransformed with each of the 5 other IVc subgroup bHLH TF genes.

Furthermore, a transient expression assay was conducted in transgenic apple calli to check the function of the interaction to

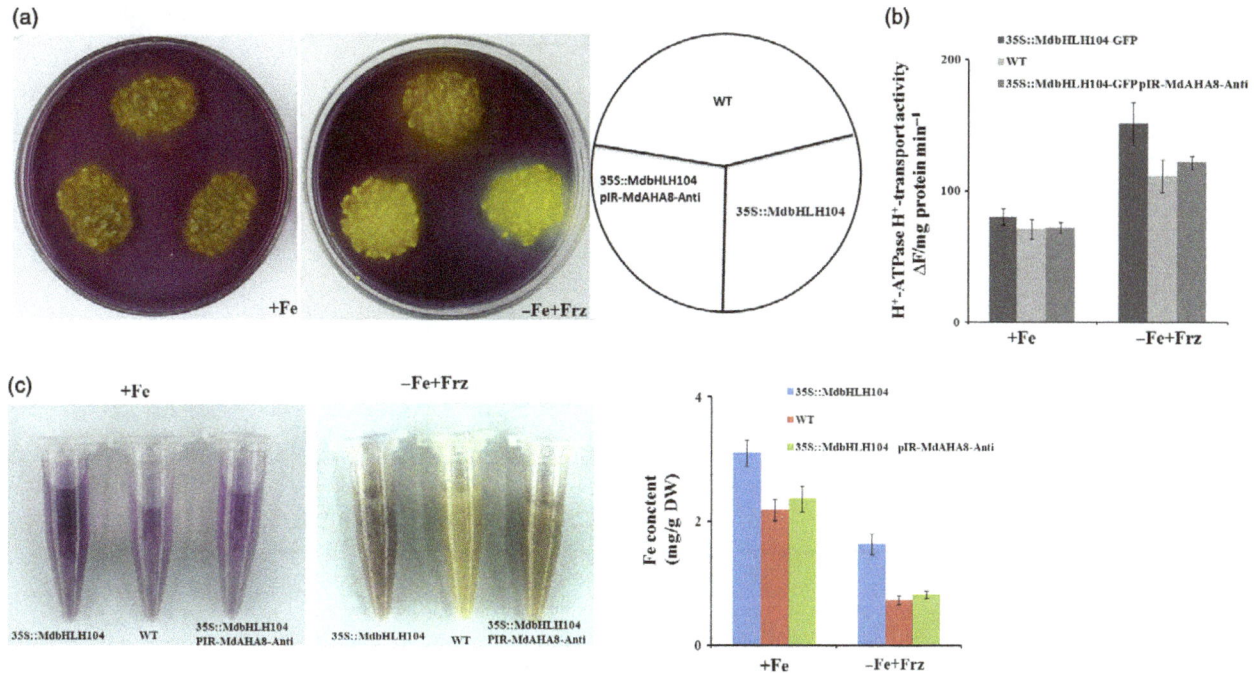

Figure 4 *MdAHA8* is required for MdbHLH104-mediated acidification and iron contents of responding iron deficient. (a) Acidification of wild-type, *35S::MdbHLH104-GFP* and *35S::MdbHLH104-GFP*/pIR-*MdAHA8-Anti* transgenic apple calli. (b) PM H⁺-ATPase activity in vesicles isolated from wild-type, *35S::MdbHLH104-GFP* and *35S::MdbHLH104-GFP*/pIR-*MdAHA8-Anti* transgenic apple calli treated with (+Fe) or without (−Fe + Frz) iron for 7 days. (c) Visualization of ferrous of Fe-sufficient and Fe-deficient conditions in wild-type, *35S::MdbHLH104-GFP* and *35S::MdbHLH104-GFP*/pIR-*MdAHA8-Anti* transgenic apple calli by FerroZine. The resulting Fe(II) is trapped by FerroZine to produce a red product. Fe content of wild-type, *35S::MdbHLH104-GFP* and *35S::MdbHLH104-GFP*/pIR-*MdAHA8-Anti* grown on Fe-sufficient (+Fe) or Fe-deficient (−Fe + Frz) media for 7 days. The data represent the means ± SD of three independent experiments. DW, dry weight.

modulate the activity of the MdAHA8 gene promoter. The coding sequences of 5 IVc subgroup bHLH TF genes were inserted into pIR viral vector, resulting in 5 transient expression viral vectors, that is pIR-*MdPYE*, pIR-*MdbHLH105*, pIR-*MdbHLH115*, pIR-*MdbHLH11* and pIR-*MdbHLH121*. The constructs were transiently transformed into P_{MdAHA8}::*GUS* plus *35S::MdbHLH104-GFP* transgenic apple calli background. IL60-1 was used as a control. The results indicated that the cotransformation of each pIR-MdbHLHs vector showed much higher GUS activity than the controls, that is IL60-1/pIR calli and P_{MdAHA8}::*GUS* plus *35S::MdbHLH104-GFP* (Figure 6c). Therefore, MdbHLH104 interacts with other IVc subgroup bHLH TFs to control the activity of the MdAHA8 gene.

The real-time RT-PCR analysis showed that the co-expression of each IVc subgroup bHLH TF together with *MdbHLH104* remarkably increased the transcript level of the *MdAHA8* gene compared to *MdbHLH104* alone (Figure 6d), demonstrating that the interaction between MdbHLH104 and each other IVc subgroup bHLH TF enhanced the expression of the MdAHA8 gene in apple calli. Finally, the PM H⁺-ATPase activity and the iron accumulation were detected in various transgenic apple calli, including *35S::MdbHLH104-GFP* and *35S::MdbHLH104-GFP* plus *35S::MdPYE*, *35S::MdbHLH105*, *35S::MdbHLH115*, *35S::MdbHLH11* or *35S::MdbHLH121*. The result showed that the calli that contained *35S::MdbHLH104-GFP* plus an interacting IVc subgroup TF gene exhibited higher PM H⁺-ATPase activity, pumped out more H⁺ into the medium and accumulated more iron than that containing *35S::MdbHLH104-GFP* alone (Figures 7 and 8). Therefore, MdbHLH104 functions together with other apple IVc subgroup

bHLH proteins to enhance PM H⁺-ATPase activity and promote iron uptake and accumulation in transgenic apple calli.

Discussion

Iron is essential for plants due to its various roles in life processes. Plant roots excrete protons mediated by PM H⁺-ATPase leading to the acidification of the rhizosphere, which in turn makes iron soluble and available in soil for uptake. In addition to proton extrusion and the associated electrochemical gradient, PM H⁺-ATPase supplies energy for iron uptake and transportation. Therefore, PM H⁺-ATPase plays a crucial role in iron acquisition and homeostasis in plants (Guerinot and Yi, 1994; Palmgren, 2001). In *Arabidopsis*, there are 11 PM H⁺-ATPases, which are encoded by genes AHA1 to AHA11 (Palmgren, 2001). Some of them are involved in Fe acquisition and homeostasis, particularly AHA2, which is crucial for the acidification of the rhizosphere (Baxter *et al.*, 2003; Haruta *et al.*, 2010). In addition, several PM H⁺-ATPase genes have been characterized by their involvement in the response to iron deficiency and the uptake of iron elements in different plant species, such as cucumber (Santi *et al.*, 2005). In this study, it was found that MdAHA8, which is one of the closest homologs among 18 apple PM H⁺-ATPases to *Arabidopsis* AHA2, transcriptionally responds to Fe deficiency. The overexpression of *MdAHA8* promoted the proton excretion of the transgenic apple calli and increased the H⁺-ATPase activity in response to iron deficiency (Figure 3).

Several types of TFs, such as bHLH, MYB and AP2 TFs, are involved in the regulation of Fe acquisition and homeostasis

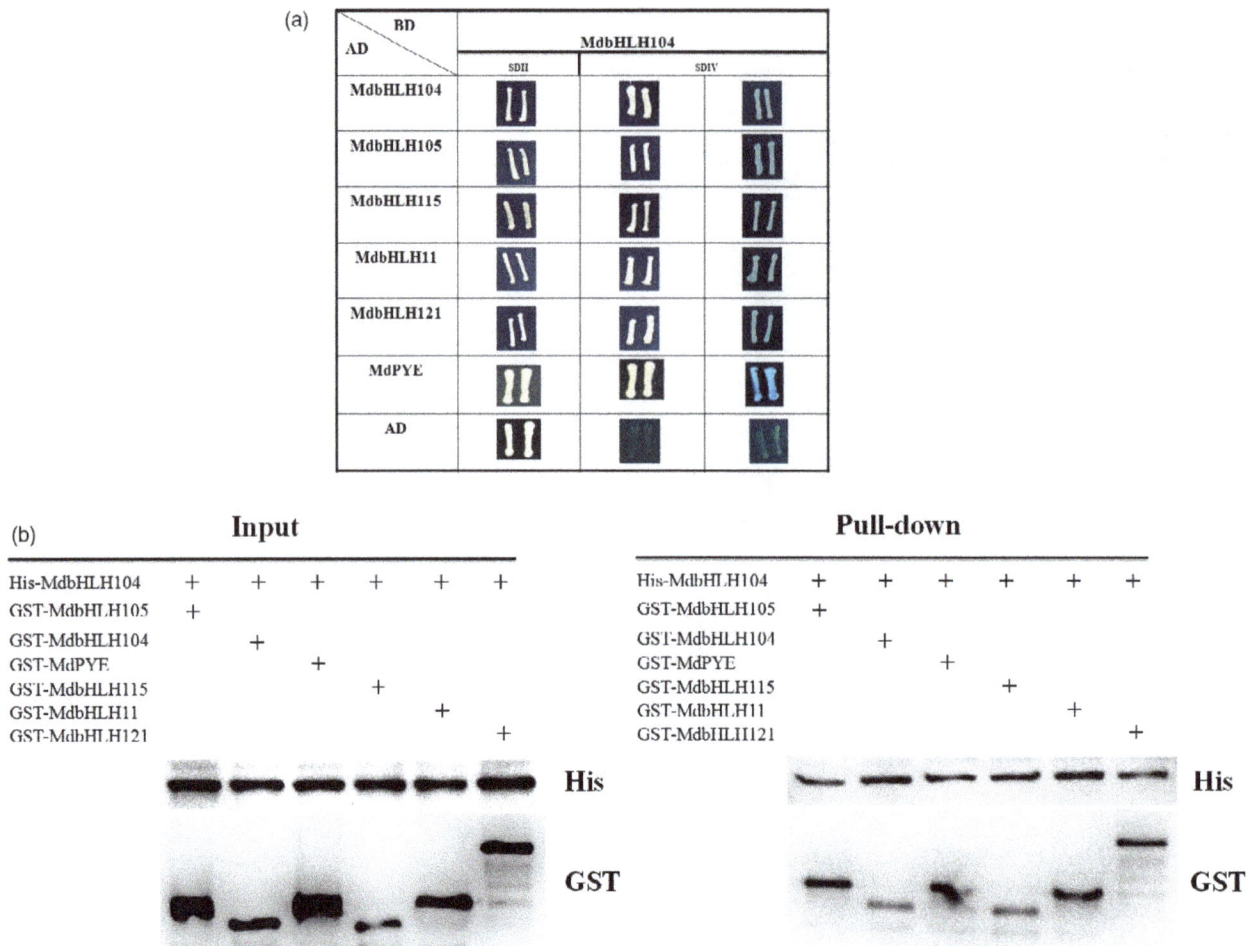

Figure 5 MdbHLH104 interacts with other IVc subgroup bHLH transcription factors. (a) MdbHLH104 interacts with other IVc subgroup bHLH transcription factors in yeast two-hybrid assays. The empty vector AD plus BD-MdbHLH104 was used as controls. The yeast cells were grown on SD/II and SDIV media. The x-α-gal assay was used to further confirm the positive interactions. (b) Pull-down assay showed the interaction of His-MdbHLH104 with GST-MdbHLHs. The His-MdbHLH104, GST-MdbHLHs and GST were expressed in BL21, and then total proteins were pulled down by Ni-agarose and detected using anti-His and anti-GST antibodies, respectively. '+' indicates presence.

(Hindt and Guerinot, 2012; Kobayashi and Nishizawa, 2012). Among them, bHLH TFs, particularly Ib and IVc subgroups bHLH TFs, play a central role in modulating the expressions of the major Fe acquisition genes (Yuan et al., 2005). The first bHLH TF, which is characterized by its function in response to iron deficiency, is an Ib subgroup bHLH TF FER in tomato (Ling et al., 2002). FIT is its Arabidopsis homolog. This bHLH and other Ib subgroup bHLHs such as bHLH38 and bHLH39 (the homologs of MdbHLH38 and MdbHLH39) act together to directly regulate the expressions of FRO2 and IRT1 genes under Fe-deficient and Fe-sufficient conditions (Ivanov et al., 2012; Wang et al., 2012; Yuan et al., 2008). However, it seems not to directly modulate the expression of PM H+-ATPase genes (Ivanov et al., 2012; Santi and Schmidt, 2009).

IVc subgroup bHLHs are also involved in Fe acquisition and homeostasis in Arabidopsis. Among them, PYE and bHLH104 regulate the acidification of rhizospheres under Fe-deficient conditions (Long et al., 2010; Selote et al., 2015; Zhang et al., 2015). CmbHLH1 is a chrysanthemum homolog of bHLH105. It regulates rhizosphere acidification and Fe uptake by enhancing the transcription of an H+-ATPase gene CmHA under iron-starvation conditions (Zhao et al., 2014). However, it is unclear

whether and how IVc bHLH TFs regulate the genes encoding PM H+-ATPases. In this study, an IVc subgroup bHLH TF MdbHLH104 is up-regulated by Fe starvation in apple (Figure S1). It directly binds to the promoter of MdAHA8 and promotes its expression (Figure 2b,c), resulting in the enhanced PM H+-ATPase activity (Figure 4). It is well known that PM H+-ATPase plays a crucial role in proton excretion from the plant root to the rhizospheric soil and in energy-promoted Fe uptake (Santi and Schmidt, 2009). MdbHLH104 transgenic apple plantlets and calli showed an enhanced proton excretion and an improved tolerance to Fe deficiency (Figure 1). Therefore, MdbHLH104 positively regulates the activity of PM H+-ATPase and the uptake and homeostasis of iron.

In Arabidopsis, Ib subgroup bHLH FIT is induced and directly up-regulates the expression of FRO2 and IRT1 under iron deficiency; however, its overexpression does not result in the strong induction of FRO2 and IRT1 under iron sufficiency (Colangelo and Guerinot, 2004; Jakoby et al., 2004). This is likely because these bHLH transcription factors within a subgroup form homodimer or heterodimer complexes with their family members, and both partners may be required for their regulatory function to the target genes (Heim et al., 2003). To be active as a

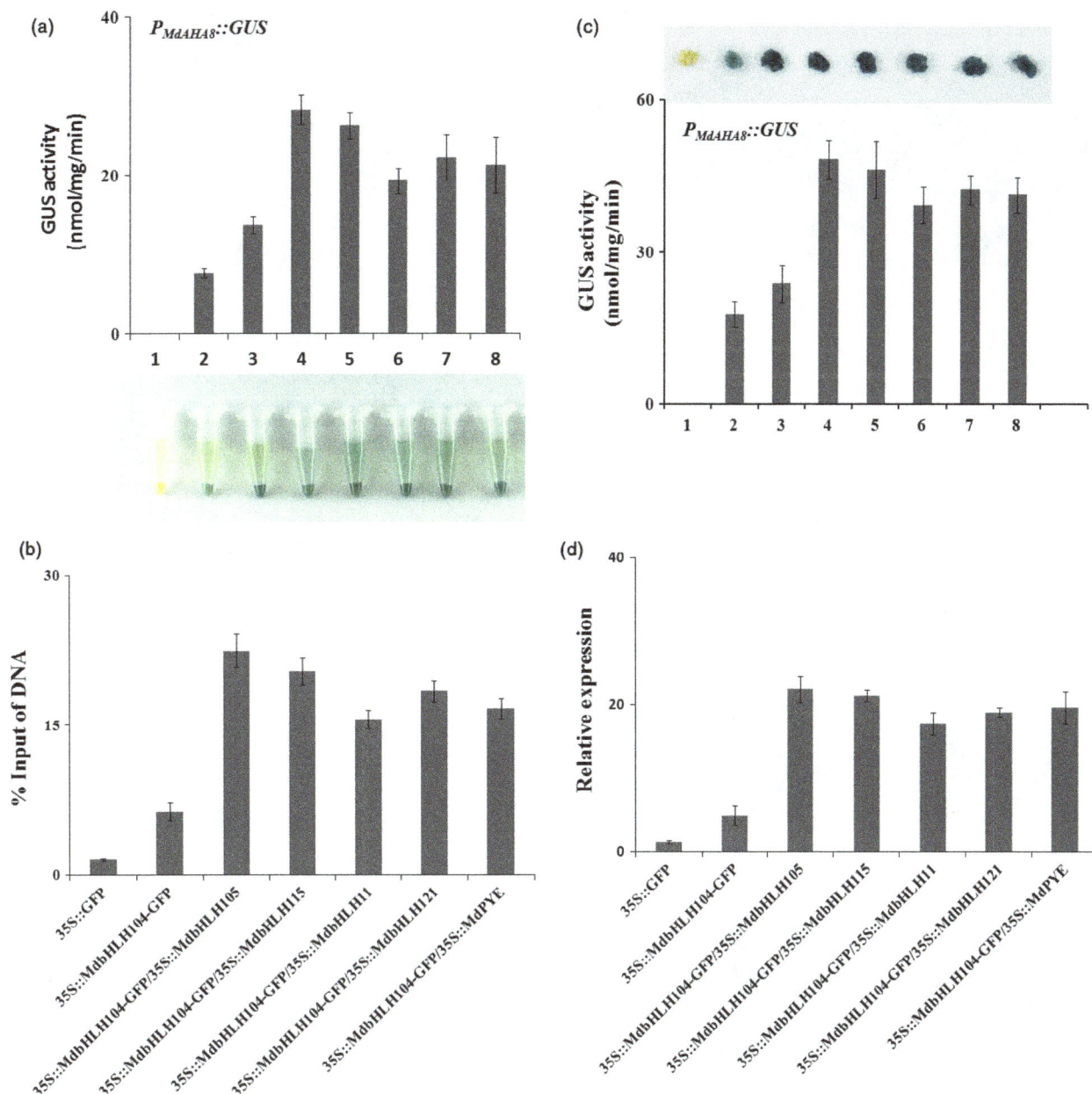

Figure 6 The Interaction with other MdbHLHs proteins affects the function of MdbHLH104 in the activation of *MdAHA8* promoter. (a) Transcription activation assay of *GUS* reporter gene in yeast cells. A series of transformant yeast cells containing different plasmid combinations, indicated as 1–8, were used. 1, pAD/pBD-*GUS*; 2, pAD/pBD-P_{MdAHA8}::*GUS*; 3, pAD/pBD-MdbHLH104-P_{MdAHA8}::*GUS*; 4–8, transformant yeast cells containing pBD-*MdbHLH104*-P_{MdAHA8}::*GUS* plus pAD-MdbHLHs plasmids, including 4, pAD-*MdbHLH105*; 5, pAD-*MdbHLH115*; 6, pAD-*MdbHLH11*; 7, pAD-*MdbHLH121*; 8, pAD-*MdPYE*. (b) The interaction enhances the binding of the MdbHLH104 protein to the promoter fragment of the *MdAHA8* gene. The immunoprecipitated DNAs were quantified through qPCR using specific primers of candidate fragments containing the E-box *cis* element. The results were quantified as the percentage of total input DNA by qPCR. (c) The interaction enhances the transcriptional activity of the MdbHLH104 to the *MdAHA8* promoter, as indicated by GUS staining and the activity test. The WT calli were labelled as 1; the transgenic calli, 2. After the expression vector *35S::MdbHLH104-GFP* was cotransformed into the calli 2, transgenic calli P_{MdAHA8}::*GUS+35S::MdbHLH104-GFP* were obtained and labelled as 3. The transgenic calli 3 were used as the background for transient co-expression with 5 viral vectors pIR-*MdbHLHs* containing other IVc subgroup bHLH genes, labelled as 4–8. 4, pIR-*MdbHLH105*; 5, pIR-*MdbHLH115*; 6, pIR-*MdbHLH11*; 7, pIR-*MdbHLH121*; 8, pIR-*MdPYE*. (d) Expression levels of *MdAHA8* gene in the WT, *35S:: MdbHLH104-GFP* and co-expression of bHLHs transgenic apple calli, as determined with qPCR.

transcriptional regulator, AtFIT forms heterodimers with other Ib subgroup member such as AtbHLH38, AtbHLH39, AtbHLH100 or AtbHLH101 to activate the expression of *FRO2* and *IRT1* genes (Ivanov *et al.*, 2012; Wang *et al.*, 2012; Yuan *et al.*, 2008). Therefore, the co-overexpression of *FIT/AtbHLH38*, *FIT/*

AtbHLH39, *FIT/AtbHLH100* or *FIT/AtbHLH101* noticeably promotes iron uptake and enhances the tolerance to iron deficiency in transgenic plants (Wang *et al.*, 2012; Yuan *et al.*, 2008). Similarly, AtbHLH104 interacts with another IVc subgroup bHLH protein such as AtbHLH105 (ILR3), AtbHLH115 or AtPYE to form

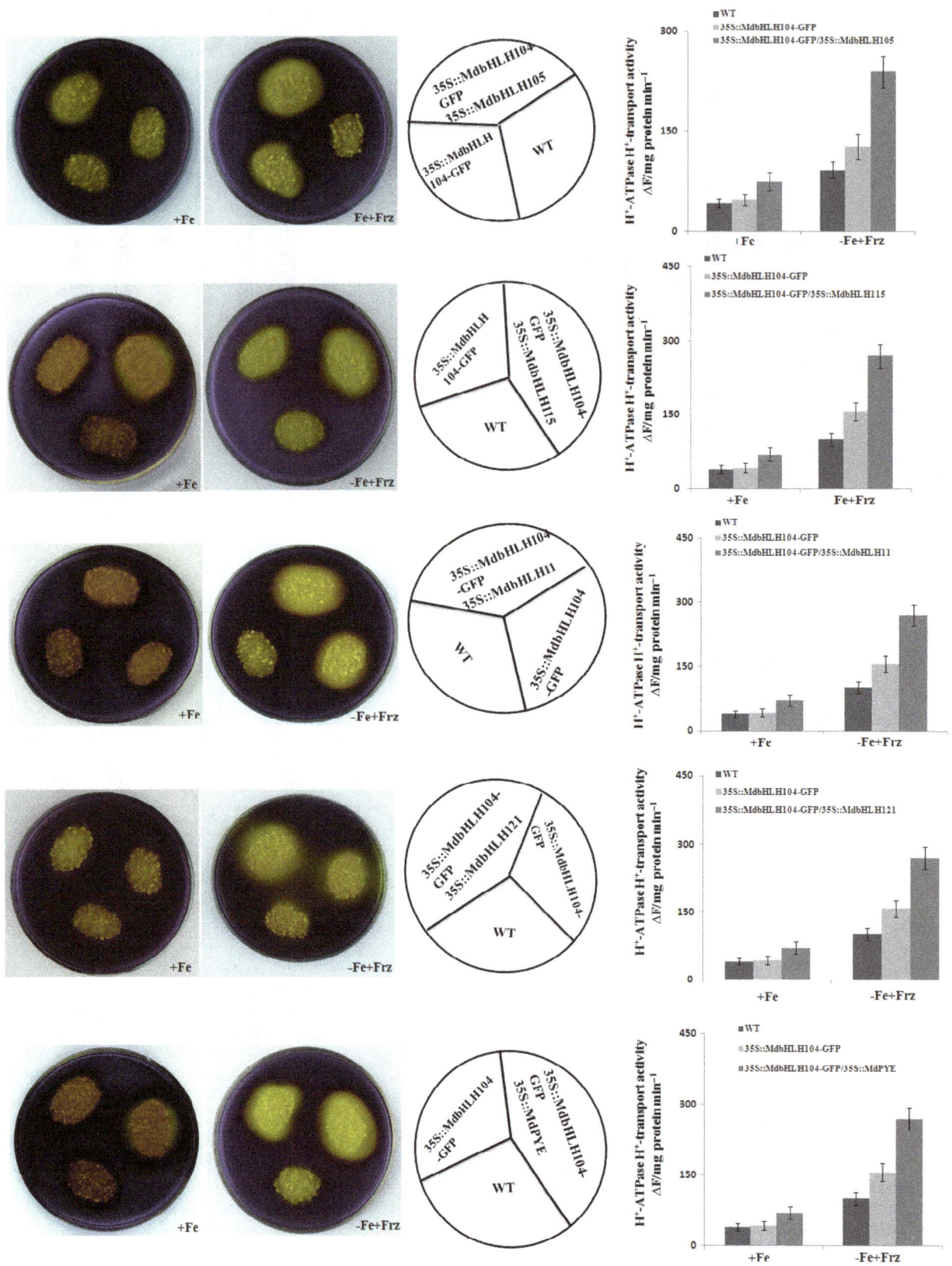

Figure 7 The Interaction with other MdbHLHs proteins affects the function of MdbHLH104 in the activation of PM H⁺-ATPase. Acidification assay with the pH indicator bromocresol purple around the different apple calli as indicated (the left panels). PM H⁺-ATPase activity in the vesicles isolated from different apple calli as indicated (the right panels).

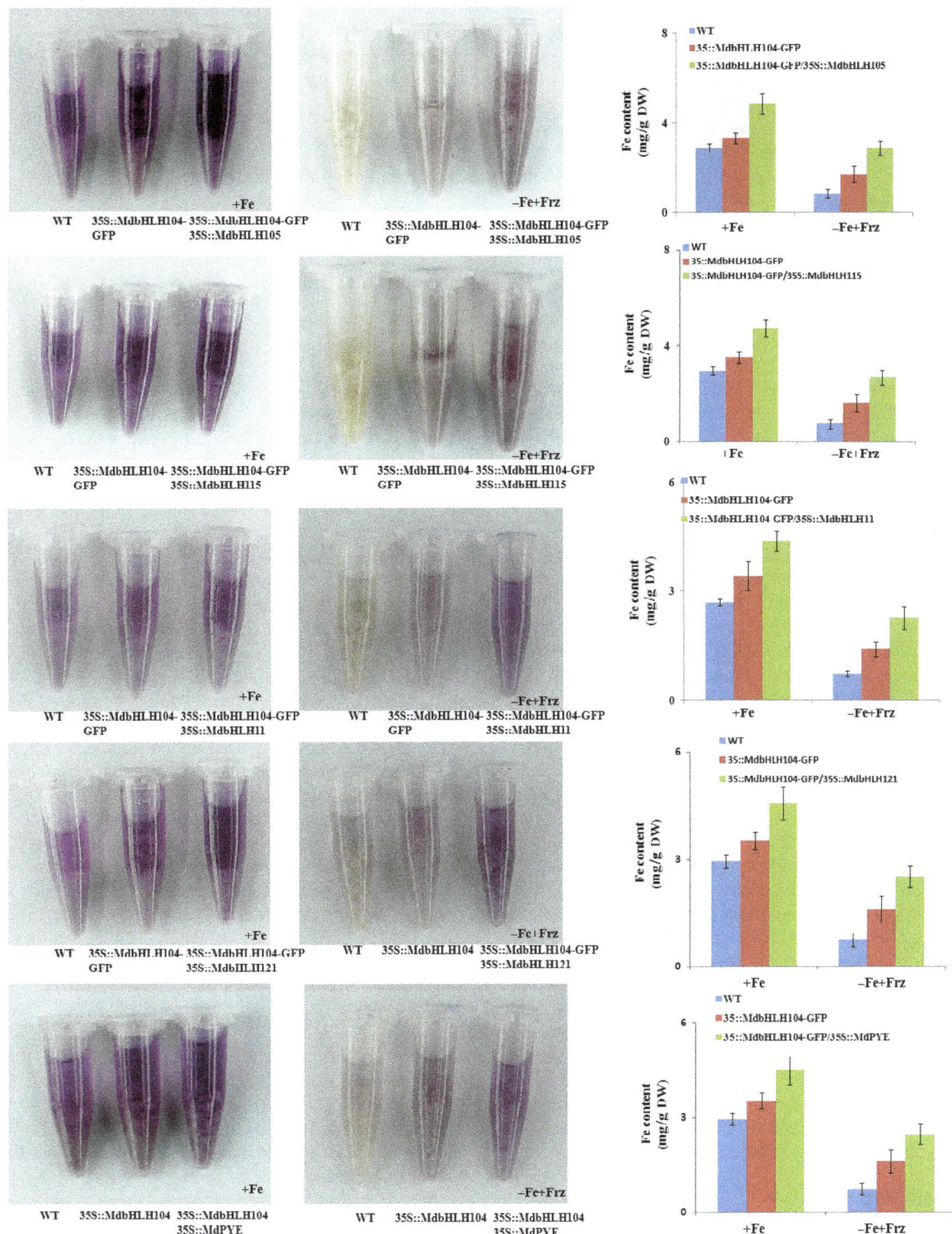

Figure 8 The Interaction with other MdbHLHs proteins affects the function of MdbHLH104 in iron uptake. Visualization of ferrous under Fe-sufficient and Fe-deficient conditions in different apple calli, as indicated. Fe contents in different apple calli, as indicated. The data represent the means ± SD of three independent experiments. DW, dry weight.

heterodimers (Selote *et al.*, 2015; Zhang *et al.*, 2015). In this study, it was found that MdbHLH104 interacted with another IVc subgroup member such as MdbHLH105, MdbHLH115, MdbHLH11, MdbHLH121 or MdPYE to increase the expression of the *MdAHA8* gene (Figures 5 and 6). Just like in *Arabidopsis* (Heim *et al.*, 2003), the interacting partners may be required for the function of MdbHLH104 under Fe deficiency. Therefore, MdbHLH104 transgenic apple plants showed higher Fe content and ATPase activity than the WT control only under Fe-deficient conditions, but not under Fe-sufficient conditions (Figure 1c,d). Taken together, MdbHLH104 works together with other IVc subgroup bHLH proteins to increase the H$^+$-ATPase activity and iron content in transgenic apple calli (Figures 7 and 8). Furthermore, MdbHLH105 and other IVc subgroup bHLH members may function in a way similar to MdbHLH104, by binding to the promoter region of *MdAHA8* gene and alleviating Fe deficiency.

In addition, Ib and IVc subgroup bHLH TFs not only regulate the H$^+$-ATPase activity but also modulate other ion transports or transcription factors. ILR3 (bHLH105) plays an important role in the metal ion-mediated auxin sensing of roots and controls metal uptake by regulating the expression of intracellular iron transport genes, such as *VIT1* (Kim *et al.*, 2006; Rampey *et al.*, 2006). PYE directly targets several genes such as *OPT3*, *FRD3*, *NRAMP4*, *ZIF1*, *NAS4* and *FRO3*, which are implicated in long-distance iron transport (Long *et al.*, 2010). In *Arabidopsis*, PYE directly regulates ANR1 and indirectly regulates other transcription factors, such as Ib bHLH subgroup TFs bHLH39 and bHLH101 (Long *et al.*, 2010; Yuan *et al.*, 2008). Additionally, bHLH104 and bHLH105 (ILR3) bind directly to the promoters of Ib subgroup bHLH genes such as *bHLH38*, *bHLH39*, *bHLH100* and *bHLH101* and to the promoter of Ib subgroup bHLH gene *POPEYE* (*PYE*; Zhang *et al.*, 2015). In apple, there are four Ib and six IVc subgroup members. Among them, MdbHLH104 directly binds to the promoters of two Ib subgroup bHLH genes *MdbHLH38* and *MdbHLH39*, and to that of an IVc subgroup one *MdPYE* (Figure S4). Therefore, bHLH104 plays pivotal roles in the regulation of Fe deficiency responses via targeting *AHA* genes, Ib or IVc subgroup bHLH genes.

Iron deficiency often results in chlorosis, which affects photosynthesis and respiration (Kosegarten *et al.*, 1998). Fruit trees are among the crops most affected by iron deficiency, which significantly decreases fruit yield and quality (Tagliavini *et al.*, 2001). It is well known that most fruit trees take up Fe nutrients through their rootstock. Therefore, high Fe-efficient rootstock is a desirable trait for fruit production (Gonzalo *et al.*, 2011). Numerous genes are involved in Fe response and homeostasis. However, just few of them are characterized and used for genetic improvement in fruit rootstock. In an apple rootstock *Malus xiaojinensis*, *MxFIT* is induced by iron deficiency. It ectopic expression confers improved tolerance to iron deficiency in transgenic *Arabidopsis* (Yin *et al.*, 2014). Our findings regarding the regulatory mechanism involved in the iron response and homeostasis are likely to favour the development of novel biotechnological tools for the generation of rootstocks for fruit trees with the enhanced ability of nutrient-use efficiency and adaptation to nutrient-poor habitats.

Materials and methods

Plant materials and growth conditions

Tissue cultures of apple (*Malus* × *domestic* cv. 'Royal Gala') were subcultured at a 1-month interval on an MS medium supplemented with 0.5 mg/L 6-BA, 0.2 mg/L NAA and 0.1 mg/L GA at 25 ± 1 °C for a 16/8-h light/night period (100 mmol/m^2/s), whereas 'Orin' apple calli were subcultured at a 3-week interval on an MS medium containing 1.5 mg/L 2,4-D and 0.4 mg/L 6-BA at 25 ± 1 °C under dark conditions. For Fe treatment, the calli-grown subculture medium was an iron-sufficient (+Fe) MS medium supplemented with 1.5 mg/L 2,4-D and 0.4 mg/L 6-BA

for 'Orin' apple calli. The iron-deficient (−Fe + Frz) medium was the same, without Fe-EDTA and with 300 mM FerroZine, an iron indicator. For transgenic apple rooting plantlets, Hoagland's nutrient solution with or without iron was used.

Plasmid construction and genetic transformation in apple and apple calli

The details are provided in the Data S1.

Gene expression analysis

The details are provided in the Data S1.

Southern blot analysis

The details are provided in the Data S1.

Phenotypic analyses

Chlorophyll content was measured in WT and transgenic apple lines under iron-deficient conditions. The young leaves were collected and ground into powder in liquid nitrogen. The powder was resuspended in 80% acetone and centrifuged at 10 000 *g* for 5 min. Chlorophyll concentrations were calculated from spectroscopy absorbance measurements at 663.2, 646.8 and 470 nm.

Acidification assays were performed as described by Yi *et al.* (1994). Wild-type and transgenic apple calli or apple lines were grown on iron-sufficient media for 10 days and then transferred to iron-deficient media for 5 days. They were finally transferred to a 1% agar plate containing 0.006% bromocresol purple and 0.2 mM CaSO$_4$ (pH adjusted to 6.5 with NaOH) for 24–48 h. Acidification is indicated by the yellow colour around the apple roots or calli.

FerroZine reagent forms a red-coloured complex with ferrous, but not with ferric iron, and the Fe(II) is trapped by FerroZine to produce a red product (Stookey, 1970).

Transgenic apple lines and calli were dried for 1–2 days at 80 °C and then wet-ashed with 1.5 mL of 13.4 M HNO$_3$ and 1.5 mL of 8.8 M H$_2$O$_2$ for 60 min at 220 °C using a muffle furnace. Iron concentration measurement was carried out as described by Kobayashi *et al.* (2013).

Chromatin immunoprecipitation (ChIP)-PCR analysis

The details are provided in the Data S1.

Electrophoretic mobility shift assay (EMSA)

MdbHLH104 ORF was amplified with primers MdbHLH104-F: 5′-ATGGGGGAATGGATAGAGTAT-3′ and MdbHLH104-R: 5′-AGCAGCAGGGGGCCTAAG-3′ containing *Bam*HI and *Sal*I restriction sites, respectively. Then, the gene was inserted into the expression vector *pET32a* after digestion with *Bam*HI and *Sal*I. The resultant expression vector was transformed into *BL21*. MdbHLH104 proteins were prepared according to the instruction manual. The 36-bp *MdAHA8* promoter probes containing a G-box element was synthesized and labelled with biotin at the 3′ end (Sangon, Shanghai, China). Unlabelled competitor probes were generated from the dimerized oligos of the *MdAHA8* promoter regions containing E-box motifs. The EMSA was carried out as described in the instruction manual (Thermo Scientific, Rockford). The double-stranded oligonucleotides wt (TTYAG-TYYGGGATTA***CAAATG***CAATACGGTCWTTCT) were used as probes and competitors for the EMSAs. The mut (TTYAGTYYGG-GATTA***ACAAGT***CAATACGGTCWTTCT) was used as the mutated competitor.

Protein extraction and western blotting

The details are provided in the Data S1.

Transcription activation analysis in yeast cells

The details are provided in the Data S1.

Transcriptional activation assays in apple calli

The details are provided in the Data S1.

GUS analysis

The details are provided in the Data S1.

Yeast two-hybrid (Y2H) assay

Y2H assays were performed as described by Xie *et al.* (2012). The MdbHLH104-coding sequence was cut with *Bam*HI and *Sal*I double digestion and cloned into pGBT9 to generate an in-frame fusion with the GAL4 activation domain. The full-length cDNAs of IVc subgroup *MdbHLHs* genes were cut by *Eco*RI and *Sal*I double digestion and cloned into pGAD424 to generate an in-frame fusion with the GAL4 DNA-binding domain. The plasmids of pGAD424-MdbHLHs and pGBT9-MdbHLH104 were cotransformed into yeast. The yeast clones were grown on SD/-Trp-Leu and SD/-Trp-Leu-His-Ade media. A selection medium supplemented with -Leu/-Trp was used as a transformation control, whereas for interaction studies, -Leu/-Trp/-His/-Ade with or without 5-bromo-4-chloro-3-indolyb-d-galactopyranoside acid (x-α-gal) was used to test for possible interactions.

Pull-down assay

The assays were carried out as described by Xie *et al.* (2012). The MdbHLH104-coding sequence was cut with *Bam*HI and *Sal*I double digestion and cloned into *pET32a*, and the full-length cDNAs of IVc subgroup MdbHLHs were cut by *Eco*RI and *Sal*I double digestion and cloned into *pGEX*. The plasmids of *pGEX*-MdbHLHs and *pET*-MdbHLH104 were transformed into *Escherichia coli* BL21 (DE3; Transgene, Beijing, China). For pull-down analysis with GST- and His-tagged proteins, GST-MdbHLH105/115/11/121/PYE proteins were first eluted from glutathione–agarose beads before being incubated with His-MdbHLH104, which that remained attached to tetradentated-chelated nickel resin. In general, proteins were incubated at least 4 h at 4 °C under shaking conditions before being centrifuged. Precipitates were washed no fewer than three times to remove unspecific bindings and boiled (10 min, 100 °C). Then, the precipitates were further analysed by SDS–PAGE and protein gel blotting using standard procedures.

Plasma membrane H⁺-ATPase isolation

Transgenic apple calli and apple lines or WT controls were grown on Fe-sufficient medium and then transformed to a Fe-deficient medium. Plasma membranes were isolated with a buffer consisting of 15 mM Tris–Cl (pH 7.5), 0.5 M sucrose, 1 mM EGTA, 1 mM EDTA, 6% (w/v) PVP, 0.1% (w/v) BSA, 0.1 mM DTT and 1 mm PMSF. Microsomal pellets were obtained from the homogenate as described (Yang *et al.*, 2010). All steps were performed at 4 °C or on ice. The homogenate was filtered through four layers of gauzes and centrifuged at 13 000 g for 10 min. The supernatant then was centrifuged for 50 min at 80 000 g to obtain a microsomal pellet that was resuspended in a buffer containing 6.4% (w/w) dextran T-500, 6.4% (w/w) PEG 3350 (Sigma-Aldrich, St Louis, MO), 5 mM phosphate buffer

titrated to pH 7.8 with KOH, 3 mM KCl, 0.1 mM EDTA, 1 mM DTT, 1 mM PMSF, 1× protease inhibitor and 0.33 M sucrose. The final upper phases were collected, diluted with a resuspension buffer containing 0.33 M sucrose, 10% (w/v) glycerol, 0.1% (w/v) BSA, 0.1 mM EDTA, 2 mM DTT, 1× protease inhibitor and 20 mM HEPES-KOH (PH 7.5) and centrifuged for 50 min at 100 000 g. The final membrane pellets were resuspended with a resuspension buffer containing 1 mM EDTA (Yang *et al.*, 2010).

PM H⁺-ATPase activity assays

For PM H⁺-ATPase activity measurement, H⁺ transport activity was measured as described (Yang *et al.*, 2010). An inside-acid pH gradient (ΔpH) was formed in the vesicles by the activity of the H⁺-ATPase and measured as a decrease (quench) in the fluorescence of quinacrine (a pH-sensitive fluorescent probe). The assays (2 mL) contained 10 μM quinacrine, 3 mM MgSO₄, 100 mM KCl, 25 mM BTP-Mes-HEPES (Sigma-Aldrich), pH 6.5, 250 mM mannitol and 50 mg/mL of a PM protein. The reactions were mixed by inversion several times and then placed in a dark chamber in a fluorescence spectrophotometer (Hitachi, Ltd., Tokyo, Japan). The reactions were initiated by the addition of ATP to a final concentration of 3 mM, and the formation of ΔpH was measured at the wavelengths of $E_x = 430$ nm and $E_x = 500$ nm. At the end of each reaction, 10 μM m-chlorophenylhydrazone (CCCP) was added to stop any remaining pH gradient. Specific activity was calculated by dividing the change in fluorescence by the mass of PM protein in the reaction per unit time (ΔF/min per mg of protein).

Acknowledgements

This work was supported by grants from NSFC (31430074), Ministry of Agriculture of China (201203075-3), Ministry of Education of China (IRT15R42) and Shandong Province (SDAIT-03-022-03).

We thank Prof Ilan Sela of Hebrew University of Jerusalem, Israel, for IL-60-BS and pIR binary vectors and Dr Takaya Moriguchi at National Institute of Fruit Tree Science, Japan, for 'Orin' apple calli.

References

Baxter, I., Tchieu, J., Sussman, M.R., Boutry, M., Palmgren, M.G., Gribskov, M., Harper, J.F. *et al.* (2003) Genomic comparison of P-type ATPase ion pumps in *Arabidopsis* and rice. *Plant Physiol.* **132**, 618–628.

Briat, J.F. and Lobréaux, S. (1997) Iron transport and storage in plants. *Trends Plant Sci.* **2**, 187–193.

Colangelo, E.P. and Guerinot, M.L. (2004) The essential basic helix-loop-helix protein FIT1 is required for the iron deficiency response. *Plant Cell*, **16**, 3400–3412.

Dell'Orto, M., Santi, S., De Nis, i P., Cesco, S., Varanini, Z., Zocchi, G. and Pinton, R. (2000) Development of Fe-deficiency responses in cucumber (*Cucumis sativus* L.) roots: involvement of plasma membrane H⁺-ATPase activity. *J. Exp. Bot.* **51**, 695–701.

Fisher, F. and Goding, C.R. (1992) Single amino acid substitutions alter helix-loop-helix protein specificity for bases flanking the core CANNTG motif. *EMBO J.* **11**, 4103.

Gonzalo, M.J., Moreno, M.A. and Gogorcena, Y. (2011) Physiological response and differential gene expression in *Prunus* rootstocks under iron deficiency conditions. *J. Plant Physiol.* **168**, 887–893.

Guerinot, M.L. and Yi, Y. (1994) Iron: nutritious, noxious, and not readily available. *Plant Physiol.* **104**, 815.

Haruta, M., Burch, H.L., Nelson, R.B., Barrett-Wilt, G., Kline, K.G., Mohsin, S.B., Young, J.C. *et al.* (2010) Molecular characterization of mutant *Arabidopsis* plants with reduced plasma membrane proton pump activity. *J. Biol. Chem.* **285**, 17918–17929.

Heim, M.A., Jakoby, M., Werber, M., Martin, C., Weisshaar, B. and Bailey, P.C. (2003) The basic helix-loop-helix transcription factor family in plants: a genome-wide study of protein structure and functional diversity. *Mol. Biol. Evol.* **20**, 735–747.

Hindt, M.N. and Guerinot, M.L. (2012) Getting a sense for signals: regulation of the plant iron deficiency response. *Biochim. Biophys. Acta*, **1823**, 1521–1530.

Ivanov, R., Brumbarova, T. and Bauer, P. (2012) Fitting into the harsh reality: regulation of iron-deficiency responses in dicotyledonous plants. *Mol. Plant*, **5**, 27–42.

Jakoby, M., Wang, H.Y., Reidt, W., Weisshaar, B. and Bauer, P. (2004) FRU (BHLH029) is required for induction of iron mobilization genes in *Arabidopsis thaliana. FEBS Lett.* **577**, 528–534.

Kim, S.A., Punshon, T., Lanzirotti, A., Li, L., Alonso, J.M., Ecker, J.R., Kaplan, J. *et al.* (2006) Localization of iron in *Arabidopsis* seed requires the vacuolar membrane transporter VIT1. *Science*, **314**, 1295–1298.

Kobayashi, T. and Nishizawa, N.K. (2012) Iron uptake, translocation, and regulation in higher plants. *Annu. Rev. Plant Biol.* **63**, 131–152.

Kobayashi, T., Nagasaka, S., Senoura, T., Itai, R.N., Nakanishi, H. and Nishizawa, N.K. (2013) Iron-binding haemerythrin RING ubiquitin ligases regulate plant iron responses and accumulation. *Nat. Commun.* **4**, 2792.

Korcak, R.F. (1988) Nutrition of blueberry and other calcifuges. *Horticultural Rev.* **10**, 183–227.

Kosegarten, H., Wilson, G.H. and Esch, A. (1998) The effect of nitrate nutrition on iron chlorosis and leaf growth in sunflower (*Helianthus annuus* L.). *Eur. J. Agron.* **8**, 283–292.

Li, X.X., Duan, X.P., Jiang, H.X., Sun, Y.J., Tang, Y.P., Yuan, Z., Guo, J.K. *et al.* (2006) Genome-wide analysis of basic/helix-loop-helix transcription factor family in rice and *Arabidopsis. Plant Physiol.* **141**, 1167–1184.

Ling, H.Q., Bauer, P., Bereczky, Z., Keller, B. and Ganal, M. (2002) The tomato *fer* gene encoding a bHLH protein controls iron-uptake responses in roots. *Proc. Natl Acad. Sci. USA*, **99**, 13938–13943.

Long, T.A., Tsukagoshi, H., Busch, W., Lahner, B., Salt, D.E. and Benfey, P.N. (2010) The bHLH transcription factor POPEYE regulates response to iron deficiency in *Arabidopsis* roots. *Plant Cell*, **22**, 2219–2236.

Marschner, H. (1995) Functions of mineral nutrients: macronutrients. *Miner Nutr Higher Plants*, **2**, 379–396.

Marschner, H. and Römheld, V. (1986) Different strategies in higher plants in mobilization and uptake of iron. *J. Plant Nutr.* **9**, 695–713.

Niu, X., Narasimhan, M.L., Salzman, R.A., Bressan, R.A. and Hasegawa, P.M. (1993) NaCl regulation of plasma membrane H^+-ATPase gene expression in a glycophyte and a halophyte. *Plant Physiol.* **103**, 713–718.

Palmgren, M.G. (2001) Plant plasma membrane H^+-ATPases: powerhouses for nutrient uptake. *Annu. Rev. Plant Biol.* **52**, 817–845.

Rampey, R.A., Woodward, A.W., Hobbs, B.N., Tierney, M.P., Lahner, B., Salt, D.E. and Bartel, B. (2006) An *Arabidopsis* basic helix-loop-helix leucine zipper protein modulates metal homeostasis and auxin conjugate responsiveness. *Genetics*, **174**, 1841–1857.

Römheld, V. and Marschner, H. (1986) Evidence for a specific uptake system for iron phytosiderophores in roots of grasses. *Plant Physiol.* **80**, 175–180.

Santi, S. and Schmidt, W. (2009) Dissecting iron deficiency-induced proton extrusion in Arabidopsis roots. *New Phytol.* **183**, 1072–1084.

Santi, S., Cesco, S., Varanini, Z. and Pinton, R. (2005) Two plasma membrane H^+-ATPases genes are differentially expressed in iron-deficient cucumber plants. *Plant Physiol. Biochem.* **43**, 287–292.

Schmidt, W. (2003) Iron solutions: acquisition strategies and signaling pathways in plants. *Trends Plant Sci.* **8**, 188–193.

Schubert, S. and Yan, F. (1997) Nitrate and ammonium nutrition of plants: effects on acid/base balance and adaptation of root cell plasmalemma H^+-ATPase. *Z. Ptlanzenemaehr und Bodenkd*, **160**, 275–281.

Selote, D., Samira, R., Matthiadis, A., Gillikin, J.W. and Long, T.A. (2015) Iron-binding E3 ligase mediates iron response in plants by targeting bHLH transcription factors. *Plant Physiol.* **167**, 273–286.

Stookey, L.L. (1970) Ferrozine-a new spectrophotometric reagent for iron. *Anal. Chem.* **42**, 779–781.

Tagliavini, M. and Rombolà, A.D. (2001) Iron deficiency and chlorosis in orchard and vineyard ecosystems. *Eur. J. Agron.* **15**, 71–92.

Wang, N., Cui, Y., Liu, Y., Fan, H., Du, J., Huang, Z. and Ling, H.Q. (2012) Requirement and functional redundancy of Ib subgroup bHLH proteins for iron deficiency responses and uptake in *Arabidopsis thaliana. Mol. Plant*, **6**, 503–513.

Xie, X.B., Li, S., Zhang, R.F., Zhao, J., Chen, Y.C., Zhao, Q. and Hao, Y.J. (2012) The bHLH transcription factor MdbHLH3 promotes anthocyanin accumulation and fruit colouration in response to low temperature in apples. *Plant, Cell Environ.* **35**, 1884–1897.

Yan, F., Feuerle, R., Schäffer, S., Fortmeier, H. and Schubert, S. (1998) Adaptation of active proton pumping and plasmalemma ATPase activity of corn roots to low root medium pH. *Plant Physiol.* **117**, 311–319.

Yang, Y., Qin, Y., Xie, C., Zhao, F., Zhao, J., Liu, D. and Guo, Y. (2010) The *Arabidopsis* chaperone J3 regulates the plasma membrane H^+-ATPase through interaction with the PKS5 kinase. *Plant Cell*, **22**, 1313–1332.

Yi, Y., Saleeba, J.A. and Guerinot, M.L. (1994) Iron uptake in Arabidopsis thaliana. In Biochemistry of Metal Micronutrients in the Rhizosphere, J. Manthey, D. Luster and D.E. Crowley, eds (Lewis Publishers, Inc: Chelsea, MI), pp. 295–307.

Yin, L., Wang, Y., Yuan, M., Zhang, X., Xu, X. and Han, Z.H. (2014) Characterization of *MxFIT*, an iron deficiency induced transcriptional factor in *Malus xiaojinensis. Plant Physiol. Biochem.* **75**, 89–95.

Yuan, Y.X., Zhang, J., Wang, D.W. and Ling, H.Q. (2005) AtbHLH29 of *Arabidopsis thaliana* is a functional ortholog of tomato FER involved in controlling iron acquisition in strategy I plants. *Cell Res.* **15**, 613–621.

Yuan, Y., Wu, H., Wang, N., Li, J., Zhao, W., Du, J. and Ling, H.Q. (2008) FIT interacts with AtbHLH38 and AtbHLH39 in regulating iron uptake gene expression for iron homeostasis in *Arabidopsis. Cell Res.* **18**, 385–397.

Zhang, J., Liu, B., Li, M.S., Feng, D.R., Jin, H.L., Wang, P., Liu, J. *et al.* (2015) The bHLH transcription factor bHLH104 interacts with IAA-LEUCINE RESISTANT3 and modulates iron homeostasis in *Arabidopsis. Plant Cell*, **27**, 787–805.

Zhao, M., Song, A., Li, P., Chen, S., Jiang, J. and Chen, F. (2014) A bHLH transcription factor regulates iron intake under Fe deficiency in chrysanthemum. *Sci. Rep.* **4**, 6694.

15

Differential expression of microRNAs and other small RNAs in barley between water and drought conditions

Michael Hackenberg[1], Perry Gustafson[2], Peter Langridge[3] and Bu-Jun Shi[3,*]

[1]*Computational Genomics and Bioinformatics Group, Genetics Department, University of Granada, Granada, Spain*
[2]*USDA-ARS, University of Missouri, Columbia, MO, USA*
[3]*Australian Centre for Plant Functional Genomics, The University of Adelaide, Urrbrae, SA, Australia*

Correspondence

email bujun.shi@adelaide.edu.au

Keywords: barley, drought, microRNA, small RNA, differential expression.

Summary

Drought is a major constraint to crop production, and microRNAs (miRNAs) play an important role in plant drought tolerance. Analysis of miRNAs and other classes of small RNAs (sRNAs) in barley grown under water and drought conditions reveals that drought selectively regulates expression of miRNAs and other classes of sRNAs. Low-expressed miRNAs and all repeat-associated siRNAs (rasiRNAs) tended towards down-regulation, while tRNA-derived sRNAs (tsRNAs) had the tendency to be up-regulated, under drought. Antisense sRNAs (putative siRNAs) did not have such a tendency under drought. In drought-tolerant transgenic barley overexpressing DREB transcription factor, most of the low-expressed miRNAs were also down-regulated. In contrast, tsRNAs, rasiRNAs and other classes of sRNAs were not consistently expressed between the drought-treated and transgenic plants. The differential expression of miRNAs and siRNAs was further confirmed by Northern hybridization and quantitative real-time PCR (qRT-PCR). Targets of the drought-regulated miRNAs and siRNAs were predicted, identified by degradome libraries and confirmed by qRT-PCR. Their functions are diverse, but most are involved in transcriptional regulation. Our data provide insight into the expression profiles of miRNAs and other sRNAs, and their relationship under drought, thereby helping understand how miRNAs and sRNAs respond to drought stress in cereal crops.

Introduction

Drought is a major constraint to crop production and occurs every year in many places in the world. Due to global warming, improving crops tolerant to drought is now becoming a primary objective of plant breeding. Barley (*Hordeum vulgare* L.) is an important cereal crop, ranking fourth among all grains in terms of quantity produced and area cultivated. Compared to its close wheat relative, barley is more tolerant to drought. In addition, barley has a short growing season with a high degree of natural and easily inducible variation. Therefore, barley would be an excellent model plant for investigating the genetic basis of drought tolerance.

MicroRNAs (miRNAs) are single-stranded small RNAs (sRNAs) with 20–24 nucleotides (nt) in length and encode no protein. miRNAs are generated from hairpin precursors (pre-miRNAs) that are formed from miRNA primary transcripts (pri-miRNAs), which are transcribed from genomic DNA. Other classes of sRNAs, including natural antisense transcript-derived siRNAs (natsiRNAs), repeat-associated siRNAs (rasiRNAs), long siRNAs (lsiRNA), heterochromatin siRNAs, secondary siRNAs, tiny noncoding (nc) RNAs (tncRNAs), 21U-RNAs, scan RNAs (scnRNAs), promoter/termini-associated sRNAs (PASRs/TASRs), transcription initiation RNAs (tiRNAs), transcription start site-associated RNAs (TSSa RNAs), splice site RNAs (spliRNAs) and sRNAs derived from rRNAs, snoRNAs, tRNAs and chloroplasts (Hackenberg *et al.*, 2012; references therein), are also about 20–24 nt in size, but generated from long linear double-stranded RNAs rather than hairpin

structures. Some dsRNAs can be generated from single-stranded sense transcripts by RDR6, a member of the RNA-dependent RNA polymerase (RdRP) family. In addition, secondary siRNAs can be further subdivided into 'phased' or 'trans-acting' (tasiRNAs) and natsiRNAs, which can act in cis (cis-natsiRNAs) or in trans (trans-natsiRNAs) (Axtell, 2013). miRNAs are predominantly involved in targeted mRNA degradation or translational repression (Huntzinger and Izaurralde, 2011), while siRNAs primarily mediate transcriptional silencing of genome loci crucial in chromatin remodelling and the maintenance of genome integrity or heterochromatic state (Elbashir *et al.*, 2001; Pontes *et al.*, 2008). The other classes of sRNAs are functionally diverse or unknown. Nevertheless, all of these classes of sRNAs are important in the regulation of biological processes (Jones-Rhoades *et al.*, 2006).

There is extensive evidence indicating that miRNAs are regulated by drought in various plant species, including rice (Zhao *et al.*, 2007; Zhou *et al.*, 2010), populous (Li *et al.*, 2011; Lu *et al.*, 2008; Shuai *et al.*, 2013), Arabidopsis (Li *et al.*, 2008, 2011; Liu *et al.*, 2008; Sunkar and Zhu, 2004), wheat (Kantar *et al.*, 2011; Yao *et al.*, 2010), maize (Li *et al.*, 2011; Xu *et al.*, 2011), soybean (Kulcheski *et al.*, 2011), *Medicago truncatula* (Trindade *et al.*, 2010; Wang *et al.*, 2011), *Phaseolus vulgaris* (Arenas-Huertero *et al.*, 2009), cassava (Ballen-Taborda *et al.*, 2013), cowpea (Barrera-Figueroa *et al.*, 2011), *Prunus persica* (Eldem *et al.*, 2012), potato (Zhang *et al.*, 2014), switchgrass (Xie *et al.*, 2014), cotton (Wang *et al.*, 2013), sugarcane (Ferreira *et al.*, 2012) and tobacco (Frazier *et al.*, 2011). In return, miRNAs turn on many genes in response to drought stress. Recent studies showed that overexpression of drought-regulated miRNAs leads to transgenic plants tolerant to drought (Li *et al.*, 2008; Zhang *et al.*, 2011). miRNAs thus have great potential to be used as a tool for improving drought tolerance in barley and other cereal crops. However, so far, little information is available on drought-regulated miRNAs or other sRNAs in barley, especially identified through genome-wide high-throughput sequencing. In barley, only dehydration-regulated miRNAs were identified using a computer-based approach and publicly available barley ESTs (Kantar *et al.*, 2010). On the other hand, genome-wide identification and analysis of drought-responsive miRNAs and other sRNAs are very helpful in identifying regulatory networks between miRNAs and other sRNAs in plants. In this work, we analysed expression profiles of both miRNAs and other sRNAs in barley under water and drought conditions. We found that a lot of miRNAs and other sRNAs were regulated by drought. While low-expressed miRNAs and rasiRNAs were down-regulated, tsRNAs were up-regulated under drought conditions in barley, suggesting that different classes of sRNAs respond differently to drought, some of which are likely to be transcriptionally regulated by each other. Our data provide a valuable resource of miRNAs and other sRNAs for potential use in improving drought tolerance in barley and other cereal crops.

Results

Deep sequencing of sRNAs from Golden Promise barley under water and drought conditions

To identify drought-regulated miRNAs and other sRNAs in barley which might serve as future targets for improving drought tolerance of cereal crops, we deeply sequenced sRNAs in Golden Promise (GP) barley grown under water and drought conditions. Under drought conditions, GP [designated as GP(−w)] showed severe wilt symptoms, while under water conditions GP [designated as GP(w)] revealed no symptoms (Figure 1). A total of 7 113 852 sequencing reads from the GP(w) sample and a total of 7 185 002 sequencing reads from the GP(−w) sample were obtained (Table 1). After removal of the adaptor sequences, low quality and short reads <16 nt, 6 671 598 and 6 664 821 sequencing reads remained in the GP(w) and GP(−w) samples, respectively (Table 1). The total number of reads is highly similar in both samples, which allows a meaningful comparison (especially of the number of unique reads which strongly depends on the total number of reads). After collapsing the reads into unique reads, 1 207 726 and 773 114 sequences were obtained in the GP(w) and GP(−w) samples, respectively (Table 1).

Cross-comparisons showed that 191 277 sequences are shared among both samples, while 1 282 989 were specific to the GP(w) and 836 578 to the GP(−w) sample. Among the shared reads, 16 827 were differentially expressed between the two samples ($|\log_2| > 1$), of which 5809 were up-regulated ($\log_2 \geq +1$), while 11 018 were down-regulated ($\log_2 \leq −1$) under drought conditions.

Length distribution of sRNA reads from water and drought-treated barley

The analysis of the read lengths of 18–25 nt gives valuable hints on the different sRNA species present in the samples. 20-nt sRNAs are the most frequent in GP(w) and GP(−w), followed by 21 nt (Figure S1A). Compared to GP(w), more 20-nt sRNAs are present in GP(−w). 21-nt sRNAs are the other way around. At a unique

Figure 1 Phenotype of GP barley under water and drought conditions. The plants were watered for 3 weeks after germination. After that one plant was stopped for watering for 5 days, at which time the photograph was taken.

read level, the 24-nt peak is dominant, followed by 21 nt in both samples (Figure S1B). 24-nt reads have been proposed to be often generated from transposable elements (TEs) (Hackenberg *et al.*, 2012). The 21-nt peak contains less unique reads but reads with higher read count in GP(w) compared to GP(−w). Shorter reads than 21 nt are more frequent in GP(−w) than in GP(w) in both read count and unique read number.

Classification of sRNAs from water and drought-treated barley

The mapping statistics to the barley nuclear and chloroplast genomes (International Barley Genome Sequencing Consortium, 2012; Saski *et al.*, 2007) showed several differences (Table 1). We observed both higher read count and more unique reads mapped to the nuclear genome compared to the chloroplast in GP(w). On the contrary, we observed higher number of unique reads from the chloroplast in GP(−w) (Table 1). sRNAs from the GP(w) and GP(−w) samples were further grouped by mapping the reads to specific databases such as the RFAM (Gardner *et al.*, 2009, release 10.0), miRBase version 19 (Griffiths-Jones *et al.*, 2008), the TIGR repeat database (Ouyang and Buell, 2004), the TREP repeat database (Wicker *et al.*, 2002), RepBase (Jurka *et al.*, 2005) and tRNA databases. We observed that 79% and 87% of the reads from the GP(w) and GP(−w) samples, respectively, could be assigned to any of the database reference sequences (Table S1). The remaining unmapped reads may be due to unavailable genome or EST sequences, or due to sequencing errors. In Table S1, the higher mean read count in GP(−w) compared to GP(w) indicates that a higher number of low copy reads must exist in GP(w). The influence of sequencing quality can be ruled out as mapped reads have been used for the calculation.

About 12% in GP(w) and 8.3% in GP(−w) were mapped to miRNAs, 0.73% in GP(w) and 0.47% in GP(−w) were mapped to repetitive elements, 5% in GP(w) and 6.9% in GP(−w) were mapped to tRNAs, and 0.83% in GP(w) and 0.53% in GP(−w) were mapped to ncRNAs from the RFAM database (Table S1). Approximate 9.5% of the total sRNAs in GP(w) and 6.7% in GP(−w) were mapped to un-annotated genomic regions (Table S1). The antisense strands of the above genomic features were also detected. 39.6% of the total sRNAs in GP(w) and 51.6% in GP(−w) were mapped to the antisense strands of barley genes (from the HVGI database) whose proportion is the highest among all

Table 1 The sequencing yield and outcome of processing

sRNA library	Raw reads	After filtration	Unique reads (UR)	Chloroplast read count (UC)	Chloroplast UR	Nuclear UC	Nuclear UR
GP(w)	7 113 852	6 671 598	1 207 726	2 752 340	64 836	3 755 898	1 091 816
GP(−w)	7 185 002	6 664 821	773 114	3 596 877	55 474	2 853 614	667 832

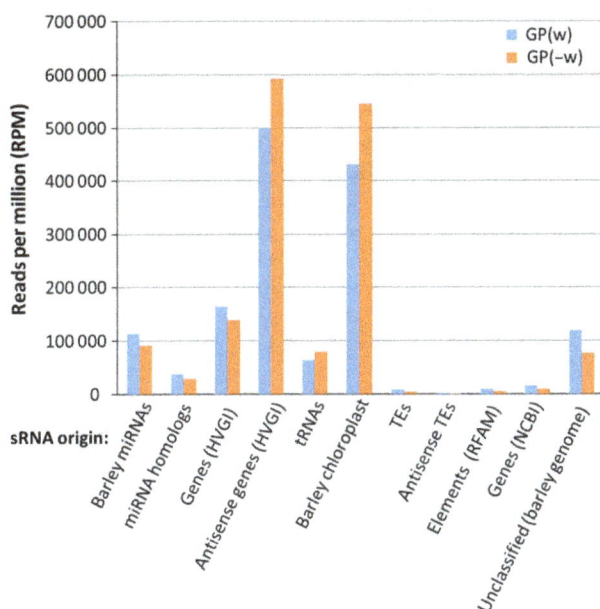

Figure 2 RPM values of different classes of sRNAs. RPM-reads per million; miRNA homologs-miRNAs from non-barley species in the miRBase; HVGI-barley EST database; Unclassified (barley genome)-mapped genome sequences are not annotated.

the mapped sRNAs (Table S1). Likewise, sRNAs originating from the chloroplast genome were also mapped to the sense and antisense strands in both samples. Significantly, a chloroplast-derived sRNA in both samples, which mapped to a chloroplast-encoded trnH-GUG gene, is the most frequent, accounting for 28.9% of the total sRNAs in GP(w) and 36.7% of the total sRNAs in GP(−w). In general, miRNAs, sRNAs derived from mRNAs, ncRNAs (RFAM) and those mapped to the genome but not classified are more frequent in GP(w), while antisense sRNAs (designated all antisense sRNAs as putative siRNAs) and sRNAs derived from tRNAs (tsRNAs) tend to be more frequent in GP(−w). For an unbiased comparison, the sRNA read counts are normalized to reads per million (RPM), calculated by dividing the read count by the total reads in the dataset and then multiplying 10^6. This result is shown in Figure 2.

Length distribution of sRNAs in water and drought-treated barley

Figure S2 shows length distribution of sRNAs (18–25 nt). The patterns are similar between water and drought conditions. Only a subset of sRNAs derived from barley genes (HVGI) are smearly distributed. This proportion of sRNAs may be generated by degradation pathways rather than from a regulated process. Reads mapped to barley genes and chloroplast-derived reads have a dominant size at 20 nt. In contrast, the dominant size of miRNAs is 21 nt followed by 20 nt, which is consistent with DCL cleavage products. However, the dominant 21 nt vary in amount between water and drought conditions. While 74.6% of all

miRNA reads have a length of 21 nt under water conditions, under drought conditions this percentage drops to 61.2% (Figure S3). This bias is in fact due to miR156. Firstly, miR156 is stronger expressed under drought conditions. Secondly, the by far most frequent miR156-derived read is 20 nt long and not 21 nt as the annotated sequence in miRBase. Note that the canonical miRBase sequence is also present but with an over two magnitudes lower read count compared to the most frequent read. In contrast to the above sRNAs, ncRNAs (from the RFAM database), repeat-associated sRNAs (in both sense and antisense directions) and the 'genome mapped' sRNAs, all have a dominant size of 24 nt (Figure S2). Of all the sRNAs, miRNAs and the reads that map the antisense direction of transcripts (designated as siRNAs) have the sharpest distributions. In the case of miRNAs, 99.1% and 98.8% of the reads are between 20 and 21 nt long in the GP(w) and GP(−w) samples, respectively. This confirms that these reads were processed, but not degradation products.

Differential expression of miRNAs between water and drought conditions

A total of 33 barley miRNAs from GP(w) and 31 from GP(−w) were detected using miRBase (Table S2). All the barley miRNAs identified in GP(−w) exist in GP(w). The two miRNAs missed in GP(−w) are hvu-miR399 and hvu-miR6177, which are both of low copies in GP(w). Among the 31 common miRNAs, 13 were significantly down-regulated ($\log_2 \leq -1$), while only one (hvu-miR5049b) was significantly up-regulated ($\log_2 \geq +1$) under drought conditions (Table S2). However, most of those miRNAs have low copy numbers. When a threshold of 10 RPM is adopted, only five, which were all down-regulated miRNAs, remain. It is worth noting that hvu-miR5048a is listed in the table as the reads do not map to this miRNA when using the strict criterion for known miRNAs, but when allowing one mismatch, this miRNA can be detected.

After removal of the barley miRNA-mapping reads, the remaining reads were further mapped (without mismatch) to miRNAs from other species in miRBase. A total of 416 putative homologous miRNAs from each of the GP(w) and GP(−w) samples were identified. However, when a threshold of 10 RPM was applied, only 74 putative homologous miRNAs from each sample remain (Table S3), of which only one (gma-miR6300) was significantly up-regulated under drought conditions (Table S3). In contrast, under the same conditions, 20 putative homologous miRNAs belonging to 10 miRNA families were significantly down-regulated ($\log_2 \leq -1$) (Table S3). This expression profile is similar to that of the previously known barley miRNAs under the same drought conditions.

Novel miRNA identification and their differential expression between water and drought conditions

To identify novel miRNAs, we combined previously described criteria (Hackenberg et al., 2011) and alignment patterns of pre-miRNAs to the barley sRNA datasets. We hereby identified three novel miRNAs, designated as hvu-miRX33, hvu-miRX34 and hvu-miRX35, respectively. However, their miRNA* was not found in the datasets. hvu-miRX33 and hvu-miRX34 are of 22 nt, while hvu-miRX35 is of 21 nt (Table 2). All of these three miRNAs have higher negative minimal free energy index (MFEI) than other different type of RNAs, which is >0.85 (Zhang et al., 2006). hvu-miRX34 and hvu-miRX35 start with U at their 5' ends, which is consistent with the preference of most miRNAs described previously (Kuang et al., 2009; Zhang et al., 2009). Another miRNA, hvu-miRX33, starts with G at its 5' end. hvu-miRX33 was only expressed under water conditions. The other two miRNAs were down-regulated under drought conditions. hvu-MIRX35 gene (Accession No. dbj|AK252755.1) was found to contain another miRNA discovered previously (Schreiber et al., 2011). Homologous sequences of hvu-miRX33 and hvu-miRX35 are also present in wheat, but hvu-miRX34 only exists in barley.

Taken together, 9% of the reads from water conditions and 8% of the reads from drought conditions are mapped to miRNAs. The lower relative frequency of miRNAs under drought conditions may indicate that drought suppresses miRNA expression in barley.

Differential expression of ncRNAs between water and drought conditions

We further mapped all reads to the RFAM database and identified 185 ncRNAs in each of the GP(−w) and GP(w) samples (Table S4). Of these ncRNAs, five were found to be significantly up-regulated ($\log_2 \geq +1$), while 21 were significantly down-regulated ($\log_2 \leq -1$), under drought. snoRNAs, which primarily guide chemical modifications of rRNAs, tRNAs and small nuclear RNAs (snRNA), were dominantly regulated by drought. When using a threshold of 10 RPM, only 12 drought down-regulated ncRNAs and one (derived from RNaseP_arch) drought up-regulated ncRNA remained. Figure S4 shows the absolute read counts, relative fractions and unique read fractions for the different analysed sRNA species as a function of the regulation class (up-, down- or non-regulated). Drastic differences can be seen between the different RNA species; while all tsRNAs seem to be regulated, the vast majority of chloroplast-derived reads are non-regulated.

Differential expression of tsRNAs between water and drought conditions

Mapping all reads to tRNA databases showed that all 61 codons of tRNAs derived sRNAs in each sample, of which sRNAs derived from 19 codons of tRNAs were up-regulated ($\log_2 \geq +1$), while sRNAs derived from two codons of tRNAs were down-regulated ($\log_2 \leq -1$) under drought conditions (Table S5). When a threshold of 10 RPM was applied, only up-regulated tsRNAs remained at an anticodon level under drought conditions ($\log_2 \geq +1$). This expression pattern contrasts to that of miRNAs or ncRNAs. Whether there is a correlation among tsRNAs, ncRNAs and miRNAs under drought conditions is interesting.

Differential expression of repeat-associated sRNAs between water and drought conditions

Alignment of all reads to the TREP repeat database, which contains 1716 TEs, revealed that 811 TEs derived sRNAs in GP (−w), while 1028 TEs derived sRNAs in GP(w) (Table S6). Of these TE-derived sRNAs, 771 were common between the two samples, 257 were specific to GP(w), while 40 were specific to GP(−w). In addition, a high proportion of sRNAs were found to be derived from the antisense strand of TEs. For example, 421 and 652 TEs derived antisense sRNAs in GP(−w) and GP(w), respectively. These antisense reads are likely to be transcribed from the antisense strand of the genome and thus may not be siRNA/piwiRNA-like sRNAs that can regulate the repeat expression. However, we cannot exclude the possibility that some of the antisense reads might originate from a repeat that inserted in antisense into any other transcribed element. In this case, these antisense reads may be able to function in the suppression of repeat expression. Nevertheless, all of these hypotheses need to be confirmed experimentally.

Among the identified antisense TE-derived sRNAs, 393 were common between the two samples, and 259 and 28 were specific to GP(w) and GP(−w), respectively. Of the common TE-derived sRNAs, only 12 sense and 9 antisense TE-derived sRNAs were significantly up-regulated ($\log_2 \geq +1$) under drought conditions. In contrast, 406 sense and 200 antisense TE-derived sRNAs were significantly down-regulated ($\log_2 \leq -1$) under the same conditions. When a threshold of 10 RPM was adopted, only 31 TE-derived sRNAs were significantly down-regulated, and no TE-derived sRNA was significantly up-regulated under drought conditions.

Table 3 shows the expression values as a function of different TE types. In general, there are no TE-specific differences between GP(w) and GP(−w), but a general down-regulation of all TE types under drought conditions, which is different from the regulation of tsRNAs under the same conditions. The majority of the mapped sense TE-derived sRNAs belong to retrotransposons (Class I TEs), accounting for 76% of the total mapped TE-derived sRNAs in GP (−w) and 73.5% in GP(w). TE-derived sRNAs belonging to DNA transposons (Class II TEs) account for 19.9% and 21.9% in GP (−w) and GP(w), respectively. There are 4.1% of TE-derived sRNAs from GP(−w) and 4.7% from GP(w) not being classified.

Table 2 Identified novel miRNAs in water and drought-treated barley

miRNA	miRNA sequence	Read count (RC) GP(w)	RC GP(−w)	Precursor sequence
hvu-miRX33	CGGUAGGGCUGUAUGAUGGCGA	159	0	UCGCCAUCAUACGCCCAACCGUGCAUUUGAU AUGCAUAUAUAUGCAUCACGAGCCA**CGGUA GGGCUGUAUGAUGGCGA**
hvu-miRX34	UGAGAAGGUAGAUCAUAAUAGC	279	2	CUCAGA**UGAGAAGGUAGAUCAUAAUAGC**UAAAA GAUAGUAUUCUCCGCAUCUCAUCUAAG
hvu-miRX35	UAAUCUUCUGGAAAUAUGCUU	26	12	AUAUCUGUCGGAUCUCUCUUAAUUUUUG**UAAUCU UCUGGAAAUAUGCUU**AGGUGUAAU

Bolded are mature miRNA sequences.

Table 3 Expression of TREP repetitive elements

Name	GP(w) RC	GP(w) RPM	GP(−w) RC	GP(−w) RPM	log$_2$[GP(−w)/GP(w)]
Gypsy	4509	466.6	2574	258.6	−0.85
Harbinger	378	41.6	160	20.7	−1.01
Mutator	146	23.1	69	10.5	−1.13
Unknown	960	128.4	460	60.3	−1.09
Mariner	713	92.1	358	39.6	−1.22
CACTA	1724	242.4	893	116.9	−1.05
Copia	5152	510.2	2930	294.8	−0.79
Name	**GP(w) RC**	**GP(w) RPM**	**GP(−w) RC**	**GP(−w) RPM**	**log$_2$[GP(−w)/GP(w)]**
LINE	228	36.5	120	17.8	−1.04
LTR	9750	990.4	5531	557.8	−0.83
TIR	2970	400.8	1481	187.9	−1.09
Unknown	636	76.9	307	37.2	−1.05

The table depicts two different levels, more specific (top) and more general (bottom).

Antisense TE-derived sRNAs gave a similar result as above (Table 4). Similar results were also observed from alignment of the reads to other repeat databases such as Repbase, the TIGR rice repeat database and the TIGR barley repeat database (data not shown).

Differential expression of chloroplast-derived sRNAs between water and drought conditions

Putative chloroplast-derived reads are strongly present in both GP(−w) and GP(w) (Table 1). Further analysis revealed that these sRNAs were distributed across the whole chloroplast genome. The comparison of chloroplast-derived sRNAs from the GP(−w) and GP(w) samples showed that the majority of chloroplast-derived sRNAs were not regulated by drought. However, tsRNAs were found again to tend to be up-regulated by drought.

Differential expression of antisense sRNAs between water and drought conditions

Antisense sRNAs (putative siRNAs) were identified via two steps: firstly, all the reads were mapped to the reverse strand of barley genes available in the databases, and then, the 'antisense' reads were mapped to the forward strand of all other libraries (tRNA, genes, repeats, etc.) in order to detect the RNA species from which the siRNAs are generated. In this way, a total of 5747 siRNA were identified to be common between water and drought conditions, and 914 siRNAs were found to be differentially expressed (|log$_2$| ≥ 1) (Table S7). Among them, 172 siRNAs were up-regulated (log$_2$ ≥ +1), while 642 were down-regulated (log$_2$ ≤ −1) under drought conditions (Table S7). The up-regulated siRNAs were attributed to antisense tsRNAs, which,

like sense tsRNAs, were also up-regulated under drought conditions.

Co-regulated miRNAs and other sRNAs in drought-treated and DREB3 transcription factor overexpressing transgenic barley

DREB transcription factors are among the first transcription factors discovered to be regulated under drought conditions and to enhance drought tolerance (Morran et al., 2011). We previously identified that overexpression of a DREB3 transcription factor from wheat (TaDREB3) in barley strongly affects the expression of sRNAs (including miRNAs) in the plants (Hackenberg et al., 2012). To identify truly drought-responsive miRNAs and siRNAs in barley under drought stress, we compared three sRNA datasets from the transgenic barley, GP(−w) and GP(w) samples. All of these three sRNA datasets were obtained under uniform conditions, for example the plants were grown at the same time in the same size of pots and same soil under the same glasshouse, and the same concentration of sRNAs was used for sequencing, which was performed at the same time in the same flowcell and generated a similar number of raw sequencing reads. Thus, these three datasets are comparable. For simplicity, we define consistent drought regulation as up-regulation in both GP(−w) and GP (TaDREB3) or down-regulation in both GP(−w) and GP(TaDREB3) when compared to GP(w). Any other combinations like up-regulation in GP(TaDREB3) but down-regulation in GP(−w) are considered as inconsistent regulation. As expected, both sample-specific and shared sRNA species were present in the datasets. Sample-specific sRNA species are generally of low read count and can be of several origins including sequencing errors and sample fluctuations. More importantly, we found many regulated

Table 4 Reads that map in antisense direction to TREP repeats

Name	GP(w) RC	GP(w) RPM	GP(−w) RC	GP(−w) RPM	log$_2$[GP(−w)/GP(w)]
Gypsy	865	115.5	469	57.6	−1.00
Unknown	194	27.8	88	14.0	−0.99
CACTA	451	64.1	234	31.4	−1.03
Copia	2810	282.6	1658	171.0	−0.72

($|log_2| \geq 1$) sRNA species, both consistently and inconsistently regulated. The consistently regulated sRNAs are shown in Table S8.

Figure S5 shows that the proportion of consistent and inconsistently regulated sRNAs fluctuates strongly as a function of RNA species. TE-derived sRNAs are 100% consistently regulated, which are all down-regulated compared to GP(w). All other sRNA species show both consistent and inconsistent regulation. Among them, tsRNAs are the most variable class of sRNAs, which have a proportion of 65% tsRNAs that are inconsistently regulated between the transgenic and GP(−w) samples. As tsRNAs tended to be up-regulated under drought condition, this result confirms that the expression of tsRNAs is drought-regulated because the transgenic barley was not treated by drought. On the other hand, it also confirms our previous result that TaDREB3 strongly affects the expression of sRNAs in plant. Whether the transcription factor overcomes drought stress by regulating some tsRNA expression is interesting. All the significantly co-regulated sRNAs among the transgenic GP(−w) and GP(w) samples were listed in Table 5. miRNAs account for the most, followed by tsRNAs and other ncRNAs. Most of the co-regulated miRNAs and ncRNAs are down-regulated, while all the co-regulated tsRNAs are up-regulated compared to their corresponding RNAs in GP(w) (Table 5).

Examination of co-regulated antisense sRNAs among the transgenic GP(−w) and GP(w) samples showed that only small portion of antisense sRNAs were up-regulated in both the transgenic and GP(−w) samples. Most antisense sRNAs were down-regulated in both samples (Table S7). Further analysis showed that the co-regulated antisense sRNAs were generated from all sources of elements including tRNAs, repeats, nuclear and chloroplast, but the majority were derived from nuclear-encoded genes (Table S7).

Experimental validation of drought-regulated miRNAs and siRNAs

To experimentally confirm whether the above co-regulated miRNAs are indeed regulated by drought, Northern hybridization was applied using total RNAs isolated from leaf and root tissues of GP treated by water and drought conditions, respectively. Of 14 selected co-regulated miRNAs (11 conserved and three novel miRNAs, Table S9), three (hvu-miR444b, hvu-miR5049a and hvu-miRX35) were not detected in any of the tissues or conditions (data not shown), which may be due to low expression levels and/or low sensitivity of Northern hybridization. Of the detected miRNAs, most did not give significant change in expression level between water and drought conditions, for example hvu-miR168-5p and hvu-miRX34 (Figure 3). Only hvu-miR159b, hvu-miR166a, ath-miR172a, osa-miR393a and hvu-miR5048 were obviously differentially expressed between water and drought conditions. Both hvu-miR159b and hvu-miR166a were up-regulated in leaf tissues, but down-regulated in root tissues, under drought conditions. ath-miR172a and hvu-miR5048 were down-regulated in both tissues under drought conditions. osa-miR393a was scarce in root tissues under both water and drought conditions, but appeared to be up-regulated under drought conditions. hvu-miR156, hvu-miR5048 and hvu-miRX33 were generally expressed less in leaf tissues than in root tissues under both water and drought conditions. hvu-miR159b and hvu-miR166a were expressed higher in root tissues only under water conditions. Under drought conditions, these miRNAs were expressed slightly higher in leaf tissues than in root tissues. ath-

miR172a and osa-miR393a were expressed higher under water conditions than under drought conditions in both leaf and root tissues. Intriguingly, hvu-miR5052 was expressed higher in leaf tissues under water conditions, but under drought conditions, this miRNA was reversely expressed. hvu-miRX33 was up-regulated, while ath-miR172, osa-miR393a, hvu-miR5052 and hvu-miR5048 were down-regulated, under drought conditions.

Expression of the selected drought-regulated miRNAs was also examined in another barley cultivar, WI4330, by Northern hybridization. WI4330 is a breeding line and not highly tolerant to drought according to our unpublished data. The result showed that most of the miRNAs were expressed similarly as in GP (Figure 3), indicating that miRNA expression tends to be conserved among barley cultivars. Only ath-miR169b was expressed slightly differently between the two cultivars. This miRNA acted to be down-regulated in both leaf and root tissues under drought conditions in WI4330.

The expression levels of the drought-regulated miRNAs were further quantitated by qRT-PCR. Of the selected 14 miRNAs (Table S9), only hvu-miR5052 was not regulated in either leaf or root tissues (Figure S6A). All other tested miRNAs were regulated to some degree by drought. hvu-miR156 and hvu-miR159b are expressed at high levels and are up-regulated under drought conditions in both tissues. Moderately expressed hvu-miR166a was up-regulated in leaves, but down-regulated in roots under drought conditions. ath-miR169b, ath-miR172a, hvu-miR444b, hvu-miR5048, hvu-miR5049a, hvu-miRX33 and hvu-miRX34 had a similar expression pattern as hvu-miR166a. In contrast, hvu-miR168-5p was only drought up-regulated in leaves. osa-miR393a and hvu-miRX35 appeared to be only expressed in leaf tissues, but not in root tissues. However, while osa-miR393a was up-regulated by drought conditions, hvu-miRX35 was down-regulated under the same conditions. Overall, the qRT-PCR result is consistent with that from the Northern hybridization or from the deep sequencing data.

Co-regulated siRNAs were also examined by Northern hybridization. Twelve co-regulated siRNAs designated as hvu-siRNA1, 2, 3, 4, 5, 6, 7, 8, 9, 10, 11 and 12, respectively, were selected (Table S9). siRNA4, siRNA7 and siRNA8 were derived from tRNAs, and siRNA1, siRNA2, siRNA3, siRNA10 and siRNA12 were derived from barley genes, while the rest were derived from unknown elements. It is notable that most of these siRNAs especially siRNA2 exist in multiple copies (Table S8). Only siRNA11 exists in a single copy, and siRNA3 and siRNA8 exist in two copies (Table S8). Northern hybridization showed that all siRNAs except for siRNA4, which was not detectable, were detected in at least one tissue and/or condition. siRNA10 was up-regulated while siRNA8 and siRNA9 were down-regulated in leaf and root tissues under drought conditions (Figure 4). siRNA6 and siRNA12 were unchanged in expression level in both tissues under drought conditions. Unlike siRNA10, siRNA1 and siRNA2 were only up-regulated in leaf tissues under drought conditions. In root tissues, these two siRNAs as well as siRNA7 were almost undetectable. siRNA11 was not detectable in leaf tissues. siRNA3 was down-regulated in leaf tissues, but up-regulated in root tissues, under drought conditions.

Northern analysis of WI4330 showed that the expression pattern of above siRNAs was the same as in GP under the same conditions (Figure 4). This indicates that the expression of siRNAs is also conserved in barley. We tried to use qRT-PCR to quantitate the expression levels of siRNAs, but failed for all siRNAs except siRNA11 which was present only in root tissues, being up-

Table 5 Significantly co-regulated elements from all classes of sRNAs

Name	GP(w) RC	GP(w) RPM	GP(−w) RC	GP(−w) RPM	GP(TaDREB3) RC	GP(TaDREB3) RPM	$\log_2$[GP(−w)/GP(w)]	$\log_2$[GP(TaDREB3)/GP(w)]
hvu microRNA								
hvu-miR5049a	546	103.2	102	17.6	116	23.4	−2.55	−2.14
hvu-miR5052	11	2.1	2	0.3	3	0.6	−2.59	−1.78
hvu-miR5053	23	4.3	6	1.0	5	1.0	−2.07	−2.11
hvu-miR1130	32	6.0	9	1.6	8	1.6	−1.96	−1.91
hvu-miR6191	9	1.7	2	0.3	3	0.6	−2.30	−1.49
hvu-miR1120	371	70.1	167	28.8	86	17.3	−1.28	−2.02
hvu-miR5048a								
hvu-miR5048b	4534	857.1	1722	296.8	2103	424.0	−1.53	−1.02
hvu-miR444b	64	12.1	29	5.0	30	6.0	−1.28	−1.00
Homologous microRNAs								
osa-miR827a	1618	305.9	736	126.9	165	33.3	−1.27	−3.20
bdi-miR5054	1325	250.5	226	39.0	610	123.0	−2.68	−1.03
ath-miR169b	1834	346.7	516	88.9	667	134.5	−1.96	−1.37
ptc-miR169s	1665	314.8	486	83.8	613	123.6	−1.91	−1.35
gma-miR169v	1642	310.4	486	83.8	613	123.6	−1.89	−1.33
rgl-miR5139	302	57.1	106	18.3	109	22.0	−1.64	−1.38
hvu-miR5048a								
hvu-miR5048b	235	44.4	75	12.9	92	18.5	−1.78	−1.26
gma-miR393h	9009	1703.1	5139	885.8	1654	333.4	−0.94	−2.35
osa-miR5072	243	45.9	114	19.7	72	14.5	−1.23	−1.66
bdi-miR5064	274	51.8	148	25.5	72	14.5	−1.02	−1.84
tRNA (anticodon)								
ValAAC	37 385	7067.6	113 732	19604.7	78 418	15808.7	1.47	1.16
RFAM								
C0719	495	93.6	182	31.4	138	27.8	−1.58	−1.75
MIR812	3445	651.3	1878	323.7	1046	210.9	−1.01	−1.63
RNaseP_bact_b	148	28.0	79	13.6	60	12.1	−1.04	−1.21
tRNA (chloroplast)								
trnS-GCU	5078	960.0	21 892	3773.7	11 765	2371.8	1.97	1.30
trnS-UGA	2100	397.0	6176	1064.6	4822	972.1	1.42	1.29
trnfM-CAU	2633	497.8	6432	1108.7	6454	1301.1	1.16	1.39

regulated by drought (Figure S6B). The failure of qRT-PCR for these siRNAs may be due to that they exist as clusters in the cells (Table S8) and hence cannot be amplified into products with a uniform size.

Prediction and validation of targets of drought-regulated miRNAs and siRNAs

To analyse the molecular function of drought-regulated miRNAs, their targets were predicted using psRNAtarget (http://plantgrn. noble.org/psRNATarget/). For most drought-regulated miRNAs, target genes can be predicted, but only some target genes are functionally known (Table S10). The known functions are mostly related to stress tolerance and developments, which emphasizes the importance of drought-regulated miRNAs. Notably, a considerable number of targets are transcription factors. Some predicted targets also encode housekeeping enzymes such as carboxylase and ribonuclease. In addition, all miRNAs have more than one target. Intriguingly, some genes are targeted by more than one miRNA as previously reported for other miRNAs' targets. Gene Ontology (GO) analysis showed that three GO terms (ADP binding, defence response and response to stress) were significantly enriched (Table S10), and likely, these genes play important roles in the plant for response of drought stress.

To verify these predicted targets, a degradome library was constructed using leaves and roots of GP plants treated by water and drought. The advantage of this degradome library is that it requires no priori miRNA target prediction and can test all targets at once. Searching this library for miRNA targets showed that only nine predicted target genes could be verified (yellow highlight in Table S10). This small number may result from the limited barley EST sequences that were used to predict or map targets of the miRNAs. In addition, it could also be possible that the miRNAs whose cleavage products were not found in the library may function in certain particular tissues or developmental stage or via the transcriptional or translational repression mechanisms according to previous studies (Shamimuzzaman and Vodkin, 2012). Furthermore, it cannot be ruled out that most of the predicted targets may not be true. Some degradome reads were aligned to noncanonical (10–11) positions, suggesting that miRNA-mediated cleavage sites may not be restricted to positions 10–11.

To further confirm the predicted targets verified by the degradome library, qRT-PCR was applied on the basis of inverse relationship between miRNA and target, that is increasing miRNA level will reduce its target level or vice versa. However, this standard does not apply for the miRNAs that function in inhibiting transcription or translation of their targets. Eight genes targeted

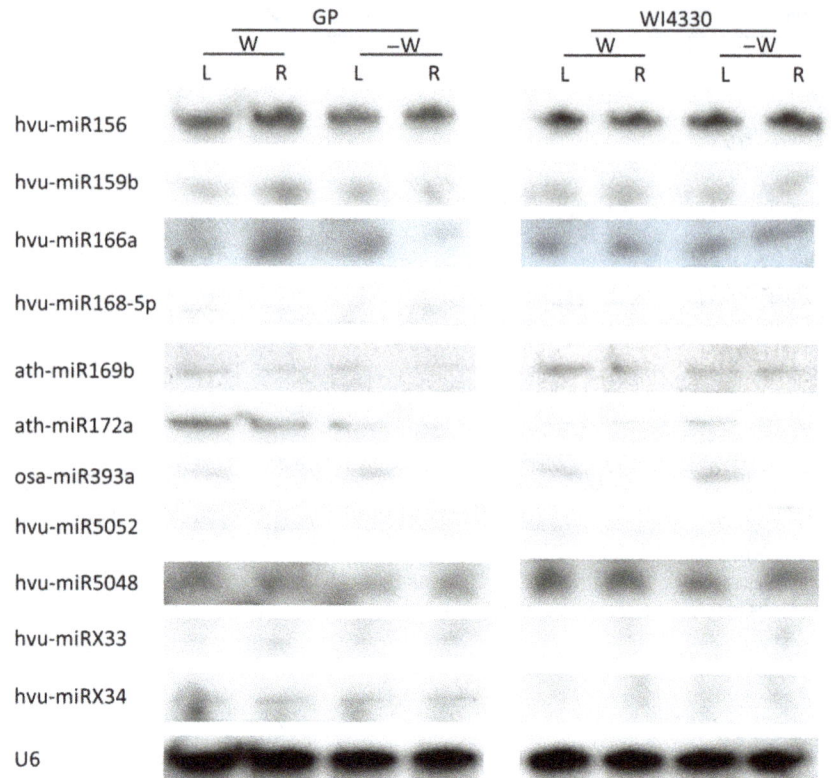

Figure 3 Northern analysis of miRNAs in leaf and root tissues of GP and WI4330 barley under water and drought conditions.

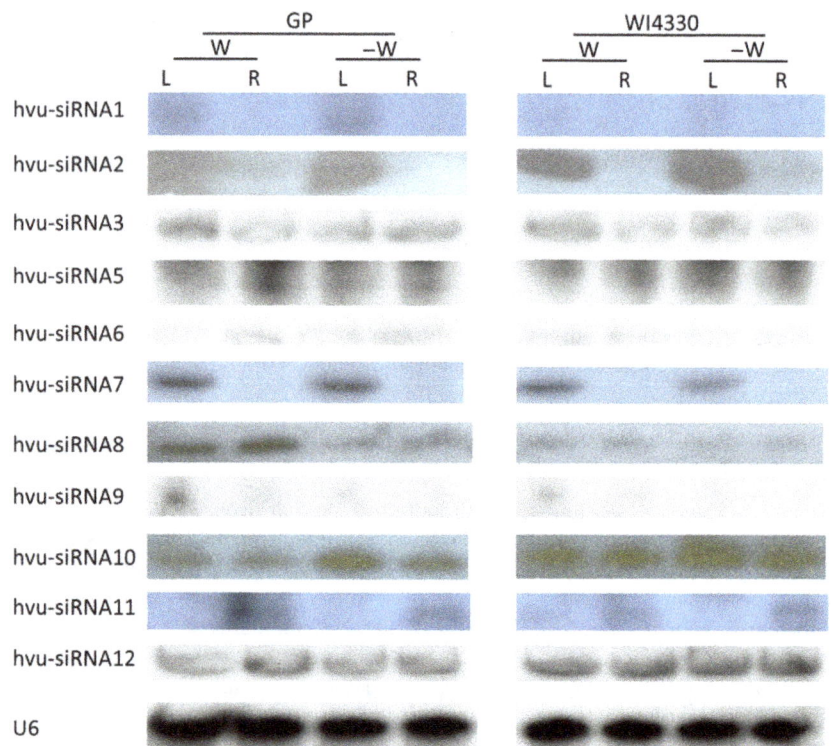

Figure 4 Northern analysis of siRNAs in leaf and root tissues of GP and WI4330 barley under water and drought conditions.

by five miRNAs were chosen (Table S10). We did not choose miR156's target like Squamosa promoter-binding-like transcription factor, because it has been well defined previously. Among the chosen targets, two are functionally unknown, while the others encode various functions (Table S10). The qRT-PCR result showed that most of these targets have a reverse expression level to their miRNAs in both leaf and root tissues (compare Figures

S6A and S7). However, osa-miR393a shows a non-inverse relationship to one of its target genes, suggesting that this target may not be a bona fide target or it is regulated via a noncleavage mechanism.

Targets of drought-regulated siRNAs were also predicted using psRNAtarget. Unlike the miRNAs above, most siRNAs found no predicted targets (Table S10). For the predicted targets, most of them are not characterized. For the characterized predicted targets, their functions are quite diverse. One target was found to contain a BTB/POZ domain and is a regulator of hormone-responsive gene expression in barley (Woodger *et al.*, 2004), suggesting that siRNAs may be involved in signalling pathways. In view of the fact that siRNAs function in the same way as miRNAs, we used the existing barley degradome library to verify the predicted targets of siRNAs (Table S11). The results showed that many targets had their cleavage products present in the library. Some targets even had their cleavage products occurring at different positions as did some of miRNA's targets. Unexpectedly, targets of the only single-copy siRNA11 in the cells were not found in the degradome library. We used qRT-PCR to examine two predicted targets (Table S10, Accession No: DN160496 and TC252151) of siRNA11, but failed as both targets were expressed at levels below the detection limits of qRT-PCR (data not shown). We did not pursue confirmation by quantifying the expression levels of predicted targets of the other siRNAs, because these siRNAs could not generate products with a uniform size by qRT-PCR.

Discussion

In this study, we used deep sequencing technology to analyse miRNAs and other sRNAs in barley under water and drought conditions. Seventy-seven per cent and 87% of sRNAs from the GP(w) and GP(−w) samples, respectively, were assigned to known entities from barley databases. The difference of the mapped sRNAs between the two samples is mainly attributed to tsRNAs, especially the most abundant trnH-GUG-derived sRNAs from the chloroplast genome, which were more prevalent in the GP(−w) sample than in the GP(w) sample. Another major contributor is siRNAs derived from antisense genes. Other sRNAs also contribute to the difference but not significantly. About 50% of rasiRNAs and those that mapped repeats [retro (transposons)] in an antisense orientation were differentially regulated between water and drought conditions. All repeat classes were down-regulated. Only a small fraction of barley-specific miRNAs are regulated by drought, most of which were down-regulated under drought stress. Overall, siRNAs and tsRNAs were up-regulated, and other sRNAs were down-regulated by drought (Figure 5). The distributions of sRNAs within each sample are also different. While miRNAs account for 8.3%–12%, TE-derived sRNAs only account for 0.47%–0.73%, of the total sRNAs. This phenomenon has been observed in Arabidopsis where miRNAs and TE-derived sRNAs account for 14.4% and 16.9% of the total sRNAs, respectively, while intergenic region-derived sRNAs account for 49.1% of the total sRNAs (Lu *et al.*, 2012). Intergenic region-derived sRNAs in barley only account for <9.5%. These results indicate that different sRNAs respond differently to drought stress, and sRNA generation is condition dependent. However, we found that the length distribution of sRNAs was not significantly affected by drought, despite that some proportion of sRNAs may be generated by degradation pathways. This suggests that sRNA length may be controlled independently of drought stress.

All tsRNAs were up-regulated under drought stress. In animals, tsRNAs also increase under stresses (Saikia *et al.*, 2012), suggesting that a common mechanism might exist between plant and animal for tsRNA generation under stresses. Curiously, tsRNAs derived from the chloroplast were also up-regulated under drought stress. This suggests that (i) part of the chloroplast

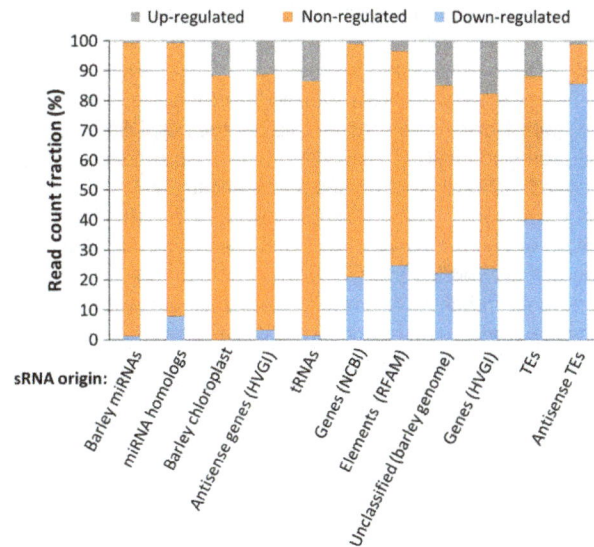

Figure 5 Distribution of overall sRNAs (siRNAs inclusive) of up-, down- and nonregulation under drought stress.

tsRNAs might be from the nucleus and/or (ii) tsRNAs in barley are genome-wide regulated (regulation is not limited to some particular organelles). The latter raises the interesting question on how two different sources of tsRNAs are coordinately regulated in plants. Perhaps, the processing of tsRNAs in the genome and chloroplast shares the same mechanism. In animals, tsRNAs have been found in the RISC and play a critical role in many biological processes (Lee *et al.*, 2009). We believe that drought-regulated tsRNAs would have roles in drought tolerance. However, the molecular mechanism of tsRNA regulation by drought desires and merits further investigation.

Intriguingly, tsRNAs and repeat classes are drought-regulated oppositely. In addition, some cis-natsiRNAs were found to overlap with some trans-natsiRNAs, thereby forming double-stranded RNAs against each function. Furthermore, cleavage products of some siRNAs' predicted targets were detected. These, combined with the fact that miRNAs can trigger siRNAs, lead us to speculate that regulatory networks may exist among sRNAs. The sRNAs may regulate or be regulated directly or indirectly via transcriptional silencing, post-transcriptional silencing and/or translational inhibition. It is likely that post-transcriptional silencing could be a dominant mechanism for all functional sRNAs because of the detection of many siRNAs' cleavage products. In contrast, transcriptional silencing could mainly be conducted by rasiRNAs and miRNAs through methylation and chromatin remodelling to control the transcription of repeat sequences (i.e. transposons and retrotransposons) and genes that derive siRNAs, respectively. Translational inhibition would involve a third factor like transcription factors for controlling the transcription of genes that derive siRNAs. The role of such regulatory networks is to fine-tune the expression of miRNAs and siRNAs in order to coordinate with the expression of other genes in response to drought stress. However, we need to mention here that we could not find inverse expression of the natsiRNA pairs by analysing our sRNA datasets, suggesting that the functional mechanism of natsiRNAs remains elusive in barley.

All up-regulated miRNAs were of low copies, suggesting that these miRNAs may be subject to weak or no selection according to previous studies (Liang and Li, 2009). This would in turn mean that they tend to turn over quickly in evolution (Liang and Li, 2009). Nevertheless, low-expressed miRNAs are generally thought to have limited biological importance, and their effect on potential target mRNAs would, in most cases, be weak or negligible. However, we cannot rule out the possibility that these low-expressed miRNAs may participate in processes that can tolerate functional variation under drought stress. Apart from low expression, some miRNAs such as osa-miR393a were found to be significantly differentially expressed between leaf and root tissues. It is surprising that miR169 was not significantly regulated by drought in barley. This result is in agreement with that from Trindade et al. (2010) but not with that from Wang et al. (2011) in M. truncatula. Wang et al. (2011) showed that miR169 is down-regulated in M. truncatula under drought conditions. The discrepancy has been proposed to result from different drought treatments. In Arabidopsis, miR169 is also down-regulated under drought (Li et al., 2008), but in rice, this miRNA is up-regulated by drought (Zhao et al., 2007; Zhou et al., 2010). The up-regulation of miR169 in rice may be attributed by its upstream dehydration-responsive element (DRE), which is associated with abiotic stresses including drought stress (Zhao et al., 2007). The conse-quence of the unchanged or slightly up-regulated miR169 in barley under drought is yet unknown, but presumably, it would have little impact on plant response to drought stress. However, the ultimate proof of this hypothesis will come from its transgenic experiments overexpressing and knocking-down miR169.

Our previous study showed that overexpression of TaDREB3 in barley significantly affects the expression of various classes of sRNAs including miRNAs (Hackenberg et al., 2012). Comparing sRNAs from drought-treated barley and those from the transgenic barley reveals that many sRNAs including miRNAs were co-regulated. In other words, these sRNAs were consistently up- or down-regulated in both the drought-treated and transgenic plants. This provides an opportunity to identify true drought-responsive sRNAs, which might be used for improvement of plant drought tolerance in the future. However, further analysis finds that most of these co-regulated sRNAs are derived from protein-coding genes and hence unlikely to be generated via the miRNA/siRNA pathway. Therefore, these sRNAs may not play an important role in drought tolerance especially if those genes turn out to be not related to drought response. Why are some sRNAs co-regulated in plants? One possible explanation is that the expression of the protein-coding genes, from which the sRNAs are derived, may be controlled by other drought or TaDREB3-regulated genes. In Arabidopsis, the roles of different classes of sRNAs have been proposed: siRNAs maintain genome stability, whereas miRNAs mediate gene expression diversity (Ha et al., 2009). It is interesting to see if these properties are conserved in other species.

Experimental procedures

Plant materials, RNA isolation and sRNA sequencing

Golden Promise plants were grown in 6-inch pots in coco-peat soil in a glasshouse at 22–23 °C day/16 °C night with a 12-h day/night light cycle under water condition. Three weeks after germination, leaf and root tissues were harvested and further harvested after water was withheld for 5 days. Total RNA was extracted using TRIzol reagent and used for both Northern hybridization, reverse transcription PCR (RT-PCR) and quantitative real-time PCR (qRT-PCR). 18–30-nt sRNAs isolated from leaf tissues via 15% polyacrylamide gel were sequenced using the 36-base Illumina platform.

Processing of sRNA reads and alignment to reference sequences

The raw sequencing reads were processed by trimming reads at the first base that has a Phred score quality ≤ 2, eliminating the adapter sequence, removing reads with a strong compositional bias (frequency of the most base ≥ 0.9), and collapsing identical sequences into unique sequences, thereby assigning a read count/unique read (the number of times in a given RNA molecule was sequenced).

All reads were mapped to the nuclear and chloroplast genome sequences without mismatches and labelled as nuclear and chloroplast upon their inferred origin. Known miRNAs were detected by mapping all reads to miRBase without mismatches. Other classes of sRNAs were detected by mapping all reads to TIGR repeat database, Triticeae repeat sequence database (TREP) and all RepBase repeats, RFAM, HVGI and NCBI without mismatches. The reads mapped to a given library were not used for next library mapping. Antisense sRNAs (putative siRNAs) were detected by mapping all reads to the antisense strand of the elements.

Prediction of novel miRNAs and targets

Prediction of novel miRNAs was performed with miRDeep (Friedlander et al., 2008). Pre-miRNA stem–loop structures were examined with MFOLD (Zuker, 2003). Only pre-miRNAs meeting a minimum free energy lower than −16 kcal/mol were retained for next analysis. miRNA targets were predicted using psTarget (http://plantgrn.noble.org/psRNATarget). Gene annotation was obtained from the same source.

Construction of degradome library

Degradome library was constructed according to Addo-Quaye et al. (2008). Poly(A) RNAs isolated using the Oligotex Kit (Qiagen, Santa Clarita, CA) were ligated with a 5′ RNA adaptor containing a MmeI restriction site using T4 RNA ligase. Following reverse transcription, second-strand synthesis, MmeI digestion, ligation of a 3′ dsDNA adaptor and gel-purification, the cDNA products were amplified by PCR and sequenced with the Illumina HiSeq platform.

Differential expression analysis of sRNAs and their predicted targets by Northern hybridization and qRT-PCR

For Northern hybridization, 50 μg total RNA was electrophoresed on 15% polyacrylamide gel containing 7 M urea and transferred to Hybond-N membrane using 20 × SSC (3 M NaCl and 0.3 M $C_6H_9Na_3O_9$). The membrane was hybridized with ^{32}P-labelled oligonucleotide probe reverse complementary to miRNA sequence. A U6 snRNA-specific probe served as a loading control. Prehybridization and hybridization were performed at 37 °C in 125 mM Na_2HPO_4 (pH 7.2), 250 mM NaCl, 7% SDS and 50% formamide. Washing was performed at 37 °C using 2 × SSC (twice) and 1 × SSC (once). The membrane was subjected to autoradiography at −80 °C for about 7 days.

For qRT-PCR, total RNA samples were polyadenylated and reverse-transcribed using the NCode™ VILO™ miRNA cDNA Synthesis Kit (Invitrogen, Carlsbad, CA) containing poly(A) poly-merase, ATP, SuperScript™III RT and a universal RT primer. cDNA was amplified with a miRNA- or target-specific forward primer and the RT primer using the following condition: 3 min at 95 °C followed by 45 cycles of 1 s at 95 °C, 1 s at 55 °C, 30 s at 72 °C (fluorescence reading acquired) and 15 s at 81 °C. Normalization was performed using three biological replicates and four control genes encoding glyceraldehyde 3-Pi dehydrogenase, heat-shock protein 70, cyclophilin and α-tubulin (Burton *et al.*, 2004).

Acknowledgements

The authors wish to thank Ursula Langridge, Hui Zhou, Yuan Li and Bryce Shi for technical assistance and Patricia Warner for reading the manuscript.

References

Addo-Quaye, C., Eshoo, T.W., Bartel, D.P. and Axtell, M.J. (2008) Endogenous siRNA and miRNA targets identified by sequencing of the *Arabidopsis* degradome. *Curr. Biol.* **18**, 758–762.

Arenas-Huertero, C., Pérez, B., Rabanal, F., Blanco-Melo, D., De la Rosa, C., Estrada-Navarrete, G., Sanchez, F., Covarrubias, A.A. and Reyes, J.L. (2009) Conserved and novel miRNAs in the legume *Phaseolus vulgaris* in response to stress. *Plant Mol. Biol.* **70**, 385–401.

Axtell, M.J. (2013) Classification and comparison of small RNAs from plants. *Annu. Rev. Plant Biol.* **64**, 137–159.

Ballen-Taborda, C., Plata, G., Ayling, S., Rodriguez-Zapata, F., Becerra Lopez-Lavalle, L.A., Duitama, J. and Tohme, J. (2013) Identification of cassava MicroRNAs under abiotic stress. *Int. J. Genomics*, **2013**, 857986.

Barrera-Figueroa, B.E., Gao, L., Diop, N.N., Wu, Z.G., Ehlers, J.D., Roberts, P.A., Close, T.J., Zhu, J.K. and Liu, R. (2011) Identification and comparative analysis of drought- associated microRNAs in two cowpea genotypes. *BMC Plant Biol.* **11**, 127.

Burton, R.A., Shirley, N.J., King, B.J., Harvey, A.J. and Fincher, G.B. (2004) The CesA gene family of barley (*Hordeum vulgare*): quantitative analysis of transcripts reveals two groups of co-expressed genes. *Plant Physiol.* **134**, 224–236.

Elbashir, S., Lendeckel, W. and Tuschl, T. (2001) RNA Interference is mediated by 21- and 22- nucleotide RNAs. *Genes Dev.* **15**, 188–200.

Eldem, V., Celikkol Akcay, U., Ozhuner, E., Bakir, Y., Uranbey, S. and Unver, T. (2012) Genome-wide identification of miRNAs responsive to drought in peach (*Prunus persica*) by high-throughput deep sequencing. *PLoS One*, **7**, e50298.

Ferreira, T.H., Gentile, A., Vilela, R.D., Costa, G.G., Dias, L.I., Endres, L. and Menossi, M. (2012) microRNAs associated with drought response in the bioenergy crop sugarcane (*Saccharum* spp.). *PLoS One*, **7**, e46703.

Frazier, T.P., Sun, G.L., Burklew, C.E. and Zhang, B.H. (2011) Salt and drought stresses induce the aberrant expression of microRNA genes in tobacco. *Mol. Biotechnol.* **14**, 159–165.

Friedlander, M.R., Chen, W., Adamidi, C., Maaskola, J., Einspanier, R., Knespel, S. and Rajewsky, N. (2008) Discovering microRNAs from deep sequencing data using miRDeep. *Nat. Biotechnol.* **26**, 407–415.

Gardner, P.P., Daub, J., Tate, J.G., Nawrocki, E.P., Kolbe, D.L., Lindgreen, S., Wilkinson, A.C., Finn, R.D., Griffiths-Jones, S., Eddy, S.R. and Bateman, A. (2009) Rfam: updates to the RNA families database. *Nucleic Acids Res.* **37**, D136–D140.

Griffiths-Jones, S., Saini, H.K., van Dongen, S. and Enright, A.J. (2008) miRBase: tools for microRNA genomics. *Nucleic Acids Res.* **36**, D154–D158.

Ha, M., Lu, J., Tian, L., Ramachandran, V., Kasschau, K.D., Chapman, E.J., Carrington, J.C., Chen, X., Wang, X.J. and Chen, Z.J. (2009) Small RNAs serve as a genetic buffer against genomic shock in Arabidopsis interspecific hybrids and allopolyploids. *Proc. Natl Acad. Sci. USA*, **106**, 17835–17840.

Hackenberg, M., Rodriguez-Ezpeleta, N. and Aransay, A.M. (2011) miRanalyzer: an update on the detection and analysis of microRNAs in high-throughput sequencing experiments. *Nucleic Acids Res.* **39**, W132–W138.

Hackenberg, M., Shi, B.J., Gustafson, P. and Langridge, P. (2012) A transgenic transcription factor (TaDREB3) in barley affects the expression of microRNAs and other small non-coding RNAs. *PLoS One*, **7**, e42030.

Huntzinger, E. and Izaurralde, E. (2011) Gene silencing by microRNAs: contributions of translational repression and mRNA decay. *Nat. Rev. Genet.* **12**, 99–110.

Jones-Rhoades, M.W., Bartel, D.P. and Bartel, B. (2006) MicroRNAs and their regulatory roles in plants. *Ann. Rev. Plant Biol.* **57**, 19–53.

Jurka, J., Kapitonov, V.V., Pavlicek, A., Klonowski, P., Kohany, O. and Walichiewicz, J. (2005) Repbase Update, a database of eukaryotic repetitive elements. *Cytogenet. Genome Res.* **110**, 462–467.

Kantar, M., Unver, T. and Budak, H. (2010) Regulation of barley miRNAs upon dehydration stress correlated with target gene expression. *Funct. Integr. Genomics*, **10**, 493–507.

Kantar, M., Lucas, S.J. and Budak, H. (2011) miRNA expression patterns of *Triticum dicoccoides* in response to shock drought stress. *Planta*, **233**, 471–484.

Kuang, H., Padmanabhan, C., Li, F., Kamei, A., Bhaskar, P.B., Ouyang, S., Jiang, J., Buell, C.R. and Baker, B. (2009) Identification of miniature inverted-repeat transposable elements (MITEs) and biogenesis of their siRNAs in the Solanaceae: new functional implications for MITEs. *Genome Res.* **19**, 42–56.

Kulcheski, F.R., de Oliveira, L.F., Molina, L.G., Almerao, M.P., Rodrigues, F.A., Marcolino, J., Barbosa, J.F., Stolf-Moreira, R., Nepomuceno, A.L., Marcelino-Guimaraes, F.C., Abdelnoor, R.V., Nascimento, L.C., Carazzolle, M.F., Pereira, G.A. and Margis, R. (2011) Identification of novel soybean microRNAs involved in abiotic and biotic stresses. *BMC Genomics*, **12**, 307.

Lee, Y.S., Shibata, Y., Malhotra, A. and Dutta, A. (2009) A novel class of small RNAs: tRNA-derived RNA fragments (tRFs). *Genes Dev.* **23**, 2639–2649.

Li, W.X., Oono, Y., Zhu, J., He, X.J., Wu, J.M., Iida, K., Lu, X.Y., Cui, X., Jin, H. and Zhu, J.K. (2008) The Arabidopsis NFYA5 transcription factor is regulated transcriptionally and posttranscriptionally to promote drought resistance. *Plant Cell*, **20**, 2238–2251.

Li, B., Qin, Y., Duan, H., Yin, W. and Xia, X. (2011) Genome-wide characterization of new and drought stress responsive microRNAs in *Populus euphratica*. *J. Exp. Bot.* **62**, 3765–3779.

Liang, H. and Li, W.H. (2009) Lowly expressed human microRNA genes evolve rapidly. *Mol. Biol. Evol.* **26**, 1195–1198.

Liu, H.H., Tian, X., Li, Y.J., Wu, C.A. and Zheng, C.C. (2008) Microarray-based analysis of stress-regulated microRNAs in *Arabidopsis thaliana*. *RNA*, **14**, 836–843.

Lu, S., Sun, Y.H. and Chiang, V.L. (2008) Stress-responsive microRNAs in Populus. *Plant J.* **55**, 131–151.

Lu, J., Zhang, C., Baulcombe, D.C. and Chen, Z.J. (2012) Maternal siRNAs as regulators of parental genome imbalance and gene expression in endosperm of Arabidopsis seeds. *Proc. Natl Acad. Sci. USA*, **109**, 5529–5534.

Morran, S., Eini, O., Pyvovarenko, T., Parent, B., Singh, R., Ismagul, A., Eliby, S., Shirley, N., Langridge, P. and Lopato, S. (2011) Improvement of stress tolerance of wheat and barley by modulation of expression of DREB/CBF factors. *Plant Biotechnol. J.* **9**, 230–249.

Ouyang, S. and Buell, C.R. (2004) The TIGR Plant Repeat Databases: a collective resource for the identification of repetitive sequences in plants. *Nucleic Acids Res.* **32**, D360–D363.

Pontes, O., Li, C., Nunes, P., Haag, J., Ream, T., Vitins, A., Jacobsen, S. and Pikaard, C. (2008) The Arabidopsis chromatin-modifying nuclear siRNA pathway involves a nucleolar RNA processing center. *Cell*, **126**, 79–92.

Saikia, M., Krokowski, D., Guan, B.J., Ivanov, P., Parisien, M., Hu, G.F., Anderson, P., Pan, T. and Hatzoglou, M. (2012) Genome-wide identification and quantitative analysis of cleaved tRNA fragments induced by cellular stress. *J. Biol. Chem.* **287**, 42708–42725.

Saski, C., Lee, S.B., Fjellheim, S., Guda, C., Jansen, R., Tomkins, J., Rognli, O.A., Daniell, H. and Clarke, J.L. (2007) Complete chloroplast genome sequences of *Hordeum vulgare*, *Sorghum bicolor*, and *Agrostis stolonifera*, and comparative analyses with other grass genomes. *Theor. Appl. Genet.* **115**, 571–590.

Schreiber, A.W., Shi, B.J., Huang, C.Y., Langridge, P. and Baumann, U. (2011) Discovery of barley miRNAs through deep sequencing of short reads. *BMC Genomics*, **12**, 129.

Shamimuzzaman, M. and Vodkin, L. (2012) Identification of soybean seed developmental stage-specific and tissue-specific miRNA targets by degradome sequencing. *BMC Genomics*, **13**, 310.

Shuai, P., Liang, D., Zhang, Z., Yin, W. and Xia, X. (2013) Identification of drought-responsive and novel *Populus trichocarpa* microRNAs by high-throughput sequencing and their targets using degradome analysis. *BMC Genomics*, **14**, 233.

Sunkar, R. and Zhu, J.K. (2004) Novel stress-regulated microRNAs other small RNAs from Arabidopsis. *Plant Cell*, **16**, 2001–2019.

The International Barley Genome Sequencing Consortium (Mayer, K.F.X., Waugh, R., Langridge, P., Close, T.J., Wise, R.P., Graner, A., Matsumoto, T., Sato, K., Schulman, A., Ariyadasa, R., Schulte, D., Poursarebani, N., Zhou, R., Steuernagel, B., Mascher, M., Scholz, U., Shi, B., Madishetty, K., Svensson, J.T., Bhat, P., Moscou, M., Resnik, J., Muehlbauer, G.J., Hedley, P., Liu, H., Morris, J., Frenkel, Z., Korol, A., BergÄš, H., Taudien, S., Felder, M., Groth, M., Platzer, M., Himmelbach, A., Lonardi, S., Duma, D., Alpert, M., Cordero, F., Beccuti, M., Ciardo, G., Ma, Y., Wanamaker, S., Cattonaro, F., Vendramin, V., Scalabrin, S., Radovic, S., Wing, R., Morgante, M., Nussbaumer, T., Gundlach, H., Martis, M., Poland, J., Pfeifer, M., Moisy, C., Tanskanen, J., Zuccolo, A., Spannagl, M., Russell, J., Druka, A., Marshall, D., Bayer, M., Swarbreck, D., Sampath, D., Ayling, S., Febrer, M., Caccamo, M., Tanaka, T., Wannamaker, S., Schmutzer, T., Brown, J.W.S., Fincher, G.B. and Stein, N.) (2012) A physical, genetic and functional sequence assembly of the barley genome. *Nature*, **491**, 711–717.

Trindade, I., Capitao, C., Dalmay, T., Fevereiro, M.P. and Santos, D.M. (2010) miR398 and miR408 are up-regulated in response to water deficit in *Medicago truncatula*. *Planta*, **231**, 705–716.

Wang, T.Z., Chen, L., Zhao, M.G., Tian, Q.Y. and Zhang, W.H. (2011) Identification of drought-responsive microRNAs and their targets in *Medicago truncatula* by genome-wide high-throughput sequencing and degradome analysis. *BMC Genomics*, **12**, 367.

Wang, M., Wang, Q. and Zhang, B. (2013) Response of miRNAs and their targets to salt and drought stresses in cotton (*Gossypium hirsutum* L.). *Gene*, **30**, 26–32.

Wicker, T., Matthews, D.E. and Keller, B. (2002) TREP: a database of Triticeae repetitive elements. *Trends Plant Sci.* **7**, 561–562.

Woodger, F.J., Jacobsen, J.V. and Gubler, F. (2004) GMPOZ, a BTB/POZ domain nuclear protein, is a regulator of hormone responsive gene expression in barley aleurone. *Plant Cell Physiol.* **45**, 945–950.

Xie, F., Stewart, C.N., Taki, F.A., He, Q., Liu, H. and Zhang, B. (2014) Highthroughput deep sequencing shows that microRNAs play important roles in switchgrass responses to drought and salinity stress. *Plant Biotechnol. J.* **12**, 354–366.

Xu, Z., Zhong, S., Li, X., Li, W., Rothstein, S.J., Zhang, S., Bi, Y. and Xie, C. (2011) Genome-wide identification of microRNAs in response to low nitrate availability in maize leaves and roots. *PLoS One*, **6**, e28009.

Yao, Y., Ni, Z., Peng, H., Sun, F., Xin, M., Sunkar, R., Zhu, J.K. and Sun, Q. (2010) Noncoding small RNAs responsive to abiotic stress in wheat (*Triticum aestivum* L.). *Funct. Integr. Genomics*, **10**, 187–190.

Zhang, B.H., Pan, X.P., Cannon, C.H., Cobb, G.P. and Anderson, T.A. (2006) Conservation and divergence of plant microRNA genes. *Plant J.* **46**, 243–259.

Zhang, J., Xu, Y., Huan, Q. and Chong, K. (2009) Deep sequencing of Brachypodium small RNAs at the global genome level identifies microRNAs involved in cold stress response. *BMC Genomics*, **10**, 449.

Zhang, X., Zou, Z., Gong, P., Zhang, J., Ziaf, K., Li, H., Xiao, F. and Ye, Z. (2011) Over-expression of microRNA169 confers enhanced drought tolerance to tomato. *Biotechnol. Lett.* **33**, 403–409.

Zhang, N., Yang, J., Wang, Z., Wen, Y., Wang, J., He, W., Liu, B., Si, H. and Wang, D. (2014) Identification of novel and conserved microRNAs related to drought stress in potato by deep sequencing. *PLoS One*, **9**, e95489.

Zhao, B.T., Liang, R.Q., Ge, L.F., Li, W., Xiao, H.S., Lin, H.X., Ruan, K.C. and Jin, Y.X. (2007) Identification of drought-induced microRNAs in rice. *Biochem. Biophys. Res. Commun.* **354**, 585–590.

Zhou, L., Liu, Y., Liu, Z., Kong, D., Duan, M. and Luo, L. (2010) Genome-wide identification and analysis of drought-responsive microRNAs in *Oryza sativa*. *J. Exp. Bot.* **61**, 4157–4168.

Zuker, M. (2003) Mfold web server for nucleic acid folding and hybridization prediction. *Nucleic Acids Res.* **31**, 3406–3415.

16

Main regulatory pathways, key genes and microRNAs involved in flower formation and development of moso bamboo (*Phyllostachys edulis*)

Wei Ge[1,†], Ying Zhang[1,2,†], Zhanchao Cheng[1,†], Dan Hou[1], Xueping Li[1] and Jian Gao[1,*]

[1]Key Laboratory of Bamboo and Rattan Science and Technology of the State Forestry Administration, International Centre for Bamboo and Rattan, Beijing, China
[2]China National Engineering Research Center for Information Technology in Agriculture, Beijing, China

*Correspondence

email gaojianicbr@163.com
[†]These authors contributed equally to this work.

Keywords: moso bamboo, flower formation and development, regulatory pathway, Dof, MADS-box, microRNA.

Summary

Moso bamboo is characterized by infrequent sexual reproduction and erratic flowering habit; however, the molecular biology of flower formation and development is not well studied in this species. We studied the molecular regulation mechanisms of moso bamboo development and flowering by selecting three key regulatory pathways: plant–pathogen interaction, plant hormone signal transduction and protein processing in endoplasmic reticulum at different stages of flowering in moso bamboo. We selected *PheDof1*, *PheMADS14* and six microRNAs involved in the three pathways through KEGG pathway and cluster analysis. Subcellular localization, transcriptional activation, Western blotting, *in situ* hybridization and qRT-PCR were used to further investigate the expression patterns and regulatory roles of pivotal genes at different flower development stages. Differential expression patterns showed that *PheDof1*, *PheMADS14* and six miRNAs may play vital regulatory roles in flower development and floral transition in moso bamboo. Our research paves way for further studies on metabolic regulatory networks and provides insight into the molecular regulation mechanisms of moso bamboo flowering and senescence.

Introduction

Moso bamboo (*Phyllostachys edulis*) is a kind of large woody bamboo with great ecological, cultural and economic values of all bamboos in Asia covering about 3 000 000 hm², accounting for 20% of the total forest area in the world (Li *et al.*, 2012). In China, it covers 6.01 million hm², and the total annual forest production was valued at 5 billion US dollars in 2013. Moso bamboo is a perennial plant characterized by rapid growth and a long vegetative stage that lasts for decades or even longer before flowering (Lin *et al.*, 2010). However, it always flowers synchronously followed by widespread death in a large area, limiting the development of the moso bamboo industry. Peng *et al.* (2013) had pointed out that 30% of flowering-related genes including DOF and MADS-box families were heat shock protein genes, stress-related genes and transcription factors. An increase expression of Dof under drought stress induced MADS-box expression, which in turn promoted flowering in moso bamboo. As Dof also affected the expression of the floral integrator *Heading date3a* (*Hd3a*) gene, Gao *et al.* (2014) suggested that an active Dof-Hd3a-MADS-flowering pathway might significantly affect flowering in Moso bamboo. Moreover, we have successfully collected full-scale moso bamboo flowering samples through several year endeavour of our research group and used the paraffin section technique to study moso bamboo inflorescence morphogenesis for the first time. A large number of differentially expressed microRNAs and their targets participated in diverse primary biological pathways and play significant regulatory roles in moso bamboo flowering (Gao *et al.*, 2015). However, it is different from the classic flowering regulation pathway based on our research data, such as no significant change in expression of *CO* during moso bamboo flowering. To propagate moso bamboo plants, it is imperative to determine the signalling pathways involved in flowering and the respective genes.

Dof (DNA binding with one zinc finger) transcription factors are a group of plant-specific transcription factors. The cDNA sequence of the first Dof was obtained from *Zea mays* (Yanagisawa and Izui, 1993). Since then, many Dof genes have been cloned from various plant species including 37 Dof genes from *Arabidopsis thaliana* (Ward *et al.*, 2005), 30 from *Oryza sativa* (Li *et al.*, 2009; Lijavetzky *et al.*, 2003) and 28 Dof genes from *Glycine max* (Wang *et al.*, 2007). Dof transcription factors are involved in many biological processes in plant growth and development. *JcDof3* is a biological clock gene that regulates the flowering time in *Jatropha carcass* (Yang *et al.*, 2011). The overexpression of *OsDof12* in *O. sativa* induces early flowering under long-day or drought stress conditions (Li *et al.*, 2009).

The MADS-box family members that were identified originally as floral homeotic genes are significant transcription factors for plant development (De Folter and Angenent, 2006; Kaufmann *et al.*, 2005). These genes were initially characterized from *Saccharomyces cerevisiae* (Passmore *et al.*, 1988) and *Arabidopsis* (Yanofsky *et al.*, 1990). MADS-box genes participate in floral organ development and flowering time regulation (Theissen *et al.*, 2000). Thirty-four MADS-box genes were identified in *O. sativa*, 15 of which regulate flower development (Lee *et al.*, 2003). *OsMADS18* regulates the differentiation of shoot apical meristem and causes early flowering in *O. sativa* (Fornara *et al.*, 2004) while *OsMADS3* and *OsMADS58* control carpel formation (Dreni *et al.*, 2007). Mutants of MADS-box genes such as *cfo1* (Sang *et al.*, 2012) can change the morphology of floral organs.

miRNAs are small regulatory RNAs that are involved in a number of processes including growth and development control, cell differentiation, phytohormone signals, abiotic and biotic stress (Liu and Chen, 2009; Zhu et al., 2008). Many miRNAs target transcription factors have been implicated in key regulatory pathways (Jones-Rhoades et al., 2006). Yu et al. (2012) showed that Gibberellins regulate flowering via miR156 targeted squamosa promoter binding-like (SPL) transcription factors. Millar and Gubler (2005) suggested that miR159 was related to the development and morphology of clinandriums. miR164 targeted NAM, ATAF1/2 and CUC2 domain containing transcription factors (NAC) and specify extraordinary cell types at the later stages of flower development (Nag and Jack, 2010). The expression levels of miR164a showed significant down-regulated trend from no flowering leaves to flowering samples in moso bamboo, negatively correlated with that of its target (Gao et al., 2015). Other flowering-related miRNAs include miR166, which affects the morphogenesis of flowers (Jung and Park, 2007) and miR167, which regulates the development and maturity of stamens and pistils (Nagpal et al., 2005). In this study, we aimed to select key regulatory pathways at different stages of floral development in moso bamboo. We used bioinformatics analysis to identify three regulatory pathways, transcription factors and the miRNAs involved in the regulatory pathways and studied the expression patterns of key genes using Western blotting, in situ hybridization and qRT-PCR. Our research reveals the molecular regulation mechanism of moso bamboo flowering and provides important information for selecting significant genes to study their function in floral development.

Results

Pathway function annotation of differentially expressed genes (DEGs)

Our results following differential gene expression (DGE) analysis showed that 11 703 differentially expressed genes changed significantly during the four stages of flower formation and development. These genes were annotated with KEGG pathway and were involved in 128 metabolic pathways. The RPKM of each metabolic pathway was calculated, and cluster analysis was carried out with the logarithm of RPKM (Figure 1). We used cluster analysis to identify the two major gene groups that were up- and down-regulated. The highest expressional pathways in the up-regulated groups were plant–pathogen interaction, plant hormone signal transduction and protein processing in endoplasmic reticulum.

Our results showed that 269 differentially expressed genes were involved in plant–pathogen interaction (Figure 2). Plant disease resistance genes FLS2, PR1, RPS2 and RPS5 showed high expression levels during the whole flower formation and developmental process. CML, PBS1 and BAK1 showed higher expression at later stages of flower formation and development compared with that at early stages (Figure 2a). The expression of WRKY gene which is related to plant senescence increased gradually with the development of the flower. The expression of WRKY was the highest when moso bamboo entered the death stage on completion of embryo development (Figure 2a). The high expression levels indicated that the process of flower formation and development might be a phenotype of moso bamboo senescence. We analysed the targets of differentially expressed miRNAs using the KEGG pathway, and miRNAs that putatively play significant regulatory roles in plant–pathogen interaction based on their targets were used to perform cluster analysis (Figure 3a). It showed that miR390a, miR5139 and miR5821 expression levels were lower in flowering samples than those in nonflowering samples (Figure 3a).

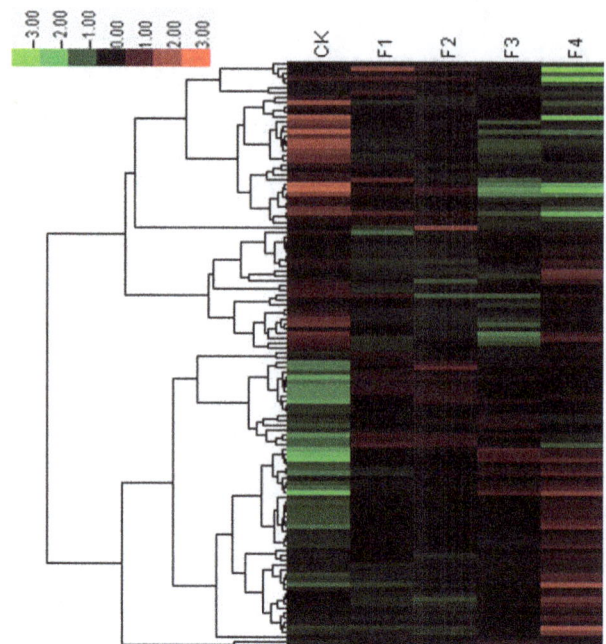

Figure 1 Hierarchical cluster analysis of DEG gene expression based on log ratio RPKM data.

Flower formation and development in plants is a cue to enter the reproductive development, and the perception and response of plant hormones plays vital roles in the regulation of flower formation and development. Our results showed that 301 differentially expressed genes were involved in plant hormone signal transduction. Cluster analysis indicated that the expression of IAA was higher in the early stage of flower formation and development and presented down-regulated trend with flower development (Figure 2b). GA metabolism genes related to flower development such as GID1 and GID2 showed high expression levels at four different periods of flower development. Thus, the genes related to GA metabolism played a crucial role in whole flower formation and development. Moreover, the expression of genes involved in jasmonic acid and salicylic acid metabolism and related to stress, including JAZ, transcription factor MYC2 and TGA, was significantly up-regulated at floral bud formation stage, bloom stage and embryo formation stage. These results showed that the stress pathways and hormone regulatory pathways related to stress resistance were mutually connected, highly expressed and jointly regulated flower development in moso bamboo. However, a number of miRNAs involved in plant hormone signal transduction, for example miR160a, miR171b and miR393, were down-regulated in the four periods (Figure 3b).

Our results showed that 170 differentially expressed genes were involved in protein processing in the endoplasmic reticulum (ER). Cluster analysis indicated that the expressions of MADS-box and MYB were markedly up-regulated during the four flower developmental stages (Figure 2c). Many transcription factors associated with stress resistance were found in protein processing in the ER such as HSP90, which showed significant up-regulation. HSP90 is associated with hormone receptors, other transcription factors (such as HSP20), stress signalling pathways, intracellular transport and kinase activity regulation. The high expression of HSP90 indicated that the process of flower development in moso

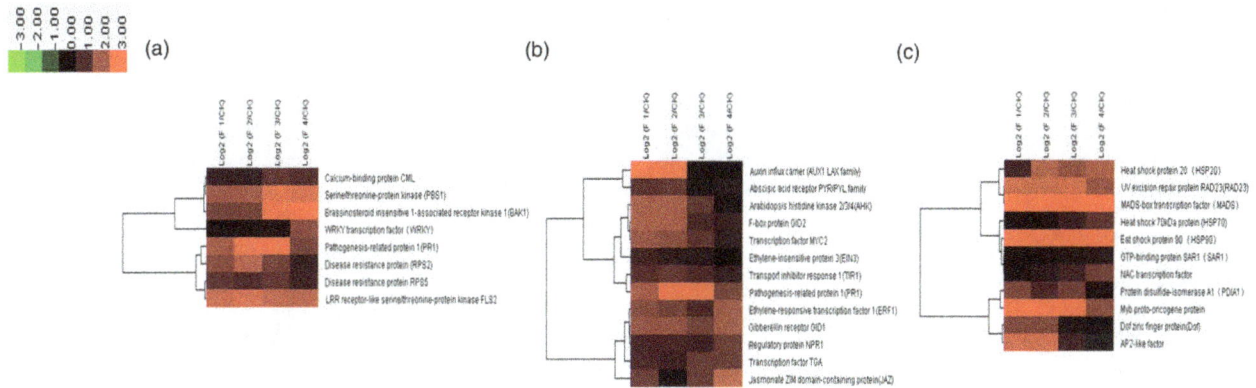

Figure 2 Differential expression analysis of related genes involved in main regulatory pathways. Heatmap for clustering analysis of related genes. The bar represents the scale of the expression levels of the genes (log 2). CK: nonflowering moso bamboo leaves, F1: floral bud formation stage, F2: inflorescence growing stage, F3: bloom stage, F4: embryo formation stage. (a) Differential expression analysis of related genes involved in plant–pathogen interaction. (b) Differential expression analysis of related genes involved in plant hormone signal transduction. (c) Differential expression analysis of related genes involved in protein processing in endoplasmic reticulum.

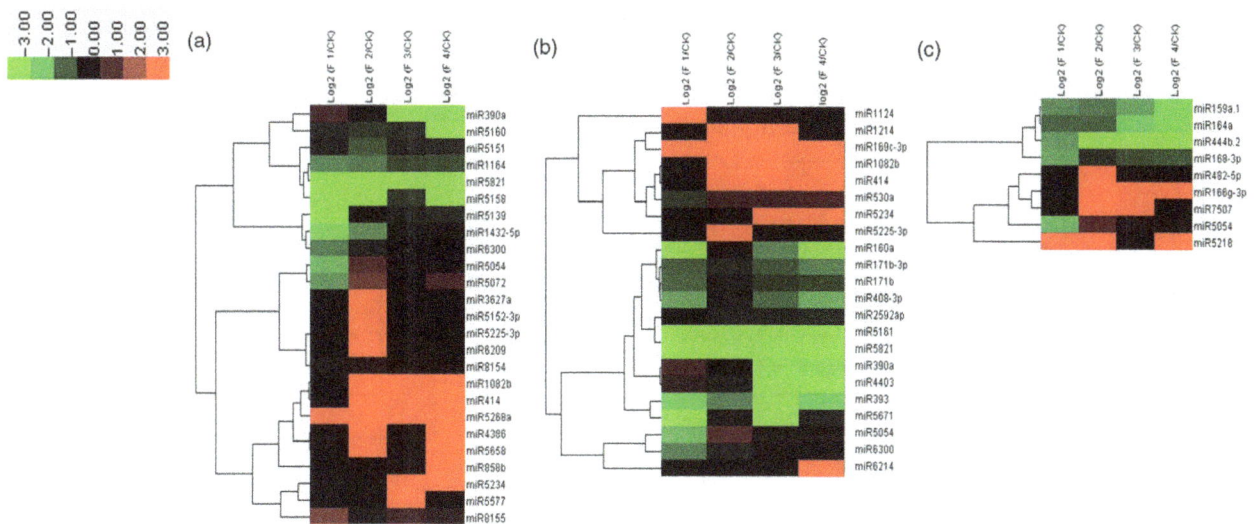

Figure 3 Differential expression analysis of related miRNAs involved in main regulatory pathways. Heatmap for clustering analysis of related miRNAs. The bar represents the scale of the expression levels of the genes (log 2). (a) Differential expression analysis of related miRNAs involved in plant–pathogen interaction. (b) Differential expression analysis of related miRNAs involved in plant hormone signal transduction. (c) Differential expression analysis of related miRNAs involved in protein processing in endoplasmic reticulum.

bamboo might be related to stress. Dof, the other important stress-related gene was highly expressed at the floral bud formation stage and inflorescence growing stage, while gradually decreased with the flowers withered and reached the minimum at embryo formation stage (Figure 2c). Therefore, we inferred that Dof mainly played roles at early stages of flower formation and development in moso bamboo; it also participated in the regulation of floral bud formation and elongation. Moreover, the expression levels of AP2, NAC and RAD23 were also significantly up-regulated in the flower development process (Figure 2c). The putative ER protein processing miRNAs such as miR159a.1, miR164a and miR168-3p showed lower expression levels in flowering samples were than those in nonflowering leaves (Figure 3c).

Phylogenetic analysis of PheDof1 and PheMADS14

We found that 15 Dof transcription factors and 34 MADS transcription factors that were involved in protein processing in

the ER in moso bamboo inflorescence transcripts. *PheDof1* (PH01000664G0640) and *PheMADS14* (PH01000222G1190) showed significantly high expression levels in the flowering period. Thus, *PheDof1* and *PheMADS14* were selected for further research.

To explore the phylogenetic relationships among the *Dof* genes in moso, *Arabidopsis* and rice, we used MEGA 4.1 to construct a phylogenetic tree using full-length CDS of *PheDof1* (918 bp, encoding 305 amino acids and 33.66 kD (Figure 4a). *PheDof1* (PH01000664G0640) was clustered in the same group with *OsDof1* and *OsDof29*, indicating that *PheDof1* was a member of the MCOG C subfamily.

We also constructed a phylogenetic tree to explore the relationship of *PheMADS14* with *Arabidopsis* and rice MADS-box genes. The full-length CDS of *PheMADS14* was 1077 bp, encoding 358 amino acids and a molecular weight of 39.49 kD (Figure 4b). *PheMADS14* (PH01000222G1190) was clustered in the same group as *OsMADS14*, *OsMADS15* and *OsMADS18*.

Thus, *PheMADS14* was a member of the AP/AGL9 subfamily and participated in the formation of floral organs.

Protein sequence motifs analysis showed that *PheDof1* (PH01000664G0640) and *PheMADS14* (PH01000222G1190) belonged to the MCOG C and AP/AGL9 subfamilies, respectively (Figure 5a and b), which was consistent with our results of the MEGA4.1 analysis.

Identification of PheDof1 and PheMADS14

The fluorescent protein-tagging method was used to investigate whether PheDof1 and PheMADS14 were nuclear-localized proteins like the transcription factors. The results showed that while green fluorescent protein (GFP) alone presented a dispersed cytoplasmic distribution, GFP-tagged *PheDof1* and *PheMADS14* were located in the nucleus as per their functions as transcription factors (Figure 6). To further ascertain whether they were able to activate transcription, PheDof1 and PheMADS14 were each fused with GAL4 DNA-binding domain (GAL4DB) and tested in yeast using a reporter construct to study whether PheDof1 and PheMADS14 had transcriptional activation activity. It was found that PheDof1 and PheMADS14 could activate the expression of the His-3 and β-Gal reporter genes (Figure 7), indicating that they are transcriptional activators.

Western blot

Western blots performed using floral buds, whole inflorescences and full blooms showed the presence of PheDof1 (35 kD) while no PheDof1 signal was detected in nonflowering leaves and embryo formation stages (Figure 8). PheMADS14 (40 kD) signal was detected at four different flower development stages. The signal was weak at the floral bud formation stage and relatively deep with the flower development. However, no PheMADS14 signal was detected in nonflowering leaves (Figure 8).

Expression patterns of PheDof1 and PheMADS14

Antisense RNA probes of *PheDof1* and *PheMADS14* were used to perform *in situ* hybridization on paraffin sections of moso bamboo flowers at different developmental stages to explore the temporal and spatial expression patterns of *PheDof1* and *PheMADS14* in the process of moso bamboo flower development. Our results showed that *PheDof1* was expressed at the early stages of moso bamboo flower development and was

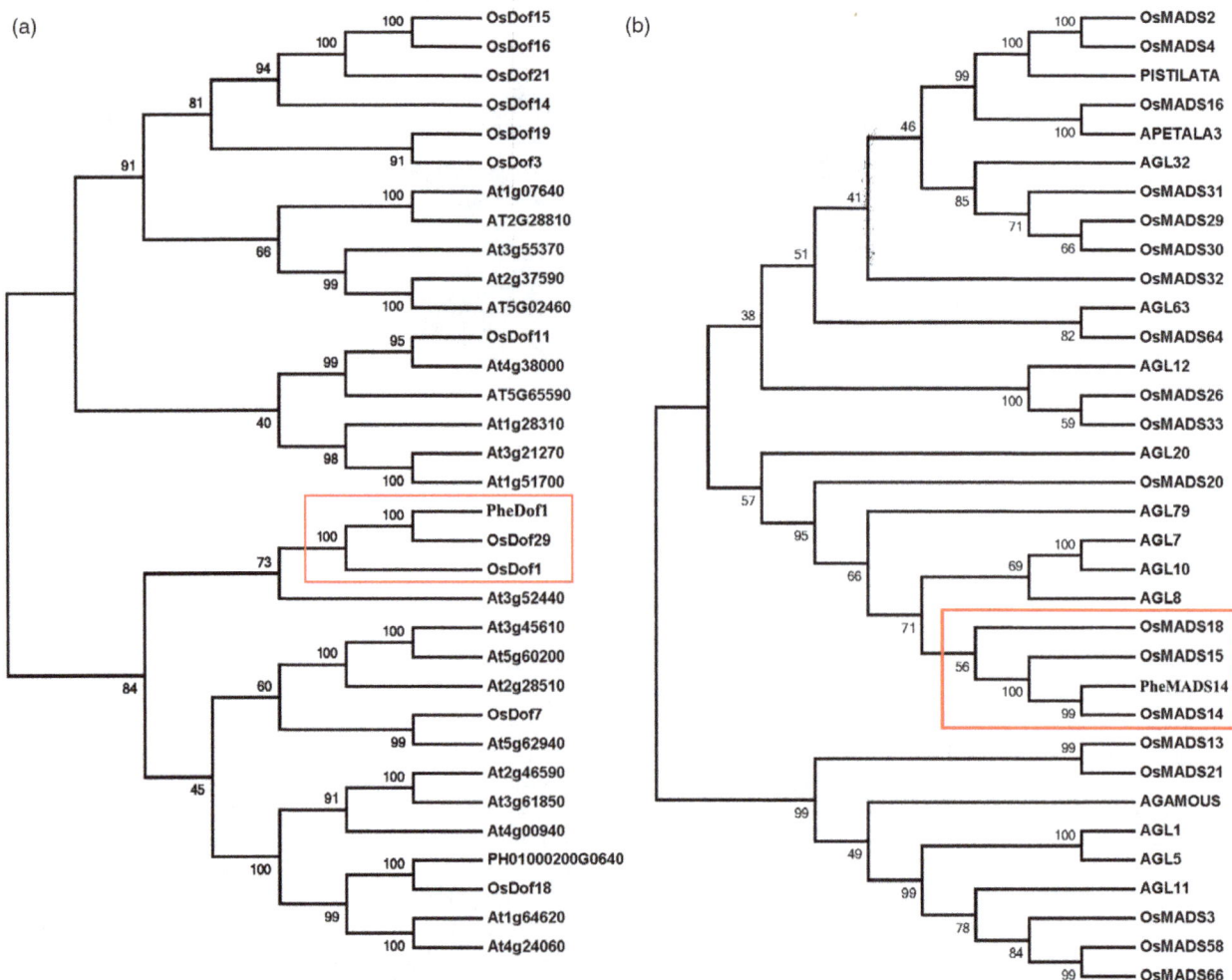

Figure 4 Phylogenetic analysis of *PheDof1* and *PheMADS14*. (a) Joined phylogenetic tree of moso bamboo *PheDof1* (PH01000200G0640) and Dof transcription factors of *Arabidopsis* and *O. sativa*. (b) Joined phylogenetic tree of moso bamboo *PheMADS14* (PH01000222G1190) and MADS transcription factors of *Arabidopsis* and *O. sativa*.

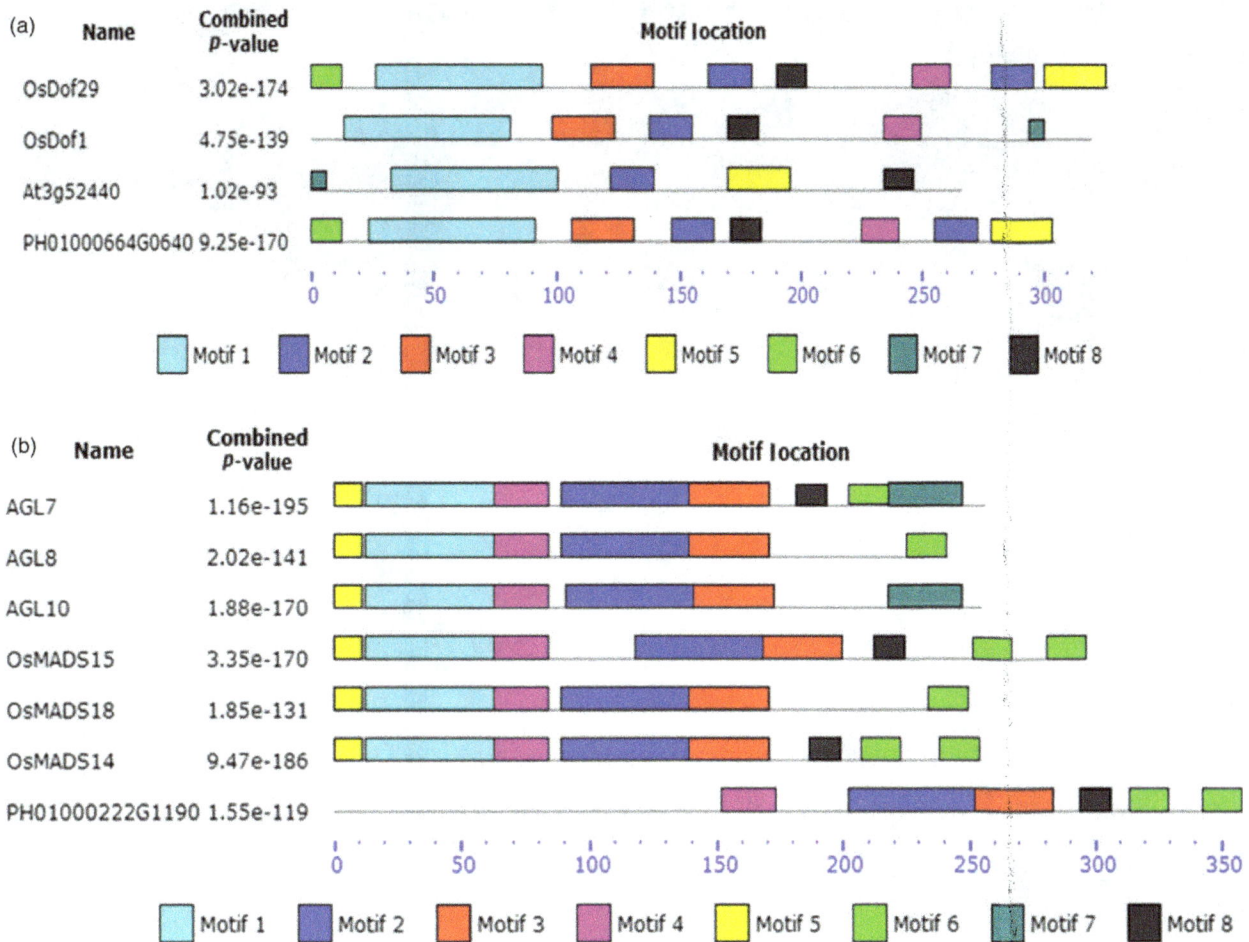

Figure 5 Distribution of the conserved motifs along the Dof and MADS transcription factors using the MEME software. (a) Dof transcription factors. (b) MADS transcription factors.

strongly expressed in the apical growth cone and inflorescence axis at the floral bud formation stage (Figure 9). Moreover, *PheDof1* expression signals were also observed in bracts. Glumes and paleas were formed after the formation of pistil and stamen primordia. *PheDof1* was positively expressed in glumes and bracts primarily while almost no expression signal was observed in stamen and pistil primordia.

PheMADS14 was expressed in different parts of moso bamboo flowers and was mainly concentrated in the parts with strong cell division such as pollen tube, bracts, pistil and stamen primordia (Figure 10). Strong expression of *PheMADS14* was observed during the differentiation of pistil and stamen primordia, which was the rapidly proliferating stage of pistil and stamen cells. *PheMADS14* was strongly expressed in stamens while no expression signal was detected in pistils, embryos and endosperms. In the process of embryo formation, the *PheMADS14* expression was detected only in glumes, paleas and lodicules. These results suggest that *PheMADS14* plays significant roles in the processes of pistil and stamen primordia formation, but is not involved in embryo formation of moso bamboo.

Expression analyses of miRNAs and target genes

Six miRNAs, miR159a.1, miR160a, miR168-3p, miR390a, miR393 and miR5139 were selected for qRT-PCR validation experiments (Figure 11). The miRNAs showed significant down-regulation that progressed through the flowering stages in the nonflowering

samples, which was consistent with our sequencing data. The six miRNAs possibly play vital regulatory roles in the moso bamboo blooming process especially the vegetative growth stage. We speculate that miR390a and miR5139 may be involved in plant–pathogen interactions while miR160a and miR393 may participate in plant hormone signal transduction. Besides, the significantly down-regulated expressions of miR159a.1 and miR168-3p indicated that they might play vital regulatory roles in protein processing in the endoplasmic reticulum.

We measured the expression levels of six targets: PH01000009G0060 (MYB family transcription factor, miR159a.1 target), PH01002685G0120 (auxin response factor, miR160a target), PH01002233G0190 (DEK C-terminal domain containing protein, miR168-3p target), PH01000001G1870 (leucine-rich repeat family protein, miR390a target), PH01000016G0700 (OsFBL16-F-box domain and LRR containing protein, miR393 target) and PH01000245G0100 (geminivirus Rep-interacting motor protein, miR5139 target). These miR targets were measured during the development of moso bamboo flowers to verify that the target genes were actually regulated by the corresponding miRNAs. The relative expression levels of PH01000009G0060, PH01002685G0120, PH01002233G0190, PH01000001G1870, PH01000016G0700 and PH01000245G0100 increased from the nonflowering samples to the flowering samples; however, those of miR159a.1, miR160a, miR168-3p, miR390a, miR393 and miR5139 were inhibited (Figure 12). The

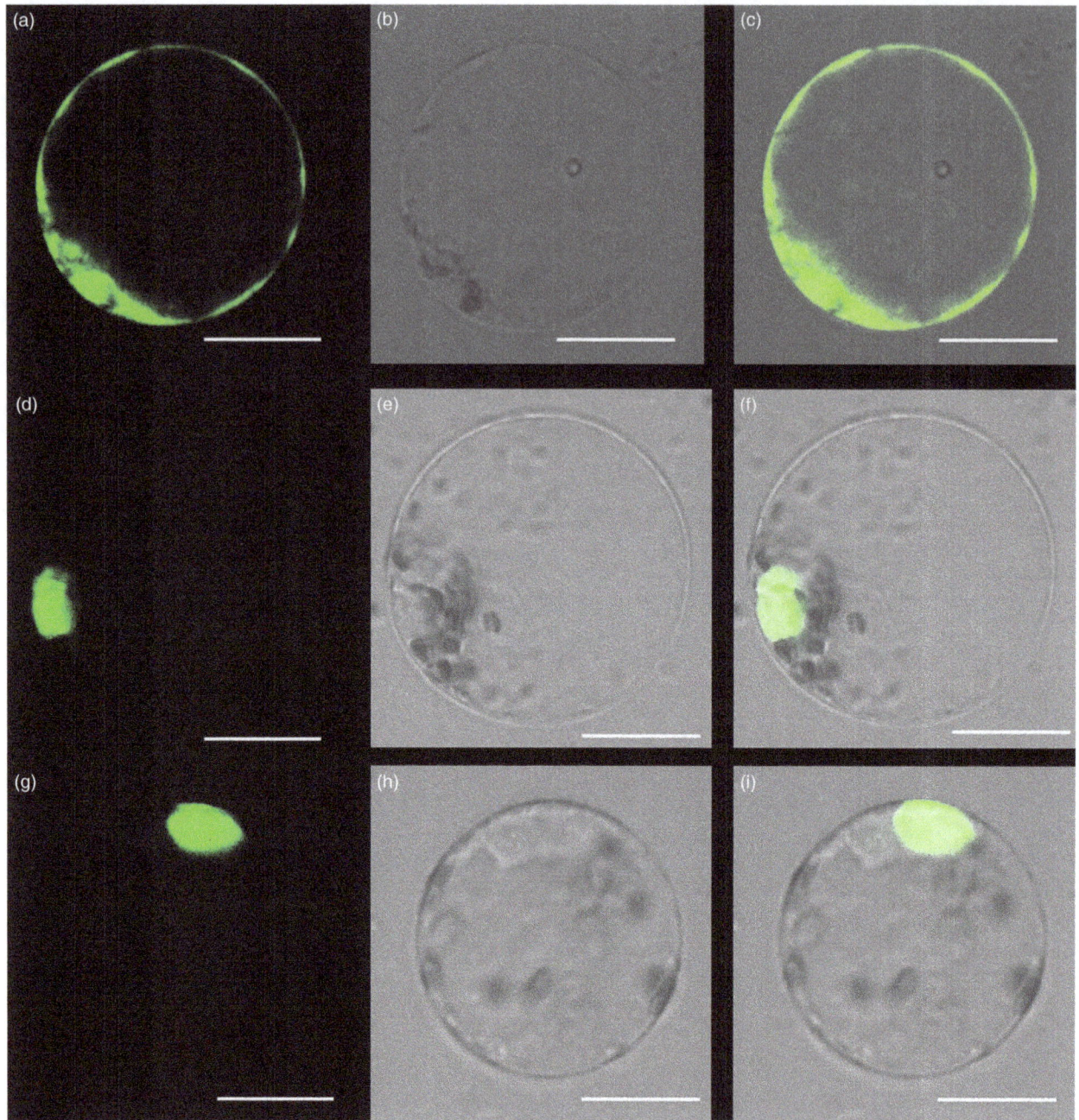

Figure 6 Subcellular localization of PheDof1 and PheMADS14. (a–c) Rice protoplasts expressing 35S-GFP. (d–f) Rice protoplasts expressing PheDof1-GFP. (g–i) Rice protoplasts expressing PheMADS14-GFP. Bar = 10 μm.

results indicated that the variation in miRNAs abundance had a negative effect on the abundance of their target genes. Thus, the expression profiles of miRNAs and their target genes were complementary.

Overexpression of pre-miR164a in Arabidopsis plants (wild-type) delayed flowering time

As miR164a might be involved in protein processing in the endoplasmic reticulum and predicted to target NAC transcription factor indicating its role in the regulation of flowering process, we overexpressed the pre-miR164a in *Arabidopsis* WT. The resulting transgenic plants showed a delayed flowering pheno-type (Figure 13a and b). We further investigated its expression in

the T_3 generation to ascertain the effects of this construct (Figure 13c).

Discussion

We showed that 11 703 differentially expressed genes changed significantly during the four flower development. They were also annotated with KEGG pathway and found to be involved in 128 metabolic pathways. Using cluster analysis, we identified two major groups: the up-regulated group and the down-regulated group. Moreover, 13 051 differentially expressed genes were found between different flower organs and nonflowering leaves of moso bamboo (unpublished data), indicating the reproducibility of the

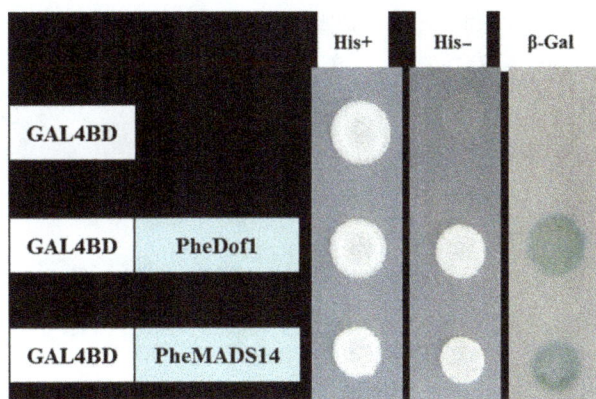

Figure 7 Transcriptional activation analysis of PheDof1 and PheMADS14.

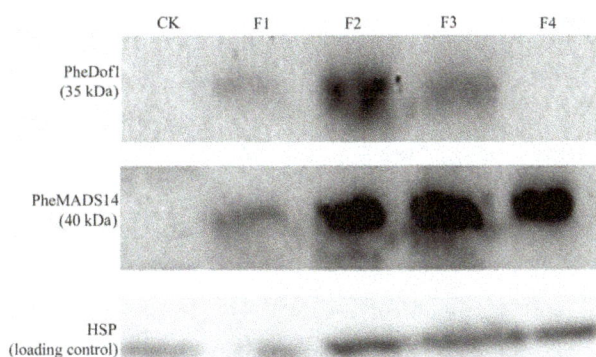

Figure 8 Western blot analysis of PheDof1 protein and PheMADS14 protein. HSP was used as the loading control. CK: nonflowering moso bamboo leaves, F1: floral bud formation stage, F2: inflorescence growing stage, F3: bloom stage, F4: embryo formation stage.

two group of sequencing data. The expression levels of genes related to glycometabolism, photosynthesis and nitrogenous metabolism pathways showed a down-regulated trend in the process of flower development. Previous studies showed that nitrogen content and nutritional ingredients gradually decreased in the leaves, stems, roots and other organs with the flowering process (Gao et al., 2002; Zhan and Li, 2007). Plant flowering is believed to be an ageing phenomenon in plants where the nutrient consumption is increased while the ability to synthesize energy materials decreases. Therefore, the growth is inhibited and 'hunger death' appears, resulting in a large number of deaths after the flowering of moso bamboo (Chai et al., 2006). Moreover, the expressions of genes related to auxin (IAA) and cytokinin (CTK) synthesis are significantly down-regulated during flower development consistent with the withering leaves and dried culms dried, whereas many genes such as PheTFL1 related to moso bamboo flowering exhibited high expression levels in inflorescence and inflorescence axis (Qi et al., 2013a,b). Besides, the content of IAA at bloom stage significantly decreased while GA_3 presented an opposite trend (Qi et al., 2013a,b). The highest expressional pathways in the up-regulated groups are plant–pathogen interaction, plant hormone signal transduction and protein processing in the endoplasmic reticulum, which suggests that related genes play significant roles in moso bamboo flower formation and development.

A large number of transcription factors participate in the pathways of plant–pathogen interaction and protein processing in the endoplasmic reticulum. Hundreds of transcription factors have been isolated from advanced plants. They play important roles in plant growth, morphogenesis and resistance to environmental stress. Peng et al. (2013) implicated stress such as drought as the main cause of moso bamboo flowering. Our DGE analysis showed that 766 transcription factors were highly expressed in the moso bamboo flower formation and development, including WRKY (110), MADS-Box (38), bHLH (34), Dof (15), NAC (61), HSF (4), HMG (4), bZIP (6), HSP20 (28), HSP70 (13), HSP90 (3), MYB (143), AP2/DREBP (32), RAD23 (14), GTA (46), CML (65), PR1 (11), RPS2 (35), RPS5 (9), FLS2 (41), PBS1 (31) and BAK1 (23). These transcription factors could play vital roles in the flower formation and development process of moso bamboo.

Plant hormones participate in all stages of plant growth and development (Richards et al., 2001). Putative homologs of gibberellin (GA) signalling pathway genes are highly expressed, including the gibberellin response modulator (GAMYB) and gibberellin receptors (GID1, GID2). They lead to the degradation of DELLAs and up-regulation of MYB21, MYB24 and MYB57, which are important for stamen filament growth (Alexandrov et al., 2009; Tsuji et al., 2006). High expression levels of these genes suggest that GA may play crucial roles in flower formation and development in moso bamboo. Jasmonic acid (JA) not only helps plants resist pathogens, but also regulates plant growth and development in all aspects including the flower formation and development, and fruit ripening (Oh et al., 2013). JA also plays significant roles in regulating flower morphogenesis. A synthetic gene mutant of JA could induce flowering in O. sativa, Triticum aestivum and Sorghum bicolor (Scott et al., 2004; Stintzi and Browse, 2000). Besides, GA could improve the activity of JA, resulting in the up-regulation expression of MYB to promote stamen filament growth (Cheng et al., 2009). According to DGEs analysis, the up-regulation of JA like genes in moso bamboo indicated that JA could interact with other hormones to jointly regulate moso bamboo flowering.

Plant–pathogen interaction, plant hormone signal transduction and protein processing in endoplasmic reticulum interact with each other and function together to regulate moso bamboo flowering (Figure 14). The related transcription factors of stress and hormone signal transduction genes identified in our study may be significant regulation factors in flower formation and development in moso bamboo. Research on the expression and regulation of these factors can help identify key genes involved in flower formation and development.

Dof (DNA binding with one zinc finger) transcription factors are a group of plant-specific transcription factors containing a single Cys2/Cys2-type zinc-finger-like Dof motif (Yanagisawa, 2002) that function in regulating flowering in response to photoperiod (Fornara et al., 2009). Ahmad et al. (2013) had found that the overexpression of AtDof4.1 could delay the flowering time of Arabidopsis and affect the development of floral organs. Besides, AtDof1.5 participated in the signal transduction pathway of phytochrome (Park et al., 2003) while AtDof5.2, AtDof5.5 and AtDof3.3 function in the photoperiod pathway (Imaizumi et al., 2005). Putative Dof transcription factors were detected in panicles, which were homologs of Dof3, Dof4, Dof5, Dof12 and CDF (Cycling DOF Factor) family, revealing similar roles of Dof3 and Dof12 genes during the flower. Our sequence analysis studies showed that PheDof1 was clustered with OsDof1 and OsDof29 and belonged to the MCOG C subfamily. We found

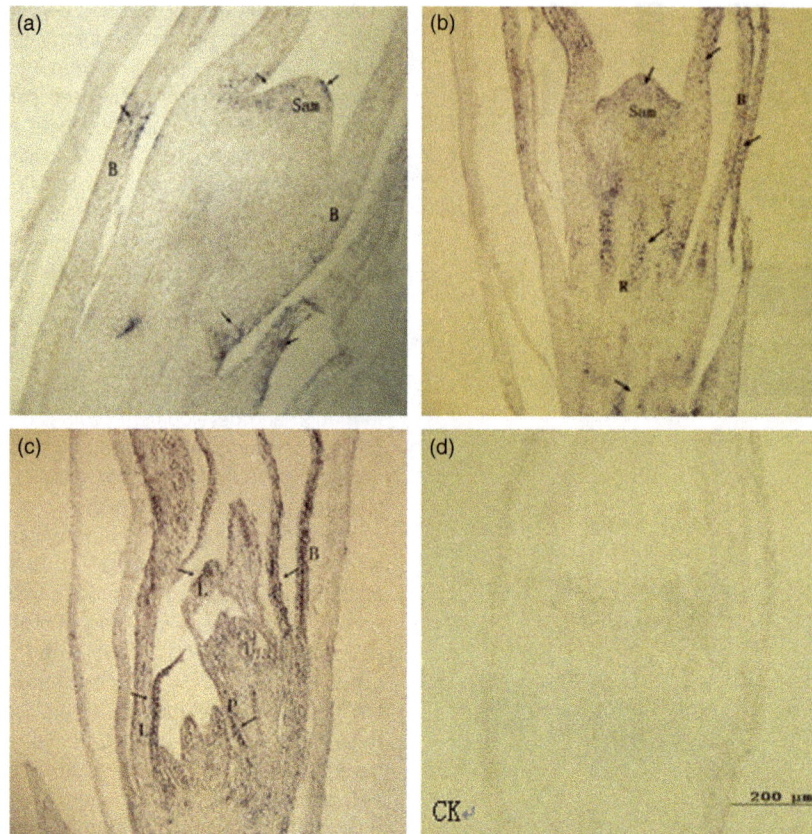

Figure 9 The expression of *PheDof1* in moso bamboo flower formation of different developmental periods. B = bract, R = rachis, Sam = shoot apical meristem, L = lemma, P = Palea. (a–b) The floral bud formation stage. (c) Bloom stage. (d) The expression of sense probe after *in situ* hybridization.

PheDof1 was located in the nucleus through the subcellular localization. PheDof1 was detected at early stages of flower formation and development, indicating that *PheDof1* might play a more significant role in the floral bud formation, inflorescence growth and bloom stages of moso bamboo. The strong expression of *PheDof1* was found in the apical growth cone and inflorescence axis at the floral bud formation stage through *in situ* hybridization, representing that *PheDof1* might take part in the start-up of moso bamboo flowering. Gao *et al.* (2014) had identified *Hd3a*, a master floral developmental regulator in rice (with apparent paralogous counterparts in *Arabidopsis* called *FT*). Drought or other environmental stresses could regulate *MADS14* in the flowering stage of bamboo (Peng *et al.*, 2013). Thus, a regulatory pathway of Dof-Hd3a-MADS-flowering could play a vital role in moso bamboo flowering.

MADS-box transcription factors contain a DNA-binding domain conserved among eukaryotes (Airoldi and Davies, 2012). Many MADS family members had been described in model plants with the floral organ identity and development proteins falling into the MIKC clade such as *MADS5*, *MADS14* and *MADS15* (Beth and Jennifer, 2005; Parenicova *et al.*, 2003; Theiben, 2001), which played significant roles in flower morphogenesis. *OsMADS14*, *OsMADS15* and *OsMADS18* belong to AP/AGL9 family, the expression patterns of which are similar to those of AP1 in *Arabidopsis*. They could determine the characteristics of floral organ meristem and are mainly expressed in glumes, paleas and lodicule primordia of *O. sativa* (Shinozuka *et al.*, 1999). Kyozuka *et al.* (2000) showed the expression of *OsMADS15* in the apical meristem at the early development of spikelets and found it

gradually located in lodicules and paleas with spikelet differentiation. *OsMADS14* is mainly expressed in glumes and paleas at the early spikelet development stage. When spikelets become mature, *OsMADS14* accumulates in stamens and carpels (Jeon *et al.*, 2000; Pelucchi *et al.*, 2002). Thirty-eight MADS-box transcription factors were identified in moso bamboo flowers, which were homologs of *MADS1*, *MADS2*, *MADS4*, *MADS7*, *MADS11*, *MADS14*, *MADS15*, *MADS17*, *MADS27*, *MADS31* and *MADS58* from *Z. mays*, *O. sativa* and *Brachypodium distachyon* (Gao *et al.*, 2014). *PheMADS14* was clustered with *OsMADS14*, *OsMADS15* and *OsMADS18* and grouped in the AP/AGL9 subfamily. We found *PheMADS14* was also located in the nucleus through the subcellular localization. PheMADS14 exhibited high expressions at later stages of flower formation and development, while a lower expression at the floral bud formation stage. Strong expression signals of *PheMADS14* were detected in pistil and stamen primordia through *in situ* hybridization, indicating that *PheMADS14* participated in the flower morphogenesis of moso bamboo and could play significant regulatory roles in moso bamboo flowering.

miRNAs function in regulating almost all aspects of plant growth and development. Our study showed that miR159a.1, miR160a, miR168-3p, miR390a, miR393 and miR5139 were involved in plant–pathogen interaction, plant hormone signal transduction and protein processing in endoplasmic reticulum based on the DGE analysis of their targets. They showed differential expression patterns from nonflowering leaves to flowering samples. Cluster analysis indicated the six miRNAs belonged to a down-regulated group. The differential expression

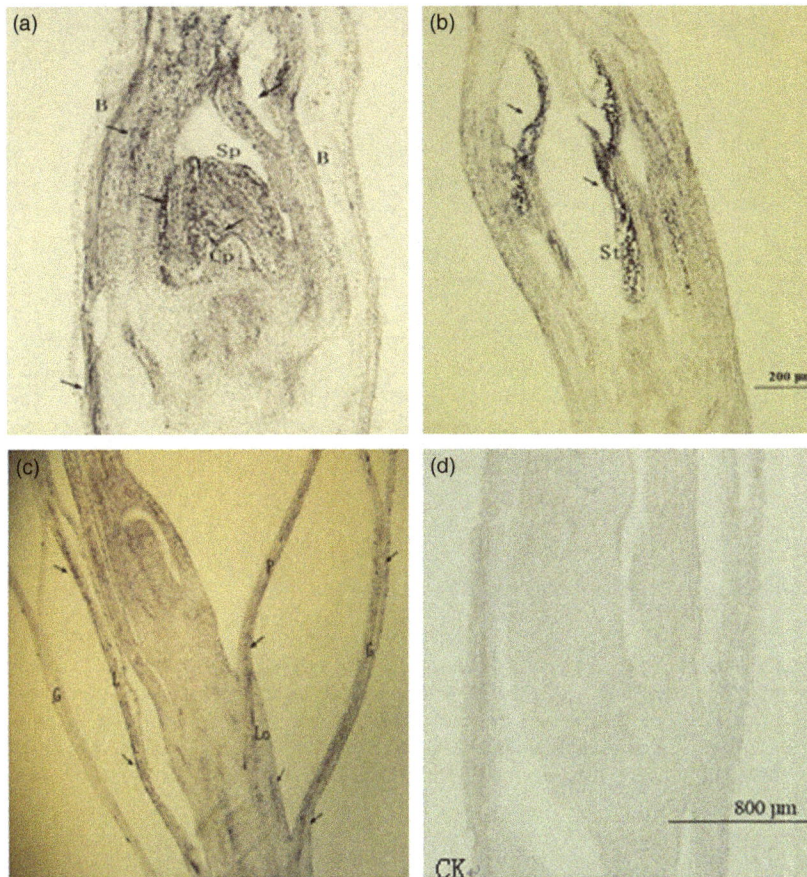

Figure 10 The expression of *PheMADS14* in moso bamboo flower formation of different developmental periods. G = glume, L = lemma, P = Palea, Lo = lodicule, B = bract, Cp = calyx primordia, Sp = stamen primordia, St = stamen. (a) Inflorescence growing stage. (b) Bloom stage. (c) Embryo formation stage. (d) The expression of sense probe after *in situ* hybridization.

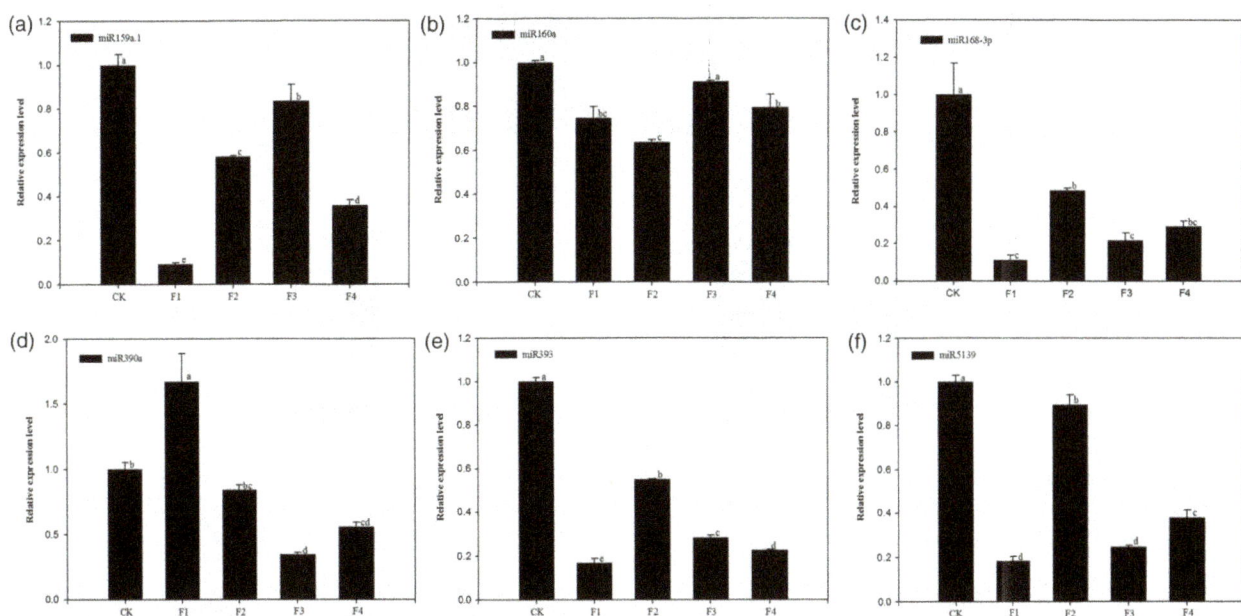

Figure 11 Expression patterns of 6 selected miRNAs among nonflowering leaves (CK) and different flower developmental stages (F1, F2, F3 and F4) in moso bamboo as determined by qRT-PCR. U6 snRNA was used as a reference in qRT-PCR. The level of every miRNA in the control was set at 1.0. Bars represent mean values of three replicates ± standard error. (a) Expression patterns of miR159a.1. (b) Expression patterns of miR160a. (c) Expression patterns of miR168-3p. (d) Expression patterns of miR390a. (e) Expression patterns of miR393. (f) Expression patterns of miR5139.

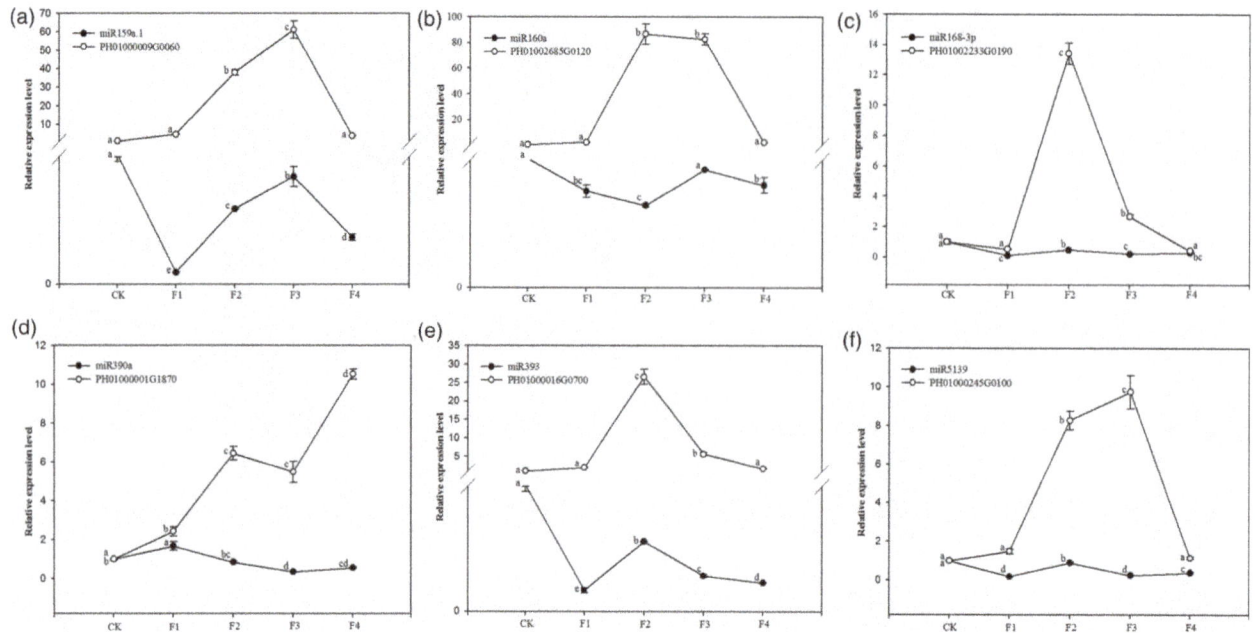

Figure 12 Expression patterns of 6 target genes and their corresponding miRNAs among nonflowering leaves (CK) and different flower developmental stages (F1, F2, F3 and F4) in moso bamboo as determined by qRT-PCR. *TIP41* was chosen as the internal housekeeping gene control for target genes. The level of every gene in the control was set at 1.0. Bars represent mean values of three replicates ± standard error. (a) Expression patterns of PH01000009G0060 and miR159a.1. (b) Expression patterns of PH01002685G0120 and miR160a. (c) Expression patterns of PH01002233G0190 and miR168-3p. (d) Expression patterns of PH01000001G1870 and miR390a. (e) Expression patterns of PH01000016G0700 and miR393. (f) Expression patterns of PH01000245G0100 and miR5139.

Figure 13 Analysis of a delayed flowering phenotype by overexpression of PhemiR164a in *Arabidopsis*. (a) The flowering phenotype of wild-type and transgenic plants at 23 °C under long-day conditions. (b) Flowering time of wild-type and transgenic plants at 23 °C under long-day conditions. (c) Relative expression of PhemiR164a in wild-type and transgenic plants were elevated by qRT-PCR. Error bars indicated the standard deviation.

patterns of miRNAs, rather than the miRNA composition, might play greater roles in regulating moso bamboo blossom. The expression patterns of miRNAs in moso bamboo could provide significant clues to their biological function and regulatory mechanism. We further corroborated the relative expression levels of six selected miRNAs through qRT-PCR. The results showed that relative expression levels of miR159a.1, miR160a, miR168-3p, miR390a, miR393 and miR5139 in nonflowering leaves were higher than those of the four flowering samples indicating their roles in moso bamboo flowering. miR390a and miR5139 may be involved in plant–pathogen interaction. miR390a could inhibit the expressions of *ARF2/ARF3/ARF4* by

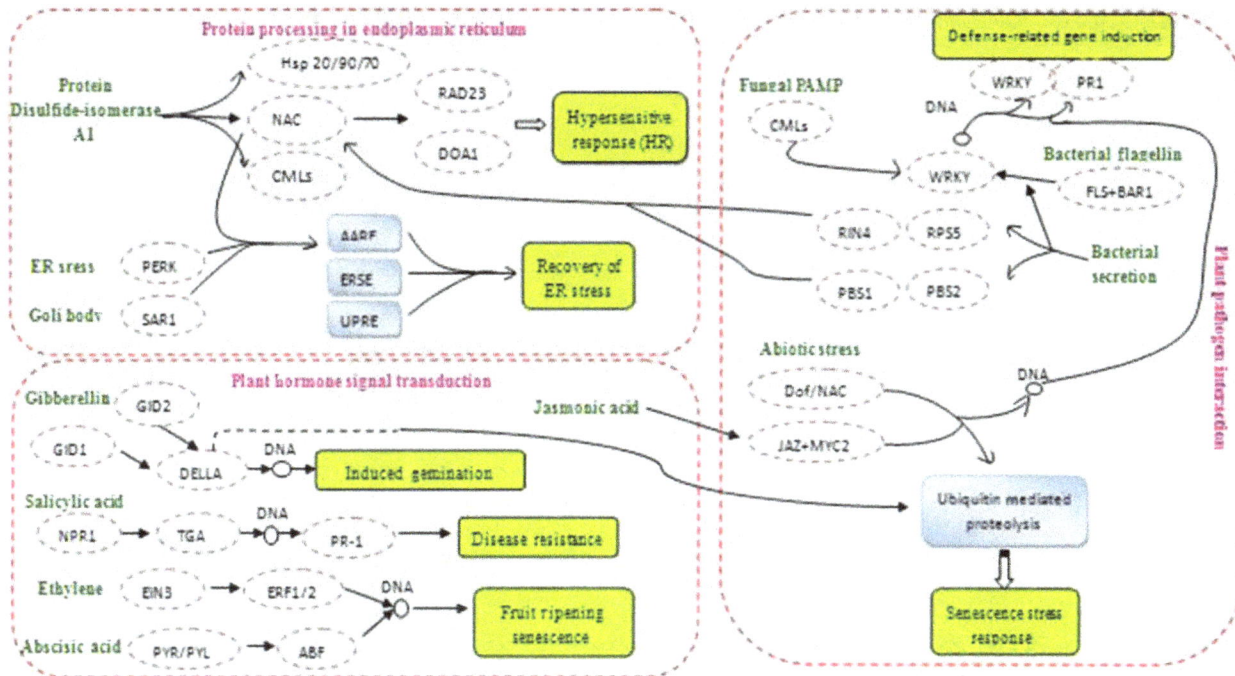

Figure 14 Different genes express in plant–pathogen interaction, plant hormone signal transduction and protein processing in endoplasmic reticulum pathways.

regulating target genes and cause a series of phenotypic changes. miR390a plays an important regulatory role in organs development (Garcia *et al.*, 2006) and growth cycle transition (Cho *et al.*, 2012). The over expression of miR390a may affect the transition from vegetative growth to reproductive growth. The target of miR5139 is geminivirus Rep-interacting motor protein. The expression of miR5139 is down-regulated under stress, so moso bamboo flowering may be related to stress. miR160a and miR393 could play potential vital roles in plant hormone signal transduction. The targets of miR160a are ARF (Rhoades *et al.*, 2002). miR160a participates in auxin signal transduction by targeting ARF and affecting vegetative growth and reproductive growth of plants (Mallory *et al.*, 2005). The targets of miR393 belong to F-box family (Navarro *et al.*, 2006), and it regulates the plant perception to auxin increasing auxin sensitivity (Parry *et al.*, 2009). miR159a.1 and miR168-3p may participate in protein processing in the endoplasmic reticulum. miR159a.1 mainly regulates MYB and directly degrades *MYB* mRNA (Patrick *et al.*, 2004), leading to a flowering delay. miR168 is important for the differentiation of plant tissues and the response to stress. Moreover, miR168 is the only miRNA that can be determined to be involved in the cross-border regulation of animals and plants (Zhang *et al.*, 2012). Thus, the over expressions of miR159a.1, miR160a, miR168-3p, miR390a, miR393 and miR5139 might inhibit moso bamboo flowering. Our study showed that their relative expressions were down-regulated during the moso bamboo flowering period. Our research indicated that the expression profiles of miRNAs were negatively correlated with those of their targets respectively, which further validated the regulation roles of miRNAs on their target genes and moso bamboo flowering.

miR164a targets *CUC2* which belongs to the NAC family and prompts the development of floral organ (Takada *et al.*, 2001). miR164a negatively regulates *CUC1* and *CUC2* to inhibit the meristem formation and floral development, influencing the establishment of organ primordia boundary (Mallory *et al.*, 2004). miR164a could participate in protein processing in the endoplasmic reticulum and affect plant flowering by regulating flower development. We report the isolation and characterization of pre-miR164a from moso bamboo. In our research, the phenotype of delaying flowering time was also observed in pCAMBIA2300-miR164a transgenic plants. The expression level of miR164a was up-regulated in transgenic plants as compared with WT plants. The results led us to suspect that miR164a delayed flowering time by regulating NAC in transgenic *Arabidopsis*. Kim *et al.* (2009) thought that miR164 could delay plant senescence. Our findings suggested that miR164a might play significant roles in moso bamboo flowering.

We selected significant regulatory pathways and analysed with miRNAs at different flower developmental stages of moso bamboo. Some candidate genes involved in plant–pathogen interaction, plant hormone signal transduction and protein processing in endoplasmic reticulum were selected and investigated. Our findings indicated that *PheDof1*, *PheMADS14* and several miRNAs might be significant regulators in moso bamboo flowering and flower development. The study will provide vital information for further functional research and key genes selection and help to explore the molecular mechanism in moso bamboo flowering and senescence.

Experimental procedures

Plant materials

Flowering moso bamboo samples at different stages and nonflowering moso bamboo leaves (CK) were collected from Guilin (E 110°17′–110°47′; E 25°04′–25°48′) in the Guangxi Zhuang Autonomous Region. The samples were stored in stationary FAA (18: 1: 1 Formaldehyde, Glacial Acetic acid and

Table 1 Primers used in miRNAs and targets qRT-PCR

Name	Primer	Sequence
miR159a.1	RT Primer	CTCAACTGGTGTCGTGGAGTCGGCAATTCAGTTGAGCAGAGCTC
	Forward Primer	ACGGGCTTTGGATTGAAGGGA
miR160a	RT Primer	CTCAACTGGTGTCGTGGAGTCGGCAATTCAGTTGAGTGGCATAC
	Forward Primer	AGCCCTGCCTGGCTCCCTGT
miR164a	RT Primer	CTCAACTGGTGTCGTGGAGTCGGCAATTCAGTTGAGTGCACGTG
	Forward Primer	AGCCCTGGAGAAGCAGGGCA
miR168-3p	RT Primer	CTCAACTGGTGTCGTGGAGTCGGCAATTCAGTTGAGATTCACTT
	Forward Primer	AGCCCGCCCGCCTTGCACCAA
miR390a	RT Primer	CTCAACTGGTGTCGTGGAGTCGGCAATTCAGTTGAGGGCGCTAT
	Forward Primer	ACGGGAAGCTCAGGAGGGAT
miR393	RT Primer	CTCAACTGGTGTCGTGGAGTCGGCAATTCAGTTGAGATCAATGC
	Forward Primer	ACGGGCTCCAAAGGGATCGC
miR5139	RT Primer	CTCAACTGGTGTCGTGGAGTCGGCAATTCAGTTGAGTGGTATCA
	Forward Primer	ACGGGCAACCTGGCTCTGATA
	Universal Reverse Primer	CTCAACTGGTGTCGTGGAGTC
U6	Forward Primer	GGACATCCGATAAAATTGGAACGATACAG
	Reverse Primer	AATTTGGACCATTTCTCGATTTATGCGTGT
PH01000009G0060	Forward Primer	TTCACTGAATCCACCCCTCC
	Reverse Primer	ATGAGGTGAAGTGTCCCCAG
PH01002685G0120	Forward Primer	CTGCAGTCGAGCCTAGTGTA
	Reverse Primer	CAGACCGCCAGAATTGCTAC
PH01002233G0190	Forward Primer	ACAACTAGCAAGGGTCCACA
	Reverse Primer	TGATGATTCCCCTTTGCCCT
PH01000001G1870	Forward Primer	GTCCACCAACAGCGATGATC
	Reverse Primer	TTCTGCTCATCCCCACCATT
PH01000016G0700	Forward Primer	CGGGGAATTTGACGGATGAC
	Reverse Primer	ACGTTGGAGTTTCACACAGC
PH01000245G0100	Forward Primer	GTCCCCAGAACTTCAGTCCA
	Reverse Primer	TCCTCAATCCCACGAAGCTT
TIP41	Forward Primer	AAAATCATTGTAGGCCATTGTCG
	Reverse Primer	ACTAAATTAAGCCAGCGGGAGTG

70% alcohol). The four stages were defined according to the anatomical structure of floral organs: F1 (floral bud formation, during which a plant transits from vegetative stage to reproductive stage), F2 (inflorescence axis continued to stretch, and lateral buds started to differentiate), F3 (bloom stage, flowers with both pistils and stamens emerging from glumes), F4 (embryo formation) (Gao et al., 2015). Besides the above, a large number of flowering moso bamboo samples at different stages and nonflowering moso bamboo samples were flash frozen in liquid nitrogen and stored at −80 °C to do further study.

DGEs analysis and small RNA analysis

Tag library preparation for samples of the four different flowering developmental periods (F1, F2, F3 and F4) was performed in parallel using the Illumina gene expression sample preparation kit. Transcriptome data of nonflowering moso bamboo leaves (CK) were regarded as the reference gene database, and the four libraries were sequenced using the Illumina high-seq 2000 at Beijing Genomics Institute (BGI) (Shenzhen, Guangdong Province, China) (Gao et al., 2014).

Total RNA was extracted from the frozen samples (CK, F1, F2, F3 and F4) using the Trizol reagent (Invitrogen, Carlsbad, California, USA), according to the manufacturer's instructions. Five small RNA libraries were constructed for moso bamboo, and the Illumina high-seq 2000 sequencing was carried out by the Beijing Genomics Institute (BGI) (Shenzhen, Guangdong Province, China) (Gao et al., 2015).

KEGG pathway analysis was performed after we obtained the sequencing data. Key regulation pathways and genes were selected according to expression abundance. Hierarchical clustering of expressional data was carried out using the Cluster 3.0 and Treeview programs (Eisen et al., 1998).

Bioinformatics analysis of PheDof1 and PheMADS14

The Dof and MADS transcription factors of moso bamboo, *Arabidopsis* and *Oryza sativa* were used to construct phylogenetic tree through MEGA 4.1. We used MEME 4.9 (http://nbcr-222.ucsd.edu/opal-jobs/) to determine the conserved motifs of protein sequence.

Subcellular localization and transcriptional activation

The subcellular localization of PheDof1 and PheMADS14 was determined by transfecting GFP-tagged PheDof1 and PheMADS14 into rice stem and sheath protoplasts (Zhang et al., 2011). The full-length cDNAs of *PheDof1* and *PheMADS14* were fused in frame with the GFP cDNA and ligated between the CaMV 35S promoter and the nopaline synthase terminator. The fluorescence signals in transfected protoplasts were examined using a confocal laser scanning microscope (Leica Microsystems).

The transcriptional activation activity of PheDof1 and Phe-MADS14 was studied by transforming the pGBKT7 construct (that had the fusion of these genes and the GAL4 DNA-binding domain) into the yeast strain PJ. The yeast strain contained the *His-3* and *LacZ* reporter genes. The transformed yeast cells were grown on synthetic defined plates (with or without *His*) and were assayed for β-galactosidase activity.

Western blotting

To generate antibodies against PheDof1 and PheMADS14, we purified the two fusion proteins from *E. coli* and used them to immunize a New Zealand white rabbit. The PheDof1 and PheMADS14 antibodies were immunopurified against nitrocellulose-bound antigen. We separated the 20 μg crude, soluble extract on 5%–10% polyacrylamide gels followed by blotting onto nitrocellulose membrane. Blots were incubated with affinity-purified PheDof1 and PheMADS14 antibodies followed by goat anti-rabbit IgG-HRP (Sigma) and detected by enhanced chemiluminescence using Super Signal West Fempto Maximum Sensitivity Substrate (Thermo Scientific).

In situ hybridization

In situ hybridization was performed as previously described (Hord et al., 2006). Flowering moso bamboo samples at different stages were fixed in the stationary liquid, FAA (18: 1: 1 Formaldehyde, Glacial Acetic acid and 70% alcohol). The samples were fixed for 48 h, dehydrated and embedded in paraffin, and 10-μm-thick sections were prepared with microtome (Leica, Germany) and mounted onto slides. The slides were dehydrated and baked, followed by dewaxing of all slides with dimethylbenzene. The antisense and sense RNAs of *PheDof1* and *PheMADS14* were labelled with digoxigenin by *in vitro* transcription of linearized pGEM-T-PheDof1 and pGEM-T-PheMADS14, which carried fragments of the *PheDof1* cDNA amplified with gene-specific primers YWZJ-Dof-F (5'-GCTGGCGAGCACAGAGGT) and YWZJ-Dof-R (5'-CGCTGCTCCAAACATCG), and *PheMADS14* cDNA amplified with gene-specific primers YWZJ-MADS-F (5'-AGGTAAAGTTT GTAAGTGGGC) and YWZJ-MADS-R (5'-AGTTGCTGCTCTAGCTG CTGA). Antidigoxigenin antibodies coupled with alkaline phosphatase and nitro blue tetrazolium were used to detect the hybridization signal.

Quantitative real-time PCR (qRT-PCR) analysis of miRNAs and target genes in moso bamboo

Total RNA was extracted from the frozen samples (CK, F1, F2, F3 and F4) and used for cDNA synthesis using the miRNA-specific stem–loop RT primer. The reaction was incubated at 16 °C for 30 min, 42 °C for 30 min, 85 °C for 5 min and 4 °C for 5 min. Real-time PCR was carried out on a Light Cycler 480 machine (Roche) using a SYBR Green I Master Kit (Roche, Switzerland). The final volume was 20 μL, containing 10 μL 2 × SYBR Premix Ex Taq, 7.2 μL of nuclease-free water, 0.4 μL of each primer (10 μM) and 2 μL of cDNA. The amplification was carried out as follows: initial denaturation at 95 °C for 10 min, followed by 43 cycles at 95 °C for 10 s, 58 °C for 20 s and 72 °C for 10 s. The melting curves were adjusted as 95 °C for 5 s and 58 °C for 1 min and then cooled to 40 °C for 30 s (Unver and Budak, 2009). All reactions were performed in triplicates. U6 snRNA were chosen as internal control for the miRNAs (Ding et al., 2011). The stem–loop reverse transcription primers were designed following the method described by Chen et al. (2005). PCR primers including a miRNA-specific forward primer and a reverse primer were used to amplify the PCR products.

The expression analysis of several target genes was also examined by quantitative RT-PCR. Reverse transcription reactions were performed using 2 μg of RNA by M-MLVRT (Promega, Madison, Wisconsin, USA) according to the manufacturer's instructions. We obtained the sequences of four selected target genes from the moso bamboo genome database (http://www.ncgr.ac.cn/bamboo). We selected *TIP41* as the internal housekeeping gene control (Fan et al. 2013). Real-time PCR was carried out on a Light Cycler 480 machine (Roche) using a SYBR Green I Master Kit (Roche, Switzerland). The 20 μL reaction mixture contained 10 μL 2 × SYBR Premix Ex Taq, 7.2 μL of nuclease-free water, 0.4 μL of each primer (10 μM) and 2 μL of cDNA. Amplification reactions were performed as the following: 95 °C for 10 s, 60 °C for 10 s and 72 °C for 20 s. All reactions were carried out in triplicate. The primers used in all quantitative RT-PCR experiments are listed in Table 1.

Constructs and plant transformation

For pCAMBIA2300-miR164a construct, a fragment of pre-miR164a from moso bamboo was obtained by PCR amplification using the pre-miR164a-F (5'-TAGGATCCCAAACCGTGCTGGA-GAAGC) and pre-miR164a-R (5'-CGAAGCTTGAAGCCATGGTG-GAGAAGGAG) primers. The construct was introduced into wild-type *Arabidopsis* plants (Columbia-0). Electroporation transformation was performed as described by *Agrobacterium*-mediated transformation. Transgenic plants were selected on kanamycin, and the first generation of transgenic plants was examined for their phenotypes. At least 10 independent transgenic plants exhibiting severe phenotypes were selected for phenotypic characterization and examined.

Acknowledgements

This project was supported by the National Natural Science Foundation of China (grant No. 31570673) and National High Technology Research and Development Program of China 'Moso Bamboo Functional Genomics Research' (grant No. 2013AA102607-4). The authors hope to express their appreciation to the reviewers for this manuscript.

References

Ahmad, M., Rim, Y., Chen, H. and Kim, J.Y. (2013) Functional characterization of *Arabidopsis* Dof transcription factor AtDof4.1. *Russian. J. Plant Physiol.* **60**, 116–123.

Airoldi, C.A. and Davies, B. (2012) Gene duplication and the evolution of plant MADS-box transcription factors. *J. Genet. Genomics*, **39**, 157–165.

Alexandrov, N., Brover, V., Freidin, S., Troukhan, M., Tatarinova, T., Tatarinova, T., Zhang, H. et al. (2009) Insights into corn genes derived from large-scale cDNA sequencing. *Plant Mol. Biol.* **69**, 179–194.

Beth, A.K. and Jennifer, C.F. (2005) Molecular mechanisms of flower development: an armchair guide. *Nature*, **6**, 688–698.

Chai, J., Qin, Y., Hua, X., Wang, Z. and Wang, Q. (2006) Advance of studies on bamboo flowering causes. *J. Zhengjiang Forest. Sci. Technol.* **26**, 53–57.

Chen, C., Ridzon, D.A., Broomer, A.J., Zhou, Z., Lee, D.H., Nguyen, J.T., Barbisin, M. et al. (2005) Real-time quantification of microRNAs by stem-loop RT-PCR. *Nucleic Acids Res.* **33**, e179.

Cheng, H., Song, S., Xiao, L., Soo, H.M., Cheng, Z., Xie, D. and Peng, J. (2009) Gibberellin acts through jasmonate to control the expression of *MYB21*, *MYB24*, and *MYB57* to promote stamen filament growth in *Arabidopsis*. *PLoS Genet.* **5**, e1000440.

Cho, S.H., Coruh, C. and Axtell, M.J. (2012) miR156 and miR390 regulate tasiRNA accumulation and developmental timing in *Physcomitrella patens*. *Plant Cell*, **24**, 4837–4839.

De Folter, S. and Angenent, G.C. (2006) Transmeets cis in MADS science. *Trends Plant Sci.* **11**, 224–231.

Ding, Y., Chen, Z. and Zhu, C. (2011) Microarray-based analysis of cadmium responsive microRNAs in rice (*Oryza sativa*). *J. Exp. Bot.* **62**, 3563–3573.

Dreni, L., Jacchia, S., Fornara, F., Fornari, M., Ouwerkerk, P.B., An, G., Colombo, L. *et al.* (2007) The D-lineage MADS-box gene *OsMADS13* controls ovule identity in rice. *Plant J.* **52**, 690–699.

Eisen, M.B., Spellman, P.T., Brown, P.O. and Botstein, D. (1998) Cluster analysis and display genome-wide expression patterns. *PNAS*, **95**(25), 14863–14868.

Fan, C.J., Ma, J.M., Guo, Q.R., Li, X.T., Wang, H. and Lu, M.Z. (2013) Selection of Reference Genes for Quantitative Real-Time PCR in Bamboo (Phyllostachys edulis). *PLoS One*, **8**(2), e56573.

Fornara, F., Parenicova, L., Falasca, G., Pelucchi, N., Masiero, S., Ciannamea, S., Lopez-Dee, Z. *et al.* (2004) Functional characterization of *OsMADS18*, a member of AP/AGL9 subfamily of MADS box genes. *Plant Physiol.* **135**, 2207–2219.

Fornara, F., Panigrahi, K.C., Gissot, L., Sauerbrunn, N., Rvhl, M., Jarillo, J.A. and Couland, G. (2009) *Arabidopsis* DOF transcription factors act redundantly to reduce CONSTANS expression and are essential for a photoperiodic flowering response. *Dev. Cell* **17**, 75–86.

Gao, P., Zheng, Y., Lin, Z., Qiu, L. and Chen, L. (2002) Study on the bio-physiological behaviors of Dendrocalamopsis oldhami, s flowering. *J. Bamboo Res.* **21**, 70–75.

Gao, J., Zhang, Y., Zhang, C., Qi, F., Li, X., Mu, S. and Peng, Z. (2014) Characterization of the floral transcriptome of Moso bamboo (*Phyllostachys edulis*) at different flowering developmental stages by Transcriptome Sequencing and RNA-seq analysis. *PLoS One*, **9**, e98910.

Gao, J., Ge, W., Zhang, Y., Cheng, Z., Li, L., Hou, D. and Hou, C. (2015) Identification and characterization of microRNAs at different flowering developmental stages in moso bamboo (*Phyllostachys edulis*) by high-throughput sequencing. *Mol. Genet. Genomics*, **290**, 2335–2353.

Garcia, D., Collier, S.A., Byrnc, M.E. and Martienssen, R.A. (2006) Specification of leaf polarity in *Arabidopsis* via the trans-acting siRNA pathway. *Curr. Biol.* **16**, 933–938.

Hord, C.L., Chen, C., Deyoung, B.J., Clark, S.E. and Ma, H. (2006) The BAM1/BAM2 receptor-like kinases are important regulators of *Arabidopsis* early anther development. *Plant Cell*, **18**, 1667–1680.

Imaizumi, T., Schultz, T.F., Harmon, F.G., Ho, L.A. and Kay, S.A. (2005) FKF1 F-box protein mediates cyclic degradation of a repressor of CONSTANS in *Arabidopsis*. *Science*, **309**, 293–297.

Jeon, J.S., Lee, S., Jung, K.H., Yang, W.S., Yi, G.H., Oh, B.G. and An, G. (2000) Production of transgenic rice plants showing reduced heading date and plant height by ectopic expression of rice MADS-box genes. *Mol. Breeding*, **6**, 581–592.

Jones-Rhoades, M.W., Bartel, D.P. and Bartel, B. (2006) MicroRNAs and their regulatory roles in plants. *Annu. Rev. Plant Biol.* **57**, 19–53.

Jung, J.H. and Park, C.M. (2007) MIR166/165 genes exhibit dynamic expression patterns in regulating shoot apical meristem and floral development in *Arabidopsis*. *Planta*, **225**, 1327–1338.

Kaufmann, K., Melzer, R. and Theissen, G. (2005) MIKC-type MADS-domain proteins: structural modularity, protein interactions and network evolution in land plants. *Gene*, **347**, 183–198.

Kim, J.H., Woo, H.R., Kim, J., Lim, P.O., Lee, I.C., Choi, S.H., Hwang, D. *et al.* (2009) Trifurcate feed-forward regulation of age-dependent cell death involving miR164 in *Arabidopsis*. *Science*, **323**, 1053–1057.

Kyozuka, J., Ikegami, A., Morita, M. and Shimamoto, K. (2000) Spatially and temporally regulated expression of rice MADS box genes with similarity to Arabidopsis class A, B and C genes. *Plant Cell Physiol.* **41**, 710–718.

Lee, S., Kim, J., Son, J.S., Nam, J., Jeong, D.H., Lee, K., Jang, S.J. *et al.* (2003) Systematic reverse genetic screening of T-DNA tagged genes in rice for

functional genomic analyses: MADS-box genes as a test case. *Plant Cell Physiol.* **44**, 1403–1411.

Li, D., Yang, C., Li, X., Gan, Q., Zhao, X. and Zhu, L. (2009) Functional characterization of rice *OsDof12*. *Planta*, **229**, 1159–1169.

Li, J.H., Yue, J.J. and Li, H.T. (2012) Evaluation of economic and ecosystem services of moso bamboo stands. *Xiandai Horticulture*, **18**, 6–7.

Lijavetzky, D., Carbonero, P. and Vicente-Carbajosa, J. (2003) Genome-wide comparative phylogenetic analysis of the rice and *Arabidopsis* Dof gene families. *BMC Evol. Biol.* **3**, 1–11.

Lin, X.C., Chow, T.Y., Chen, H.H., Liu, C.C., Chou, S.J., Huang, B.L., Kuo, C.I. *et al.* (2010) Understanding bamboo flowering based on large-scale analysis of expressed sequence tags. *Genet. Mol. Res.* **9**, 1085–1093.

Liu, Q. and Chen, Y.Q. (2009) Insights into the mechanism of plant development: interactions of miRNAs pathway with phytohormone response. *Biochem. Biophys. Res. Commun.* **384**, 1–5.

Mallory, A.C., Dugas, D.V., Bartel, D.P. and Bartel, B. (2004) MicroRNA regulation of NAC-domain targets is required for proper formation and separation of adjacent embryonic, vegetative, and floral organs. *Curr. Biol.* **14**, 1035–1046.

Mallory, A.C., Bartel, D.P. and Bartel, B. (2005) MicroRNA-directed regulation of *Arabidopsis* AUXIN RESPONSE FACTOR17 is essential for proper development and modulates expression of early auxin response genes. *Plant Cell*, **17**, 1360–1375.

Millar, A.A. and Gubler, F. (2005) *Arabidopsis* GAMYB-1ike genes, MYB33 and MYB65 are microRNA-regulated genes that redundantly facilitate anther development. *Plant Cell*, **17**, 705–721.

Nag, A. and Jack, T. (2010) Sculpting the flower; the role of microRNAs in flower development. *Curr. Top. Dev. Biol.* **91**, 349–378.

Nagpal, P., Ellis, C.M., Weber, H., Ploense, S.E., Barkawi, L.S., Guilfoyle, T.J., Hagen, G. *et al.* (2005) Auxin response factors ARF6 and ARF8 promote jasmonic acid production and flower maturation. *Development*, **132**, 4107–4118.

Navarro, L., Dunoyer, P., Jay, F., Arnold, B., Dharmasiri, N., Estelle, M., Voinnet, O. *et al.* (2006) A plant miRNA contributes to antibacterial resistance by repressing auxin signaling. *Science*, **312**, 436–439.

Oh, Y., Baldwin, I.T. and Galis, I. (2013) A jasmonate ZIM-domain protein NaJAZd regulates floral jasmonic acid levels and counteracts flower abscission in nicotiana attenuata plants. *PLoS One*, **8**, e57868.

Parenicova, L., De Folter, S., Kieffer, M., Horner, D.S., Favalli, C., Busscher, J., Cook, H.E. *et al.* (2003) Molecular and phylogenetic analyses of the complete MADS-box transcription factor family in *Arabidopsis*: new openings to the MADS world. *Plant Cell*, **15**, 1538–1551.

Park, D.H., Lim, P.O., Kim, J.K., Cho, D.S., Hong, S.H. and Nam, H.G. (2003) The *Arabidopsis* COG1 gene encodes a Dof domain transcription factor and negatively regulates phytochrome signaling. *Plant J.* **34**, 161–171.

Parry, G., Calderon-Villalobos, L.I., Prigge, M., Peret, B., Dharmasiri, S., Itoh, H., Lechner, E. *et al.* (2009) Complex regulation of the TIR1/AFB family of auxin receptors. *Proc. Natl Acad. Sci. USA*, **106**, 22540–22545.

Passmore, S., Maine, G.T., Elble, R., Christ, C. and Tye, B.K. (1988) Saccharomyces cerevisiae protein involved in plasmid maintenance is necessary for mating of MAT alpha cells. *J. Mol. Biol.* **204**, 593–606.

Patrick, A., Alan, H., Baulcombe, D.C. and Harberd, N.P. (2004) Modulation of floral development by a gibberellin-regulated microRNA. *Development*, **131**, 3357–3365.

Pelucchi, N., Fornanra, F., Favalli, C., Masiero, S., Lago, C., Pe, E.M., Colombo, L. *et al.* (2002) Comparative analysis of rice MADS-box genes expressed during flower development. *Sex. Plant Reprod.* **15**, 113–122.

Peng, Z.H., Lu, Y., Li, L.B., Zhao, Q., Feng, Q., Gao, Z.M., Lu, H.Y. *et al.* (2013) The draft genome of the fast-growing non-timber forest species moso bamboo (*Phyllostachys heterocycla*). *Nat. Genet.* **45**, 456–461.

Qi, F.Y., Hu, T., Peng, Z.H. and Gao, J. (2013a) Screening of reference genes used in qRT-PCR and expression analysis of *PheTFL1* gene in moso bamboo. *Acta Botanica Boreali-Occidentalia Sinica*, **33**, 48–52.

Qi, F.Y., Peng, Z.H., Hu, T. and Gao, J. (2013b) Change of endogenous hormones in different organs during the flowering phase of moso bamboo. *Forest Res.* **26**, 332–336.

Rhoades, M.W., Reinhart, B.J., Lim, L.P., Burge, C.B., Bartel, B. and Bartel, D.P. (2002) Prediction of plant microRNA targets. *Cell*, **110**, 513–520.

Richards, D.E., King, K.E., Ait-ali, T. and Harberd, N.P. (2001) How gibberellin regulates plant growth and development: a molecular genetic analysis of gibberellin signaling. *Annu. Rev. Plant Physiol. Plant Mol. Biol.* **52**, 67–88.

Sang, X., Li, Y., Luo, Z., Ren, D., Fang, L., Wang, N., Zhao, F. *et al.* (2012) *CHIMERIC FLORAL ORGANS1*, encoding a monocot-specific MADS box protein, regulates floral organ identity in rice. *Plant Physiol.* **160**, 788–807.

Scott, J.R., Spielman, M. and Dickinson, H.G. (2004) Stamen structure and function. *Plant Cell*, **16**, S46–S60.

Shinozuka, Y., Kojima, S., Shomura, A., Ichimura, H., Yano, M., Yamamoto, K. and Sasaki, T. (1999) Isolation and characterization of rice MADS-box gene homologues and their RFLP mapping. *DNA Res.* **6**, 123–129.

Stintzi, A. and Browse, J. (2000) The *Arabidopsis* male-sterile mutant, opr3, lacks the 12-oxophytodienoic acid reductase required for jasmonate synthesis. *Proc. Natl Acad. Sci.* **97**, 10625–10630.

Takada, S., Hibara, K.I., Ishida, T. and Tasaka, M. (2001) The *CUP-SHAPEDCOTYLEDON1* gene of *Arabidopsis* regulates shoot apical meristem formation. *Development*, **128**, 1127–1135.

Theiben, G. (2001) Development of floral organ identity, stories from the MADS house. *Curr. Opin. Plant Biol.* **4**, 75–85.

Theissen, G., Becker, A., Di Rosa, A., Kanno, A., Kim, J.T., Munster, T., Winter, K.U. *et al.* (2000) A short history of MADS-box genes in plants. *Plant Mol. Biol.* **42**, 115–149.

Tsuji, H., Aya, K., Ueguchi-Tanaka, M., Shimada, Y., Nakazono, M., Watanabe, R., Nishizawa, N.K. *et al.* (2006) GAMYB controls different sets of genes and is differentially regulated by microRNA in aleurone cells and anthers. *Plant J.* **47**, 427–444.

Unver, T. and Budak, H. (2009) Conserved microRNAs and their targets in model grass species Brachypodium distachyon. *Planta*, **230**, 659–669.

Wang, H.W., Zhang, B., Hao, Y.J., Huang, J., Tian, A.G., Liao, Y., Zhang, J.S. *et al.* (2007) The soybean Dof-type transcription factor genes, *GmDof4* and *GmDof11*, enhance lipid content in the seeds of transgenic *Arabidopsis* seeds. *Plant J.* **52**, 716–729.

Ward, J.M., Cufr, C.A., Denzel, M.A. and Neff, M.M. (2005) The Dof transcription factor OBP3 modulates phytochrome and cryptochrome signaling in *Arabidopsis*. *Plant Cell*, **17**, 475–485.

Yanagisawa, S. (2002) The Dof family of plant transcription factors. *Trends Plant Sci.* **7**, 555–560.

Yanagisawa, S. and Izui, K. (1993) Molecular cloning of two DNA-binding proteins of maize that are structurally different but interact with the same sequence motif. *J. Biol. Chem.* **268**, 16028–16036.

Yang, J., Yang, M.F., Wen, P.Z., Fan, C. and Shen, S.H. (2011) A putative flowering-time-related Dof transcription factor gene, *JcDof3*, is controlled by the circadian clock in *Jatropha curcas*. *Plant Sci.* **181**, 667–674.

Yanofsky, M.F., Ma, H., Bowman, J.L., Drews, G.N., Feldmann, K.A. and Meyerowitz, E.M. (1990) The protein encoded by the *Arabidopsis* homeotic gene agamous resembles transcription factors. *Nature*, **346**, 35–39.

Yu, S., Galvao, C.V., Zhang, Y.C., Horrer, D., Zhang, T.Q., Hao, Y.H., Feng, Y.Q. *et al.* (2012) Gibberellin regulates the *Arabidopsis* floral transition through miR156-targeted *SQUAMOSA PROMOTER BINDING–LIKE* transcription factors. *Plant Cell*, **24**, 3320–3332.

Zhan, A. and Li, Z. (2007) Dynamic of mineral elements in umbrella bamboo (*Fargesia murielae*) before and after flowering. *World Bamboo Rattan*, **5**, 12–15.

Zhang, Y., Su, J., Duan, S., Ao, Y., Dai, J., Liu, J., Wang, P. *et al.* (2011) A highly efficient rice green tissue protoplast system for transient gene expression and studying light/chloroplast-related processes. *Plant Methods*, **7**, 30.

Zhang, L., Hou, D.X., Chen, X., Li, D.H., Zhu, L.Y., Zhang, Y.J., Li, J. *et al.* (2012) Exogenous plant MIR168a specifically targets mammalian *LDLRAP1*: evidence of cross-kingdom regulation by microRNA. *Cell Res.* **22**, 107–126.

Zhu, Q.H., Spriggs, A., Matthew, L., Fan, L., Kennedy, G., Gubler, F. and Helliwell, C. (2008) A diverse set of microRNAs and microRNA-like small RNAs in developing rice grains. *Genome Res.* **18**, 1456–1465.

17

Improving cold storage and processing traits in potato through targeted gene knockout

Benjamin M. Clasen[1], Thomas J. Stoddard[1], Song Luo[1], Zachary L. Demorest[1], Jin Li[1], Frederic Cedrone[2], Redeat Tibebu[1], Shawn Davison[1], Erin E. Ray[1], Aurelie Daulhac[1], Andrew Coffman[1], Ann Yabandith[1], Adam Retterath[1], William Haun[1], Nicholas J. Baltes[1], Luc Mathis[1], Daniel F. Voytas[1] and Feng Zhang[1,*]

[1]Cellectis plant sciences Inc., New Brighton, MN, USA
[2]Cellectis SA, Paris, France

*Correspondence
email fzhang@cellectis.
com

Keywords: transcription activator-like effector nucleases, gene editing, vacuolar invertase, cold-induced sweetening, acrylamide reduction, potato.

Summary

Cold storage of potato tubers is commonly used to reduce sprouting and extend postharvest shelf life. However, cold temperature stimulates the accumulation of reducing sugars in potato tubers. Upon high-temperature processing, these reducing sugars react with free amino acids, resulting in brown, bitter-tasting products and elevated levels of acrylamide—a potential carcinogen. To minimize the accumulation of reducing sugars, RNA interference (RNAi) technology was used to silence the vacuolar invertase gene (*VInv*), which encodes a protein that breaks down sucrose to glucose and fructose. Because RNAi often results in incomplete gene silencing and requires the plant to be transgenic, here we used transcription activator-like effector nucleases (TALENs) to knockout *VInv* within the commercial potato variety, Ranger Russet. We isolated 18 plants containing mutations in at least one *VInv* allele, and five of these plants had mutations in all *VInv* alleles. Tubers from full *VInv*-knockout plants had undetectable levels of reducing sugars, and processed chips contained reduced levels of acrylamide and were lightly coloured. Furthermore, seven of the 18 modified plant lines appeared to contain no TALEN DNA insertions in the potato genome. These results provide a framework for using TALENs to quickly improve traits in commercially relevant autotetraploid potato lines.

Introduction

Potato (*Solanum tuberosum*) is the third most important food crop, with worldwide production estimated at 376 million metric tons in 2013 (http://faostat3.fao.org/browse/rankings/commodities_by_regions/E). The majority of potatoes (61% of the 2010 crop in the United States) are used by processors for potato chips, French fries and other processed products. Because potatoes are harvested only once a year, it is necessary to cold store the tubers to ensure a year-round supply of high-quality potatoes for processing. Without cold storage, potatoes have a shelf life of about 6 months, after which they rapidly deteriorate in quality (Bianchi *et al.*, 2014). In addition to prolonged storage, cold temperatures also reduce sprouting, losses due to shrinkage and the spread of disease. One undesirable consequence of cold storage is cold-induced sweetening (CIS), in which reducing sugars accumulate in the tubers. When processed at high temperatures, reducing sugars form dark-pigmented products that are bitter and unacceptable to consumers (Dale and Bradshaw, 2003; Kumar *et al.*, 2004; Mottram *et al.*, 2002; Sowokinos, 2001; Stadler *et al.*, 2002). In the United States, CIS causes up to 15% of potatoes being rejected at processing plants every year (Bhaskar *et al.*, 2010).

In addition to producing bitter-tasting products, heat processing causes reducing sugars (e.g. glucose and fructose) in potato tubers to react with free amino acids (e.g. asparagine) to form the potential cancer-causing agent, acrylamide, via the nonenzymatic Maillard reaction (Tareke *et al.*, 2002). Acrylamide is particularly prevalent in heat-processed potatoes that have undergone CIS, due to their high levels of reducing sugars. Acrylamide in potato

chips and French fries has caused global food safety concerns, and legal measures have been sought to limit acrylamide levels in fried potato products (Grob, 2007; Medeiros Vinci *et al.*, 2012). Thus, methods that reduce acrylamide are being sought by the potato processing industry, and one effective way is to decrease reducing sugars in cold-stored tubers (Bhaskar *et al.*, 2010; Matsuura-Endo *et al.*, 2006).

Accumulation of reducing sugars during CIS is influenced by several metabolic processes, including starch synthesis, starch degradation, glycolysis, hexogenesis and mitochondrial respiration (Sowokinos, 2001). A portion of the sucrose is cleaved to produce the reducing sugars, fructose and glucose, by a family of ubiquitous enzymes termed invertases (Roitsch and Gonzalez, 2004). There are three types of invertase isoenzymes classified by their solubility, subcellular localization, pH-optima and isoelectric point: cell wall-bound invertases, neutral invertases and vacuolar invertases (VInv) (Roitsch and Gonzalez, 2004). VInv is localized in the vacuole and plays a particularly important role in the production of reducing sugars in cold-stored tubers (Kumar *et al.*, 2004; Matsuura-Endo *et al.*, 2006; Sowokinos, 2001). Transgenic RNA interference (RNAi) approaches have confirmed that knocking down *VInv* expression lowers reducing sugars and dark-pigmented nonenzymatic browning in cold-stored potatoes (Bhaskar *et al.*, 2010; Wu *et al.*, 2011; Ye *et al.*, 2010).

Here we will report the generation of potato varieties with knockout mutations in all alleles of the *VInv* gene through precise genome engineering (Voytas, 2013). This was accomplished by transiently expressing transcription activator-like effector nucleases (TALENs) designed to bind and cleave specific DNA sequences in the *VInv* locus. The double-stranded breaks (DSBs) created by the TALENs were repaired by nonhomologous end joining (NHEJ), an error prone mechanism that introduces indel (insertion/deletion) mutations that compromised *VInv* gene function. Due to the high levels of heterozygosity in the potato genome, the task of simultaneously targeting multiple alleles required careful TALEN design and optimization (Draffehn *et al.*, 2010). In contrast to previous RNAi work, TALENs achieved complete knockout lines without the need to incorporate foreign DNA. As a result, the new potato lines have significantly lower levels of reducing sugars and acrylamide in heat-processed products.

Results

TALEN design and activity assessment at the VInv target site

To design TALENs targeting the *VInv* gene in cultivated potato, exon 1 of *VInv* was sequenced from *Solanum tuberosum* cv Ranger Russet. Ranger Russet was chosen because this variety is widely used for frying, and processing is hindered due to cold-induced sweetening. As most cultivated potato varieties are tetraploid with high levels of heterozygosity, 454 deep sequencing was used to assess allele diversity. Sequencing primers were designed based on conserved regions identified through a multiple DNA sequence alignment of the *VInv* genes described in previous studies (Slugina *et al.*, 2014). A region near the start codon (600 bp) was amplified from Ranger Russet and subjected to 454 pyro-sequencing. From a total of 8268 reads, three distinct allele types were identified (designated as A1, A2 and A3). The ratio of read numbers for each allele approximated 1 : 2 : 1 (A1, 1814 reads; A2, 4265 reads; A3, 2189 reads) suggesting that there is one copy of A1, two copies of A2 (designated as A2(1) and A2(2)) and one copy of the A3 in Ranger Russet. Three TALEN pairs were designed to target within the first 200 bp of coding sequence using TALEN™ hit software (Figure 1b) (Haun *et al.*, 2014). The most conserved TALEN-binding sites are identical in sequence between the alleles A2(1), A2(2) and A3, but contain one, two, or three single nucleotide polymorphisms (SNPs) in A1 (Figure 1c). TALENs were synthesized based on the consensus target sequences among the three *VInv* alleles.

TALEN activity was tested at endogenous target sites by expressing the TALENs in protoplasts and surveying the target sites for mutations introduced by NHEJ (Figure 2a). TALEN-encoding plasmids (VInv_T1, VInv_T2 and VInv_T3) were individually introduced into Ranger Russet protoplasts by polyethylene glycol (PEG)-mediated transformation (Figure S1). Protoplasts were sacrificed 48 h post-transformation and genomic DNA was isolated. The genomic DNA was subjected to PCR amplification, producing a 272-bp fragment encompassing the TALEN recognition sites. The PCR product was then subjected to 454 pyro-sequencing, and the resulting sequencing reads were evaluated for NHEJ-induced mutations. The frequency of mutated sequences in each sample was used to estimate the activity of the corresponding TALENs.

The *VInv* TALEN pairs, encoded by plasmids VInv_T1, VInv_T2 and VInv_T3, induced NHEJ mutations at frequencies of 3.6%, 9.5% and 9.9%, respectively (Figure 2b). All three TALEN pairs induced mutations in all *VInv* alleles, despite the DNA sequence polymorphisms. TALEN pairs were also tested in protoplasts isolated from three other commercial varieties, namely Russet

Burbank, Atlantic and Shepody. The observed mutation frequency for each TALEN pair in each variety is summarized in Figure S2. Of the three TALEN pairs, VInv_2 TALENs performed consistently well across all four varieties, with mutation frequencies ranging from 2.1% to 15.9%.

Generating potato lines with knockout mutations in *VInv*

To create potato lines harbouring mutations in the *VInv* alleles, Ranger Russet protoplasts were transformed with TALEN-encoding plasmids and regenerated into whole plants (Figure 3a). Given the previous pyro-sequencing results, we focused our efforts on VInv_T2, as this TALEN pair performed consistently well across all four potato varieties and contained the fewest SNPs in the TALE-binding regions among the three different alleles. For each experiment, 100 000–200 000 protoplasts were transformed with a plasmid VInv_T2 using PEG. Transformed protoplasts were transferred to nonselective regeneration media and allowed to regenerate. Shoots that appeared from calli (approximately 12 weeks post-transformation) were excised and transferred to rooting medium. To reduce the likelihood of recovering a chimeric potato plant (containing a mixture of WT and modified cells), several meristematic subcultures were propagated. At the time that plantlets were transferred to rooting medium, a small leaf sample was removed and DNA was isolated for genotyping of the *VInv* locus.

Plants were initially screened for mutations by directly sequencing PCR amplicons obtained using primers that recognize all three allele types, A1, A2 and A3. Sequencing chromatograms with two or more peaks near the TALEN target sites suggested that there were mutations in one or more *VInv* alleles (Figure S3). From a total of approximately 600 regenerated shoots, 18 plants were identified that likely harbour mutations in at least one *VInv* allele. To further characterize mutations within the *VInv* alleles, PCR amplicons encompassing the *VInv* TALEN-binding sites were cloned, and 24–48 clones were sequenced. Of the 18 candidate plants, six contained mutations in one *VInv* allele, five contained mutations in two *VInv* alleles, two contained mutations in three *VInv* alleles, and five contained mutations in all four *VInv* alleles (Table 1).

From transformation group St116, nine mutants were isolated, seven of which had 1 or 2 mutations within the *VInv* alleles. Two plants were recovered with mutations in all *VInv* alleles (St116_1 and St116_9). Notably, plant St116_1 contained a four-base-pair deletion in A1, an identical four-base-pair deletion in both A2(1) and A2(2), and a 17-bp deletion in A3. Because all mutations within plant St116_1 resulted in frame shifts, *VInv* gene function was predicted to be completely knocked out. For this reason, plant St116_1 was advanced to phenotypic characterization.

It is possible that transformation of protoplasts with plasmid DNA results in stable integration of DNA into the plant genome. To determine whether VInv_T2 plasmid DNA was present in the mutants, PCR was performed using primer pairs designed to amplify three regions of the TALEN-expression cassette (Figure S4A). Of 18 mutant plants, 7 (39%) had no detectable TALEN vector sequence. A representative example of a PCR to detect VInv_T2 plasmid sequence is shown in Figure S4B. Taken together, our results demonstrate the usefulness of TALENs and protoplast transformation/regeneration to create modified, nontransgenic potato lines in a single generation.

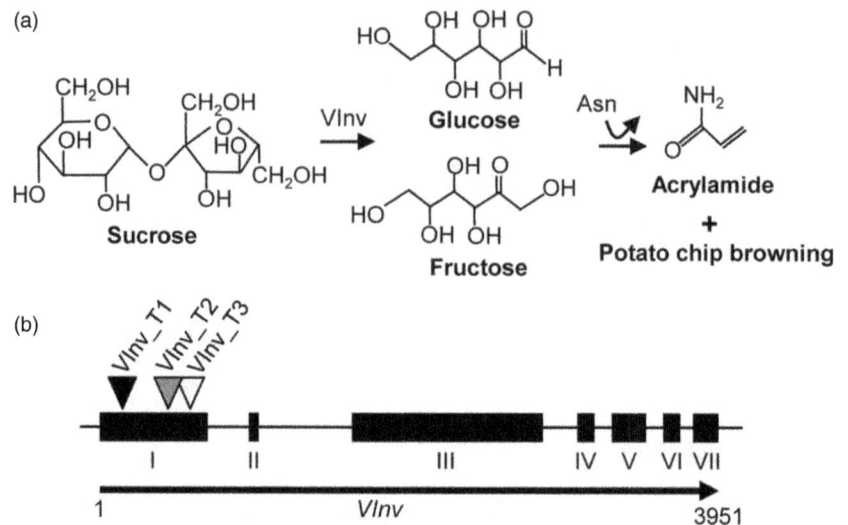

Figure 1 Targeting the *Solanum tuberosum* cv Ranger Russet *VInv* gene with transcription activator-like effector nucleases (TALENs). (a) During cold storage, potato tubers accumulate acrylamide through a nonenzymatic Maillard reaction, which uses reducing sugars (primarily glucose and fructose) and free amino acids (asparagine [Asn]). Reducing sugars accumulate through hydrolysis of sucrose by vacuolar acid invertase (VInv). In addition to accumulating acrylamide, cold-stored potatoes also produce brown- to black-pigmented products. (b) Schematic of the *VInv* gene. TALEN target sites are indicated with black, grey and white triangles (VInv_T1, VInv_T2, VInv_T3, respectively). (c) TALEN target sites within exon 1. Single nucleotide polymorphisms are indicated by lowercase bold letters. Underlined letters indicate TALEN-binding sites. A1, allele 1; A2(1), copy 1 of allele 2; A2(2) copy 2 of allele 2; A(3), allele 3.

Assessment of sugar levels from cold-stored potato tubers from *VInv*-knockout lines

To assess levels of reducing sugars in cold-stored tubers, potatoes from wild-type Ranger Russet, St116_8 (containing a mutation in one *VInv* allele) and St116_1 (containing mutations in all *VInv* alleles), were cold-stored at 4 °C for 14 days. Following cold storage, tubers were homogenized, and the levels of sucrose, glucose and fructose were quantified using high-pressure liquid chromatography. As shown in Figure 4a, the levels of fructose, glucose and sucrose in wild-type tubers were 1.4, 1.5 and 1.1 mg/g (% of fresh weight), respectively. In tubers from St116-8, the sugar levels were roughly comparable to wild type with 1.2, 1.2 and 2 mg/g of fructose, glucose and sucrose, respectively. Tubers from the full knockout mutant, St116-1, exhibited significantly decreased levels of fructose and glucose that were below the limit of detection for this assay (<0.1 mg/g), and increased levels of sucrose (3.3 mg/g) (Figure 4a). To further assess sugar levels in our mutant potato plants, three additional lines, St116_5, St116_3 and St123_1, containing three wild-type *VInv* alleles, two wild-type *VInv* alleles and two wild-type *VInv* alleles, respectively, were assessed for levels of fructose, glucose and sucrose (Figure S5). Compared to cold-stored tubers from wild-type plants and St116_1, all additional lines had intermediate sugar levels.

Acrylamide levels and quality assessment of potato chips produced from cold-stored tubers from *VInv*-knockout lines

Acrylamide levels were assessed in potato chips from wild-type Ranger Russet and the St116_1 mutant. Chips were prepared from tubers stored at 4 °C for 14 days. Acrylamide levels were quantified using high-pressure liquid chromatography. Acrylamide content in chips from St116_1 tubers was 1820 µg/kg, 73.3% less than that of chips from wild-type tubers (6820 µg/kg) (Figure 4b). Potato chips from the wild-type and the mutant plants were also assessed for colour. Chips processed from cold-stored wild-type and St116_1 mutant tubers are shown in Figure 4c. As predicted, chips from St116_1 were lighter-coloured than those from wild type. The chip colour, as determined by the Hunter score, was substantially better for line St116_1 compared to wild type (Figure 4c).

Discussion

Potatoes are an integral part of the world's food supply; however, trait-improvement efforts are met with challenges due to the complexity of the autotetraploid genome and the high degree of heterozygosity (Barrell *et al.*, 2013). As a result, classic improve-

(a)

S. tuberosum 'Ranger Russet'

Isolate protoplasts

Transform protoplasts with plasmids encoding TALEN pairs

PCR exon 1 and deep sequence

VInv

(b)

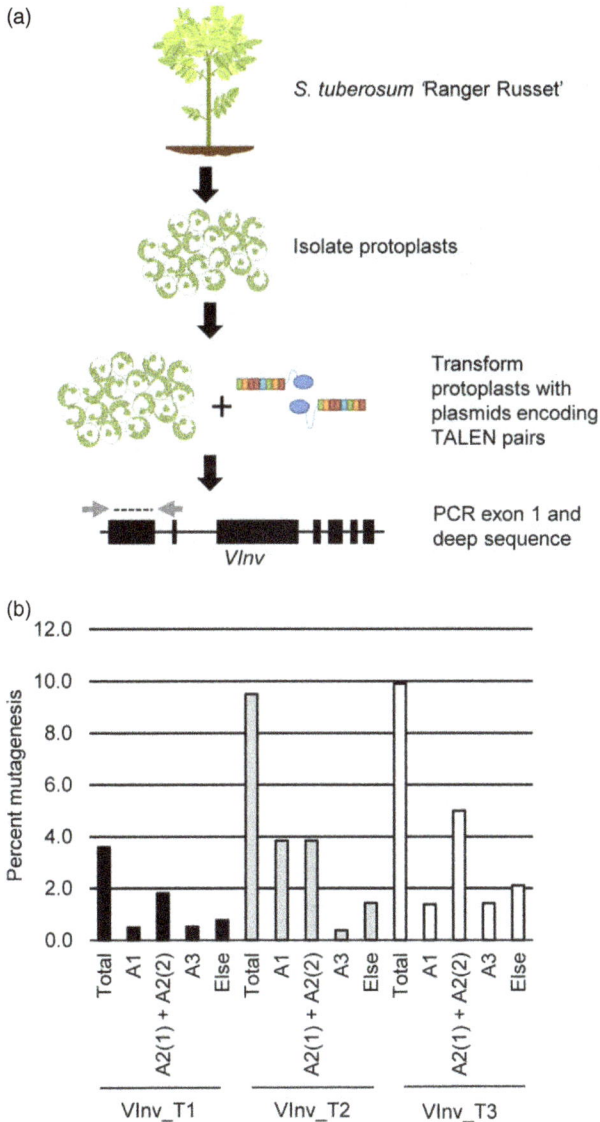

Figure 2 Assessment of transcription activator-like effector nuclease (TALEN) activity in *Solanum tuberosum* cv Ranger Russet protoplasts. (a) Protoplasts from leaves on 3-week-old potato plants (Ranger Russet) were isolated and transformed with plasmids encoding TALEN pairs. Following transformation, exon 1 was amplified by PCR and mutations were assessed by 454 pyro-sequencing. (b) Percentage of vacuolar invertase sequences containing TALEN-induced mutations. Total refers to the combined percentage of mutations in all four alleles. Else refers to sequences that could not be assigned an allele type due to TALEN-induced mutations that removed the allele-defining single nucleotide polymorphisms.

mation. As a result, the expression of transgenes is not predictable and largely depends on the location, copy number and configuration of the transgenes. Thus, the characterization of a large population of independently transformed lines is often required to identify those with the desired expression of the transgene (Barrell and Conner, 2009; Barrell *et al.*, 2013; Jacobs *et al.*, 2009). TILLING is also time-consuming and laborious, and numerous mutations occur outside the target gene that can only be removed by extensive backcrossing regimes (Elias *et al.*, 2009; Muth *et al.*, 2008). Recent advances in technologies and methods that enable precise gene modification hold promise to overcome these challenges and accelerate trait improvement in potato (Voytas, 2013).

To circumvent the difficulties inherent in working with a highly heterozygous autopolyploid species, the method of targeted gene modification needs to be highly efficient, such that mutations can be introduced simultaneously in multiple loci and multiple alleles. In this study, we demonstrated that TALEN-mediated gene editing creates a valuable trait in a commercial potato cultivar with high efficiency. Sequence-specific TALENs were designed to cleave multiple alleles of a target gene and were delivered to potato protoplasts with over 80% efficiency. Potato plants were then regenerated from single-cell protoplasts, and about 3% of the plants were found to contain mutations in one to four *VInv* alleles. The ability to generate multiple mutant lines is particularly valuable for potato trait development, as somaclonal variation commonly occurs in plants derived from tissue culture leading to abnormal phenotypes (Barrell *et al.*, 2013). In addition to the cultivar Ranger Russet, TALENs were found to be highly efficient mutagens in protoplasts derived from other commercial varieties. As plants can be readily regenerated from these transformed protoplasts, the method described in this study can be used as a framework for efficiently introducing targeted mutations into a gene-of-interest in a single generation in elite potato varieties.

Previous studies have indicated that off-target mutagenesis can occur when using TALENs for genome modification (Frock *et al.*, 2015; Mali *et al.*, 2013). Furthermore, we have shown that TALEN pairs could tolerate up to 3–4 mismatches in the recognition sites (Juillerat *et al.*, 2014). Therefore, it may be possible that the TALEN reagents used in this study bind and cleave unintended target sites. However, in an effort to minimize the chance of off-target mutations, during TALEN design and synthesis, we searched publicly available potato genome sequences for potential TALEN-binding sites—no potential target sequences were found with four or less mismatches. To completely rule out the possibility of off-target mutations in the VInv mutant lines, it is possible to perform whole-genome sequencing.

Previous research demonstrated that silencing of the *VInv* gene by RNAi decreases the reducing sugar and acrylamide levels in cold-stored, processed potato tubers and potato chips (Bhaskar *et al.*, 2010; Wu *et al.*, 2011). No accumulation of the reducing sugars, glucose and fructose, was detected in the RNAi lines that suppressed *VInv* gene expression by more than 97%, whereas partial suppression did not control CIS effectively. This observation suggested that the *VInv* gene has to be almost completely silenced to confer the CIS-resistant trait. In this study, we tested this hypothesis through targeted mutagenesis. By generating a series of mutants with 1–4 *VInv* alleles knocked out, our results showed a positive correlation between the number of wild-type alleles and the levels of reducing sugars in

ment schemes require long breeding cycles and screening of exceptionally large populations to accumulate desired alleles (Bradshaw *et al.*, 2006). In recent years, new technologies, such as transgenesis/cisgenesis, RNAi and TILLING, have been developed to overcome these challenges (Barrell *et al.*, 2013; Elias *et al.*, 2009). The major shortcoming of these technologies is the lack of precision to produce desired traits. For example, transgenesis/cisgenesis and RNAi technologies deliver and integrate transgenes in a random fashion into the potato genomes by either direct DNA uptake or *Agrobacterium*-mediated transfor-

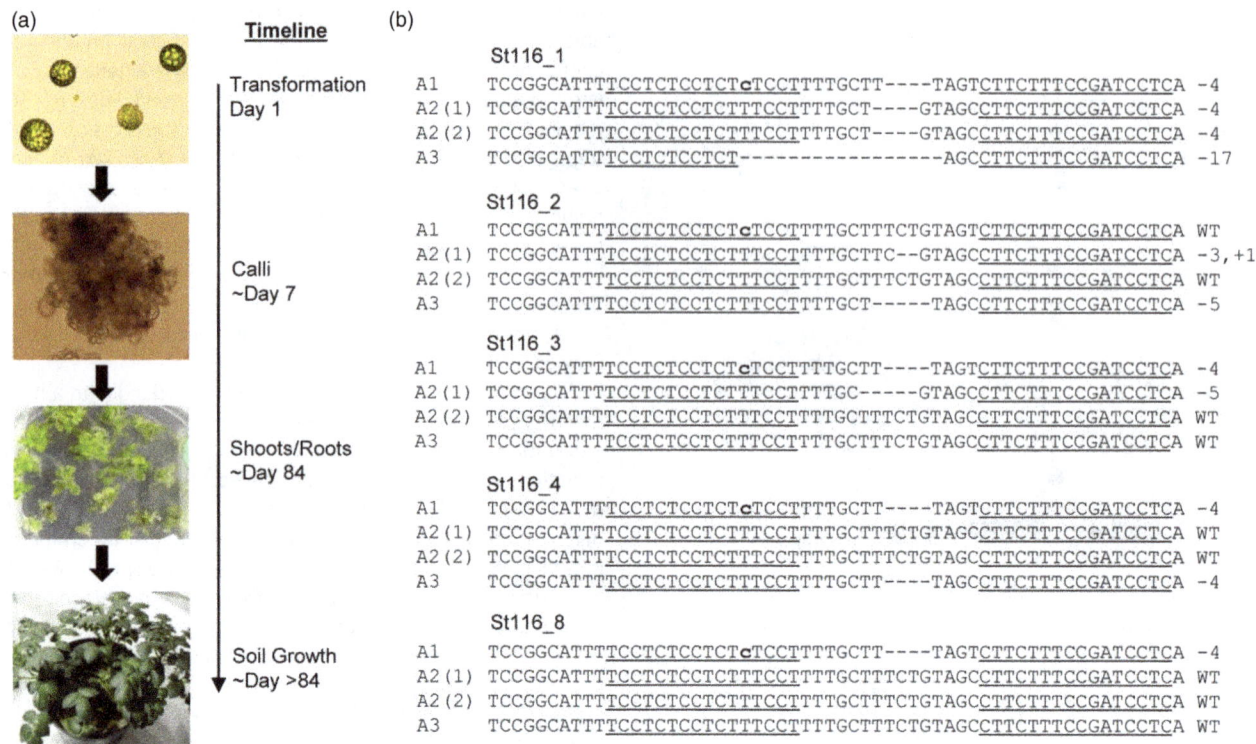

Figure 3 Recovery of potato lines carrying mutations within vacuolar invertase (*VInv*). (a) Approach and timeline to regenerate plants with mutations in *VInv*. Protoplasts were transformed with plasmids encoding the VInv_T2 transcription activator-like effector nucleases pair and were cultured in nonselective regeneration medium. Following shoot and root formation, potato plantlets were transferred to soil. (b) Examples of plant lines carrying mutations in one or more of the *VInv* alleles. WT, wild type.

Table 1 Characterization of vacuolar invertase mutant plants

Number of mutant alleles	Number of plants	Plants without transgene
4	5 (28%)	2
3	2 (11%)	1
2	5 (28%)	2
1	6 (33%)	2
Total	18	7 (39%)

cold-stored potato tubers. When all four *VInv* alleles were mutated, reducing sugars were undetectable (<0.1 µg/kg), which is consistent with previous RNAi research. The low level of acrylamide generated within the tubers of our complete knockout line suggests the presence of reducing sugars, albeit at undetectable levels. These reducing sugars are most likely derived from invertase activity in the cytoplasm and cell wall (Bhaskar *et al.*, 2010; Roitsch and Gonzalez, 2004). Therefore, it may be possible to further reduce acrylamide content by targeting additional, nonessential invertases. The agronomic performance of the full *VInv*-knockout lines remains to be assessed in field trails in the coming years.

In the past two decades, transgenic or cisgenic approaches have been used to create potato varieties with new traits (Barrell *et al.*, 2013; Holme *et al.*, 2013; Rommens, 2007). Although a few new varieties have been deregulated by the USDA, transgenic potatoes have not been well accepted by the market because of their GM status (Fernandez-Cornejo *et al.*, 2014). In this study, several *VInv*-knockout lines were created that lack foreign DNA (e.g. the TALEN-encoding constructs) in their genome.

Regulatory authorities around the world are currently considering how to handle plants that lack transgenes and carry targeted mutations such as those created by NHEJ (Kuzma and Kokotovich, 2011). It is interesting to consider that the potato lines created in this study have no foreign DNA and thus are not different from naturally occurring varieties.

Materials and methods

Identification of TALEN target sites in the *VInv* gene

To completely inactivate or knockout the *VInv* gene in *S. tuberosum* cv Ranger Russet, sequence-specific nucleases were designed that target the protein coding region in the vicinity of the start codon. Three TALEN target sites were identified (VInv_T1, VInv_T2 and VInv_T3) within the first 200 bp of the coding sequence using TALEN™ Hit software. The locations of the TALENs and their corresponding sequences are shown in Figure 1c. TALENs were synthesized using methods similar to those described elsewhere (Beurdeley *et al.*, 2013; Cermak *et al.*, 2011).

Protoplast isolation and transformation

Protoplast isolation and transformation were carried out as previously described with slight modifications (Craig *et al.*, 2005). Shoots derived from tubers (grown in vermiculite at 22 °C under 120 µmol/m²/s for 3 weeks) were surface-sterilized and cultured on StProp medium (MS basal salts plus vitamins, 2% sucrose, 0.7% agar, pH 5.8). Subcultures were made by single node propagation at biweekly intervals, and leaves from these subcultures served as the source material for protoplast isolation as previously described (Yoo *et al.*, 2007; Zhang *et al.*, 2013).

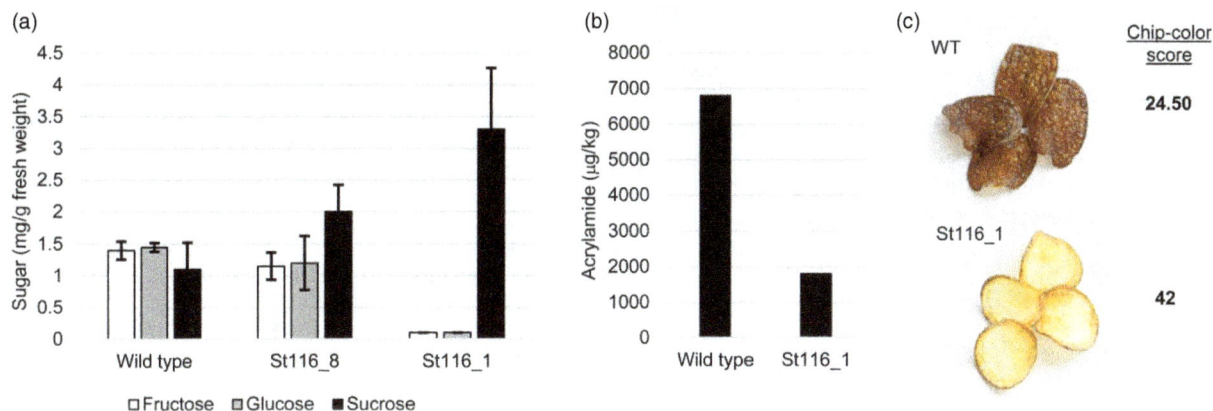

Figure 4 Quality assessment of mutant potato lines. (a) Analysis of sugar content within potato tubers stored at 4 °C for 14 days. Error bars represent standard deviation. (b) Analysis of acrylamide content in potato chips that were processed from tubers that were stored at 4 °C for 14 days. (c) Images of potato chips after being processed from tubers stored at 4 °C for 14 days. The colorimetric score is listed to the right of the image.

Protoplasts were suspended in transfection buffer at a cell density of 100 000 protoplasts/mL in a final volume of 200 μL and placed in a 2-mL Eppendorf tube containing 15 μg of the plasmid of interest. Immediately following suspension, 200 μL of 40% PEG4000 was added to the protoplast suspension. The tube was gently vortexed and incubated in the dark at room temperature for 30 min. After incubation, 1 mL of wash buffer was added to each tube. Tubes were gently vortexed and cells were pelleted at 500 g for 5 min followed by resuspension in plating medium. Transformation efficiencies were monitored by transforming protoplasts with a plasmid encoding YFP.

Validation of TALENs in potato protoplasts

To assess TALEN activity, protoplasts from varieties Ranger Russet, Atlantic, Russet Burbank and Shepody were transformed with plasmid DNA encoding VInv_T1, VInv_T2 or VInv_T3. Two days after transformation, protoplasts were sacrificed and subjected to DNA extraction using the hexadecyltrimethylammonium bromide (CTAB)-based method (Zhang et al., 2013). Purified DNA was used as a template for PCR using primers designed to amplify VInv exon 1 (5′-CACCCAGTACCATTCCAGTTATG-3′ and 5′-TTTTTGA GGTTGAAAATGGTAAGCA-3′). The resulting amplicons were subjected to 454 pyro-sequencing (Roche GS Junior), and TALEN-induced mutations were quantified.

Regeneration of mutant plants

Immediately after transformation of protoplasts with a plasmid encoding VInv_T2, cells were cultured using methods and media previously described by Gamborg et al. with slight modifications (Gamborg and Shyluk, 1981). Protoplasts were suspended in P-medium and stored at 25 °C in the dark. Two weeks after transformation, after the majority of the protoplasts had divided at least once, the protoplast culture was diluted twofold in a suspension of P-medium. Two weeks after dilution, cultures were plated on a solid reservoir of CUL medium in a 10-cm petri plate (Haberlach et al., 1985). Six weeks after plating, protoplast-derived calli (p-calli) were transferred to a solid reservoir of DIF medium (Haberlach et al., 1985). P-calli were transferred to fresh DIF medium at biweekly intervals. As shoots formed from the calli, they were excised, sampled for DNA isolation and placed into a solid reservoir of R-medium (Gamborg and Shyluk, 1981). As the shoots elongated, they were subcloned at individual nodes and separated into two populations: one for propagation and the

other for tuber production. The propagation population was maintained and expanded on StProp, whereas the tuber population was transferred to a soilless potting substrate (SunGro, Agawam, MA, USA, All Purpose Planting Mix) and grown to maturity in a growth chamber.

Genotyping VInv mutant plantlets

DNA was isolated from fresh in vitro leaf material using methods described by Aljanabi and Martinez (Aljanabi and Martinez, 1997). Mutants were identified using a two-step screening process. For the first step, DNA was isolated and subjected to PCR to amplify the locus of interest using primers designed for VInv (forward primer, 5′-CACCCAGTACCATTCCAGTTATG-3′; reverse primer, 5′-CATTGGACCACGCATAAGAA-3′). The resulting pool of amplicons were directly sequenced using Sanger sequencing (Eurofins MWG Operon, Huntsville, AL, USA). We predicted that if a plant contained insertions or deletions within one or more VInv alleles, the resulting sequencing reads would contain unordered traces near the predicted TALEN cut site. Sequence results were scored by hand, to check for a disordered trace. Candidate plants were subjected to further analysis: the PCR product was cloned using the CloneJet PCR Cloning Kit (Thermo Scientific, Waltham, MA, USA), and clones were subjected to Sanger sequencing. Given the highly heterozygous nature of the potato genome, a large number of clones (n ≥ 10) were sequenced to recover all allele types. Sequencing files associated with the clones were aligned with the S. tuberosum VInv gene (NCBI Accession JN661860) using the discontinuous megablast algorithm with default parameters with a standalone version of NCBI-BLAST+ for Windows.

Sugar profiling, acrylamide analysis and colorimetry of fried potato chips

Sugar profile analysis was carried out by Rtech Laboratories (Land O'Lakes, MN) using AOAC official methods (Method 982.14; Glucose, Fructose, Sucrose, and Maltose in Presweetened Cereals; http://www.eoma.aoac.org/). The analysis was performed on 30 g (mass was determined after 14 days of storage at 4 ± 1 °C) of homogenized tuber tissue comprised of the medial one centimetre of two to three independent mature tubers from a single plant. Potato chips were processed using the method previously described (Bhaskar et al., 2010). Acrylamide analysis was performed on chips prepared from tubers that were stored at

4 ± 1 °C for 14 days. Analysis was performed by Covance laboratories (United States) using a modified method described by the United States Food and Drug Administration (Detection and Quantitation of Acrylamide in Foods; http://www.fda.gov/). The colour intensity of potato chips was measured using Hunter scores. Chip colour measurements were determined on a subset of transversely sliced chips (15 g) from the medial one centimetre using a D25LT colorimeter (HunterLab) (Bhaskar et al., 2010).

References

Aljanabi, S.M. and Martinez, I. (1997) Universal and rapid salt-extraction of high quality genomic DNA for PCR-based techniques. *Nucleic Acids Res.* **25**, 4692–4693.

Barrell, P.J. and Conner, A.J. (2009) Expression of a chimeric magainin gene in potato confers improved resistance to the phytopathogen Erwinia carotovora. *Open Plant Sci. J.* **3**, 14–21.

Barrell, P.J., Meiyalaghan, S., Jacobs, J.M. and Conner, A.J. (2013) Applications of biotechnology and genomics in potato improvement. *Plant Biotechnol. J.* **11**, 907–920.

Beurdeley, M., Bietz, F., Li, J., Thomas, S., Stoddard, T., Juillerat, A., Zhang, F., Voytas, D.F., Duchateau, P. and Silva, G.H. (2013) Compact designer TALENs for efficient genome engineering. *Nat. Commun.* **4**, 1762.

Bhaskar, P.B., Wu, L., Busse, J.S., Whitty, B.R., Hamernik, A.J., Jansky, S.H., Buell, C.R., Bethke, P.C. and Jiang, J. (2010) Suppression of the vacuolar invertase gene prevents cold-induced sweetening in potato. *Plant Physiol.* **154**, 939–948.

Bianchi, G., Scalzo, R.L., Testoni, A. and Maestrelli, A. (2014) Nondestructive analysis to monitor potato quality during cold storage. *J. Food Qual.* **37**, 9–17.

Bradshaw, J.E., Bryan, G.J. and Ramsay, G. (2006) Genetic resources (including wild and cultivated *Solanum* species) and progress in their utilisation in potato breeding. *Potato Res.* **49**, 49–65.

Cermak, T., Doyle, E.L., Christian, M., Wang, L., Zhang, Y., Schmidt, C., Baller, J.A., Somia, N.V., Bogdanove, A.J. and Voytas, D.F. (2011) Efficient design and assembly of custom TALEN and other TAL effector-based constructs for DNA targeting. *Nucleic Acids Res.* **39**, e82.

Craig, W., Gargano, D., Scotti, N., Nguyen, T., Lao, N., Kavanagh, T., Dix, P. and Cardi, T. (2005) Direct gene transfer in potato: a comparison of particle bombardment of leaf explants and PEG-mediated transformation of protoplasts. *Plant Cell Rep.* **24**, 603–611.

Dale, M.F. and Bradshaw, J.E. (2003) Progress in improving processing attributes in potato. *Trends Plant Sci.* **8**, 310–312.

Draffehn, A., Meller, S., Li, L. and Gebhardt, C. (2010) Natural diversity of potato (*Solanum tuberosum*) invertases. *BMC Plant Biol.* **10**, 271.

Elias, R., Till, B., Mba, C. and Al-Safadi, B. (2009) Optimizing TILLING and Ecotilling techniques for potato (*Solanum tuberosum* L). *BMC Res. Notes*, **2**, 141.

Fernandez-Cornejo, J., Wechsler, S., Livingston, M. and Mitchell, L. (2014) *Genetically Engineered Crops in the United States*. Washington D.C: US Department of Agriculture Economic Research Service.

Frock, R.L., Hu, J., Meyers, R.M., Ho, Y.J., Kii, E. and Alt, F.W. (2015) Genome-wide detection of DNA double-stranded breaks induced by engineered nucleases. *Nat. Biotechnol.* **33**, 179–186.

Gamborg, O.L. and Shyluk, J.P. (1981) *Nutrition, Media and Characteristics of Plant Cell and Tissue Cultures*. In: Plant Tissue Culture (Thorpe, T.A. ed) pp. 21–44. San Diego, CA: Academic Press.

Grob, K. (2007) Options for legal measures to reduce acrylamide contents in the most relevant foods. *Food Addit. Contam.* **24**, 71–81.

Haberlach, G., Cohen, B., Reichert, N., Baer, M., Towill, L. and Helgeson, J.P. (1985) Isolation, culture and regeneration of protoplasts from potato and several related *Solanum* species. *Plant Sci.* **39**, 67–74.

Haun, W., Coffman, A., Clasen, B.M., Demorest, Z.L., Lowy, A., Ray, E., Retterath, A., Stoddard, T., Juillerat, A., Cedrone, F., Mathis, L., Voytas, D.F. and Zhang, F. (2014) Improved soybean oil quality by targeted mutagenesis of the fatty acid desaturase 2 gene family. *Plant Biotechnol. J.* **12**, 934–940.

Holme, I.B., Wendt, T. and Holm, P.B. (2013) Intragenesis and cisgenesis as alternatives to transgenic crop development. *Plant Biotechnol. J.* **11**, 395–407.

Jacobs, J.M., Takla, M.F., Docherty, L.C., Frater, C.M., Markwick, N.P., Meiyalaghan, S. and Conner, A.J. (2009) Potato transformation with modified nucleotide sequences of the cry9Aa2 gene improves resistance to potato tuber moth. *Potato Res.* **52**, 367–378.

Juillerat, A., Dubois, G., Valton, J., Thomas, S., Stella, S., Marechal, A., Langevin, S., Benomari, N., Bertonati, C., Silva, G.H., Daboussi, F., Epinat, J.C., Montoya, G., Duclert, A. and Duchateau, P. (2014) Comprehensive analysis of the specificity of transcription activator-like effector nucleases. *Nucleic Acids Res.* **42**, 5390–5402.

Kumar, D., Singh, B.P. and Kumar, P. (2004) An overview of the factors affecting sugar content of potatoes. *Ann. Appl. Biol.* **145**, 247–256.

Kuzma, J. and Kokotovich, A. (2011) Renegotiating GM crop regulation. Targeted gene-modification technology raises new issues for the oversight of genetically modified crops. *EMBO reports* **12**, 883–888.

Mali, P., Aach, J., Stranges, P.B., Esvelt, K.M., Moosburner, M., Kosuri, S., Yang, L. and Church, G.M. (2013) CAS9 transcriptional activators for target specificity screening and paired nickases for cooperative genome engineering. *Nat. Biotechnol.* **31**, 833–838.

Matsuura-Endo, C., Ohara-Takada, A., Chuda, Y., Ono, H., Yada, H., Yoshida, M., Kobayashi, A., Tsuda, S., Takigawa, S., Noda, T., Yamauchi, H. and Mori, M. (2006) Effects of storage temperature on the contents of sugars and free amino acids in tubers from different potato cultivars and acrylamide in chips. *Biosci. Biotechnol. Biochem.* **70**, 1173–1180.

Medeiros Vinci, R., Mestdagh, F. and De Meulenaer, B. (2012) Acrylamide formation in fried potato products – present and future, a critical review on mitigation strategies. *Food Chem.* **133**, 1138–1154.

Mottram, D.S., Wedzicha, B.L. and Dodson, A.T. (2002) Acrylamide is formed in the Maillard reaction. *Nature*, **419**, 448–449.

Muth, J., Hartje, S., Twyman, R.M., Hofferbert, H.R., Tacke, E. and Prüfer, D. (2008) Precision breeding for novel starch variants in potato. *Plant Biotechnol. J.* **6**, 576–584.

Roitsch, T. and Gonzalez, M.C. (2004) Function and regulation of plant invertases: sweet sensations. *Trends Plant Sci.* **9**, 606–613.

Rommens, C.M. (2007) Intragenic crop improvement: combining the benefits of traditional breeding and genetic engineering. *J. Agric. Food Chem.* **55**, 4281–4288.

Slugina, M.A., Khrapalova, I.A., Ryzhova, N.N., Kochieva, E.Z. and Skryabin, K.G. (2014) Polymorphism of Pain-1 invertase gene in *Solanum* species. *Dokl. Biochem. Biophys.* **454**, 1–3.

Sowokinos, J. (2001) Biochemical and molecular control of cold-induced sweetening in potatoes. *Am. J. Potato Res.* **78**, 221–236.

Stadler, R.H., Blank, I., Varga, N., Robert, F., Hau, J., Guy, P.A., Robert, M.C. and Riediker, S. (2002) Acrylamide from Maillard reaction products. *Nature*, **419**, 449–450.

Tareke, E., Rydberg, P., Karlsson, P., Eriksson, S. and Tornqvist, M. (2002) Analysis of acrylamide, a carcinogen formed in heated foodstuffs. *J. Agric. Food Chem.* **50**, 4998–5006.

Voytas, D.F. (2013) Plant genome engineering with sequence-specific nucleases. *Annu. Rev. Plant Biol.* **64**, 327–350.

Wu, L., Bhaskar, P.B., Busse, J.S., Zhang, R., Bethke, P.C. and Jiang, J. (2011) Developing Cold-Chipping Potato Varieties by Silencing the Vacuolar Invertase Gene All rights reserved. No part of this periodical may be reproduced or transmitted in any form or by any means, electronic or mechanical, including photocopying, recording, or any information storage and retrieval system, without permission in writing from the publisher. Permission for printing and for reprinting the material contained herein has been obtained by the publisher. *Crop Sci.* **51**, 981–990.

Ye, J., Shakya, R., Shrestha, P. and Rommens, C.M. (2010) Tuber-specific silencing of the acid invertase gene substantially lowers the acrylamide-forming potential of potato. *J. Agric. Food Chem.* **58**, 12162–12167.

Yoo, S.D., Cho, Y.H. and Sheen, J. (2007) Arabidopsis mesophyll protoplasts: a versatile cell system for transient gene expression analysis. *Nat. Protoc.* **2**, 1565–1572.

Permissions

List of Contributors

Melissa Bredow and Barbara Vanderbeld
Department of Biology, Queen's University, Kingston, ON, Canada

Virginia K. Walker
Department of Biology, Queen's University, Kingston, ON, Canada
Department of Biomedical and Molecular Sciences and School of Environmental Studies, Queen's University, Kingston, ON, Canada

Gagan Garg, Rhonda Foley and Ling-ling Gao
CSIRO Agriculture, Wembley, WA, Australia

James K. Hane
CSIRO Agriculture, Wembley, WA, Australia
Department of Environment and Agriculture, CCDM Bioinformatics, Centre for Crop and Disease Management, Curtin University, Bentley, WA, Australia
Curtin Institute for Computation, Curtin University, Bentley, WA, Australia

Lars G. Kamphuis and Karam B. Singh
CSIRO Agriculture, Wembley, WA, Australia
UWA Institute of Agriculture, University of Western Australia, Crawley, WA, Australia

Yao Ming and Jianbo Jian
Department of Plant and Animal Genome Research, Beijing Genome Institute, Shenzhen, China

Matthew N. Nelson and Craig A. Atkins
UWA Institute of Agriculture, University of Western Australia, Crawley, WA, Australia
School of Plant Biology, University of Western Australia, Crawley, WA, Australia

Philipp E. Bayer
School of Plant Biology, University of Western Australia, Crawley, WA, Australia

David Edwards and Bhavna Hurgobin
School of Plant Biology, University of Western Australia, Crawley, WA, Australia
University of Queensland, Brisbane, Qld, Australia

Armando Bravo and Maria J. Harrison
Boyce Thompson Institute for Plant Research, Ithaca, NY, USA

Scott Bringans
Proteomics International, Nedlands, WA, Australia

Steven Cannon
Usda-Ars Corn Insects And Crop Genetics Research Unit, Crop Genome Informatics Lab, Iowa State University, Ames, IA, USA
Department of Agronomy, Iowa State University, Ames, IA, USA

Wei Huang
Department of Agronomy, Iowa State University, Ames, IA, USA

Sean Li, Annette McGrath and Jeremy Murray
Data, Csiro, Canberra, Act, Australia

Cheng-Wu Liu
John Innes Centre, Norwich Research Park, Norfolk, UK

Grant Morahan
Centre for Diabetes Research, University of Western Australia, Crawley, WA, Australia

James Weller
School of Biological Sciences, University of Tasmania, Hobart, TAS, Australia

Abdul Baten and Daniel L. E. Waters
Southern Cross Plant Science, Southern Cross University, Lismore, NSW, Australia

Kwanho Jeong, Cecile C. Julia and Terry J. Rose
Southern Cross Plant Science, Southern Cross University, Lismore, NSW, Australia
Southern Cross GeoScience, Southern Cross University, Lismore, NSW, Australia

Omar Pantoja
Southern Cross Plant Science, Southern Cross University, Lismore, NSW, Australia
Instituto de Biotecnologıa, Universidad Nacional Autonoma de Mexico, Cuernavaca, Morelos, Mexico

Matthias Wissuwa
Crop, Livestock and Environment Division, Japan International Research Center for Agricultural Sciences, Tsukuba, Ibaraki, Japan

Sigrid Heuer
University of Adelaide, School of Agriculture Food and Wine / Australian Centre for Plant Functional Genomics (Acpfg), Adelaide, SA, Australia

Tobias Kretzschmar
Genotyping Services Laboratory, International Rice Research Institute (Irri), Metro Manila, Philippines

Marco Maccaferri, Andrea Ricci, Silvio Salvi, Sara Giulia Milner, Enrico Noli and Roberto Tuberosa
Department of Agricultural Sciences (DipSA), University of Bologna, Bologna, Italy

Pier Luigi Martelli and Rita Casadio
Biocomputing Group, University of Bologna, Bologna, Italy

Eduard Akhunov
Department of Plant Pathology, Kansas State University, Manhattan, KS, USA

Simone Scalabrin, Vera Vendramin and Michele Morgante
Istituto di Genomica Applicata, Udine, Italy
Dipartimento di Scienze Agrarie e Ambientali, University of Udine, Udine, Italy

Karim Ammar
Cimmyt Carretera Mexico, Texcoco, Mexico

Antonio Blanco
Dipartimento di Biologia e Chimica Agro-forestale ed ambientale, Universita di Bari, Aldo Moro, Bari, Italy

Francesca Desiderio
Consiglio per la ricerca e la sperimentazione in agricoltura, Genomics Research Centre, Fiorenzuola d'Arda, Italy

Assaf Distelfeld,
Faculty of Life Sciences, Department of Molecular Biology and Ecology of Plants, Tel Aviv University, Tel Aviv, Israel

Jorge Dubcovsky
Department of Plant Sciences, University of California, Davis, CA, USA
Howard Hughes Medical Institute, Chevy Chase, MD, USA

Tzion Fahima and Abraham Korol
Department of Evolutionary and Environmental Biology, Institute of Evolution, Faculty of Science and Science Education, University of Haifa, Haifa, Israel

Justin Faris and Steven Xu
USDA-ARS Cereal Crops Research Unit, Fargo, ND, USA

Andrea Massi
Societá Produttori Sementi Bologna (Psb), Argelato, Italy

Anna Maria Mastrangelo
Consiglio per la ricerca e la sperimentazione in agricoltura, Cereal Research Centre, Foggia, Italy

Curtis Pozniak and Amidou N'Diaye
Crop Development Centre and Department of Plant Sciences, University of Saskatchewan, Saskatoon, SK, Canada

Yulia A. Meshcheriakova, Pooja Saxena and George P. Lomonossoff
Department of Biological Chemistry, John Innes Centre, Norwich, UK

Laura Pascual, Nelly Desplat, Aurore Desgroux, Jean-Paul Bouchet and Mathilde Causse
Inra, UR1052, Génétique et Amélioration des Fruits et Légumes, Montfavet, France

Bevan E. Huang
Computational Informatics and Food Futures Flagship, Csiro, Dutton Park, Qld, Australia

Laure Bruguier, Betty Chauchard and Philippe Verschave
Vilmorin, Centre de La Costiére, Ledenon, France

Quang H. Le
Vilmorin and Cie, Route d'Ennezat, Chappes, France

Ricardo H. Ramirez-Gonzalez and Mario Caccamo
The Genome Analysis Centre, Norwich, UK

Vanesa Segovia and Nicholas Bird
John Innes Centre, Norwich, UK

Cristobal Uauy
John Innes Centre, Norwich, UK
National Institute of Agricultural Botany, Cambridge, UK

Paul Fenwick and Simon Berry
Limagrain UK Ltd, Rothwell, UK

Sarah Holdgate and Peter Jack
RAGT Seeds, Saffron Walden, UK

Ji Tian, Zhen Peng, Jie Zhang, Tingting Song, Huihua Wan, Meiling Zhang and Yuncong Yao
Beijing Key Laboratory for Agricultural Application and New Technique, Department of Plant Science and Technology, Beijing University of Agriculture, Beijing, China

Qian Wang, Deepa Panicker, Hui-Zhu Mao, Nadimuthu Kumar, Chakravarthy Rajan, Prasanna Nori Venkatesh and Rajani Sarojam
Temasek Life Sciences Laboratory, Research Link, National University of Singapore, Singapore City, Singapore

Vaishnavi Amarr Reddy
Temasek Life Sciences Laboratory, Research Link, National University of Singapore, Singapore City, Singapore
Department of Biological Sciences, National University of Singapore, Singapore City, Singapore

Nam-Hai Chua
Laboratory of Plant Molecular Biology, The Rockefeller University, New York, NY, USA

Qing-Jie Wang, Hong Sun, Qing-Long Dong, Tian-Yu Sun, Zhong-Xin Jin and Yu-Xin Yao
State Key Laboratory of Crop Biology, College of Horticulture Science and Engineering, Shandong Agricultural University, Tai-An, Shandong, China

Huibin Xu, Yidong Wei, Yongsheng Zhu, Ling Lian, Hongguang Xie, Qiuhua Cai, Huaan Xie and Jianfu Zhang
Rice Research Institute, Fujian Academy of Agricultural Sciences, Fuzhou, China
Incubator of National Key Laboratory of Fujian Crop Germplasm Innovation and Molecular Breeding between Fujian and Ministry of Sciences and Technology, Fuzhou, China
Key Laboratory of Germplasm Innovation and Molecular Breeding of Hybrid Rice for South China, Ministry of Agriculture, Fuzhou, China
South-China Base of National Key Laboratory of Hybrid Rice of China, Fuzhou, China
National Engineering Laboratory of Rice, Fuzhou, Fujian, China

Qiushi Chen
Rice Research Institute, Fujian Academy of Agricultural Sciences, Fuzhou, China
Incubator of National Key Laboratory of Fujian Crop Germplasm Innovation and Molecular Breeding between Fujian and Ministry of Sciences and Technology, Fuzhou, China
Key Laboratory of Germplasm Innovation and Molecular Breeding of Hybrid Rice for South China, Ministry of Agriculture, Fuzhou, China
South-China Base of National Key Laboratory of Hybrid Rice of China, Fuzhou, China
National Engineering Laboratory of Rice, Fuzhou, Fujian, China
Fujian Agriculture and Forestry University, Fuzhou, China

Zonghua Wang
Fujian Agriculture and Forestry University, Fuzhou, China

Zhongping Lin
School Of Life Sciences, Peking University, Beijing, China

Zhen Yue, Xiaoguang Liu, Zijing Zhou, Guangming Hou, Jinping Hua and Zhangwu Zhao
College of Agriculture and Biotechnology, China Agricultural University, Beijing, China

Dongyan Zhao and Guo-qing Song
Plant Biotechnology Resource and Outreach Center, Department of Horticulture, Michigan State University, East Lansing, MI, USA

Qiang Zhao, Yi-Ran Ren, Qing-Jie Wang, Yu-Xin Yao, Chun-Xiang You and Yu-Jin Hao
National Key Laboratory of Crop Biology, National Research Center for Apple Engineering and Technology, College of Horticulture Science and Engineering, Shandong Agricultural University, Tai-An, Shandong, China

Michael Hackenberg
Computational Genomics and Bioinformatics Group, Genetics Department, University of Granada, Granada, Spain

Perry Gustafson
USDA-ARS, University of Missouri, Columbia, Mo, USA

Peter Langridge and Bu-Jun Shi
Australian Centre for Plant Functional Genomics, The University of Adelaide, Urrbrae, SA, Australia

Wei Ge, Zhanchao Cheng, Dan Hou, Xueping Li and Jian Gao
Key Laboratory of Bamboo and Rattan Science and Technology of the State Forestry Administration, International Centre for Bamboo and Rattan, Beijing, China

Ying Zhang
Key Laboratory of Bamboo and Rattan Science and Technology of the State Forestry Administration, International Centre for Bamboo and Rattan, Beijing, China
China National Engineering Research Center for Information Technology in Agriculture, Beijing, China

Benjamin M. Clasen, Thomas J. Stoddard, Song Luo, Zachary L. Demorest, Jin Li, Redeat Tibebu, Shawn Davison, Erin E. Ray, Aurelie Daulhac, Andrew Coffman, Ann Yabandith, Adam Retterath, William Haun, Nicholas J. Baltes, Luc Mathis, Daniel F. Voytas and Feng Zhang
Cellectis plant sciences Inc., New Brighton, MN, USA

Frederic Cedrone
Cellectis SA, Paris, France

Index

www.ingramcontent.com/pod-product-compliance
Lightning Source LLC
Chambersburg PA
CBHW080644200326
41458CB00013B/4727